Flavor of Meat, Meat Products and Seafoods

Flavor of Meat, Meat Products and Seafoods

Second edition

Edited by

F. SHAHIDI
Department of Biochemistry
Memorial University of Newfoundland
St John's, Newfoundland
Canada

BLACKIE ACADEMIC & PROFESSIONAL
An Imprint of Chapman & Hall
London · Weinheim · New York · Tokyo · Melbourne · Madras

Published by Blackie Academic and Professional, an imprint of Chapman & Hall, 2–6 Boundary Row, London SE1 8HN, UK

Thomson Science, 2–6 Boundary Row, London SE1 8HN, UK

Thomson Science, 115 Fifth Avenue, New York, NY 10003, USA

Thomson Science, Suite 750, 400 Market Street, Philadelphia, PA 19106, USA

Thomson Science, Pappelallee 3, 69469 Weinheim, Germany

First edition 1994

Reprinted, 1995, 1997

Second edition 1998

Thomson Science is a division of International Thomson Publishing I(T)P®

Typeset in 10/12pt Times by Florencetype Ltd, Stoodleigh, Devon

Printed in Great Britain by St Edmundsbury Press,
Bury St Edmunds, Suffolk

ISBN 0 7514 0484 5

A catalogue record for this book is available from the British Library

Library of Congress Catalog Card Number: 98-71052

∞ Printed on acid-free text paper, manufactured in accordance with ANSI/NISO Z39.48-1992 (Permanence of Paper).

Preface

Flavor is an important sensory aspect of the overall acceptability of muscle foods. Complex flavor systems of meat, and seafoods in particular, are comprised of both taste- and aroma-active components. While taste-active constituents of muscle foods are generally non-volatile, their aroma-active components are volatile in nature. Whether we accept or reject a food depends primarily on its flavor, in the first instance aroma. Furthermore, threshold values of different flavor active compounds have an important effect on the cumulative sensory properties of all foods.

While meat and meat products, in the raw state, have little aroma and only a blood-like taste, fresh seafoods have a more readily perceivable aroma. The flavors of fresh seafoods are primarily impacted by lipoxygenase-derived lipid-based volatiles and, to a lesser extent, environmentally induced components such as halocompounds and amines. Upon heat processing, many low-molecular-weight aroma-active compounds are formed via lipid oxidation, Strecker degradation and Maillard reaction. Mode of heat processing and presence of other constituents and additives also have a profound effect on the flavor of prepared muscle foods.

The first edition of this book, published in 1994 and reprinted in 1995 and 1997, received overwhelming interest from professionals in the industry, government laboratories and academic institutions. This edition provides an updated presentation of the earlier material, but some have been rewritten completely by different scientists. The monograph has also been expanded in order to include a discussion of flavor attributes of other processed meats and seafoods. The first chapter provides an overview of the field of muscle food flavor research while Chapter 2 discusses its chemistry. Meanwhile, Chapters 3 to 8 present a concise account of the flavor of different species of muscle foods, namely beef, pork, poultry, sheep, fish and shellfish. In Chapters 9 to 15, the role of meat constituents and processing on flavor is described. The final section of the book (Chapters 16 to 18) summarizes analytical methodologies for assessing the flavor quality of meat, meat products and seafoods.

Finally, I wish to extend my sincere thanks to all authors for their cooperative efforts and commendable contributions.

Fereidoon Shahidi

Contributors

M.E. Bailey Department of Food Science and Human Nutrition, 21 Agricultural Building, University of Missouri-Columbia, Columbia, MO 65211, USA

G. Bertelsen Department of Dairy and Food Science, The Royal Veterinary and Agricultural University, Rolihedsvej 30, DK-1958 Frederiksberg C, Denmark

K.L. Bett Southern Regional Research Center, USDA, Agricultural Research Service, 1100 Robert E. Lee Boulevard, New Orleans, LA 70124, USA

T.J. Braggins MIRINZ, East Street (Ruakara Campus), PO Box 671, Hamilton, New Zealand

K.R. Cadwallader Department of Food Science and Technology, Mississippi Agricultural and Forestry Experiment Station, Mississippi State University, MS 39762-5953, USA

J. Chen Department of Food Science, Rutgers University, New Brunswick, NJ 08903, USA

C.-W. Chen Department of Food Science, Rutgers University, New Brunswick, NJ 08903, USA

T. Cheraghi Department of Food Science, Wageningen Agricultural University, Wageningen, The Netherlands

E. Durnford Department of Biochemistry, Memorial University of Newfoundland, St. John's, Newfoundland, Canada, A1B 3X9

M. Flores Instituto de Agroquimica y Tecnologia de Alimentos (CSIC), Aparto de Correos 73, 46100 Burgassot, Valencia, Spain

C.-T. Ho Department of Food Science, Rutgers University, New Brunswick, NJ 08903, USA

A.J. MacLeod Department of Chemistry, King's College London, Strand, London WC2R 2LS, UK

G. MacLeod 12 The Coppice, Rhine Field Road, Brockenhurst, Hants. SO42 7QZ, UK

J.A. Maga Department of Food Science and Human Nutrition, Colorado State University, Fort Collins, Colorado 80523, USA

A. Mikkelsen Department of Dairy and Food Science, The Royal Veterinary and Agricultural University, Rolihedsvej 30, DK-1958 Frederiksberg C, Denmark

D.S. Mottram University of Reading, Department of Food Science and Technology, Whiteknights, Reading, RG6 2AP, UK

B.S. Pan Department of Marine Food Science, Fisheries Science College, National Taiwan Ocean University, Keelung, Taiwan, Republic of China

N. Ramarathnam GENOX Corporation, 1414 Key Highway, Baltimore, MD 21230, USA

J.P. Roozen Department of Food Science, Wageningen Agricultural University, Wageningen, The Netherlands

J. Rozum Red Arrow Products Company Inc., Manitowoe, WI 54221, USA

F. Shahidi Department of Biochemistry, Memorial University of Newfoundland, St. John's, Newfoundland, Canada, A1B 3X9

L.H. Skibstead Department of Dairy and Food Science, The Royal Veterinary and Agricultural University, Rolihedsvej 30, DK-1958 Frederiksberg C, Denmark

A.M. Spanier Southern Regional Research Center, USDA, Agricultural Research Service, 1100 Robert E. Lee Boulevard, New Orleans, LA 70124, USA

S. Spurvey Department of Biochemistry, Memorial University of Newfoundland, St. John's, Newfoundland, Canada, A1B 3X9

A.J. St. Angelo Southern Regional Research Center, USDA, Agricultural Research Service, 1100 Robert E. Lee Boulevard, New Orleans, LA 70124, USA

F. Toldrá Instituto de Agroquimica y Tecnologia de Alimentos (CSIC), Aparto de Correos 73, 46100 Burgassot, Valencia, Spain

S.M. van Ruth Department of Food Science, Wageningen Agricultural University, Wageningen, The Netherlands

B.T. Vinyard Southern Regional Research Center, USDA, Agricultural Research Service, 1100 Robert E. Lee Boulevard, New Orleans, LA 70124, USA

O.A. Young MIRINZ, East Street (Ruakara Campus), PO Box 617, Hamilton, New Zealand

Contents

1 Flavour of muscle foods – an overview

F. SHAHIDI

1.1 Introduction

Flavour is an important sensory aspect of the overall acceptability of meat and seafood products. The overwhelming effect of flavour volatiles has a tremendous influence on the sensory quality of foods. However, the taste properties of high molecular weight components and contribution of nonvolatile precursors to the flavour of muscle foods should also be considered.

Although raw meat has little aroma and only a blood-like taste, it is a rich reservoir of compounds with taste tactile properties as well as aroma precursors and flavour enhancers (Crocker, 1948; Bender and Ballance, 1961). However, seafoods are somewhat different in that they may carry their own flavour in the raw state. The presence of amines in the gadoid fish and autoxidation in fatty fish such as herring and mackerel is of particular interest. The flavour of both fish and shellfish in the fresh raw state is also impacted by lipoxygenase-derived lipid-based volatiles. Furthermore, shellfish have varied flavour characteristics which arise mainly from existing differences in their nonvolatile taste-active constituents. The nonvolatile precursors of muscle flavour include free amino acids, peptides, reducing sugars, vitamins and nucleotides. The interaction of these components with one another and/or their degradation products via the Strecker degradation and Maillard reaction produces a large number of intermediates and/or volatiles which contribute to the development of desirable aroma of muscle foods during thermal processing. Lipids also play an important role in the overall flavour of meat and seafoods which is distinct and species-dependent (Mottram *et al.*, 1982; Mottram and Edwards, 1983; Shahidi and Cadwallader, 1997).

Dietary regime, metabolic pathway, and species of the animal under investigation may have an effect on the flavour quality of muscle foods. For example, cattle and poultry on a fish meat-supplemented diet are fed a cereal-based finishing feed in order to prevent occurrence of a fish-tainted flavour in their meat. Furthermore, branched fatty acids such as 4-methyloctanoic and 4-methylnonanoic acids are mutton-specific and a swine sex odour is associated with boars (Wong *et al.*, 1975; Gower *et al.*, 1981).

1.2 Flavour volatiles of muscle foods

Nearly 1000 compounds have so far been identified in the volatile constituents of meat from beef, chicken, pork and mutton (Shahidi, 1992) as well as seafoods (Pan and Kuo, 1994). These volatiles are representative of most classes of organic compounds, such as hydrocarbons, alcohols, aldehydes, ketones, carboxylic acids, esters, lactones, ethers, furans, pyridines, pyrazines, pyrroles, oxazoles, oxazolines, thiazoles, thiazolines, thiophenes, and other sulphur- and halogen-containing compounds. It is believed that the predominant contribution to aroma is made by sulphurous acyclic compounds, heterocyclic compounds containing nitrogen, oxygen and/or sulphur, and carbonyl-containing volatiles (Shahidi, 1989, 1992).

Although the chemical nature of many flavour volatiles of meat from different species is similar qualitatively, there are quantitative differences. For example, it has been reported that mutton aromas contain a higher concentration of 3,5-dimethyl-1,2,4-trithiolane and 2,4,6-trimethylperhydro-1,3,5-dithiazine (thialdine) as compared to those of other species. Other sulphur-containing compounds were also present in high concentration and were attributed to the high content of suphurous amino acids in mutton as compared with those of beef and pork. Similarly a higher concentration of alkyl-substituted heterocyclics was noted in mutton volatiles (Buttery *et al.*, 1977). Mercaptothiophenes and mercaptofurans were significant contributors to beef aroma (Macleod, 1986).

Compared to the total number of volatile compounds identified in meat from different species, only a small fraction of them have been reported to possess meaty aroma characteristics (Shahidi, 1989) and these are mainly sulphur-containing in nature. While most of the sulphurous volatiles of meat exhibit a pleasant meaty aroma at concentrations present in meat, at high levels their odour is objectionable. Therefore, both qualitative and quantitative aspects of volatiles have to be considered when assessing the flavour quality of muscle foods. In addition, possible synergisms between various aroma constituents have to be considered.

Finally, in the evaluation of flavour quality of meat, the contribution to taste by amino acids, peptides and nucleotides must be considered. These compounds not only interact with other components to produce flavour volatiles, they also contribute to sweet, salty, bitter, sour and umami sensation of muscle foods. In the production of soups and gravies, proteins are partially hydrolysed to enhance taste sensation of the molecules. Therefore, studies in this area would allow us to optimize conditions to yield products with a maximum level of acceptability.

1.3 Impact of processing and storage on muscle food flavour

Processing of meat and fish such as curing (Shahidi, 1992) and/or smoking (Maga, 1987) brings about a characteristic flavour in the products. Interaction of nitrite with muscle foods generally retards the formation of off-flavour volatiles which may mask the natural flavour of products. Therefore, process flavours contribute greatly to the desirable characteristics of a wide range of well-loved products.

Cured and salted muscle foods generally retain their flavour, but nitrite curing also modifies the aroma and allows storage of products for a reasonably long period. Progression of oxidation and meat flavour deterioration is dependent primarily on the species of meat and its lipid content. Furthermore, interaction of oxidation products with muscle food components may in turn bring about changes in their colour, texture and nutritional value (Spanier *et al.*, 1992). In gadoid fish, formation of formaldehyde from degradation of trimethylamine oxide leads to toughening of the muscle texture during frozen storage because of cross-linking of protein molecules. Therefore, control of deteriorative processes, including autoxidation, and methods to quantitate them, have received considerable attention over the last few decades.

As analytical methodologies improve, the identification of new flavour-active compounds contributing to aroma at low threshold values becomes possible. Fundamental research has helped the application of important findings to improve the quality of muscle food products and these findings will continue to have an impact in new industrial developments. As more data become available, better understanding of the mechanisms involved in flavour perception and formulation of true-to-nature flavorants may be embraced.

1.4 Methodologies and concerns

The methods used for the analysis of all muscle food flavours are essentially the same (Ho and Manley, 1993; Marsili, 1997). However, identification of characteristic and important aroma compounds in meat has been challenging due to the presence of these compounds at extremely low levels, often at sub parts-per-billion concentrations. Isolation and sampling of volatiles prior to gas chromatographic analysis can be conducted in a number of ways, including equilibrium and dynamic headspace sampling, distillation under atmospheric or vacuum conditions with subsequent solvent extraction, direct solvent extraction or sublimation *in vacuo* and direct sample injection. Each isolation technique has its own advantages and weaknesses and each will give somewhat biased results since it selects for certain groups of compounds over others and some methods are prone to artifact formation.

A good approach to account for some of the bias has been to rely on two or more of the above techniques for isolation of volatiles. The subsequent analysis of volatile extract is most often accomplished by gas chromatography (GC) and GC–mass spectrometry (GC–MS). In addition to these, GC–olfactometry (GC–O) has proven useful for the identification of character-impact compounds in meat and seafoods. The use of an electronic aroma sensor for analysis of flavours is a recent development.

References

Bender, A.E. and Ballance, P.E. (1961). A preliminary examination of the flavour of meat extract. *J. Sci. Food Agric.*, **12**, 683–687.

Buttery, R.G., Ling, L.C., Teranishi, R. and Mon, T.R. (1977). Roasted lamb fat: Basic volatile components. *J. Agric. Food Chem.*, **25**, 1227–1229.

Crocker, E.C. (1948). The flavor of meat. *Food Res.*, **13**, 179–183.

Gower, D.E., Hancok, M.R. and Bannister, L.H. (1981). In *Biochemistry of Taste and Olfaction*, eds. R.H. Cagan, and M.R. Kave, Academic Press, New York, pp. 7–31.

Ho, C-T. and Manley, C.H. eds. (1993). *Flavor Measurement.* Marcel Dekker, New York.

MacLeod, G. (1986). The Scientific and Technological Basis of Meat Flavours. In *Development in Food Flavours*, eds. G.G. Birch and M.G. Lindley. Elsevier Applied Science, London. pp. 191–223.

Maga, J.A. (1987). The flavor chemistry of wood smoke. *Food Reviews International* **3**, 139–183.

Marsili, R. ed. (1997). *Techniques for Analyzing Food Aroma.* Marcel Dekker, New York.

Mottram, D.S. and Edwards, R.A. (1983). The role of triglycerides and phospholipids in the aroma of cooked beef. *J. Sci. Food Agric.*, **34**, 517–522.

Mottram, D.S., Edwards, R.A. and MacFie, H.J.H. (1982). A comparison of the flavor volatiles from cooked beef and pork meat systems. *J. Sci. Food Agric.*, **33**, 934–944.

Pan, B.S. and Kuo, J.-M. (1994). Flavour of shellfish and kamaboko flavourants. In *Seafoods: Chemistry, Processing Technology and Quality*, eds. F. Shahidi and J.R. Botta. Blackie Academic & Professional, Glasgow, pp. 85–114.

Shahidi, F. (1989). Flavour of cooked meats. In *Flavour Chemistry: Trends and Developments*, eds. R. Teranishi, R.E. Buttery, and F. Shahidi, ACS Symp, Ser. 388, American Chemical Society, Washington, DC, pp. 188–201.

Shahidi, F. (1992). Prevention of lipid oxidation in muscle foods by nitrite and nitrite-free compositions. In *Lipid Oxidation in Food*, ed. A.J. St. Angelo, ACS Symposium Series 500, American Chemical Society, Washington, DC, pp. 161–182.

Shahidi, F., Rubin, L.J. and D'Souza, L.A. (1986). Meat flavor volatiles: A review of the composition, techniques of analysis, and sensory evaluation. *CRC Crit. Rev. Food Sci. Nutr.*, **24**, 141–243.

Shahidi, F. and Cadwallader, K.R., eds. (1997). *Flavor and Lipid Chemistry of Seafoods.* ACS Symposium Series 674, American Chemical Society, Washington, DC.

Spanier, A.M., Miller, J.A. and Bland, J.M. (1992). Lipid oxidation: Effect on meat proteins. In *Lipid Oxidation in Food*, ed. A.J. St. Angelo, ACS Symposium Series 500, American Chemical Society, Washington, DC, pp. 161–182.

Wong, E., Nixon, L.N. and Johnson, C.B. (1975). Volatile medium chain fatty acids and mutton flavor. *J. Agric. Food Chem.*, **23**, 495–498.

2 The chemistry of meat flavour

D.S. MOTTRAM

2.1 Introduction

The flavours associated with cooked meats have proved particularly difficult to characterize, both for the sensory analyst and the flavour chemist. Meat flavour is influenced by compounds contributing to the sense of taste as well as those stimulating the olfactory organ. Other sensations such as mouthfeel and juiciness will also affect the overall flavour sensation. However it is the volatile compounds of cooked meat that determine the aroma attributes and contribute most to the characteristic flavours of meat. It has been one of the most researched of food flavours, with over 1000 volatile compounds having been isolated. A survey of the volatiles found in meat shows a much larger number from beef than the other meats, but this is reflected in the much larger number of publications for beef compared with pork, sheep meat or poultry (Mottram, 1991).

Meat flavour is thermally derived, since uncooked meat has little or no aroma and only a blood-like taste. Much research has been aimed at understanding the chemistry of meat aroma and the nature of the reactions occurring during cooking which are responsible for the formation of aroma compounds. The major precursors of meat flavour can be divided into two categories: water-soluble components (amino acids, peptides, carbohydrates, nucleotides, thiamine, etc.) and lipids. The main reactions during cooking which result in aroma volatiles, are the Maillard reaction between amino acids and reducing sugars, and the thermal degradation of lipids.

2.2 Meat flavour precursors

Most of the early work on meat flavour, during the 1950s and 1960s, was concerned with identifying those components of meat which, on heating, gave the characteristic flavour (Kramlich and Pearson, 1960; Hornstein and Crowe, 1960; Macey *et al.*, 1964; Wasserman and Gray, 1965). It was concluded that meat flavour precursors are low molecular weight water-soluble components and that the high molecular weight fibrillar and sarcoplasmic proteins are unimportant. The main flavour precursors were suggested to be free sugars, sugar phosphates, nucleotide-bound sugars, free amino acids, peptides, nucleotides and other nitrogenous components such as thiamine. A number of studies examined the changes which occurred in the quantities of these water-soluble compounds on heating. Depletions in the quantities of carbohydrates and amino acids were observed, the most significant losses occurring for cysteine and ribose. Subsequent studies of the aromas produced on heating mixtures of amino acids and sugars, confirmed the important role played by cysteine in meat flavour formation and led to the classic patent of Morton *et al.* (1960) which involved the formation of a meat-like flavour by heating a mixture of cysteine and ribose. Most subsequent patent proposals for 'reaction product' meat flavourings have involved sulphur, usually as cysteine or other sulphur-containing amino acids or hydrogen sulphide (MacLeod and Seyyedain-Ardebili, 1981; MacLeod, 1986).

The role of the lipid fractions of meat, both adipose tissue and fat contained within the lean, has been the subject of some debate which still continues. Hornstein and Crowe 1960, 1963) found that aqueous extracts of beef, pork and lamb had similar aromas when heated, while heating the fats yielded species-characteristic aromas. It was suggested that lipid provides volatile compounds which give the characteristic flavours of the different species, and the lean is responsible for a basic meaty flavour common to all species. It was also recognized that autooxidation of fat could produce undesirable flavour compounds and lead to rancidity. Fat can also serve as a solvent for aroma compounds, obtained either from extraneous sources or as part of the flavour-forming reactions, and it can therefore influence the release of flavour from the meat (Wasserman and Spinelli, 1972). However, it is now realized that this is an over-simplification of the role of fat in meat flavour and that lipid-derived volatiles have an important part to play in desirable meat aroma both as aroma compounds and as intermediates to other compounds.

2.3 Reactions leading to meat aroma

A wide range of temperature conditions exist during normal cooking of meat; the centre of a rare steak may only reach 50°C, the centre of roast meat may attain 70–80°C, while the outside of grilled or roast meat will be subjected to much higher temperatures and localized dehydration of the surface will occur. On the other hand, in stewing the meat remains at a temperature of 100°C, in the presence of excess water, for several hours. It is not surprising, therefore, that a wide range of different flavour sensations is perceived in cooked meats; some meats may be relatively bland while others may have strong meaty notes and others will be distinctly roasted.

The primary reactions occurring on heating which can lead to meat flavour include pyrolysis of amino acids and peptides, caramelization of carbohydrates, degradation of ribonucleotides, thiamine degradation, interaction of sugars with amino acids or peptides, and thermal degradation of lipid (MacLeod and Seyyedain-Ardebili, 1981; Mottram, 1991). This complicated array of reactions is made even more complex by the whole host of secondary reactions which can occur between the products of the initial reactions, giving rise to the vast number of volatile compounds which contribute to meat flavour. The thermal decomposition of amino acids and peptides and the caramelization of sugars normally require temperatures over 150°C before aroma compounds are formed (Feather and Harris, 1973; Wasserman, 1979). Such temperatures are higher than those which are normally encountered during the cooking of meat, except for small areas of the surface which may dehydrate during grilling or roasting.

2.3.1 Maillard reaction

The Maillard reaction between reducing sugars and amino compounds does not require the very high temperatures associated with sugar caramelization and protein pyrolysis, and readily produces aroma compounds at the temperatures associated with the cooking of food. It is, therefore, one of the most important routes to flavour compounds in cooked foods. It occurs much more readily at low moisture levels; hence in meat, flavour compounds produced by the Maillard reaction tend to be associated with the areas of the cooking meat which have been dehydrated by the heat source (van den Ouweland *et al.*, 1978).

The Maillard reaction in relation to flavour has been the subject of a number of reviews (Hurrell, 1982; Mottram, 1994; Nursten, 1986; Tressl *et al.*, 1993) and is discussed elsewhere in this book (Bailey, 1998). The initial stages of the reaction have been studied in some detail and involve the condensation of the carbonyl group of the reducing sugar with the amino compound, to give a glycosylamine. Subsequently, this rearranges and dehydrates, via deoxyosones, to various sugar dehydration and degradation products such as furfural and furanone derivatives, hydroxyketones and dicarbonyl compounds. The reaction has been discussed in many research papers, but it is interesting to note that the mechanism proposed by Hodge in 1953 still provides the basis for our understanding of the early stages of this reaction (Hodge, 1953; Ledl and Schleicher, 1990). The subsequent stages of the Maillard reaction involve the interaction of these compounds with other reactive components such as amines, amino acids, aldehydes, hydrogen sulphide and ammonia. It is these steps which provide the aroma compounds which characterise cooked foods and, therefore, are of particular interest to the flavour chemist.

An important reaction associated with the Maillard reaction is Strecker degradation which involves the oxidative deamination and decarboxylation of an α-amino acid in the presence of a dicarbonyl compound (Figure 2.1). This leads to the formation of an aldehyde, containing one fewer

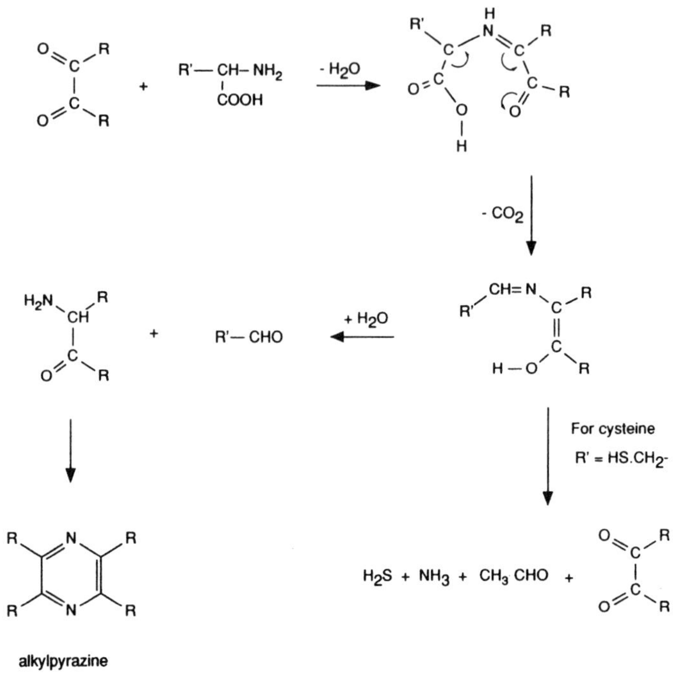

Figure 2.1 The Strecker degradation of α-amino acids showing the formation of hydrogen sulphide and ammonia in the Strecker degradation of cysteine.

carbon atom than the original amino acid, and an α-aminoketone. The Strecker degradation of cysteine yields the expected Strecker aldehyde, mercaptoacetaldehyde and an α-amino ketone, but hydrogen sulphide, ammonia and acetaldehyde are also formed from the breakdown of the intermediate mercaptoiminol. These compounds are important as reactive intermediates for the formation of many highly odoriferous compounds which play important roles in meat flavour, and this emphasizes the importance of cysteine in the development of meat flavour. The Strecker degradation of methionine is another source of sulphur-containing intermediates; in this case methional, methanethiol and 2-propenal are produced (Schutte, 1974).

All these Maillard products are capable of further reaction, and the subsequent stages of the Maillard reaction involve the interaction of furfurals, furanones and dicarbonyls with other reactive compounds such as amines, amino acids, hydrogen sulphide, thiols, ammonia, acetaldehyde and other aldehydes. These additional reactions lead to many important classes of flavour compounds including heterocyclics such as pyrazines, oxazoles, thiophenes, thiazoles and other heterocyclic sulphur compounds (Vernin and Parkanyi, 1982).

2.3.2 *Lipid degradation*

In addition to the subcutaneous fat and other fat depot tissues, triacylglycerols are present within the muscle in amounts which depend on the particular muscle and the age of the animal (Lawrie, 1992). All tissues also contain structural phospholipids. During cooking of meat, thermal degradation of lipids results in the formation of many volatile compounds, indeed more than half the volatiles reported in meat are lipid derived. One of the main routes to aroma volatiles during meat cooking is the thermally induced oxidation of the acyl chains of the lipids. Autoxidation of unsaturated fatty acid chains is also responsible for the undesirable flavours associated with rancidity which develops during the storage of fatty foods. The reactions by which volatile aroma compounds are formed from lipids follow the same general routes for both thermal oxidation and rancid oxidation, although subtle changes in the mechanisms give rise to different profiles of volatiles in the two systems.

The oxidative breakdown of the unsaturated alkyl chains of lipids involves a free radical mechanism and the formaton of intermediate hydroperoxides (Frankel, 1980). Decomposition of these hydroperoxides involves further free radical mechanisms and the formation of non-radical products including volatile aroma compounds (Forss, 1972; Grosch, 1982). The degradation of hydroperoxides (Figure 2.2) initially involves homolysis to give an alkoxy radical and a hydroxy radical; this is followed by cleavage of the fatty acid chain adjacent to the alkoxy radical. The nature of the volatile product from a particular hydroperoxide depends on the composition of the alkyl chain and the position where cleavage of the chain

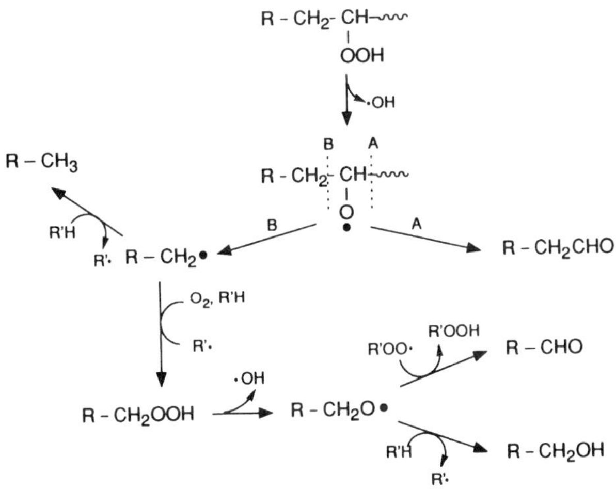

Figure 2.2 Breakdown of simple lipid hydroperoxides to give volatile products.

takes place (A or B). If the alkyl group is saturated and cleavage takes place at A, a saturated aldehyde results, while cleavage at B give an alkyl radical which can give an alkane or, alternatively, can react with oxygen to give a hydroperoxide. The latter, just like the hydroperoxide of the lipid, breaks down to give an alkoxy radical which then gives a stable non-radical product such as an alcohol or aldehyde. One or more double bonds in the alkyl chain will give analogous compounds containing the double bonds, but the final range of products is more complex because further oxidation of the unsaturated chain can occur.

Other compounds formed in these and related reactions include ketones, furans and aromatic hydrocarbons, aldehydes and ketones. Free fatty acids can be formed from thermal hydrolysis of lipids while the gamma and delta lactones which are found in meat arise from the lactonization of hydroxy fatty acids (Watanabe and Sato, 1970).

In boiled and lightly grilled or roasted meat, lipid degradation products have been found to dominate the volatile extracts (Mottram *et al.*, 1982; Mottram, 1985). However, the contribution of many of these compounds to the overall flavour of the cooked meat may be small because their odour threshold values are relatively high. Those compounds which have sufficiently low odour threshold values for them to contribute directly to meat aroma are aldehydes, unsaturated alcohols and ketones and lactones.

2.4 Compounds contributing to meat flavour

Among the many aroma characteristics that can be differentiated in meat, those which are most clearly recognized are fatty, species-related, roast, boiled-meat and the characteristic 'meaty' aroma which is associated with all meats regardless of species.

The characteristic flavour of the different meat species is generally believed to be derived from the lipid sources, although it has been suggested that interaction of lipid with other meat components may also be involved (Wasserman and Spinelli, 1972). Aldehydes, as major lipid degradation products, are probably involved in certain species' characteristics. The higher proportion of unsaturated fatty acids in the tryglycerides of pork and chicken gives more unsaturated volatile aldehydes in these meats and these compounds may be important in determining the specific aromas of these species (Noleau and Toulemonde, 1987).

Sheep meat has been found to contain a number of methyl-branched saturated fatty acids which have not been reported in other meats. These acids have been associated with the characteristic flavour of mutton which results in low consumer acceptance of sheep meat in many countries, and which the Chinese described as 'soo' flavour (Wong *et al.*, 1975). Two acids, 4-methyloctanoic and 4-methylnonanoic, are considered to be primarily responsible for this flavour. The lipids of sheep contain significant quantities of methyl-branched fatty acids, unlike other species, and these are known to arise from the metabolic process occurring in the rumen of sheep. Branched acids with methyl substituents at even-numbered carbon atoms result from fatty acid synthesis utilizing methyl malonate (arising from propionate metabolism) instead of malonate in the chain lengthening (Christie, 1978). Fatty flavours, of course, originate from the lipid, and aldehydes (e.g. 2,4-decadienal), ketones and lactones will contribute to the fatty aromas associated with cooked meat.

Recently, 12-methyltridecanal was reported to be an important component in the volatiles of cooked beef where it was found at relatively high concentrations. However, it was only found at very low levels in veal, lamb and deer and only trace amounts in pork, chicken and turkey (Grosch *et al.*, 1993; Guth and Grosch, 1993, 1995). The compound had a tallowy, beef-like aroma and it was suggested to play an important role in determining the characteristic aroma of beef. A number of other *iso*- and *anteiso*-methyl-branched aldehydes with between 11 and 17 carbon atoms have been reported in cooked beef and some have also been found in pork and chicken (Werkhoff *et al.*, 1993). It was concluded that these methyl-branched aldehydes arose from the hydrolysis of plasmalogens, which are phosphoglycerides in which one position on the glycerol moiety is linked to an aldehyde through an enol–ether link.

Roast flavours appear to be associated with heterocyclic compounds (e.g. pyrazines and thiazoles) formed in the later stages of the Maillard reaction. Many different alkyl pyrazines have been found in meat volatiles as well as two classes of interesting bicyclic compounds, 6,7-dihydro-5(*H*)-cyclopentapyrazines and pyrrolo[1,2*a*]pyrazines (Flament *et al.*, 1976, 1977). This latter class of compounds has not been reported in any other food. Alkyl-substituted thiazoles in general have lower odour thresholds than pyrazines (Petit and Hruza, 1974), although they are found at lower concentrations in meat. Both classes of compounds increase markedly with the increasing severity of the heat treatment and, in well-done grilled meat, pyrazines are the dominant class of volatiles (Mottram, 1985).

One notable feature of the volatiles from cooked meat is the preponderance of sulphur-containing compounds. The majority occur at low concentrations, but their very low odour thresholds make them important contributors to the aromas of cooked meat. A comparison of boiled and roast beef shows that many more aliphatic thiols, sulphides and disulphides have been reported in the boiled meat (Mottram, 1991). A number of heterocyclic compounds with two or three sulphur atoms in five- and six-membered rings (e.g. trithiolanes, trithianes) have also been found in boiled beef, and thiophenes are much more prevalent in boiled beef than in the roast meat. Many of these sulphur compounds have low odour thresholds with sulphurous, onion-like and, sometimes, meaty aromas

(Fors, 1983), and they probably contribute to the overall flavour by providing sulphurous notes which form part of the aroma of boiled meat.

Isolation of compounds with the characteristic 'meaty' aroma associated with all meats has been the subject of much research. These desirable characteristics of meat flavour have also been sought in the production of simulated meat flavourings which are of considerable importance in convenience and processed savoury foods. Furans (and also thiophenes) with a thiol group in the 3-position appear to have meat-like aromas, as do the disulphides formed by oxidation of furan and thiophene thiols. A number of such compounds have been found in the volatiles from heated model systems containing hydrogen sulphide or cysteine, and pentoses or other sources of carbonyl compounds, and some are reported to have meaty aromas. 2-Methyl-3-furanthiol was shown to be one of the products formed in the reaction of hydrogen sulphide with 4-hydroxy-5-methyl-3(2*H*)-furanone, which is a dehydration product of ribose (van den Ouweland and Peer, 1975). The meaty characteristics of the reaction mixture led to patents dealing with a number of related compounds with potential as meat flavourings (MacLeod, 1986). 2-Methyl-3-furanthiol, and the analogous thiophenethiol, have been found in the volatiles from the reaction of cysteine and ribose (Farmer *et al.*, 1989; Farmer and Mottram, 1990a), and from the thermal degradation of thiamine (van der Linde, *et al.*, 1979; Werkhoff *et al.*, 1990; Güntert *et al.*, 1993). These thiols can be oxidized to the corresponding disulphides, which have also been found in some of these model systems.

Although thiol-substituted furans and thiophenes and related disulphides were known to possess strong meat-like and/or roast aromas and exceptionally low odour threshold values (Evers, 1970; Evers *et al.*, 1976; van den Ouweland and Peer, 1972), it was not until relatively recently that such compounds were first reported in meat itself. MacLeod and Ames (1986) identified 2-methyl-3-(methylthio)furan in cooked beef. It has been reported to have a low odour threshold value (0.05 mg/kg) and a meaty aroma at levels below 1 mg/kg. Subsequently, 2-methyl-3-furanthiol and the corresponding disulphide, bis-(2-methyl-3-furanyl) disulphide, were identified as major contributors to the meaty aroma of cooked beef, chicken and pork (Gasser and Grosch, 1988, 1990). The odour theshold value of this disulphide has been reported as 0.02 ng/kg, one of the lowest known threshold values (Buttery *et al.*, 1984). Other related thiols and disulphides have also been found in the volatiles of heated meat systems (Figure 2.3).

From the work on model systems and from attempts to synthesize structures with meat-like aromas, it has become clear that the furanthiols and thiophenethiols and their disulphides have particularly meaty aroma characteristics. The importance of these types of compound in meat flavour has now been confirmed by their detection in meat itself.

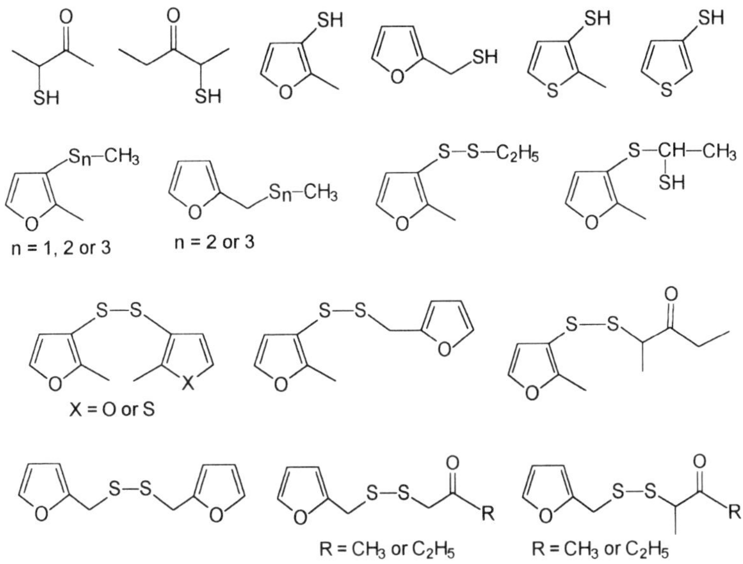

Figure 2.3 Some thiols and disulphides found in the volatiles of cooked meat which may contribute to meaty characteristics (Farmer and Patterson, 1991; Gasser and Grosch, 1988; MacLeod and Ames, 1986; Werkhoff *et al.*, 1993; Madruga and Mottram, 1995).

2.5 Pathways for the formation of some meat aroma volatiles

The Maillard reaction produces a number of pentose and texose degradation products containing carbonyl groups, such as 2-oxopropanal, 2,3-butanedione, hydroxypropanone, 3-hydroxy-2-butanone, 2-furfural, 5-methyl-2-furfural, 5-hydroxymethyl-2-furfural and 5-methyl-4-hydroxy-3(2*H*)-furanone and its dimethyl homologue. Strecker degradation of amino acids by these dicarbonyl compounds yields aldehydes, while Strecker degradation of cysteine also produces hydrogen sulphide and ammonia. These compounds provide the main reactants for the formation of the large number of different heterocyclic compounds which are found in meat volatiles.

An important route to alkyl pyrazines is from α-aminoketones, which are formed in Strecker degradation or from the reaction of α-dicarbonyls with ammonia. Condensation of two aminoketone molecules yields a dihydropyrazine which oxidizes to the pyrazine (Figure 2.1).

Thiazoles and oxazoles may be formed by related mechanisms, involving the formation of an intermediate imine from the reaction of an aldehyde (e.g. a Strecker aldehyde) with ammonia (Figure 2.4). Reaction of the imine with a dicarbonyl produces an oxazoline which may undergo oxidation to the corresponding oxazole. If hydrogen sulphide is present it may

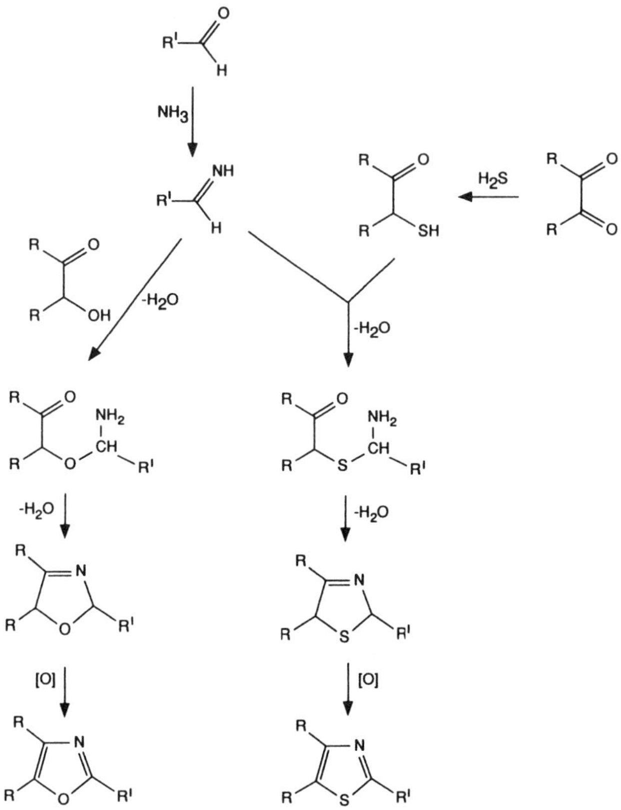

Figure 2.4 Formation of thiazoles, thiazolines, oxazoles and oxazolines by the reaction of aliphatic aldehydes with dicarbonyls, hydrogen sulphide and ammonia (Vernin and Parkanyi, 1982).

react with the dicarbonyl to give an α-mercaptoketone which then yields thiazolines and thiazoles from reaction with the imine.

Many of the important aroma volatiles in meat contain sulphur. Hydrogen sulphide, formed from cysteine by Strecker degradation or hydrolysis, appears to be the essential intermediate for these compounds. Some of the reactions involving hydrogen sulphide in the formation of sulphur-containing heterocyclics are shown in Figure 2.5. Formylthiophenes and furylmethanethiol can be formed from furfurals, while the reaction of short-chain aldehydes, such as acetaldehyde, with hydrogen sulphide has been shown to yield a number of diffrent sulphur compounds including trithiolanes and trithianes (Boelens *et al.*, 1974). One of the most important reactions for the formation of meaty aromas is the reaction of 4-hydroxy-5-methyl-3(2*H*)-furanone with hydrogen sulphide to give 2-methyl-3-furanthiol, 2-methyl-3-thiophenethiol and their dihydro homologues (van den Ouweland and Peer, 1975).

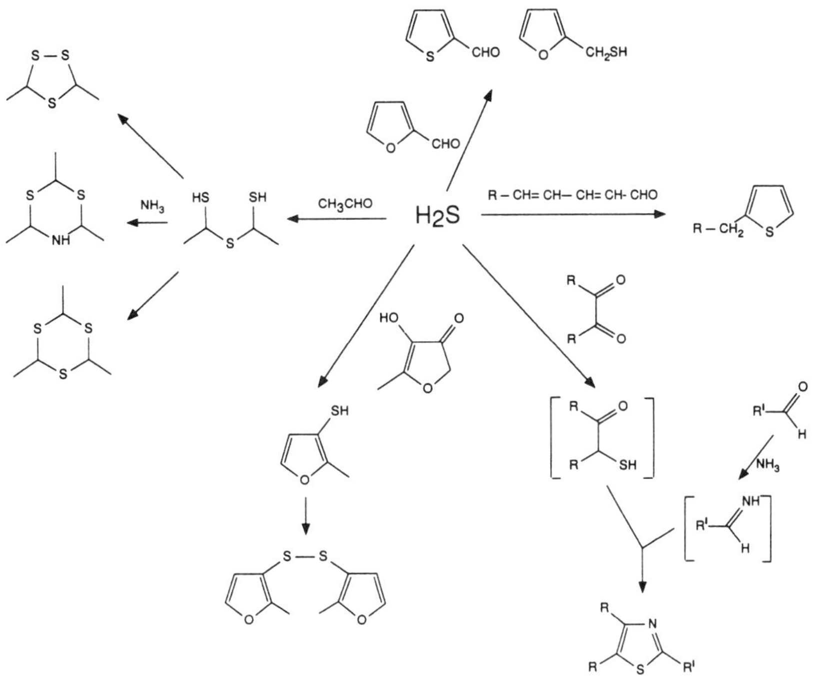

Figure 2.5 Some classes of volatile aroma compounds formed by the reaction of hydrogen sulphide with carbonyl compounds.

Oxidation of these thiols then results in mixtures of symmetrical and unsymmetrical disulphides. The extent to which the different thiols and disulphides exist in cooked meat or model meat systems will depend on a number of factors, including the presence of other thiol groups such as those occurring in proteins (Mottram *et al.*, 1996). 4-Hydroxy-5-methyl-3(2*H*)-furanone can be formed in the Maillard reaction from pentose sugars via 1-deoxypentosone. In meat, dephosphorylation and dehydration of ribose phosphate, associated with the ribonucleotides, provides an alternative route of this furanone (Figure 2.6). Although these reactions may only occur to a small extent, the exceedingly low odour threshold values of the compounds and their meaty character gives the reaction major importance for meat flavour.

Compounds produced in other reactions may also react with products of the Maillard reaction, especially hydrogen sulphide and ammonia. For example, aldehydes and other carbonyls formed during lipid oxidation could readily participate in some of the reactions described above.

2.6 Interaction of lipid with the Maillard reaction

In an examination of the contribution that lipids make to the development of aroma during the heating of meat, the phospholipids were shown to be particularly important (Mottram and Edwards, 1983). Sensory panels and consumer studies had failed to show any relationship between the flavour of lean meat and the amount of fat on the carcass, apparently confirming the early work on meat flavour in which meatiness was associated with the water-soluble flavour precursors and species characteristics with the lipid. However, the volatiles of cooked meat are often dominated by lipid-derived volatiles and such volatiles would be expected to have some effect on meaty flavour. The contribution of lipid to meat aroma was investigated by evaluating the change in aroma which occurred when lipids were extracted from meat prior to cooking. When inter- and intra-muscular triacylglycerols were removed from lean muscle using hexane, the aroma after cooking could not be differentiated from the untreated material in sensory triangle tests; both preparations were judged to be meaty. However, when a more polar solvent (chloroform–methanol) was used to extract all the lipids – phospholipids as well as triacylglycerols – a very marked difference in aroma resulted; the meaty aroma was replaced by a roast, biscuit-like aroma. Comparison of the aroma volatiles from these meat preparations showed that the control and the material extracted with hexane had similar profiles, dominated by aliphatic aldehydes and alcohols, while removal of phospholipids as well as triacylglycerols gave a very different profile; the lipid oxidation products were lost, but there was a large increase in the amounts of alkylpyrazines. This implied that in normal meat, lipids or their degradation products inhibit the formation of heterocyclic compounds by participating in the Maillard reaction.

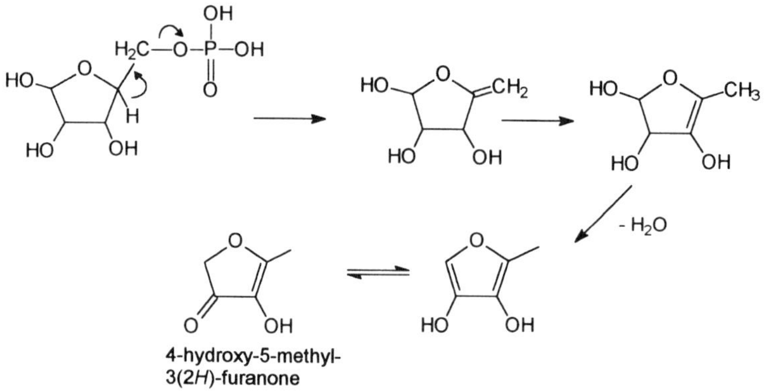

Figure 2.6 Formation of 4-hydroxy-5-methyl-3(2*H*)-furanone from ribose-5-phosphate (Peer and van den Ouweland, 1968).

2.6.1 *Model systems*

These results, showing a reduction in amounts of volatile Maillard products in defatted cooked meat, led to investigations of the effect of phospholipids on the volatile products of heated mixtures of amino acids and sugars (Whitfield *et al.*, 1988; Salter *et al.*, 1988; Farmer *et al.*, 1989; Farmer and Mottram, 1990b, 1992, 1994). Several amino acids were used, including cysteine, and ribose was chosen as the reducing sugar because of its recognized role in the formation of meat flavour, and concentrations were selected to approximate their relative concentrations in muscle. Reactions were carried out in aqueous solution buffered at pH 5.6 with phosphate under pressure in sealed glass tubes at 140°C. The effects of several phospholipid preparations, including egg-yolk phospholipids and triacylglycerols extracted from beef, on the volatiles from the reaction were examined. In the absence of lipid the reaction mixtures yielded complex mixtures of volatiles including furfurals, furanones, alkylpyrazines and pyrroles. The volatiles from reactions involving cysteine were dominated by sulphur-containing heterocyclics, particularly thiophenes, thienothiophenes, dithiolanones, dithianones, trithiolanes and trithianes, together with 2-methyl-3-furanthiol, 2-furanmethanethiol and 2-methyl-3-thiophenethiol.

In the presence of phospholipids a reduction in the amounts of many of these volatiles was observed, confirming the observations in meat that phospholipids exert a quenching effect on the quantities of heterocyclic compounds formed in the Maillard reaction. As would be expected the inclusion of phospholipids in the reaction mixtures produced many volatiles derived from lipid degradation, such as hydrocarbons, alkylfurans, saturated and unsaturated alcohols, aldehydes and ketones. In addition, the reaction mixtures contained several compounds derived from the interaction of the lipid or its degradation products with Maillard reaction intermediates. In reaction mixtures containing cysteine, ribose and phospholipid, the most abundant volatile compounds were 2-pentylpyridine, 2-pentylthiophene, 2-hexylthiophene and 2-pentyl-2*H*-thiapyran. Smaller amounts of other 2-alkylthiophenes with *n*-alkyl substituents between C_4 and C_8 were found together with 2-(1-hexenyl)thiophene, 1-heptanethiol and 1-octanethiol. All these compounds are probably formed by the reaction of lipid breakdown products with hydrogen sulphide or ammonia derived from cysteine. Figure 2.7 shows a scheme for the formation of 2-pentylpyridine, 2-hexylthiophene and 2-pentyl-2*H*-thiapyran from 2,4-decadienal which is one of the major oxidation products of polyunsaturated fatty acids.

The particular role of the phospholipids, compared with triacylglycerols, was also examined (Farmer and Mottram, 1990b). The structural phospholipids contain a high proportion of polyunsaturated fatty acids, particularly those with three or more double bonds such a arachidonic

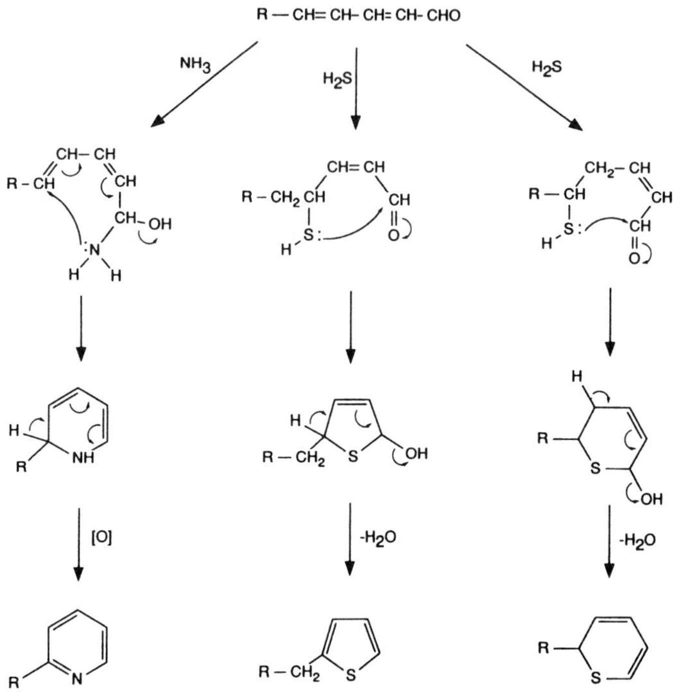

Figure 2.7 Reaction of 2,4-decadienal with hydrogen sulphide and ammonia. R = C_5H_{11} (Farmer and Mottram, 1990b).

acid (20:4), and these would be expected to break down during heating to give products which could react with Maillard products. The triacylglycerols in meat contain only a very small proportion of polyunsaturated fatty acids and this may explain the observations on defatted meat, which suggested that phospholipids rather than triacylglycerols were important for meat flavour. Reaction mixtures of cysteine and ribose were prepared with different lipids: triacylglycerols (BTG) and phospholipids (BPL) extracted from beef, and commercial egg phosphatidylcholine (PC) and phosphatidylethanolamine (PE). The aroma of the reaction mixture without any lipid was described as 'sulphurous, rubbery' but there was a distinct underlying meaty aroma. Addition of the beef triacylglycerols did not affect the aroma; however when beef phospholipids were used, the meaty aroma was more intense and the sulphurous notes less pronounced. Similarly, the addition of PC or PE gave mixtures with increased meatiness, with the mixture containing PE exhibiting the most meaty character.

The effect of these different lipids on the formation of selected volatiles was also examined (Table 2.1). The lipid preparations differed in the way they influenced the profile of volatiles from the reaction mixtures. All the phospholipids produced 2-pentylpyridine, 2-pentylthiapyran and the

Table 2.1 Relative concentration of some selected heterocyclic compounds formed from the reaction between cysteine and ribose with different lipids (Mottram and Salter, 1989; Farmer and Mottram, 1990b)

Compound	No lipid	BTG	BPL	PC	PE
2-Mercapto-3-pentanone	1	0.72	0.49	0.53	0.53
3-Mercapto-2-pentanone	1	0.77	0.47	0.50	0.49
2-Furylmethanethiol	1	0.67	0.63	0.62	0.72
2-Methyl-3-furanthiol	1	0.40	0.15	0.27	0.24
2-Thiophenethiol	1	0.32	0.03	0.46	0.14
2-Methyl-3-thiophenethiol	1	0.08	0.01	0.20	0.10
2-Pentylpyridine	0	0.09	1	18.6	1.5
2-Pentylthiophene	0	0	1	23.9	10.9
2-Hexylthiophene	0	0.15	1	6.6	2.4
2-Pentyl-2*H*-thiapyran	0	0.01	1	11.0	4.0

2-alkylthiophenes, but the quantities differed markedly. Only trace amounts of these compounds were found in the triacylglycerol-containing system. Many Maillard reaction products showed marked reductions on addition of lipid, although not all the volatiles were affected to the same extent, and the different lipids did not all behave in the same way. In general BTG showed some effect on the Maillard volatiles, but this was not as marked as the phospholipid preparations.

These results clearly demonstrated the important role that lipids, especially phospholipids, play in these Maillard reactions which are the basis of meat flavour formation. The early stages of the Maillard reaction give rise to many reactive intermediates, of which dicarbonyls, furfurals, furanones, Strecker aldehydes, ammonia and hydrogen sulphide are the most important. These intermediates provide the reactants for most of the important classes of meat aroma volatiles. The relative amounts of the different volatiles produced from these intermediates must depend on the concentrations produced by the early Maillard reaction, and the relative rates of the different reactions. The addition of lipid (especially unsaturated lipids) and lipid degradation products (such as aldehydes, ketones and alcohols) to this mixture of Maillard intermediates provides competing reactions which produce other volatiles and affect the relative proportions of compounds produced by the other reactions. Hydrogen sulphide and ammonia are extremely important intermediates in many of the aroma forming reactions in meat, and their interaction with lipid degradation products is a clear example of how lipid will influence the relative proportions of heterocyclic Maillard reaction products.

2.6.2 Volatiles in meat formed via lipid–Maillard interactions

A number of volatile compounds have been found in meat which could be formed from the interaction of lipid with the Maillard reaction (Table 2.2).

Table 2.2 Some heterocyclic volatiles identified in meat which are derived from the interaction of lipid with the Maillard reaction

Compound	Type of meat	Compound	Type of meat
Pyridines:		2-pentyl-4-methyl-	beef [8]
2-butyl-	lamb [1], chicken [2]	2-pentyl-4,5-dimethyl-	beef [8]
2-pentyl-	beef [3], pork [4], lamb [1]	2-pentyl-4-ethyl-	beef [8]
	chicken [2]	2 pentyl-4-ethyl-5-methyl-	beef [8]
3-pentyl-	lamb [1]	2 pentyl-5-ethyl-4-methyl-	beef [8]
2-pentyl-5-methyl-	lamb [1]	2-hexyl-4-methyl-	beef [8]
2-pentyl-5-ethyl-	lamb [1]	2-hexyl-4,5-dimethyl-	beef [8]
2-hexylpyridine-	lamb [1]	2-hexyl-4-ethyl-	beef [8]
Pyrroles:		2-hexyl-4-ethyl-5-methyl-	beef [8]
2-butyl-	beef [5]	2-hexyl-5-ethyl-4-methyl-	beef [8]
1-pentyl-	beef [6]	2-heptyl-4-methyl-	beef [8]
Pyrazines:		2-heptyl-4-ethyl-	beef [8]
2-butyl-	chicken [2]	2-heptyl-4-ethyl-5-methyl-	beef [8]
2-butyl-3-methyl-	chicken [2]	2-heptyl-5-ethyl-4-methyl-	beef [8]
methylpentyl-	pork [4]	2-octyl-4-methyl-	beef [8]
dimethylbutyl-	pork [4]	2-octyl-4,5-dimethyl-	beef [8]
dimethylpentyl-	pork [4]	2-octyl-4-ethyl-	beef [8]
Thiazoles:		2-octyl-4-ethyl-5-methyl-	beef [8]
2-butyl-4,5-dimethyl-	beef [5], chicken [2],	2-octyl-5-ethyl-4-methyl-	beef [8]
	bacon [7]	2-nonyl-4,5 dimethyl-	beef [8]
2-butyl-5-ethyl-4-methyl-	beef [5], chicken [2]	2-nonyl-4-ethyl-5-methyl-	beef [8]
2-pentyl-5-methyl-	beef [5]	**Thiophenes:**	
2-pentyl-4,5-dimethyl-	beef [5,8], chicken [2]	2-butylthiophene	beef [10,11], chicken [2], pork [12]
2-pentyl-4-ethyl-5-methyl-	beef [8], chicken [2]	2-pentylthiophene	beef [10,11], chicken [2], pork [12]
2-hexyl-4-ethyl-5-methyl-	beef [8]	2-hexylthiophene	beef [10], pork [12]
2-heptyl-4,5-dimethyl-	beef [8], chicken [2]	2-heptylthiophene	beef [10], pork [12]
2-heptyl-4-ethyl-5-methyl-	beef [8], chicken [2]	2-octylthiophene	beef [10,11], pork [12]
2-octyl-4,5-dimethyl-	beef [8], chicken [2]	tetradecylthiophene	beef [11]
2-octyl-4-ethyl-5-methyl-	beef [8]	2-butanoylthiophene	beef [13]
2-octyl-5-ethyl-4-methyl-	beef [8]	2-heptanoylthiophene	beef [13]
2-tridecyl-4-ethyl-	beef [9]	2-octanoylthiophene	beef [13]
2-tridecyl-4,5-dimethyl-	beef [9]	**Trithiolanes:**	
2-tetradecyl-4-methyl-	beef [9]	3-methyl-5-butyl-1,2,4-	chicken [14], pork [12]
2-tetradecyl-4-ethyl-	beef [9]	3-methyl-5-pentyl-1,2,4-	chicken [14], pork [12]
2-tetradecyl-4,5-dimethyl-	beef [9]	**Oxazoles:**	
2-pentadecyl-	beef [9], chicken [9]	2-butyl-	bacon [15]
2-pentadecyl-4-methyl-	beef [9], chicken [9]	2-methyl-5-pentyl-	bacon [15]
2-pentadecyl-4-ethyl-	beef [9], chicken [9]	2,5-dimethyl-4-butyl-	beef [5], bacon [15]
3-Thiazolines:		2,5-dimethyl-4-pentyl-	bacon [15]
2-butyl-4-methyl-	beef [8]	2,5-dimethyl-4-hexyl-	beef [5]
2-butyl-4,5-dimethyl-	beef [8]		

[1]Buttery *et al.* (1977); [2]Tang *et al.* (1983); [3]Watanabe and Sato (1971); [4]Mottram (1985); [5]Hartman *et al.* (1983); [6]Coppock and Macleod (1977); [7]Ho *et al.* (1983); [8]Elmore *et al.* (1997); [9]Farmer and Mottram (1994); [10]Min *et al.* (1979); [11]Wilson *et al.* (1973); [12]Werkhoff *et al.* (1993); [13]Hsu *et al.* (1982); [14]Hwang *et al.* (1986); [15]Ho and Carlin (1989).

The occurrence of such compounds in meat and other cooked foods has been reviewed by Whitfield (1992). These compounds are O-, N- or S-heterocycles containing long *n*-alkyl substituents (C_5–C_{15}). The alkyl groups usually derive from aliphatic aldehydes, obtained from lipid oxidation, while amino acids are the source of the nitrogen and sulphur.

2-Pentylpyridine is commonly found in the volatiles of cooked meat and is probably formed from 2,4-decadienal and ammonia (Figure 2.7). A number of other alkylpyridines have been reported in lamb fat (Buttery *et al.*, 1977). Related reactions between dienals and hydrogen sulphide may be responsible for the formation of 2-alkylthiophenes with C_4–C_8 alkyl substituents which have been reported in pressure-cooked beef (Min *et al.*, 1979; Wilson *et al.*, 1973). Other heterocyclic compounds, with long *n*-alkyl substituents, found in meat include butyl- and pentyl-pyrazines (Tang *et al.*, 1983; Mottram, 1985). It has been suggested that these could result from the reaction of pentanal or hexanal (from lipid oxidation) with a dihydropyrazine, formed by the condensation of two aminoketone molecules (Figure 2.8). The latter is a product of the Strecker degradation of amino acids with α-dicarbonyl compounds. Pentanal and hexanal also appear to be involved in the formation of 5-butyl-3-methyl-1,2,4-trithiolane and its 5-pentyl homologue, which have both been reported in fried chicken (Hwang *et al.*, 1986) and pork (Werkhoff *et al.*, 1993). Trithiolanes can be formed from aldehydes and hydrogen sulphide and the reaction of hydrogen sulphide, acetaldehyde and pentanal or hexanal has been suggested as the route to these butyl and pentyl trithiolanes (Figure 2.9).

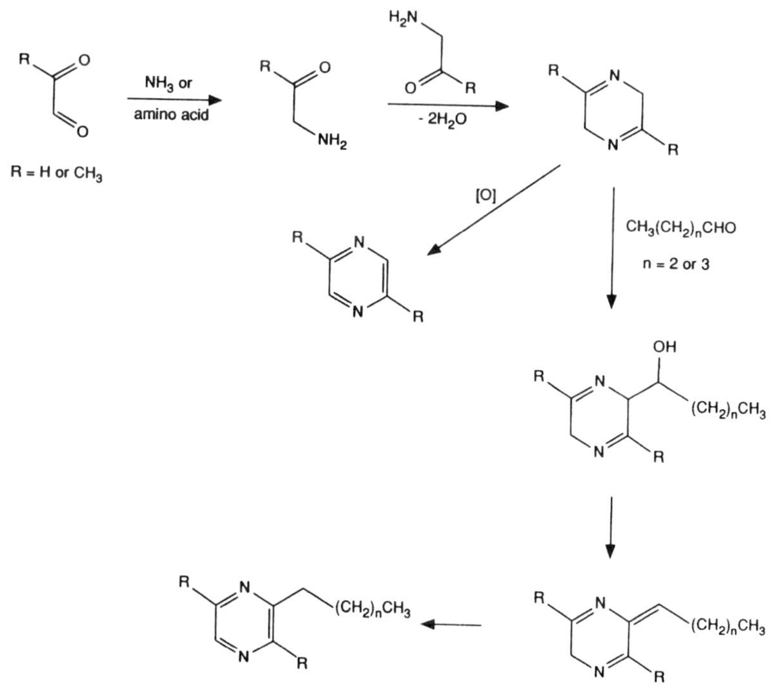

Figure 2.8 Mechanism for the formation of *n*-butyl and *n*-pentyl pyrazines (Ho *et al.*, 1987; Chiu *et al.*, 1990).

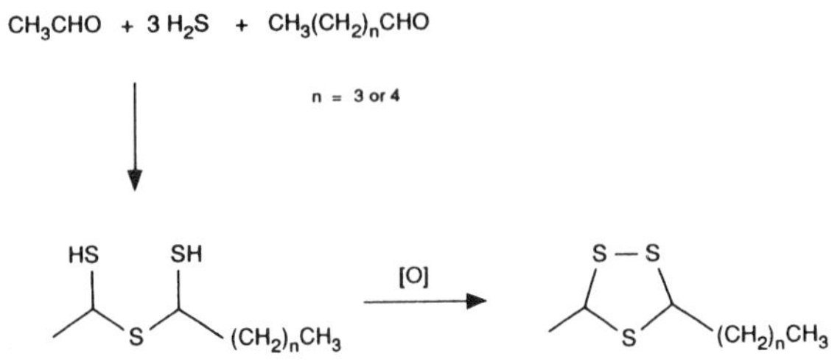

Figure 2.9 Mechanism for the formation of trithiolanes with *n*-butyl and *n*-pentyl substituents (Boelens *et al.*, 1974; Ho *et al.*, 1987).

Several thiazoles with C_4–C_8 *n*-alkyl substituents in the 2-position have been reported in roast beef (Hartman *et al.*, 1983) and fried chicken (Tang *et al.*, 1983). Other alkylthiazoles with longer 2-alkyl substituents (C_{13}–C_{15}) were found in the volatiles of heated beef and chicken with the highest concentrations in beef heart muscle (Farmer and Mottram, 1994). Aliphatic aldehydes from lipid oxidation are the likely source of the long *n*-alkyl groups in these compounds. The alkylthiazoles containing C_{13}–C_{15} alkyl groups require C_{14}–C_{16} aldehydes, and the most likely sources of these are the plasmalogens which contain long-chain alkenyl ether substituents which hydrolyse to give fatty aldehydes. Heart muscle contains higher concentrations of plasmalogens which explains the higher levels of these alkylthiazoles found in heated beef heart.

Recently, a large number of alkyl-3-thiazolines have been isolated from cooked beef (Elmore and Mottram, 1997). Most of the thiazolines contained C_5–C_9 *n*-alkyl substituents in the 2-position. Small quantities of some thiazoles, with similar substitution to the thiazolines, were also obtained. The meat was produced in a study of the quality of beef from cattle fed diets which attempted to modify the polyunsaturated fatty acid composition by feeding lipid supplements. Although most of the 3-thiazolines were present in meat from cattle fed on control diets, their concentrations were much higher in meat from the cattle fed with fish oil or linseed supplements. The cooked meat from the animals which had been fed fish oil or linseed also had considerably higher concentrations of saturated and unsaturated aldehydes than meat from the control.

The most likely routes to the 3-thiazolines and thiazoles are from α-hydroxyketones or α-diones, hydrogen sulphide, ammonia and aldehydes (Figure 2.5). Hydroxyketones and diones are sugar degradation products from the Maillard reaction, while ammonia and hydrogen sulphide may be produced from cysteine by Strecker degradation or hydrolysis. Lipid oxidation provides long chain aldehydes. This route was confirmed by heating aqueous mixtures containing α-hydroxyketones or α-dicarbonyls, *n*-alkanals and ammonium sulphide (Elmore and Mottram, 1997). 3-Thiazolines, with C_3–C_9 alkyl substituents in the 2-position and methyl or ethyl in positions

4 and/or 5, were major products of the reactions involving hydroxyketones (up to 35% of total volatiles formed) and small amounts of thiazoles were also obtained. Reactions involving diones also resulted in thiazolines, but they also gave thiazoles in similar quantities.

The aroma characteristics have only been examined for a few of these alkyl-substituted heterocyclic compounds, but those which have been reported suggest that such compounds may contribute to fatty, fried aromas (Buttery *et al.*, 1977; Ho *et al.*, 1987). Odour-port GC assessment of the alkyl-3-thiazolines and alkylthiazoles indicated that they did not possess low odour thresholds and, therefore, may not be very important odour impact compounds. However, the reactions by which these compounds are formed compete for the intermediates of the other flavour-forming Maillard reactions, and thereby may modify and control the generation of desirable aroma compounds.

2.7 Conclusion

The flavour of meat and other thermally processed foods derives from a complex series of reactions involving both the Maillard reaction and lipid degradation. Compounds contributing to the roast characteristics derive from the Maillard reaction, whereas those responsible for species characteristics are formed from the degradation of lipid. Sulphur-containing compounds are extremely important in meat flavour. They are only present in low concentrations, but the very low odour threshold values of many of these compounds results in major contributions to the aroma characteristics of cooked meat from very small quantities. Sulphur-substituted furans appear to be particularly important in determining the meaty characteristics of cooked meat. Interactions between compounds produced by the reaction of intermediates of the Maillard reaction with lipid oxidation products leads to a number of heterocyclic compounds with long-chain alkyl substituents, such as pyridines, pyrazines, thiophenes, thiazoles and thiazolines. Some of these compounds may contribute to the aroma of cooked meat, but the reactions by which they are formed will also compete with other flavour-forming Maillard reactions and, thus, influence the overall aroma profiles of cooked meat.

References

Bailey, M.E. (1998). Maillard reactions and meat flavour development. In *The Flavour of Meat and Meat Products*, 2nd edn, ed. F. Shahidi. Blackie, London, Chapter 9.

Boelens, M., van der Linde, L.M., de Valois, P.J., van Dort, H.M. and Takken, H.J. (1974). Organic sulphur compounds from fatty aldehydes, hydrogen sulfide, thiols and ammonia as flavour constituents. *J. Agric. Food Chem.*, **22**, 1071–1076.

Buttery, R.G., Lin. L.C., Teranishi, R. and Mon, T.R. (1977). Roast lamb fat: basic volatile components. *J. Agric. Food Chem.*, **25**, 1227–1229.

Buttery, R.G., Haddon, W.F., Seifert, R.M. and Turnbaugh, J.G. (1984). Thiamin odor and bis(2-methyl-3-furyl)disulphide. *J. Agric. Food Chem.*, **32**, 674–676.

Chiu, E.M., Kuo, M.C., Bruechert, L.J. and Ho, C.T. (1990). Substitution of pyrazines by aldehydes in model systems. *J. Agric. Food Chem.*, **38**, 58–61.

Christie, W.W. (1978). The composition, structure and function of lipids in the tissues of ruminant animals. *Prog. Lipid Res.*, **17**, 111–205.

Coppock, B.M. and MacLeod, G. (1977). The effect of ageing on the sensory and chemical properties of boiled beef aroma. *J. Sci. Food Agric.*, **28**, 206–214.

Elmore, J.S. and Mottram, D.S. (1997). Formation of thiazoles and thiazolines in the reactions of α-hydroxyketones, and ammonium sulfide. *J. Agric. Food Chem.*, **45**, 3595–3602.

Elmore, J.S., Mottram, D.S., Enser, M.B. and Wood, J.D. (1997). Novel thiazoles and 3-thiazolines in cooked beef aroma. *J. Agric. Food Chem.*, **45**, 3603–3607.

Evers, W.J. (1970). 3-Furylsulfides or 3-furanthiols as flavor additives for foods. British Patent 1 256 462.

Evers, W.J., Heinsohn, H.H., Mayers, B.J. and Sanderson, A. (1976). Furans substituted in the three position with sulfur. In *Phenolic, Sulfur and Nitrogen Compounds in Food Flavors*, eds. G. Charalambous and I. Katz. American Chemical Society, Washington DC, pp. 184–193.

Farmer, L.J. and Mottram, D.S. (1990a). Recent studies on the formation of meat aroma compounds. In *Flavour Science and Technology*, eds. Y. Bessiere and A.F. Thomas. Wiley, Chichester, pp. 113–116.

Farmer, L.J. and Mottram, D.S. (1990b). Interaction of lipid in the Maillard reaction between cysteine and ribose: effect of a triglyceride and three phospholipids on the volatile products. *J. Sci. Food Agric.*, **53**, 590–595.

Farmer, L.J. and Mottram, D.S. (1992). Effect of cysteine and ribose on the volatile thermal degradation products of a triglyceride and three phospholipids. *J. Sci. Food Agric.*, **60**, 489–497.

Farmer, L.J. and Mottram, D.S. (1994). Lipid–Maillard interactions in the formation of volatile aroma compounds. In *Trends in Flavour Research*, eds. H. Maarse and D.G. van der Heij. Elsevier, Amsterdam, pp. 313–326.

Farmer, L.J. and Patterson, R.L.S. (1991). Compounds contributing to meat flavour. *Food Chem.*, **40**, 201–205.

Farmer, L.J., Mottram, D.S. and Whitfield, F.B. (1989). Volatile compounds produced in Maillard reactions involving cysteine, ribose and phospholipid. *J. Sci. Food Agric.*, **49**, 347–368.

Feather, M.S. and Harris, J.F. (1973). Dehydration reactions of carbohydrates. *Adv. Carbohyd. Chem. Biochem.*, **28**, 161–224.

Flament, I., Kohler, M. and Aschiero, R. (1976). Sur l'arome de viande de boeuff grillée. II. Dihydro-6,7-5H-cyclopenta-[b]pyrazines, identification et mode de formation. *Helv. Chim. Acta*, **59**, 2308–2313.

Flament, I., Sonnay, P. and Ohloff, G. (1977). Sur l'arome de viande de boeuff grillée. III. Pyrrolo[1,2a]pyrazines, identification et synthèse. *Helv. Chim. Acta.*, **60**, 1872–1883.

Fors, S. (1983). Sensory properties of volatile Maillard reaction products and related compounds. In *The Maillard Reaction in Foods and Nutrition*, eds. G.R. Waller and M.S. Feather, American Chemical Society, Washington, DC, pp. 185–286.

Forss, D.A. (1972). Odour and flavour compounds from lipids. *Prog. Chem. Fats other Lipids*, **13**, 181–258.

Frankel, E.N. (1980). Lipid oxidation. *Prog. Lipid Res.*, **19**, 1–22.

Gasser, U. and Grosch, W. (1988). Identification of volatile flavour compounds with high aroma values from cooked beef. *Z. Lebensm. Unters Forsch.*, **186**, 489–494.

Gasser, U. and Grosch, W. (1990). Primary odorants of chicken broth – a comparative-study with meat broths from cow and ox. *Z. Lebensm. Unters. Forsch.*, **190**, 3–8.

Grosch, W. (1982). Lipid degradation products and flavours. In *Food Flavours*, eds. I.D. Morton and A.J. MacLeod. Elsevier, Amsterdam, pp. 325–398.

Grosch, W., Zeiler-Hilgart, G., Cerny, C. and Guth, H. (1993) Studies on the formation of odorants contributing to meat flavours. In *Progress in Flavour Precursor Studies*, ed. P. Schreier and P. Winterhalter. Allured Publishing, Carol Stream, IL, pp. 329–342.

Güntert, M., Bertram, H.-J., Hopp, R., Silberzahn, W., Sommer, H. and Werhoff, P. (1993). Thermal generation of flavor compounds from thiamin and various amino acids. In *Recent Developments in Flavor and Fragrance Chemistry*, eds. R. Hopp and K. Mori. VCH, Weinheim, pp. 215–240.

Guth, H. and Grosch, W. (1993). 12-Methyltridecanal, a species-specific odorant of stewed beef. *Lebensm. Wiss. Technol.*, **26**, 171–177.

Guth, H. and Grosch, W. (1995). Dependence of the 12-methyltridecanal concentration in beef on the age of the animal. *Z. Lebensm. Unters. Forsch.*, **201**, 25–26.
Hartman, G.J., Jin, Q.Z., Collins, G.J., Lee, K.N., Ho, C.T. and Chang, S.S. (1983). Nitrogen-containing heterocyclic compounds identified in the volatile flavour constituents of roast beef. *J. Agric. Food Chem.*, **31**, 1030–1033.
Ho, C.T. and Carlin, J.T. (1989). Formation and aroma characteristics of heterocyclic compounds in foods. In *Flavor Chemistry Trends and Developments*, eds. R. Teranishi, R.G. Buttery and F. Shahidi. American Chemical Society, Washington, DC, pp. 92–104.
Ho, C.T., Carlin, J.T. and Huang, T.C. (1987). Flavour development in deep-fat fried foods. In *Flavour Science and Technology*, eds. M. Martens, G.A. Dalen and H. Russwurm. Wiley, Chichester, pp. 35–42.
Ho, C.T., Lee, K.N. and Jin, Q.Z. (1983). Isolation and identification of flavour compounds in fried bacon. *J. Agric. Food Chem.*, **31**, 336–342.
Hodge, J.E. (1953). Chemistry of browning reactions in model systems. *J. Agric. Food Chem.*, **1**, 928–943.
Hornstein, I. and Crowe, P.F. (1960). Flavour studies on beef and pork. *J. Agric. Food Chem.*, **8**, 494–498.
Hornstein, I. and Crowe, P.F. (1963). Meat flavour: lamb. *J. Agric. Food Chem.*, **11**, 147–149.
Hsu, C.M., Peterson, R.J., Jin, Q.Z., Ho, C.T. and Chang, S.S. (1982). Characterization of new volatile compound in the neutral fraction of roast beef flavour. *J. Food Sci.*, **47**, 2068–2071.
Hurrell, R.F. (1982). Maillard reaction in flavour. In *Food Flavours*, eds. I.D. Morton and A.J. MacLeod. Elsevier, Amsterdam, pp. 399–437.
Hwang, S.S., Carlin, J.T., Bao, Y., Hartman, G.J. and Ho, C.T. (1986). Characterisation of volatile compounds generated from the reactions of aldehydes with ammonium sulphide. *J. Agric. Food Chem.*, **34**, 538–542.
Kramlich, W.E. and Pearson, A.M. (1960). Separation and identification of cooked beef flavour components. *Food Res.*, **25**, 712–719.
Lawrie, R.A. (1992). *Meat Science*, 5th edn. Pergamon Press, Oxford.
Ledl, F. and Schleicher, E. (1990). New aspects of the Maillard reaction in food and the human body. *Angewandte Chemie* International Edition in English, **29**, 565–706.
Macey, R.L. Jr., Naumann, N.D. and Bailey, M.E. (1964). Water-soluble flavour and odour precursors of meat. *J. Food Sci.* **29**, 136–148.
MacLeod, G. (1986). The scientific and technological basis of meat flavours. In *Developments in Food Flavours*, eds. G.G. Birch and M.G. Lindley. Elsevier, London, pp. 191–223.
MacLeod, G. and Ames, J.M. (1986). 2-Methyl-3-(methylthio)furan: a meaty character impact aroma compound identified from cooked beef. *Chem. Ind. (London)*, 175–177.
MacLeod, G. and Seyyedain-Ardebili, M. (1981). Natural and simulated meat flavours (with particular reference to beef). *CRC Crit. Rev. Food Sci. Nutr.*, **14**, 309–437.
Madruga, M.S. and Mottram, D.S. (1995). The effect of pH on the formation of Maillard-derived aroma volatiles using a cooked meat system. *J. Sci. Food Agric.*, **68**, 305–310.
Min, D.B., Ina, K., Peterson, R.J. and Chang, S.S. (1979). Preliminary identification of volatile flavour compounds in the neutral fraction of roast beef. *J. Food Sci.*, **44**, 639–642.
Morton, I.D., Akroyd, P. and May, C.G. (1960). Flavouring substances and their preparation. British Patent 836 694.
Mottram, D.S. (1985). The effect of cooking conditions on the formation of volatile heterocyclic compounds in pork. *J. Sci. Food Agric.*, **36**, 377–382.
Mottram, D.S. (1991). Meat. In *Volatile Compounds in Foods and Beverages*, ed. H. Maarse. Marcel Dekker, New York, pp. 107–177.
Mottram, D.S. (1994). Flavour compounds formed during the Maillard reaction. In *Thermally Generated Flavors. Maillard, Microwave, and Extrusion Processes*, eds. T.H. Parliment, M.J. Morello and R.J. McGorrin. American Chemical Society, Washington, DC, pp. 104–126.
Mottram, D.S. and Edwards, R.A. (1983). The role of triglycerides and phospholipids in the aroma of cooked beef. *J. Sci. Food Agric.*, **34**, 517–522.
Mottram, D.S. and Salter, L.J. (1989). Flavour formation in meat-related Maillard systems containing phospholipids. In *Thermal Generation of Aromas*, eds. T.H. Parliment, R.J. McGorrin and C.T. Ho. American Chemical Society, Washington, DC, pp. 442–451.
Mottram, D.S., Edwards, R.A. and MacFie, H.J.H. (1982). A comparison of the flavour volatiles from cooked beef and pork meat systems. *J. Sci. Food Agric.*, **33**, 934–944.

Mottram, D.S., Szauman-Szumski, C. and Dodson, A. (1996). Interaction of thiol and disulfide flavor compounds with food components. *J. Agric. Food Chem.*, **44**, 2349–2351.

Noleau, I. and Toulemonde, B. (1987). Volatile components of roast chicken fat. *Lebensm. Wiss. Technol.*, **20**, 37–41.

Nursten, H.E. (1986). Aroma compounds from the Maillard reaction. In *Developments in Food Flavours*, eds. G.G. Birch and M.G. Lindley. Elsevier, London, pp. 173–190.

Peer, H.G. and van den Ouweland, G.A.M. (1968). Synthesis of 4-hydroxy-5-methyl-2,3-dihydro-3-furanone from D-ribose-5-phosphate. *Recl. Trav. Chim. Pays-Bas*, **87**, 1017–1020.

Petit, A.O. and Hruza, D.A. (1974). Comparative study of flavour properties of thiazole derivatives. *J. Agric. Food Chem.*, **22**, 264–269.

Salter, L.J., Mottram, D.S. and Whitfield, F.B. (1988). Volatile compounds produced in Maillard reactions involving glycine, ribose and phospholipid. *J. Sci. Food Agric.*, **46**, 227–242.

Schutte, L. (1974). Precursors of sulphur-containing flavour compounds. *CRC Crit. Rev. Food Technol.*, **4**, 457–505.

Tang, J., Jin, Q.Z., Shen, G.H., Ho, C.T. and Chang, S.S. (1983). Isolation and identification of volatile compounds from fried chicken. *J. Agric. Food Chem.*, **31**, 1287–1292.

Tressl, R., Helak, B., Kersten, E. and Nittka, C. (1993). Formation of flavor compounds by Maillard reaction. In *Recent Developments in Flavor and Fragrance Chemistry*, eds. R. Hopp and K. Mori. VCH, Weinheim, pp. 167–181.

van den Ouweland, G.A.M. and Peer, H.G.F. (1972). Mercaptofurane and mercaptothiophene derivatives. British Patent 1 283 912.

van den Ouweland, G.A.M. and Peer, H.G.F. (1975). Components contributing to beef flavour. Volatile compounds produced by the reaction of 4-hydroxy-5-methyl-3(2H)-furanone and its thio analog with hydrogen sulfide. *J. Agric. Food Chem.*, **23**, 501–505.

van den Ouweland, G.A.M., Peer, H.G. and Tjan, S.B. (1978). Occurrence of Amadori and Heyns rearrangement products in processed foods and their role in flavour formation. In *Flavor in Foods and Beverages*, eds. G. Charalambous and G.E. Inglett. Academic Press, New York, pp. 131–143.

van der Linde, L.M., van Dort, J.M., de Valois, P., Boelens, H. and de Rijke, D. (1979). Volatile compounds from thermally degraded thiamin. In *Progress in Flavour Research*, eds. D.G. Land and H.E. Nursten. Applied Science, London, pp. 219–224.

Vernin, G. and Parkanyi, C. (1982). Mechanisms of formation of heterocyclic compounds in Maillard and pyrolysis reactions. In *Chemistry of Heterocyclic Compounds in Flavours and Aromas*, ed. G. Vernin. Ellis Horwood, Chichester, pp. 151–207.

Wasserman, A.E. (1979). Chemical basis for meat flavour: a review. *J. Food Sci.*, **44**, 6–11.

Wasserman, A.E. and Gray, N. (1965). Meat flavour. I. Fractionation of water-soluble flavour precursors of beef. *J. Food Sci.*, **30**, 801–807.

Wasserman, A.E. and Spinelli, A.M. (1972). Effect of some water-soluble components on aroma of heated adipose tissue. *J. Agric. Food Chem.*, **20**, 171–174.

Watanabe, K. and Sato, Y. (1970). Conversion of some saturated fatty acids, aldehydes and alcohols into γ- and δ-lactones. *Agric. Biol. Chem.*, **34**, 464–472.

Watanabe, K. and Sato, Y. (1971). Some alkyl-substituted pyrazines and pyridines in the flavour components of shallow fried beef. *J. Agric. Food Chem.*, **19**, 1017–1019.

Werkhoff, P., Brüning, J., Emberger, R., Güntert, M., Kopsel, M., Kuhn, W. and Surburg, H. (1990). Isolation and characterisation of volatile sulphur-containing meat flavour components in model systems. *J. Agric. Food Chem.*, **38**, 777–791.

Werkhoff, P., Brüning, J., Emberger, R., Güntert, M. and Hopp, R. (1993). Flavor chemistry of meat volatiles: New results on flavor components from beef, pork and chicken. In *Recent Developments in Flavor and Fragrance Chemistry*, eds. R. Hopp and K. Mori. VCH, Weinheim, pp. 183–213.

Whitfield, F.B. (1992). Volatiles from interactions of Maillard reactions and lipids. *Crit. Rev. Food Sci. Nutr.*, **31**, 1–58.

Whitfield, F.B., Mottram, D.S., Brock, S., Puckey, D.J. and Salter, L.J. (1988). The effect of phospholipid on the formation of volatile heterocyclic compounds in heated aqueous solutions of amino acids and ribose. *J. Sci. Food Agric.*, **42**, 261–272.

Wilson, R.A. Mussinan, C.J., Katz, I. and Sanderson, A. (1973). Isolation and identification of some sulphur chemicals present in pressure-cooked beef. *J. Agric. Food Chem.*, **21**, 873–876.

Wong, E., Nixon, L.N. and Johnson, C.B. (1975). Volatile medium chain fatty acids and mutton fat. *J. Agric. Food Chem.*, **23**, 495–498.

3 The flavour of beef

G. MACLEOD

3.1 Introduction

The flavour of beef has been investigated more extensively than any other meat flavour, probably because of its greater consumer popularity, and hence its commerical significance in the creation of successful simulated meat flavourings. Literature reports over the past 30 years show that the flavour of beef is highly complex. In its simplest format, it consists of taste-active compounds, flavour enhancers and aroma components.

3.2 Taste-active compounds

With regard to taste (MacLeod and Seyyedain-Ardebili, 1981; MacLeod, 1986; Kuninaka, 1981; Haefeli and Glaser, 1990), sweetness has been associated with glucose, fructose, ribose and several L-amino acids such as glycine, alanine, serine, threonine, lysine, cysteine, methionine, asparagine, glutamine, proline and hydroxyproline. Sourness stems from aspartic acid, glutamic acid, histidine and asparagine, together with succinic, lactic, inosinic, *ortho*-phosphoric and pyrrolidone carboxylic acids. Saltiness is largely due to the presence of inorganic salts and the sodium salts of glutamate and aspartate; while bitterness may be derived from hypoxanthine together with anserine, carnosine and other peptides, and also the L-amino acids histidine, arginine, lysine, methionine, valine, leucine, isoleucine, phenylalanine, tryptophan, tyrosine, asparagine and glutamine.

The umami taste has a characteristic savoury quality and is supplied by glutamic acid, monosodium glutamate (MSG), 5′-inosine monophosphate (IMP), 5′-guanosine monophosphate (GMP) and certain peptides. Although generally speaking, glutamate is the most important contributor, its presence at a lower concentration in beef than in pork or chicken, for example, gives rise to a lower perceived umami taste intensity in beef (Kato and Nishimura, 1989; Kawamura, 1990). Similarly, the effect of conditioning these three species has shown a significant increase in intensity of the savoury, brothy taste in pork and chicken after aging, yet no significant difference in conditioned beef (Nishimura *et al.*, 1988). This disparity was paralleled by the observation that the increased concentration of free amino acids and of oligopeptides on aging was significantly smaller in beef than in pork or chicken (Nishimura *et al.*, 1988). However, subsequent rates of changes in concentration of these nonvolatile components and of others on continued heating, and the associated aroma manifestations, could well alter the overall comparative conclusions on flavour improvement, in general, on aging different meat species.

3.3 Flavour enhancers

A more important sensory contribution than the taste characteristics of glutamic acid, MSG, IMP and, to a lesser extent, GMP in beef, is their flavour enhancing property. It has been proposed that our sensory receptors for flavour enhancement are independent and sterically different from the traditional basic taste receptors (Kuninaka, 1981; Kawamura, 1990). The 5′-ribonucleotides have strong flavour potentiating effects individually, but more importantly, they exhibit a potent synergistic effect when present, as in meat, in conjunction with glutamic acid or MSG (Kawamura, 1990). It appears that, in some mammals, this synergy is due to an induced increased strength of binding of glutamate to the receptor protein site, whereas in other mammals, an enhanced amount of glutamate is actually bound (Kawamura, 1990). The 5′-nucleotides are reported to enhance meaty, brothy, MSG-like, mouthfilling, dry and astringent qualities; they suppress sulphurous and HVP-like notes, while sweet, sour, oily/fatty, starchy and burnt qualities remain unchanged (Kuninaka, 1981).

Thermal decomposition of both classes of flavour enhancers may occur, with a resultant loss of activity. For example, at 121°C and a pH of 4.5–6.5 (e.g. during canning), an initial loss of the phosphate group from IMP and GMP, converting the nucleotide into the corresponding nucleoside, is followed by slow hydrolysis releasing the base – either hypoxanthine (from IMP) or guanine (from GMP) (Shaoul and Sporns, 1987). The first triggering reaction of the phosphate loss is depressed in the presence of certain divalent metals, e.g. calcium ions (Kuchiba *et al.*, 1990). Under similar heating conditions (100°C/pH 4–6), glutamic acid and MSG are converted into pyrrolidone carboxylic acid (PCA) (Gayte-Sorbier *et al.*, 1985). The reverse reaction is also possible, but is favoured by extreme pH values of <2.5 or >11 (Airaudo *et al.*, 1987). Not only is there no flavour enhancement activity from any of the decomposition products (Kuninaka, 1981; Gayte-Sorbier, 1985), but a distinct off-flavour is associated with PCA at certain concentrations.

3.4 Aroma components

Aroma components are generated in beef from nonvolatile precursors on cooking. Primary reactions occurring are (1) lipid oxidation/degradation; (2) thermal degradation and inter-reactions of proteins, peptides, amino acids, sugars and ribonucleotides; and (3) thermal degradation of thiamine. But reaction products become reactants, and the end result is a complex and intertwining network of reactions. In consequence, the most recent edition of the now classic TNO-CIVO publication *Volatile Compounds in Food* lists 880 volatile components reported from cooked beef (Maarse and Visscher, 1989). To place this figure in a better perspective, a rough breakdown of the chemical classes represented in shown in Table 3.1 (Maarse and Visscher, 1989).

With such a large number of potential contributors to the sensory perception of cooked beef aroma, the critical question is, 'What is the relative sensory significance of these volatiles?' The answer is not totally clear, but three conclusions can be drawn. First, many are relatively unimportant. Second, the term 'meatiness' can be cleanly dissected sensorially into about ten different odour qualities (Galt and MacLeod, 1983), in which case, many of the identified volatiles are acting as 'aroma modifiers' contributing buttery, caramel, roast, burnt, sulphurous, green, fragrant, oily/fatty and nutty qualities. Structure–activity correlations, or at least associations, do exist in the literature for many of these odour qualities. Third, some aroma components do contribute a truly specific 'meaty' odour, and are therefore

Table 3.1 Chemical classes of aroma components reported from cooked beef

Class of compound	Number of components reported
Aliphatic	
Hydrocarbons	103
Alcohols	70
Aldehydes	55
Ketones	49
Carboxylic acids	24
Esters	56
Ethers	7
Amines	20
Alicyclic	
Hydrocarbons	44
Alcohols	3
Ketones	18
Heterocyclic	
Lactones	38
Furans and derivatives	44
Thiophenes and derivatives	40
Pyrroles and derivatives	20
Pyridines and derivatives	21
Pyrazines and derivatives	54
Oxazol(in)es	13
Thiazol(in)es	29
Other sulphur heterocyclics	13
Benzenoids	80
Sulphur compounds (not heterocyclic)	72
Miscellaneous	7

From Maarse and Visscher (1989).

character impact compounds. Several potent, key and trace meaty compounds are present in natural cooked beef aromas and many remain to be identified. Clearly, the ultimate chemical and sensory resolution of the beef flavour complex relies primarily on defining these particular meaty/beefy compounds and the reactions which generate them.

A search of the literature for individual chemical compounds described as meaty (e.g. MacLeod, 1986; Werkhoff *et al.*, 1989, 1990, 1996; Guntert *et al.*, 1990) shows that, of the 880 cooked beef aroma components identified to date (Maarse and Visscher, 1989; Werkhoff *et al.*, 1989, 1990, 1996), only 25 have been reported to possess a meaty odour. These are presented in Figure 3.1. The meaty quality of several of these has been under attack by some workers, probably because it is a difficult quality to define with precision. Nevertheless, they are included for completeness. Also, many compounds are meaty only at certain concentrations, usually very low concentrations which are often unspecified. In the discussion below, which considers how various cooked beef aroma components are formed during cooking, the meaty compounds of Figure 3.1 are numbered **1**–**25**. All other compounds are quoted without a numerical code label.

3.4.1 Effect of heat on sugars and/or amino acids

To the seasoned flavour chemist, even a cursory glance at Table 3.1 shows that a high proportion of the total number of volatiles identified is derived from reactions which result from the effect of heat on sugars and or amino acids. Strecker degradations and Maillard reactions are critical contributors, both chemically and sensorially. Furthermore, a host of secondary reactions can occur involving the products of the above reactions (e.g. H_2S, NH_3, thiols and simple carbonyl compounds), thereby increasing quite considerably the variety of compounds which may be formed.

Strecker degradation is depicted in Figure 3.2 (MacLeod and Ames, 1988), and several Strecker aldehydes are well-known cooked beef aroma components, e.g. acetaldehyde (from alanine), methylpropanal (from valine), 2-methylbutanal (from isoleucine), 3-methybutanal (from leucine), phenylacetaldehyde (from phenylalanine) and methional, which readily decomposes into methanethiol, dimethyl sulphide, dimethyl disulphide and propenal (from methionine).

The Maillard reaction between compounds containing a free amino group (e.g. amino acids, amines, peptides, proteins, ammonia) and carbonyl compounds (e.g. aldehydes, ketones, reducing sugars) is an open-ended complex reaction consisting of a series of inter-reactions and decompositions and resulting in numberous volatile products. The first stage in the reaction of an α-amino acid with an aldose or ketose is the formation of an Amadori or Heyns compound, respectively. Both are non-volatile but they are heat labile and they decompose thermally. The second gross stage

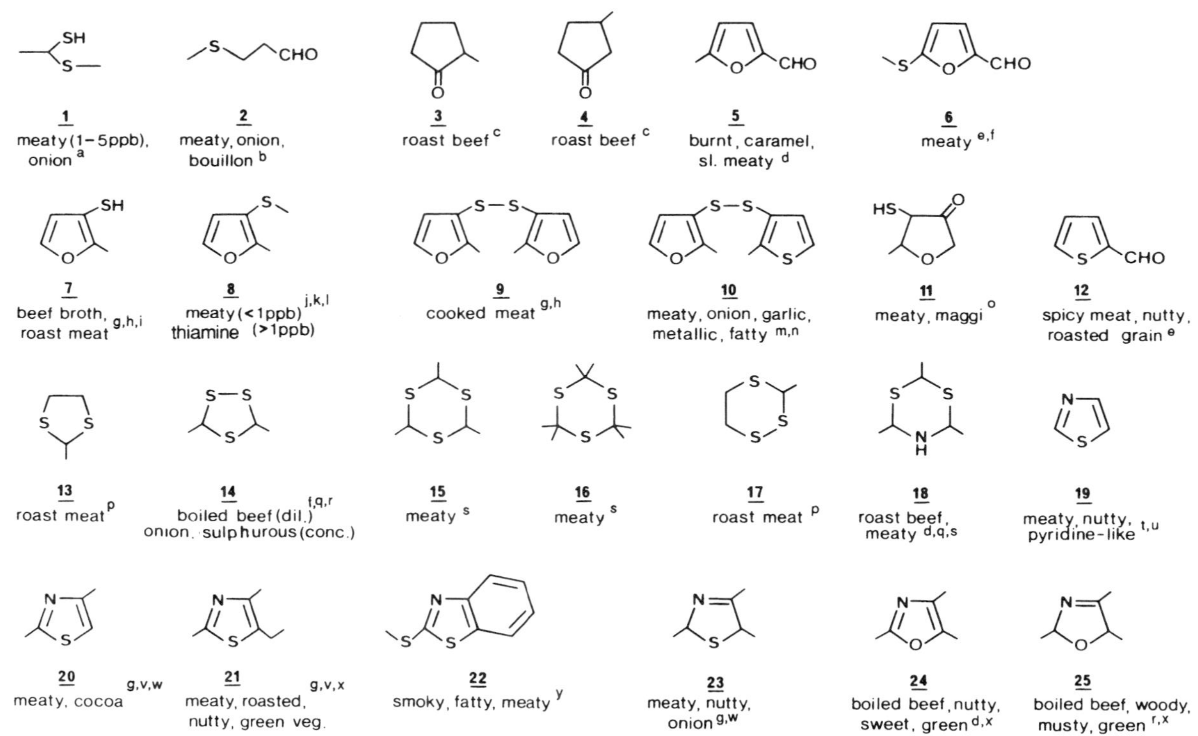

Figure 3.1 Compounds identified from cooked beef aromas (Maarse and Vischer, 1989; Werkhoff *et al.*, 1989, 1990) and reported to possess meaty odour ([b]Arctander, 1969; [t]Baltes, 1979; [a]Brinkman *et al.*, 1972; [h]Evers *et al.*, 1976; [u]Furia and Bellanca, 1975; [l]IFF Inc., 1979; [g]Katz, 1981; [q]Kubota *et al.*, 1980; Macleod, 1986; [j]MacLeod and Ames, 1986; [x]Mussinan *et al.*, 1976; [c]Nishimura *et al.*, 1980; [d]Ohloff and Flament, 1978; [w]Pittet and Hruza, 1974; [e]Roedel and Kruse, 1980; [r]Self *et al.*, 1963; [f]Shibamoto, 1980; [k]Tressl and Silwar, 1981; [p]Tressl *et al.*, 1983; [o]van den Ouweland and Peer, 1975; [i]van der Linde *et al.*, 1979; [v]Vernin, 1979, [y]1982; [n]Werkhoff *et al.*, 1989, [m]1990; [s]Wilson *et al.*, 1974).

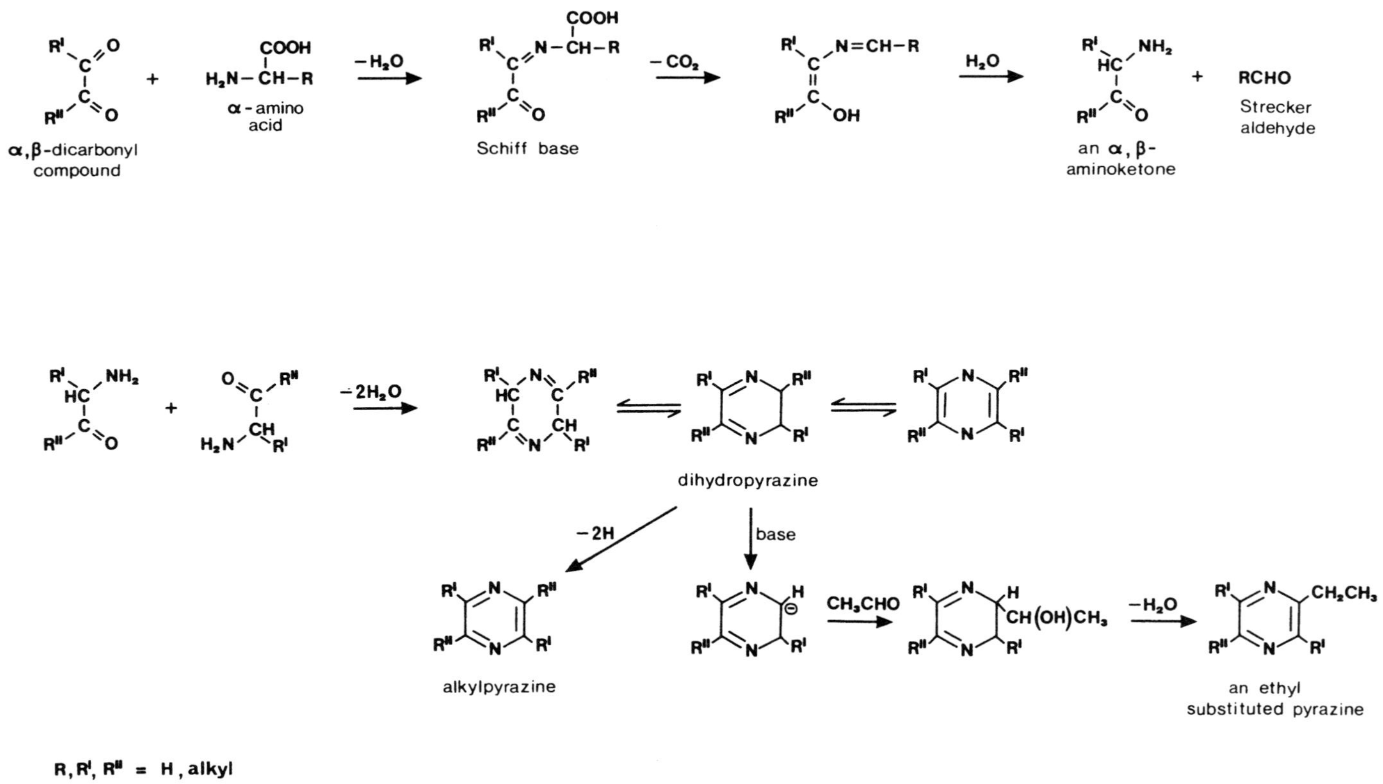

Figure 3.2 The Strecker degradation of α-amino acids and the subsequent formation of alkylpyrazines. (Reprinted with permission from MacLeod, G. and Ames, J.M. (1988) *CRC Crit. Revs Food Sci. Nutr.*, **27**, 219–400. Copyright CRC Press, Inc., Boca Raton, FL.)

involves rearrangement and subsequent decomposition of the Amadori and Heyns compounds, as exemplified in Figure 3.3 (MacLeod and Ames, 1988). Thus, 2-furaldehyde forms from pentoses, and 5-hydroxymethyl-2-furaldehyde from hexoses. Additionally, a number of dicarbonyl and hydroxycarbonyl fragmentation products can be formed, e.g. glyoxal, glycolaldehyde, glyceraldehyde, pyruvaldehyde, hydroxyacetone, dihydroxyacetone, diacetyl, acetoin and hydroxydiacetyl. These are all highly reactive compounds, and therefore the final global stage of the Maillard reaction can be considered as the decomposition and further inter-reaction of furanoids and aliphatic carbonyl compounds, formed as exemplified in Figure 3.3, with other reactive species present in the system, e.g. H_2S, NH_3, amines and aldehydes. Yayalan has recently proposed that, since the currently accepted mechanisms of 1,2- and 2,3-enolizations of the open chain form of the Amadori compounds, and their subsequent dehydration (as just described), do not adequately account for many products observed in model systems and in foods, alternative mechanisms should be considered, e.g. direct dehydrations from cyclic forms of the Amadori compounds (Yayalan, 1990). By whichever mechanism, the final stage reactions are extremely large in number and varied in nature and cannot be generalized. Examples are as follows.

Furanoids often arise, as shown in Figure 3.3, via the 1,2-enolization pathway. One exception is 2-acetylfuran which probably is derived from a 1-deoxyhexosone intermediate (Tressl *et al.*, 1979). Some other furanoids, e.g. 4-hydroxy-5-methyl-(2*H*)furan-3-one and 2,5-dimethyl-4-hydroxy-(2*H*)furan-3-one, also arise by this 2,3-enolization route. 2-Furaldehyde is an important precursor of other furanoids, and indeed of other heterocyclic compounds too, such as thiophenes and pyrroles. This is because the oxygen of the furan ring, in the presence of H_2S or NH_3, may be substituted by sulphur or nitrogen, forming the corresponding thiophenes or pyrrole derivatives. The formation of such compounds is shown in Figure 3.4 (MacLeod and Ames, 1988).

The most likely pathway for the formation of alkylpyrazines is by self-condensation of the α,β-aminoketones formed during Strecker degradation, as shown in Figure 3.2. While alkylpyrazines are frequently occurring volatiles in many heated foods, the bicyclic pyrazines, namely the alkyl-(5*H*)6,7-dihydrocyclopenta[*b*]pyrazines and the pyrrolo[1,2-*a*] pyrazines, are unique to grilled and roasted beef aromas (Maarse and Visscher, 1989). The former arise from reaction of an alkylhydroxy-cyclopentenone (e.g. cyclotene) with an α,β-dicarbonyl and NH_3 as shown in Figure 3.5 (Flament *et al.*, 1976; MacLeod and Seyyedain-Ardebili, 1981), while the latter derive from condensation of an α,β-aminoketone (from Strecker degradation) with a hydroxy α,β-dicarbonyl compound (formed from reducing sugars) as shown in Figure 3.6 (Flament *et al.*, 1977; MacLeod and Seyyedain-Ardebili, 1981).

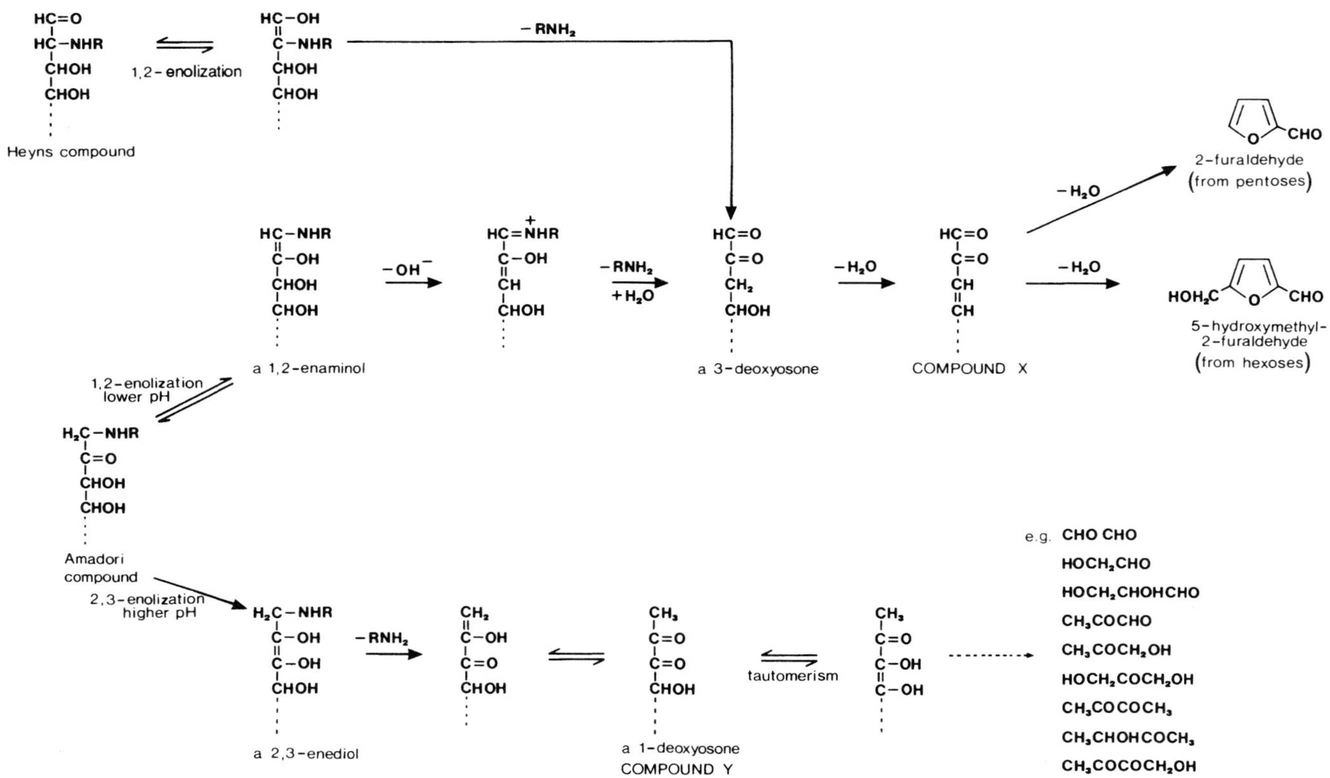

Figure 3.3 Decomposition of Amadori and Heyns compounds. (Reprinted with permission from MacLeod, G. and Ames, J.M. (1988) *CRC Crit. Revs Food Sci. Nutr.*, **27**, 219–400. Copyright CRC Press, Inc., Boca Raton, FL.)

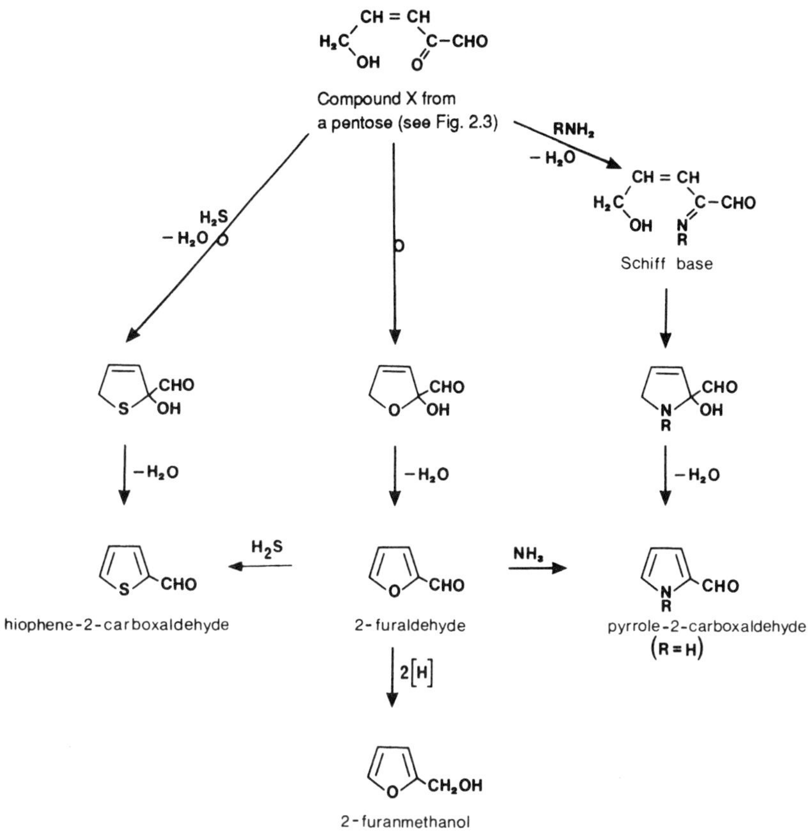

Figure 3.4 Formation of some related furan, thiophene and pyrrole derivatives from Maillard intermediates. (Reprinted with permission from MacLeod, G. and Ames, J.M. (1988) *CRC Crit. Revs Food Sci. Nutr.*, **27**, 219–400. Copyright CRC Press, Inc., Boca Raton, FL.)

Maillard type reactions of cysteine and/or cysteine have been extensively studied, and a recent review summarizes the various cysteine-specific Maillard products (Tressl *et al.*, 1989). Reaction products include carbonyl compounds, amines, benzenoids, acids, lactones, thiols, sulphides, furanoids, thiophenoids, thianes, thiolanes, pyrroles, pyridines, pyrazines, thiazoles, thiazolines and thiazolidines. The type of products formed, and their concentrations, are strongly influenced in model systems by the solvent. For example, reaction of cysteine/dihydroxyacetone in water favours the formation of sulphur products, especially mercaptopropanone and some thiophenes, whereas in triacylglycerol or glycerine systems, dehydration reactions are favoured producing preferentially various pyrazines and thiazoles (Okumura *et al.*, 1990).

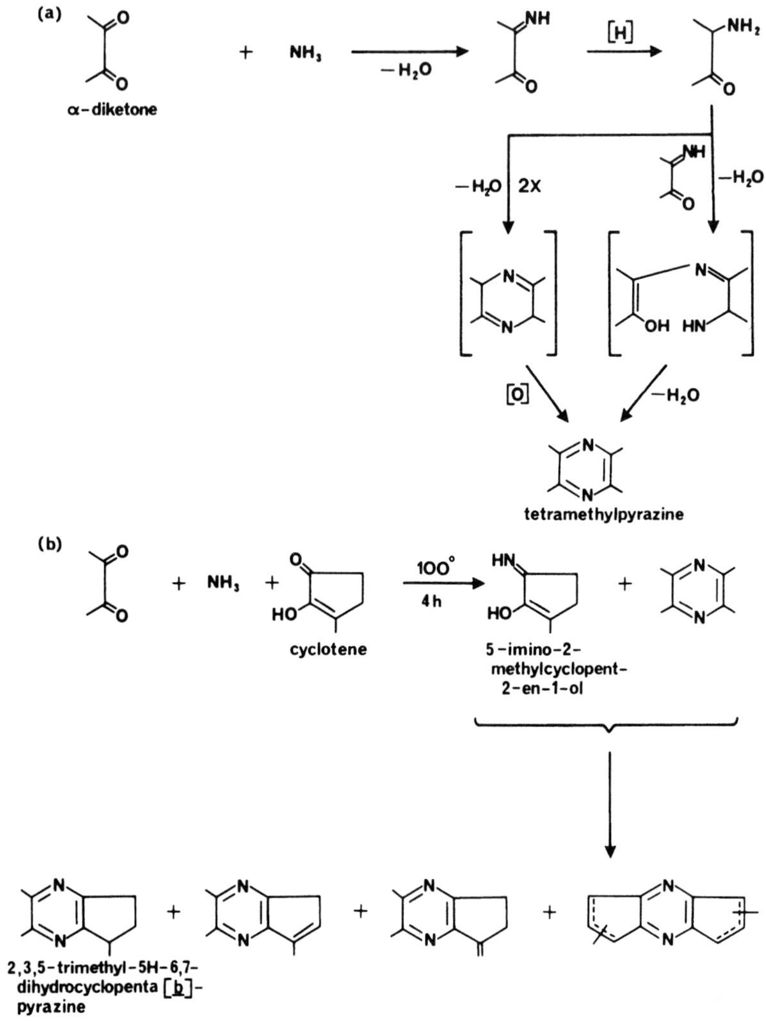

Figure 3.5 Formation of alkyl-(5*H*)6,7-dihydrocyclopenta[*b*]pyrazines by Maillard reaction. (Reprinted with permission from MacLeod, G. and Seyyedain-Ardebili, M. (1981) *CRC Crit. Revs Food Sci. Nutr.*, **14**, 309–437. Copyright CRC Press, Inc., Boca Raton, FL.)

The long list of chemical classes deriving from cysteine/cystine systems, and indeed the identification of several representatives within each class, serves to stress the considerable impact of cysteine (in particular) as a Strecker/Maillard reactant. This significance of cysteine is attributed to the very high reactivity of its initial Strecker degradation products, namely mercaptoacetaldehyde, acetaldehyde, H_2S and NH_3, all of which undergo numerous further reactions. Sheldon *et al.* (1986) recently showed that

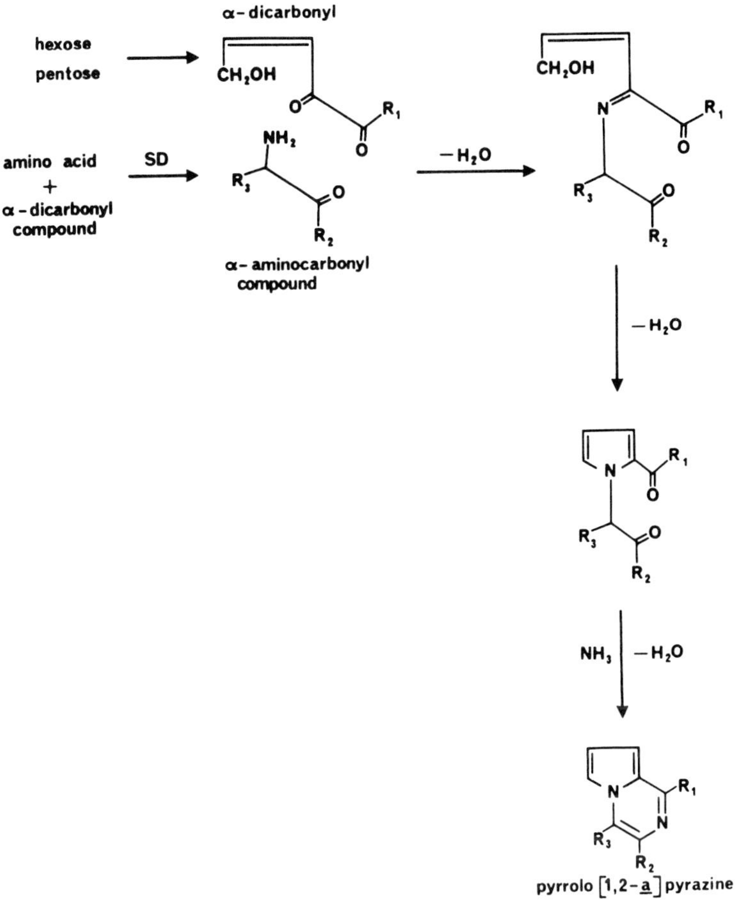

Figure 3.6 Formation of pyrrolo[1,2-*a*]pyrazines by Maillard reaction. (Reprinted with permission from MacLeod, G. and Seyyedain-Ardebili, M. (1981) *CRC Crit. Revs Food Sci. Nutr.*, **14**, 309–437. Copyright CRC Press, Inc., Boca Raton, FL.)

some Maillard reaction between cysteine and glucose occurs even at room temperature in the dark. At high temperatures approximating roasting conditions (i.e. Shigematsu conditions), the products formed reflect the fact that amino acid pyrolysis (rather than Strecker degradation) is an important reaction (de Rijke *et al.*, 1981), and cysteine is primarily transformed into six products, namely mercaptoacetaldehyde, acetaldehyde, cysteamine, ethane-1,2-dithiol, H_2S and NH_3 (Tressl *et al.*, 1983). These highly reactive compounds trigger a number of aldol and other condensation reactions with sugar degradation products and with each other. For example, the reaction of ethane-1,2-dithiol/acetaldehyde/H_2S gives rise to 2-methyl-1,3-dithiolane (**13**) and 3-methyl-1,2,4-trithiane (**17**) (Tressl *et al.*, 1983).

Further examples of Figure 3.1 compounds arising from Maillard type reactions are thiazole (**19**) from cysteine/pyruvaldehyde (Kato *et al.*, 1973), 2,4,5-trimethyloxazole (**24**) from cysteine/butanedione (Ho and Hartman, 1982), thiophene-2-carboxaldehyde (**12**) (Scanlan *et al.*, 1973), 3-methyl-1,2,4-trithiane (**17**) and thialdine (**18**) (de Rijke *et al.*, 1981) from cysteine/glucose; 3-methyl-1,2,4-trithiane (**17**) and 2,4-dimethyl-5-ethylthiazole (**21**) from cysteine/cystine–ribose (Mulders, 1973), 2-methylfuran-3-thiol (**7**), 2-methyl-3-(methylthio)furan (**8**) and 2,4-dimethyl-5-ethylthiazole (**21**) from cysteine/ribose (Whitfield *et al.*, 1988; Farmer *et al.*, 1989; Mottram and Salter, 1989; Farmer and Mottram, 1990; Mottram and Leseigneur, 1990), methional (**2**) and 2-methyl-3-(methylthio)furan (**8**) from methionine/reducing sugar (Tressl *et al.*, 1989) and the 2- and 3-methylcyclopentanones (**3**, **4**) from cylcotene/H_2S/NH_3 reaction (Nishimura *et al.*, 1980).

Frequently the effective sulphur reactant is H_2S as shown, for example in Figure 3.4, explaining the generation of some thiophenes. Thiazol(in)es derive from the reaction of an α,β-dicarbonyl compound, an aldehyde, H_2S and NH_3 as shown in Figure 3.7 (MacLeod and Ames, 1988; Takken *et al.*, 1976). In the absence of H_2S, parallel reactions yield oxazol(in)es.

On a similar theme, inter-reactions involving acetaldehyde, H_2S and NH_3 explain the formation of several other cyclic sulphur compounds identified in cooked beef aromas, e.g. 3,5-dimethyl-1,2,4-trithiolane (**14**), trithioacetaldehyde (**15**) and thialdine (**18**). These reactions are shown in Figure 3.8 (MacLeod and Seyyedain-Ardebili, 1981; Takken *et al.*, 1976). The thiadiazine is decomposed on storage into thialdine (Kawai *et al.*, 1985). Interestingly, the isomeric trithiolanes (**14**) are major products (63% of total volatiles) from heated glutathione, whereas H_2S/NH_3 interaction products such as the dithiazine and thiadiazine predominate (70% of total volatiles) from heated cysteine (Zhang *et al.*, 1988). This is explained by the fact that a very much milder degradation occurs with glutathione and it releases H_2S more readily than NH_3 (Zhang *et al.*, 1988). It is generally established that glutathione is the main H_2S precursor in meat during the early stages of cooking, but cysteine takes over this major role on prolonged heating. The formation of another reportedly important beef aroma component is shown in Figure 3.8, i.e. 1-(methylthio)ethanethiol (**1**) which derives from reaction of acetaldehyde with methanethiol and H_2S.

The reactants required for the reactions just described are liberated on heating either aqueous cysteine (Shu *et al.*, 1985b) or cystine (Shu *et al.*, 1985a) alone. Shu *et al.* (1985b) have reported the thermal degradation products of cysteine in water at 160°C/30 min at different pH values. They identified 26 volatile products of which five have been described as meaty, i.e. 2-methyl-1,3-dithiolane (**13**), 3-methyl-1,2-dithiolane, 3,5-dimethyl-1,2,4-trithiolane (**14**), 3-methyl-1,2,4-trithiane (**17**), thiazole (**19**) and 1,2,3-trithiacyclohept-5-ene. The most vigorous reaction occurred at the

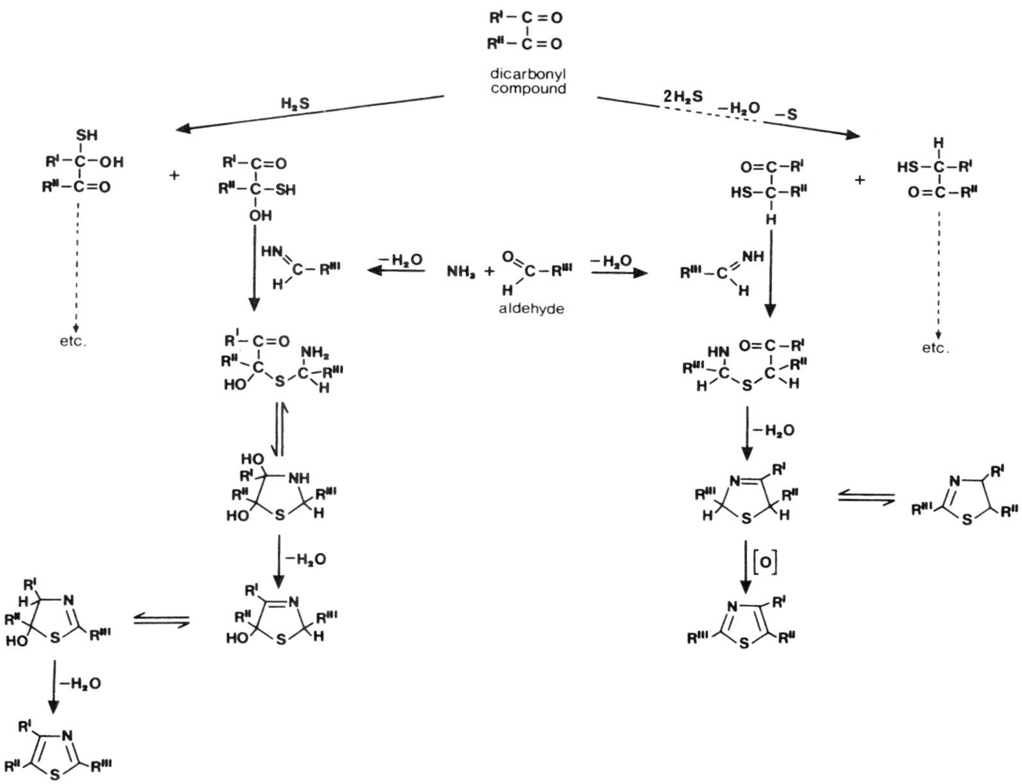

Figure 3.7 Proposed formation pathway for thiazoles from Maillard reactions. (Reprinted with permission from MacLeod, G. and Ames, J.M. (1988) *CRC Crit. Revs Food Sci. Nutr.*, **27**, 219–400. Copyright CRC Press, Inc., Boca Raton, FL.)

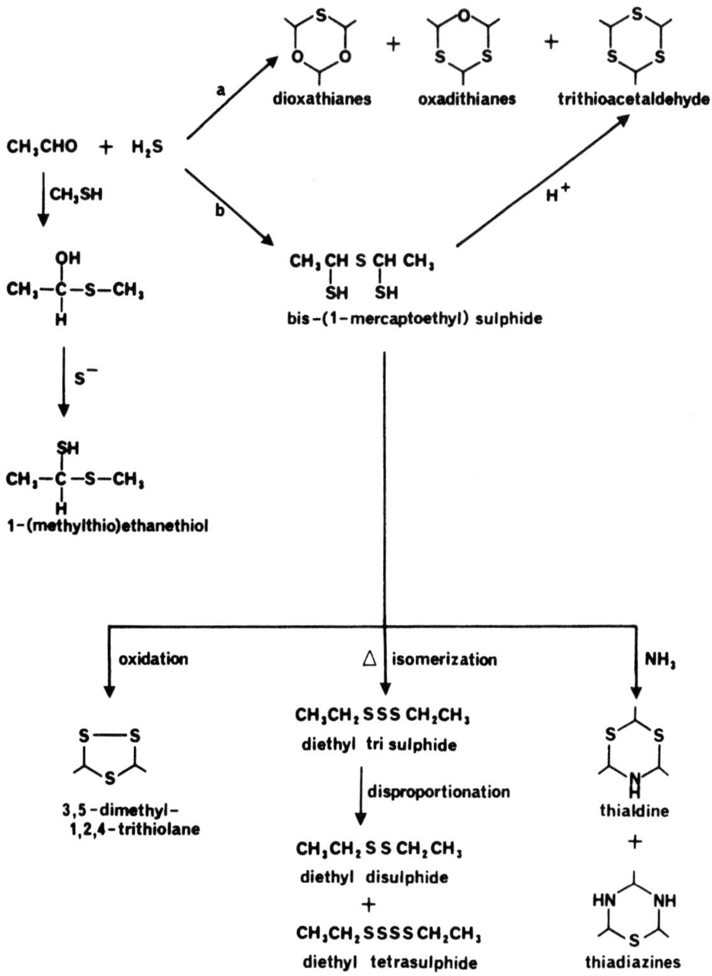

Figure 3.8 Some reactions of aldehydes with hydrogen sulphide/ammonia/methanethiol: a = atmospheric pressure; b = closed vessel + excess hydrogen sulphide. (Reprinted with permission from MacLeod, G. and Seyyedain-Ardebili, M. (1981) *CRC Crit. Revs Food Sci. Nutr.*, **14**, 309–437. Copyright CRC Press, Inc., Boca Raton, FL.)

isoelectric point (IEP) of 5.1, which is also the most relevant pH as far as beef is concerned. The volatile mixture was relatively simple and contained 55% acetone. This pH also favoured the formation of the isomeric 3,5-dimethyl-1,2,4-trithiolanes (**14**). At pH 2.2, the main volatile products were cyclic sulphur compounds containing 5-, 6- and 7-membered rings, together with thiophenes. The major component (34% of the isolate) was the novel meaty compound 1,2,3-trithiacyclohept-5-ene. Only a mild degradation occurred at pH 7.1 and, apart from butanone, only cyclic compounds were identified. Major products were the 3,5-dimethyl-1,2,4-

trithiolanes (**14**), 3-methyl-1,2,4-trithiane (**17**) and 2-propanoylthiophene (Shu *et al.*, 1985b).

From aqueous cystine treated similarly, 42 compounds were identified; they were mainly thiazoles, aliphatic sulphides, thiolanes and thianes (Shu *et al.*, 1985a). Significantly more thiazoles were present than in the corresponding cysteine system (Shu *et al.*, 1985b). Several compounds reported to be meaty were generated at the pH of meat (pH 5.5). Figure 3.1 compounds produced were 3-methylcyclopentanone (**4**), 2-methyl-1,3-dithiolane (**13**), 3,5-dimethyl-1,2,4-trithiolane (**14**), 3-methyl-1,2,4-trithiane (**17**), trithioacetaldehyde (**15**) and thiazole (**19**) (Shu *et al.*, 1985a). On the whole, cyclic sulphur compounds were generated more readily at pH 2.3, which is the converse of what was reported for the cysteine system where the thiolanes formed preferentially at the IEP rather than below it (Shu *et al.*, 1985b). It was suggested that this was related to the thermal stability of the disulphide bond of cystine at different pH values.

Often an integral part of the degradations just described, but not necessarily so, is a series of reactions of simple low molecular weight compounds, in particular aliphatic aldehydes, alkane-2,3-diones, methanethiol, H_2S and NH_3. Acetaldehyde is involved in many of them and its relevant reactions are schematically represented in Figure 3.9 (Katz, 1981; Mussinan *et al.*, 1976; Takken *et al.*, 1976), explaining the formation of the reportedly meaty 1-(methylthio)ethanethiol (**1**), 3,5-dimethyl-1,2,4-trithiolane (**14**), trithioacetaldehyde (**15**), thialdine (**18**), 2,4-dimethylthiazole (**20**), 2,4-dimethyl-5-ethylthiazole (**21**), 2,4,5-trimethyl-3-thiazoline (**23**), 2,4,5-trimethyloxazole (**24**) and 2,4,5-trimethyl-3-oxazoline (**25**).

3.4.2 *Reactions of hydroxyfuranones*

Two important cooked beef aroma precursors are 4-hydroxy-5-methyl-(2*H*)furan-3-one (HMFone) and 2,5-dimethyl-4-hydroxy-(2*H*)furan-3-one (HDFone). Both have been isolated from beef (Tonsbeek *et al.*, 1968, 1969) but not from any other meats. Natural precursors of HMFone in beef are ribose-5-phosphate and PCA or taurine (to a lesser extent) or both, and the furanone forms on heating at 100°C/2.5 h in a dilute aqueous medium of pH 5.5 (Tonsbeek *et al.*, 1969). PCA itself is readily formed from ammonia and glutamic acid or glutamine on heating (Tonsbeek *et al.*, 1969). The furanone can also be derived from 5′-ribonucleotides via ribose-5-phosphate obtained, for example, by heating at 60°C (Macy *et al.*, 1970) or during autolysis in muscle (Jones, 1969). HMFone is also generated from Maillard reaction of pentoses, e.g. ribose. Precursors of the related HDFone are hexoses, e.g. glucose or fructose under Maillard conditions. HDFone is a relatively unstable compound. Its optimum stability is at pH 4 and its decomposition increases rapidly with temperature (Hirvi *et al.*, 1980). Shu *et al.* thermally degraded HDFone in water

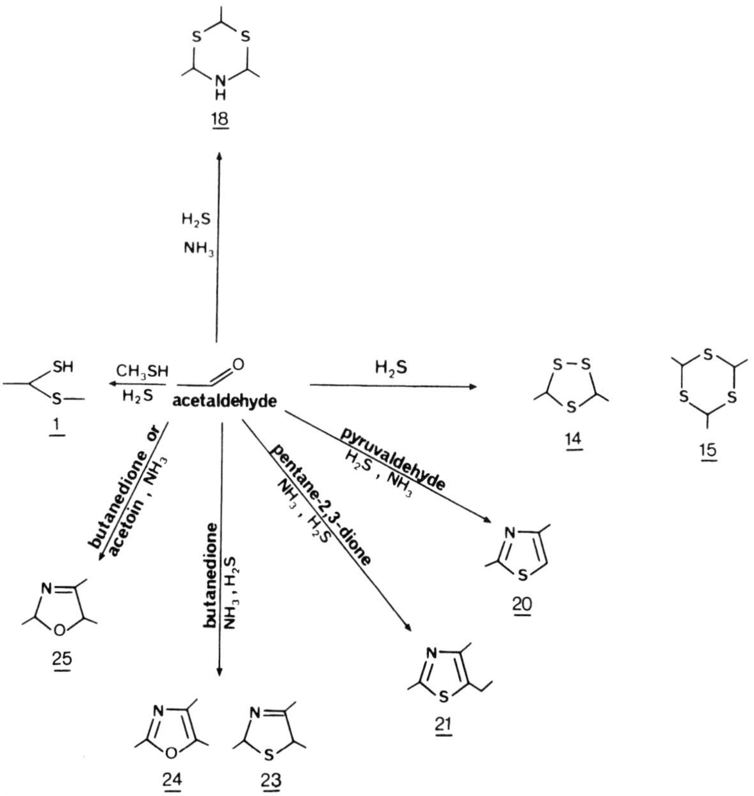

Figure 3.9 Schematic reactions of acetaldehyde-generating compounds from Figure 3.1.

at 160°C/30 min at pH 2.2, 5.1 and 7.1 (Shu *et al.*, 1985c). Degradation occurred more readily at the lower pH values. They identified 20 volatile products, most of which were reactive simple aliphatic carbonyls and dicarbonyls; the remainder of the volatile reaction product mixture consisted of alkyl-substituted (2*H*)furan-3-ones, which were favoured at higher pH values. Most of these were shown to derive from pentane-2,3-dione as intermediate, indicating that the HDFone ring opened initially on heating.

The significance of these two hydroxyfuranones in cooked beef aroma is in their reaction with hydrogen sulphide. For HMFone/H_2S, the reaction is summarized in Figure 3.10 (MacLeod and Seyyedain-Ardebili, 1981; van den Ouweland and Peer, 1975; van den Ouweland *et al.*, 1978). Again, ring opening is indicated and an initial partial substitution of the ring oxygen with sulphur occurs. The reaction product mixture possesses an overall odour of roasted meat and the majority of the reaction products are mercaptofuranoids and mercaptothiophenoids, most of which possess meaty odours (van den Ouweland and Peer, 1975; van den Ouweland

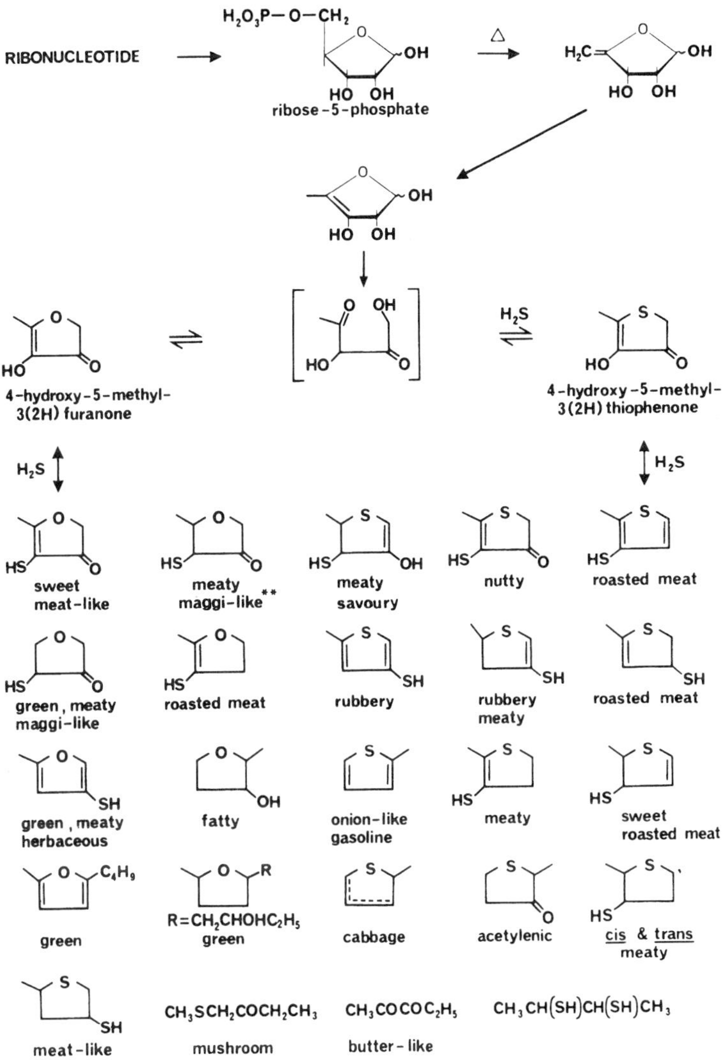

Figure 3.10 Formation from ribonucleotides of 4-hydroxy-5-methyl-(2*H*)furan-3-one, and its reaction with hydrogen sulphide. (Reprinted with permission from MacLeod, G. and Seyyedain-Ardebili, M. (1981) *CRC Crit. Revs Food Sci. Nutr.*, **14**, 309–437. Copyright CRC Press, Inc., Boca Raton, FL.)

et al., 1978). Only one of these (asterisked in Figure 3.10) has so far been identified from natural cooked beef aromas, i.e. 5-methyl-4-mercapto-tetrahydrofuran-3-one (**11**) (Ching, 1979). It is likely, however, that the remainder are present, since several of them possessed GC retention times and odour port assessments which were very similar to those of trace components detected from a natural beef broth aroma isolate (Ching, 1979). Bodrero *et al.* (1981), using surface response methodology to study the contribution of various volatiles to cooked beef aroma, showed that the HMFone/HS_2S reaction product mixture gave the highest predicted score of their entire study. HDFone reacts with H_2S in a similar manner (van den Ouweland and Peer, 1975; van den Ouweland *et al.*, 1978) producing at least two meaty compounds, i.e. 2,5-dimethylfuran-3-thiol and 2,5-dimethyl-4-hydroxy-(2*H*)thiophen-3-one, neither of which has yet been identified from beef. The chances are that other compounds identified from this reaction are meaty also, but their individual sensory properties have not been reported.

Shu *et al.* have reacted HDFone with cystine at 160°C/30 min in water at pH 2.4 (Shu *et al.*, 1985d). The volatile products changed on storage for two weeks at 4°C and were therefore analysed after holding for two weeks. When compared with the products generated from HDFone (Shu *et al.*, 1985c) and from cystine (Shu *et al.*, 1985a) treated similarly, the major difference was the presence of relatively large concentrations of hexane-2,4-dione (16%) and of three thiophenones (22.5%). One of these was the 2,5-dimethyl-4-hydroxy-(2*H*)thiophen-3-one mentioned above and previously reported from HDFone/H_2S reaction, but the other two were novel, namely 2,5-dimethyl-2-hydroxy-(2*H*)thiophen-3-one (possessing a roasted onion odour) and 2,5-dimethyl-2,4-dihydroxy-(2*H*)thiophen-3-one possessing a meaty, pot roast aroma and taste. Yet again, this meaty compound has eluded identification in natural beef. Optimum conditions for the formation of meaty compounds (including these thiophenones and the isomeric 3,5-dimethyl-1,2,4-trithiolanes (**14**)) were 160°C/aqueous medium; 75% water/pH 4.7 (Shu and Ho, 1989). More recently, all three thiophenones have been reported from cysteine/glucose reaction (Tressl *et al.*, 1989).

In a study involving HDFone/cysteine, Shu *et al.* (1986) showed that the two novel thiophenones mentioned above were only trace components. Instead, two novel thiophenes were characterized, namely 3-methyl-2-(2-oxopropyl)thiophene and 2-methyl-3-propanoylthiophene. Their odour properties were not described. Meaty compounds formed preferentially at pH 2.2 and 5.1 rather than at pH 7.1, when various secondary reactions generated a host of different compounds (Shu and Ho, 1988).

3.4.3 *Thermal degradation of thiamine*

The thermal degradation of thiamine produces some important compounds (Werkhoff *et al.*, 1989, 1990; Guntert *et al.*, 1990; van der Linde *et al.*, 1979; Hartman *et al.*, 1984; Reineccius and Liardon, 1985; Dwivedi and Arnold, 1973). van der Linde *et al.* (1979) reported five primary products, the main component being 4-methyl-5-(2-hydroxyethyl)thiazole, which is responsible for the formation of several thiazoles on further degradation. The other sulphur-containing primary product is 5-hydroxy-3-mercapto-pentan-2-one, a key intermediate compound giving rise to a number of aliphatic sulphur compounds, furans and thiophenes (van der Linde, 1979). Two of these were also present in the previously discussed reaction properties of HMFone/H_2S (van den Ouweland and Peer, 1975; van den Ouweland *et al.*, 1978), namely 2-methyltetrahydrofuran-3-one, and the meaty 2-methyl-4,5-dihydrofuran-3-thiol. Hartman *et al.* (1984) degraded thiamine at 135°C/30 min in water (pH 2.3) and also in propane-1,2-diol. Few decomposition products were formed from the diol system but several carbonyls, furanoids, thiophenoids, thiazoles and aliphatic sulphur compounds were isolated from the aqueous reaction. In addition, a novel compound was reported, i.e. 3-methyl-4-oxo-1,2-dithiane, also present (and in enhanced concentration) from a model system of cystine/ascorbic acid/thiamine (Hartman *et al.*, 1984). Reineccius and Liardon (1985) studied the volatile products from thiamine at lower temperatures (40°C, 60°C, 90°C) and at pH 5, 7 and 9 respectively. At pH 5 and 7, the meaty 2-methylfuran-3-thiol (7) and bis(2-methyl-3-furyl) disulphide (9), together with various thiophenes were the major products, but at pH 9, the meaty compounds (7) and (9) were not significant and the thiophenes predominated. Apart from compounds (7) and (9), other compounds which are reportedly meaty and which have been identified from thermally degraded thiamine are 3-mercaptopentan-2-one, 2-methyl-4,5-dihydrofuran-3-thiol, 2-methyltetrahydrofuran-3-thiol, 4,5-dimethylthiazole and 2,5-dimethylfuran-3-thiol (van der Linde *et al.*, 1979; Hartman *et al.*, 1984).

In their recent paper reporting the formation of selected sulphur-containing compounds from various meat model systems, Guntert *et al.* (1990) have proposed very comprehensive reaction schemes for the thermal degradation of thiamine. An extract of these schemes, in so far as it relates to the formation of some compounds reported to be meaty, is shown in Figure 3.11 (Guntert *et al.*, 1990). In the following, meaty compounds from thiamine are coded (**A**)–(**U**); all other compounds are uncoded. One of the primary degradation products, i.e. 4-methyl-5-(2-hydroxyethyl) thiazole, as mentioned above, is responsible for the subsequent generation of several thiazoles, such as 4,5-dimethylthiazole (**B**) and thiazole itself (**19**). Other primary degradation products are 4-amino-5-(aminomethyl)-2-methylpyrimidine (**a**), formic acid and the key intermediates H_2S and 5-hydroxy-3-mercaptopentan-2-one (**c**). The latter can form seven additional intermediates, only one of which, i.e. 3,5-dimercaptopentan-2-one (**b**), is shown in Figure 3.11. Further reactions of **b** lead to the meaty compounds 3-acetyl-1,2-dithiolane (**A**), 2-methyltetrahydrothiophene-2-thiol (**C**), 2-methyl-4,5-dihydrothiophene-3-thiol (**D**) 2-methylthiophene-3-thiol (**F**) and bis(2-methyl-3-thienyl)disulphide (**G**) as shown, whereas intermediate **c** is responsible for the formation of 2-methyl-4,5-dihydrofuran-3-thiol (**E**),

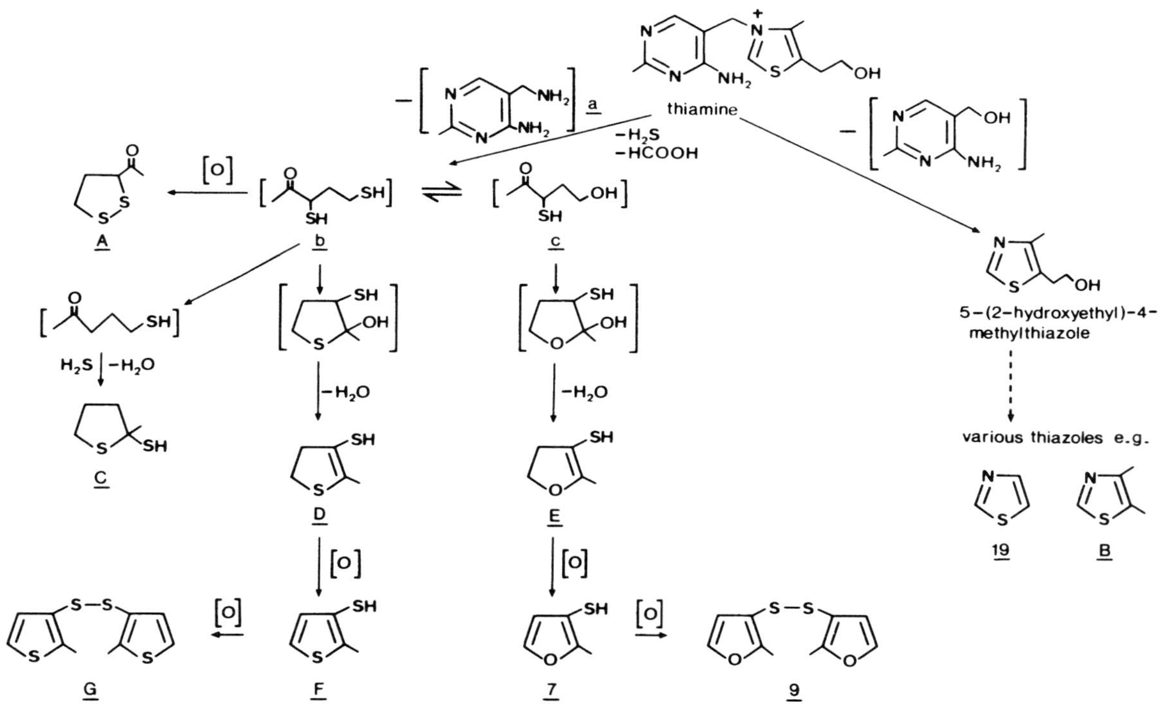

Figure 3.11 Formation of some meaty compounds from the thermal degradation of thiamine (Guntert *et al.*, 1990). Reported odour qualities are as follows: 3-acetyl-1,2-dithiolane (**A**) = meaty, onion, shiitake, liver (Guntert *et al.*, 1990); 4,5-dimethylthiazole (**B**) = meaty, roasted, nutty, green, poultry (Pittet and Hruza, 1974; Vernin, 1979); 2-methyltetrahydrothiophene-2-thiol (**C**) = meaty, sulphurous, tropical fruit, buccu, blackcurrant (Guntert *et al.*, 1990); 2-methyl-4,5-dihydrothiophene-3-thiol (**D**) = meaty (van den Ouweland and Peer, 1975); 2-methyl-4,5-dihydrofuran-3-thiol (**E**) = roast meat (van den Ouweland and Peer, 1975; 2-methylthiophene-3-thiol (**F**) = roast meat (van den Ouweland and Peer, 1975); bis(2-methyl-3-thienyl)disulphide (**G**) = sulphurous, metallic, rubbery, slightly meaty (Werkhoff *et al.*, 1989, 1990).

2-methylfuran-3-thiol (7) and bis(2-methyl-3-furyl)disulphide (9). Two further meaty compounds, which are not shown in Figure 3.11, were also identified (Guntert *et al.*, 1990). These were mercaptopropanone and tetrahydrothiophene-2-thiol. To date, only compounds 7, 9 and **19** have been identified from beef; mercaptopropanone has been reported from canned pork (Maarse and Visscher, 1989), and it is interesting to note that pork has a significantly higher thiamine content than beef.

As a sequel to this study just described, Werkhoff *et al.* (1989, 1990) have recently published their excellent work on the identification of interesting volatile sulphur compounds giving meaty notes to a model meat system consisting of cystine/thiamine/glutamate/ascorbic acid/water heated at 120°C/0.5 h at an initial pH of 5.0. They positively characterized 70 sulphur components, of which 19 possessed individual odours described as meaty. These are presented in Figure 3.12. The majority are new to the flavour literature. Four have already been identified from natural cooked beef, namely 2-methylfuran-3-thiol (7) (Gasser and Grosch, 1988), bis(2-methyl-3-furyl)disulphide (9) (Gasser and Grosch, 1988), 2-methyl-3-[2-methyl-3-thienyl)dithio]furan (**10**) (Werkhoff *et al.*, 1989, 1990) and 1-(methylthio)ethanethiol (**1**) (Maarse and Visscher, 1989). Apart from the latter, the others are all recent identities from beef, the most recent being 2-methyl-3-[(2-methyl-3-thienyl)dithio]furan (**10**), identified by retrospective use of information gained from the above model system in which it was the main component. The disulphides **9**, **10**, **G**, **H** and **J** of Figure 3.12 are derived from oxidation of the corresponding thiols. Even air oxidation of the monomers results in dimerization without effort (Werkhoff *et al.*, 1989, 1990). 1-[2-Methyl-3-thienyl)thio]ethanethiol (**T**) and 1-[(2-methyl-3-furyl)thio]ethanethiol (**U**) are totally new compounds. They have flavour threshold values in water of <0.05 $\mu g\ kg^{-1}$ (Werkhoff *et al.*, 1989, 1990). They are derived from the reaction of 2-methylthiophen-3-thiol (**F**) or 2-methylfuran-3-thiol (7) with acetaldehyde and H_2S respectively. Formation pathways from thiamine have been proposed for most of the Figure 3.12 compounds (Werkhoff *et al.*, 1989, 1990), highlighting the important role of thiamine as precursor of meaty compounds.

3.4.4 Lipid oxidation/degradation

Lipid decomposition creates a prolific number of volatiles. The main reactions involved are oxidation and degradation of both unsaturated and saturated fatty acids. The primary oxidation products, the monohydroperoxides, decompose via an intermediate alkoxy radical, forming a range of aroma volatiles. Such decompositions are summarized in Figures 3.13 and 3.14 (MacLeod and Ames, 1988), explaining the generation of many aliphatic hydrocarbons, alcohols, aldehydes, ketones, acids, lactones and 2-alkylfurans. Esters are formed from esterification of alcohols and acids.

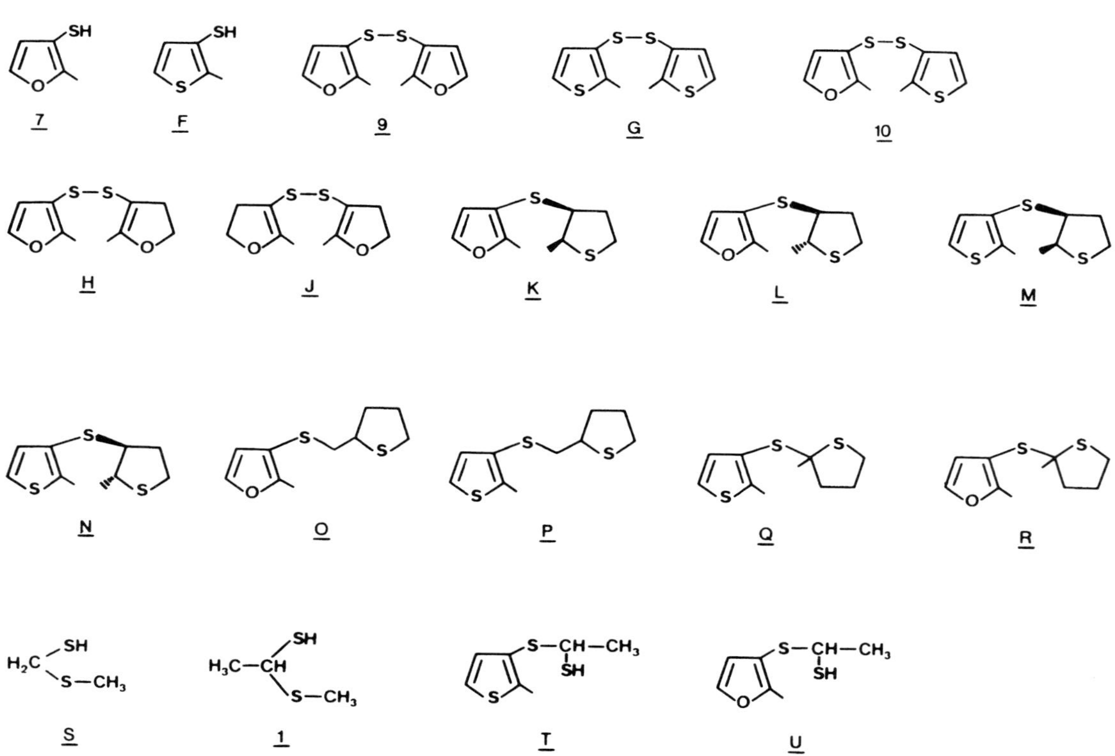

Figure 3.12 Meaty compounds from a model meat system of cystine/thiamine/glutamate/ascorbic acid/water (Werkhoff *et al.*, 1989, 1990).

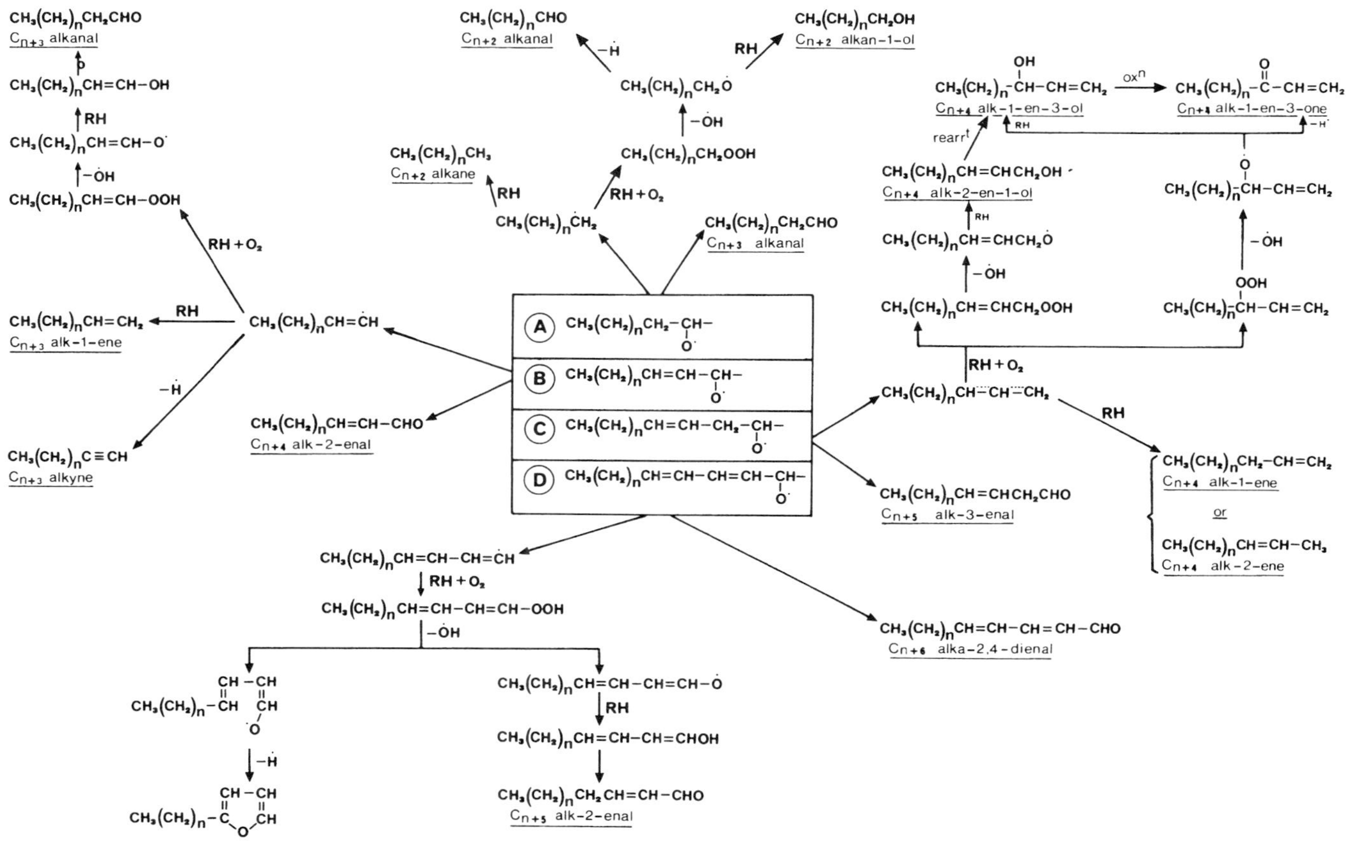

Figure 3.13 Decomposition of alkoxy radicals (RO$^\bullet$) derived from monohydroperoxides and forming aroma volatiles. Example A is RO$^\bullet$ where R is saturated; example B is RO$^\bullet$ where R has one double bond; example C is RO$^\bullet$ where R has an allylic system; and example D is RO$^\bullet$ where R is a diene. (Reprinted with permission from MacLeod, G. and Ames, J.M. (1988) *CRC Crit. Revs Food Sci. Nutr.*, **27**, 219–400. Copyright CRC Press, Inc., Boca Raton, FL.)

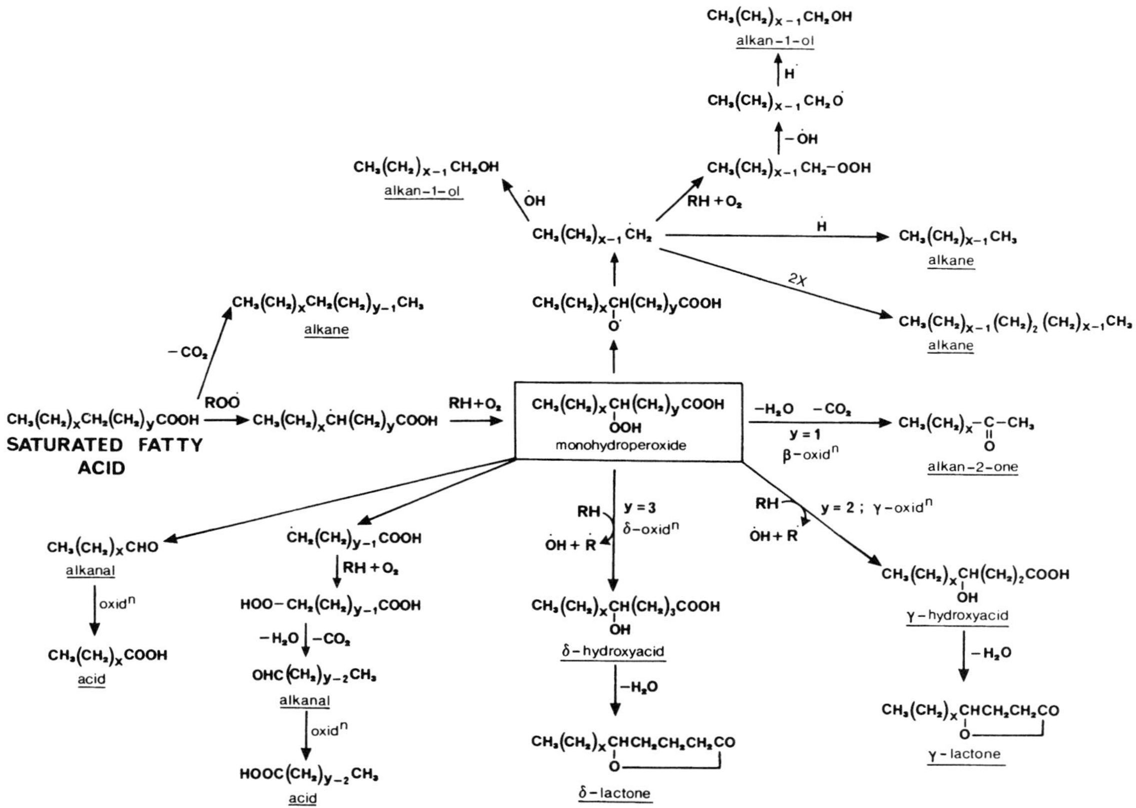

Figure 3.14 Schematic representation of some possible reaction pathways in the thermal oxidation of saturated fatty acids. (Reprinted with permission from MacLeod, G. and Ames, J.M. (1988) *CRC Crit. Revs Food Sci. Nutr.*, **27**, 219–400. Copyright CRC Press, Inc., Boca Raton, FL.)

Other cooked beef aroma components created by lipid degradation are several benzenoids, e.g. benzaldehyde, benzoic acid, alkylbenzenes and naphthalene.

Lipid oxidation starts in raw beef and continues during cooking. Even in lean muscle, the intramuscular lipids are a source of a very large number of volatiles, many of which are present in relatively high concentrations (Bailey and Einig, 1989; Buckholz, 1989). In fact, they create quite a nuisance effect, analytically speaking, because they dominate gas chromatograms of aroma isolates from lean beef and hinder the detection of trace components.

The role of lipids in beef flavour has been considerably clarified by the work of Mottram and his colleagues. They showed that the addition of adipose tissue to lean beef does not give a proportional increase in lipid-derived volatiles, indicating that the intramuscular lipids are the major source of volatile components (Mottram *et al.*, 1982). It has also been shown that species-specific flavour precursors are present in lean beef, although the addition of fat induces fat/lean interactions of some kind which enhance species differences. Intramuscular lipids consist of marbling fat (mainly triacylglycerols) and structural or membrane lipids, which are largely phospholipids, and which contain a relatively higher content of unsaturated fatty acids. Selected removal of the inter- and intramucular triacylglycerols from lean beef causes no significant chemical or sensory aroma differences, but removal of both triacylglycerols and phospholipids generates marked chemical and sensory differences (Mottram and Edwards, 1983). The aroma is less meaty and more roasted, and it contains less lipid oxidation products but higher concentrations of certain heterocyclic compounds, including some alkylpyrazines (Mottram and Edwards, 1983). This implies that, in beef, lipids (or their degradation products) may inhibit the formation of some heterocycles specifically generated from Maillard reactions (Mottram and Edwards, 1983).

This hypothesis was tested using model systems, e.g. of glycine/ribose and of cystein/ribose, both in the presence and absence of lecithin (Whitfield *et al.*, 1988; Farmer *et al.*, 1989; Mottram and Salter, 1989; Farmer and Mottram, 1990; Mottram, 1987; Salter *et al.*, 1988). The same overall effect was observed in both cases (i.e. a lower concentration of some heterocyclic compounds generated by the amino acid–ribose reaction. For some compounds in the cysteine system, e.g. 2-methylfuran-3-thiol (7) and thiophene-2-thiol, the decrease was as much as 65% or more (Mottram and Salter, 1989). This is explained by competition by lecithin degradation products for H_2S and NH_3 derived from cysteine, and this was proved to occur by the additional presence in the reaction product mixture of some different heterocycles known to arise from such reactions. The possible interaction of different lipids was also tested. For example, the addition of beef triacylglycerol had no effect on the aroma of the cysteine/ribose reaction product mixture, but the addition of beef phospholipid caused a significant increase of meaty aroma, a significantly decreased concentration of the normal Maillard reaction heterocycles and a significantly increased concentration of new heterocycles specific to lipid–Maillard interactions (Mottram and Salter, 1989). Examples of such products are 2-pentyl-pyridine (from deca-2,4-dienal/NH_3 interaction), 2-pentyl-, 2-hexyl- and 2-hex-1-enyl-thiophenes (from alka-2,4-dienals/H_2S or from 2-alkylfurans/H_2S), heptane-1-thiol and octane-1-thiol (from alkan-1-ols/H_2S) and 4,5-dimethyl-2-pentylthiazole (from hexanal/diacetyl/H_2S/NH_3). The formation of heterocyclic compounds with long-chain alkyl substituents has also been confirmed in model systems of deca-2,4-dienal and dec-2-enal with H_2S (van den Ouweland *et al.*, 1989), of deca-2,4-dienal with cysteine or glutathione (Zhang and Ho, 1989) and from acetol/NH_3/pentanal or hexanal reaction (Chiu *et al.*, 1990).

Therefore, in beef, some lipid is necessary for a full meaty aroma, but the intramuscular tissue phospholipids (which constitute only about 1% of the muscle composition) are sufficient, and the triacylglycerols are not essential (Mottram and Edwards, 1983). Interactions between water-soluble and phospholipid-derived components occur and could be important.

3.4.5 *Selected aroma components of high sensory significance*

Grosch and his co-workers have recently developed a screening procedure for highlighting the most important volatile compounds in aroma isolates. In this technique, the volatiles of an extract are 'arranged' in order of their flavour significance according to their 'flavour dilution factors' (FD factors). The FD factor of any compound is proportional to its 'aroma value' which is defined as the ratio of the concentration of the flavour compound to its odour threshold. When the technique was applied to cooked beef, Gasser and Grosch (1988) identified 35 compounds possessing FD factors $\geqslant 4$. These are listed, together with the odour qualities apportioned to each, in Table 3.2. Seventeen aroma components (of which 15 were identified) had relatively high FD factors $\geqslant 64$), thereby contributing with high aroma values to the flavour of the cooked beef. Only two compounds were described as meaty, i.e. 2-methylfuran-3-thiol (**7**) and bis(2-methyl-3-furyl)disulphide (**9**). Both possessed the highest FD factor measured (512), highlighting their considerable odour potency. The odour thresholds of those two compounds were determined as 0.0025–0.01 ng l^{-1} (air) and 0.0007–0.0028 ng l^{-1} (air) respectively (Gasser and Grosch, 1990a,b). Gasser and Grosch have also reported a virtually identical study on chicken volatiles (Gasser and Grosch, 1990a) and on commerical meat flavourings (Gasser and Grosch, 1990b). The major differences between beef and chicken were that the sulphur compounds bis(2-methyl-3-furyl)disulphide (**9**) (meaty) and methional (**2**) (cooked potato) predominated in beef whereas volatiles from oxidation of

Table 3.2 Compounds of cooked beef aroma possessing relatively high flavour dilution factors

Flavour dilution factor	Aroma component	Odour quality
512	2-Methylfuran-3-thiol	Meaty, sweet, sulphurous
	Unknown	Roasted
	Methional	Cooked potato
	Non-2(*E*)-enal	Tallowy, fatty
	Deca-2(*E*),4(*E*)-dienal	Fatty, fried potato
	β-Ionone	Violets
	Bis(2-methyl-3-furyl) disulphide	Meaty
256	2-Acetyl-1-pyrroline	Roasted, sweet
	Oct-1-en-3-one	Mushroom
	Phenylacetaldehyde	Honey, sweet
128	2-Acetylthiazole	Roasted
	Nona-2(*E*),4(*E*)-dienal	Fatty
64	Octan-2-one	Fruity, musty
	Oct-2(*E*)-enal	Fruity, fatty, tallowy
	Decan-2-one	Musty, fruity
	Unknown	Sulphurous, onion
	Dodecan-2-one	Musty, fruity
32	Hept-2(*E*)-enal	Fatty, tallowy
	Octa-1,5(*Z*)-dien-3-one	Geranium, metallic
	Unknown	Musty, fatty
	Benzothiazole	Pyridine, metallic
16	Hexanal	Green
	Hex-2(*E*)-enal	Green
	Heptan-2-one	Fruity, musty
	Heptanal	Green, fatty, oily
	Dimethyl trisulphide	Cabbage, sulphurous
	Benzylthiol	Sulphurous
	Nona-2(*E*),6(*Z*)-dienal	Cucumber
	Undecan-2-one	Tallowy, fruity
	Tridecan-2-one	Rancid, fruity, tallowy
8	Octo-1-en-3-ol	Mushroom
	Nonan-2-one	Fruity, musty
	Nonanal	Tallowy, green
	Unknown	Tallowy, cardboard
	5-Methylthiophene-2-carboxaldehyde	Mouldy, sulphurous
	Unknown	Sulphurous
	3-Acetyl-2,5-dimethylthiophene	Sulphurous
	A deca-2,4-dienal (not *E*,*E*)	Fatty
4	2-Methyl-3-(methylthio)furan	Sulphurous
	2-Acetylthiophene	Sulphurous, sweet

From Gasser and Grosch (1988).

unsaturated lipids, in particular deca-2(*E*),4(*E*)-dienal (fatty) and γ-dodecalactone (tallowy, fruity), prevailed in chicken (Gasser and Grosch, 1990a). Interestingly, while the concentration of the important beef aroma compound, bis(2-methyl-3-furyl)disulphide (**9**), was significantly lower in chicken than in beef, by contrast, its reduction product, 2-methylfuran-3-thiol (**7**), was present at approximately the same level in both. Gasser and Grosch suggest that the relatively higher level of linoleic acid in chicken captures most of the available gaseous oxygen for peroxidation reactions, thus protecting the thiol against oxidation to the disulphide. This hypothesis links with the previously mentioned findings of Whitfield *et al.* (1988) who showed a very significant decrease in concentration of 2-methylfuran-3-thiol (7) from a cysteine/ribose model system when lecithin was added. They suggested that carbonyl compounds (from the lecithin) were preferentially capturing reactants such as H_2S. Since the combined levels of 2-methylfuran-3-thiol (7) and its disulphide were much lower in chicken than in beef, this difference could well be due to the reactions proposed by Whitfield *et al.* Furthermore, the higher FD factor for methional in beef, compared with chicken volatiles, tends to indicate a partial inhibition of the Strecker degradation of methionine on heating chicken (Gasser and Grosch, 1990a).

A family of furans with sulphur substituents in the 3-position – as exemplified by compounds **7–10** of Figure 3.1 and others also discussed in this chapter, but so far unidentified in beef – are likely to be of extreme importance in cooked beef aromas. Many have been synthesized, patented and described sensorially (Werkhoff *et al.*, 1989, 1990; Evers *et al.*, 1975; Tressl and Silwar, 1981; Gasser and Grosch, 1988). These four compounds have only recently been reported from cooked beef however (Werkhoff *et al.*, 1989, 1990; MacLeod and Ames, 1986; Gasser and Grosch, 1988). van den Ouweland *et al.* have proposed that an essential structural requirement for meaty aroma is a 5- or 6-membered ring, which is more or less planar and substituted with an enol, thiol and a methyl group adjacent to the thiol, as exemplified by the compounds shown in Scheme 3.1 (van den Ouweland, 1989).

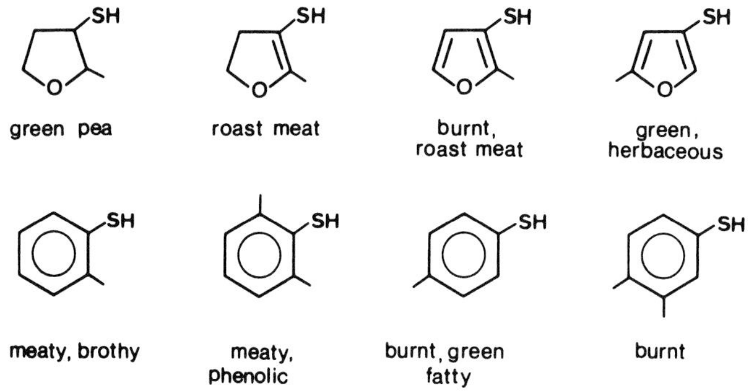

Scheme 3.1

A comprehensive investigation into structure/activity correlation in meaty compounds has been reported by Dimoglo *et al.* (1988). They concluded that the generalized molecular fragment shown in Scheme 3.2 accounts for meaty odour (see also van Wassenaar *et al.*, 1995; Wang *et al.*, 1996; Hau *et al.*, 1997).

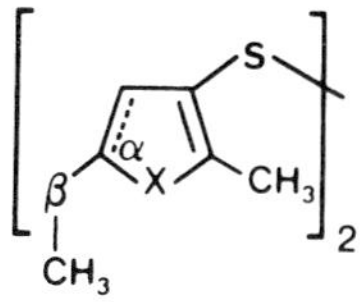

Scheme 3.2

In Scheme 3.2, X = O, S and β = a group coplanar with the α-carbon and the methyl group carbon (e.g. C=O) or an atom different from O, e.g. S. For furan and thiophene derivatives, the methyl group on C_2 plays an important role in meaty odour; additionally, two furan rings and/or two or more sulphur atoms favour meatiness. They also showed that all meaty compounds contain the general structural fragment XH_2 where X = O, N, S and H_2 = two hydrogen atoms belonging, as a rule, to a methyl group. The methyl group must rotate freely. For X = O, N and inter-atomic distance ($X–H_1$) of 0.262–0.277 nm, compounds are described as 'meaty, meat sauce-like, meat soup odour'; for X = S with $X–H_1$ distance of 0.278–0.306 nm, compounds possess a 'roast meat' odour (Scheme 3.3).

The extremely low odour thresholds for compounds 9 and 7 have already been mentioned. Related structures are likely to be potent odorants too. It follows therefore that only minute traces of these types of compounds need be present for them to be aroma effective, creating enormous analytical difficulties for their detection.

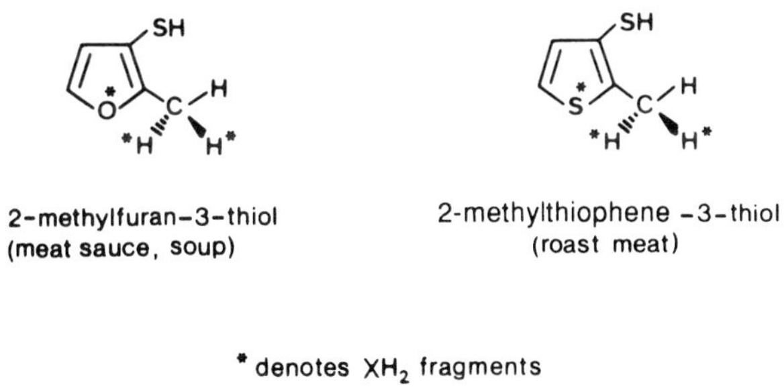

Scheme 3.3

3.5 Conclusions

Because of these difficulties, many researchers have investigated relevant model systems. Such studies have led to the characterization of important volatiles, mostly sulphur compounds, the presence of which can then be investigated retrospectively in natural cooked beef aromas. The extreme success of this philosophy is now evident, particularly so in the recent literature. Several compounds exhibiting meaty/beefy odours have been fully characterized, both chemically and sensorially, and formation mechanisms from defined precursors have been proposed. A few of these compounds have also been isolated from beef itself, with strong indications that others are inherently present in trace quantities. There is no doubt that the successes of the last decade have evolved from such approaches.

References

Airaudo, C.B., Gayte-Sorbier, A. and Armand, P. (1987). Stability of glutamine and pyroglutamic acid under model system conditions. *J. Food Sci.*, **52**, 1750–1752.

Arctander, S. (1969). *Perfume and Flavor Chemicals*, published by the author, New Jersey.

Bailey, M.E. and Einig, R.G. (1989). Reaction flavors of meat. In *Thermal Generation of Aromas*, eds. T.H. Parliment, R.J. McGorrin and C.-T. Ho. American Chemical Society, Washington, DC, pp. 421–432.

Baltes, W. (1979). Rostaromen. *Deutsch. Lebensm. Rundsch.*, **72**, 2–7.

Bodrero, K.O., Pearson, A.M. and Magee, W.T. (1981). Evaluation of the contribution of flavor volatiles to the aroma of beef by surface response methodology. *J. Food Sci.*, **46**, 26–31.

Brinkman, H.W., Copier, H., de Leuw, J.J.M. and Tjan, S.B. (1972). Components contributing to beef flavor. *J. Agric. Food Chem.*, **20**, 177–181.

Buckholz, L.L. Jr. (1989). Maillard technology as applied to meat and savory flavors. In *Thermal Generation of Aromas*, eds. T.H. Parliment, R.J. McGorrin and C.-T. Ho. American Chemical Society, Washington, DC, pp. 406–420.

Ching, J.C.-Y. (1979). Volatile Flavor Compounds from Beef and Beef Constituents. PhD Thesis, University of Missouri.

Chiu, E.-M., Kuo, M.-C., Bruechert, L.J. and Ho, C.-T. (1990) Substitution of pyrazines by aldehydes in model systems. *J. Agric. Food Chem.*, **38**, 58–61.

de Rijke, D., van Dort, J.M. and Boelens, H. (1981). Shigematsu variation of the Maillard reaction. In *Flavour '81*, ed. P. Schreier. de Gruyter, Berlin, pp. 417–431.

Dimoglo, A.S., Gorbachov, M.Y., Bersuker, I.B., Greni, A.I., Vysotskaya, L.E., Stepanova, O.V. and Lukash, E.Y. (1988). Structural and electronic origin of meat odour of organic heteroatomic compounds. *Nahrung*, **32**, 461–473.

Dwivedi, B.K. and Arnold, R.G. (1973). Chemistry of thiamine degradation in food products and model systems. *J. Agric. Food Chem.*, **21**, 54–60.

Evers, W.J., Heinsohn, H.H. Jr., Mayers, B.J. and Sanderson, A. (1976). Furans substituted at the three position with sulfur. in *Phenolic, Sulfur and Nitrogen Compounds in Food Flavors*, eds. G. Charalambous and I. Katz. American Chemical Society, Washington, DC, pp. 184–193.

Farmer, J.J. and Mottram, D.S. (1990). Recent studies on the formation of meat-like aroma compounds. In *Flavour Science and Technology*, eds. Y. Bessiere and A.F. Thomas. Wiley, Chichester, pp. 113–116.

Farmer, L.J., Mottram, D.S. and Whitfield, F.B. (1989). Volatile compounds produced in Maillard reactions involving cysteine, ribose and phospholipid. *J. Sci. Food Agric.*, **49**, 347–368.

Flament, I., Kohler, M. and Aschiero, R. (1976). Sur l'arome de viande de boeuf grille: II. Dihydro-6,7-5H-cyclopenta[*b*]pyrazines, identification et mode de formation. *Helv. Chim. Acta*, **59**, 2308–2313.

Flament, I., Sonnay, P. and Ohloff, G. (1977). Sur l'arome de boeuf grille: III. Pyrrolo-[1,2-*a*]pryazines, identification et synthèse. *Helv. Chim. Acta*, **60**, 1872–1883.

Furia, T.W. and Bellanca, N. (1975). *Fenaroli's Handbook of Flavor Ingredients*, 2nd edn, Vol. 1. CRC Press, Ohio.

Galt, A.M. and MacLeod, G. (1983). The application of factor analysis to cooked beef aroma descriptors. *J. Food Sci.*, **48**, 1354–1355.

Gasser, U. and Grosch, W. (1988). Identification of volatile flavour compounds with high aroma values from cooked beef. *Z. Lebensm. Unters. Forsch.*, **186**, 489–494.

Gasser, U. and Grosch, W. (1990a). Primary odorants of chicken broth. *Z. Lebensm. Unters. Forsch.*, **190**, 3–8.
Gasser, U. and Grosch, W. (1990b). Aroma extract dilution analysis of commercial meat flavourings. *Z. Lebensm. Unters. Forsch.*, **190**, 511–515.
Gayte-Sorbier, A., Airaudo, C.B. and Armand, P. (1985). Stability of glutamic acid and monosodium glutamate under model system conditions. *J. Food Sci.*, **50**, 350–360.
Guntert, M., Bruning, J., Emberger, R., Kopsel, M., Kuhn, W., Thielmann, T. and Werkhoff, P. (1990). Identification and formation of some selected sulfur-containing flavor compounds in various meat model systems. *J. Agric. Food Chem.*, **38**, 2027–2041.
Haefeli, R.J. and Glaser, D. (1990). Taste response and thresholds obtained with the primary amino acids in humans. *Lebensm. Wiss. Technol.*, **23**, 523–527.
Hartman, G.J., Carlin, J.T., Scheide, J.D. and Ho, C.-T. (1984). Volatile products formed from the thermal degradation of thiamine at high and low moisture levels. *J. Agric. Food Chem.*, **32** 1015–1018.
Hau, J., Cazes, D. and Fay, L.B. (1997). Comparative study of the 'beefy meaty peptide'. *J. Agric. Food Chem.*, **45**, 1351–1355.
Hirvi, T., Honkanen, E. and Pyysalo, T. (1980). Stability of 2,5-dimethyl-4-hydroxy-3(2H)furanone and 2,5-dimethyl-4-methoxy-3(2H)furanone in aqueous buffer solutions. *Lebensm. Wiss. Technol.*, **13**, 324–325.
Ho, C.-T. and Hartman, G.J. (1982). Formation of oxazolines and oxazoles in Strecker degradation of DL-alanine and L-cysteine with 2,3-butanedione. *J. Agric. Food Chem.*, **30**, 793–794.
IFF Inc. (1979) British Patent 1 543 653.
Jones, N.R. (1969). Meat and fish flavors; significance of ribomononucleotides and other metabolites. *J. Agric. Food Chem.*, **17**, 712–716.
Kato, H. and Nishimura, T. (1987). Taste components and conditioning of beef, pork and chicken. In *Umami: One of the Basic Tastes*, eds. Y. Kawamura and M.R. Kare. Marcel Dekker, New York.
Kato, S., Kurata, T. and Fujimaki, M. (1973). Volatile compounds produced by the reaction of L-cysteine or L-cystine with carbonyl compounds. *Agric. Biol. Chem.*, **37**, 539–544.
Katz, I. (1981). Recent progress in some aspects of meat flavor chemistry. In *Flavor Research: Recent Advances*, eds. R. Teranishi, R.A. Flath and H. Sugisawa. Marcel Dekker, New York, pp. 217–229.
Kawai, T., Irie, M. and Sakaguchi, M. (1985). Degradation of 2,4,6-trialkyltetrahydro-1,3,5-thiadiazines during storage. *J. Agric. Food Chem.*, **33**, 393–397.
Kawamura, Y. (1990). Umami: one of the basic tastes. *Food Technol. Inter. Europe*, 151–155.
Kubota, K., Kobayashi, A. and Yamanishi, T. (1980). Some sulfur-containing compounds in cooked odor concentrate from boiled Antarctic krills. *J. Agric. Food Chem.*, **44**, 2677–2682.
Kuchiba, M., Kaizaki, S., Matoba, T. and Hasegawa, K. (1990). Depressing effect of salts on thermal degradation of inosine 5′-monophosphate and guanosine 5′-monophosphate in aqueous solution. *J. Agric. Food Chem.*, **38**, 593–598.
Kuninaka, A. (1981). Taste and flavour enhancers. In *Flavor Research: Recent Advances*, eds. R. Teranishi and R.A. Flath. Marcel Dekker, New York, pp. 305–353.
Maarse, H. and Visscher, C.A. (1989). *Volatile Compounds in Food – Qualitative and Quantitative Data.* TNO–CIVO, Zeist, The Netherlands.
MacLeod, G. (1986). The scientific and technological basis of meat flavours. In *Developments in Food Flavours*, eds. G.G. Birch and M.G. Lindley. Elsevier Applied Science, London, pp. 191–223.
MacLeod, G. and Ames, J.M. (1986). 2-Methyl-3-(methylthio)furan: a meaty character impact compound identified from cooked beef. *Chem. Ind.*, 175–177.
MacLeod, G. and Ames, J.M. (1988). Soy flavor and its improvement. *CRC Crit. Revs Food Sci. Nutr.*, **27**, 219–400.
MacLeod, G. and Seyyedain-Ardebili, M. (1981). Natural and simulated meat flavors (with particular reference to beef). *CRC Crit. Revs Food Sci. Nutr.*, **14**, 309–437.
Macy, R.L. Jr., Naumann, H.D. and Bailey, M.E. (1970). Water soluble flavor and odor precursors of meat. *J. Food Sci.*, **35**, 81–83.
Mottram, D.S. (1987). The effect of lipid on the formation of volatile heterocyclic compounds in the Maillard reaction. In *Flavour Science and Technology*, eds. M. Martens, G.A. Dalen and H. Russwurm Jr. Wiley, Chichester, pp. 29–34.

Mottram, D.S. and Edwards, R.A. (1983). The role of triglycerides and phospholipids in the aroma of cooked beef. *J. Sci. Food Agric.*, **34**, 517–522.

Mottram, D.S. and Leseigneur, A. (1990). The effect of pH on the formation of aroma volatiles in meat-like Maillard systems. in *Flavour Science and Technology*, eds. Y. Bessiere and A.F. Thomas. Wiley, Chichester, pp. 121–124.

Mottram, D.S. and Salter, L.J. (1989). Flavor formation in meat-related Maillard systems containing phospholipids. In *Thermal Generation of Aromas*, eds. T.H. Parliment, R.J. McGorrin and C.-T. Ho. American Chemical Society, Washington, DC, pp. 442–451.

Mottram, D.S., Edwards, R.A. and MacFie, H.J.H. (1982). A comparison of the flavour volatiles from cooked beef and pork meat systems. *J. Sci. Food Agric.*, **33**, 934–944.

Mulders, E.J. (1973). Volatile components from the non-enzymic browning reaction of the cysteine/cystine–ribose system. *Z. Lebensm. Unters. Forsch.*, **152**, 193–201.

Mussinan, C.J., Wilson, R.A., Katz, I., Hruza, A. and Vock, M.H. (1976). Identification and flavor properties of some 3-oxazolines and 3-thiazolines isolated from cooked beef. In *Phenolic, Sulfur and Nitrogen Compounds in Food Flavors*, eds. G. Charalambous and I. Katz. American Chemical Society, Washington, DC, pp. 133–145.

Nishimura, O., Mihara, S. and Shibamoto, T. (1980). Compounds produced by the reaction of 2-hydroxy-3-methyl-2-cyclopenten-1-one with ammonia and hydrogen sulfide. *J. Agric. Food Chem.*, **28**, 39–43.

Nishimura, T., Rhue, M.R., Okitani, A. and Kato, H. (1988). Components contributing to the improvement of meat taste during storage. *Agric. Biol. Chem.*, **52**, 2323–2330.

Ohloff, G. and Flament, I. (1978). Heterocyclic constituents of meat aroma. *Heterocycles*, **11**, 663–695.

Okumura, J., Yanai, T., Yajima, I. and Hayashi, K. (1990). Volatile products formed from L-cysteine and dihydroxyacetone thermally treated in different solvents. *Agric. Biol. Chem.*, **54**, 1631–1638.

Pittet, A. and Hruza, D. (1974). Comparative study of flavor properties of thiazole derivatives. *J. Agric. Food Chem.*, **22**, 264–269.

Reineccius, G.A. and Liardon, R. (1985). The use of charcoal traps and microwave desorption for the analysis of headspace volatiles above heated thiamine solutions. In *Topics in Flavour Research*, eds. R.G. Berger, S. Nitz and P. Schreier. Eichhorn, Marzling–Hangenhan, pp. 125–136.

Roedel, W. and Kruse, H.P. (1980). Present problems of meat aroma research. *Nahrung*, **24**, 129–139.

Salter, L.J., Mottram, D.S. and Whitfield, F.B. (1988). Volatile compounds produced in Maillard reactions involving glycine, ribose and phospholipid. *J. Sci. Food Agric.*, **46**, 227–242.

Scanlan, R.A., Kayser, S.G., Libbey, L.M. and Morgan, M.E. (1973). Identification of volatile compounds from heated L-cysteine–HCl–D-glucose. *J. Agric. Food Chem.*, **21**, 673–675.

Self, R., Casey, J.C. and Swain, T. (1963). The low boiling volatiles of cooked foods. *Chem. Ind.*, 863–864.

Shaoul, O. and Sporns, P. (1987). Hydrolytic stability at intermediate pHs of the common purine nucleotides in food. *J. Food Sci.*, **52**, 810–812.

Sheldon, S.A., Russell, G.F. and Shibamoto, T. (1986). Photochemical and thermal activation of model Maillard reaction systems. In *Amino–Carbonyl Reactions in Food and Biological Systems*, eds. M. Fujimaki, M. Namiki and H. Kato. Elsevier Science, New York, pp. 145–154.

Shibamoto, T. (1980). Heterocyclic compounds found in cooked meats. *J. Agric. Food Chem.*, **28**, 237–243.

Shu, C.-K. and Ho, C.-T. (1988). Efffect of pH on the volatile formation from the reaction between cysteine and 2,5-dimethyl-4-hydroxy-3(2H)furanone. *J. Agric. Food Chem.*, **36**, 801–803.

Shu, C.-K. and Ho, C.-T. (1989). Parameter effects on the thermal reaction of cystine and 2,5-dimethyl-4-hydroxy-3(2H)furanone. In *Thermal Generation of Aromas*, eds. T.H. Parliment, R.J. McGorrin and C.-T. Ho. American Chemical Society, Washington, DC, pp. 229–241.

Shu, C.-K., Hagedorn, M.L., Mookherjee, B.D. and Ho, C.-T. (1985a). Volatile components of the thermal degradation of cysteine in water. *J. Agric. Food Chem.*, **33**, 438–442.

Shu, C.-K., Hagedorn, M.L., Mookherjee, B.D. and Ho, C.-T. (1985b). pH effect on the volatile components in the thermal degradation of cysteine. *J. Agric. Food Chem.*, **33**, 442–446.

Shu, C.-K., Hagedorn, M.L., Mookherjee, B.D. and Ho, C.-T. (1985c). Volatile components of the thermal degradation of 2,5-dimethyl-4-hydroxy-3(2H)furanone. *J. Agric. Food Chem.*, **33**, 446–448.

Shu, C.-K., Hagedorn, M.L., Mookherjee, B.D. and Ho, C.-T. (1985d). Two novel 2-hydroxy-3(2H)thiophenones from the reaction between cystine and 2,5-dimethyl-4-hydroxy-3(2H)furanone. *J. Agric. Food Chem.*, **33**, 638–641.

Shu, C.-K., Hagedorn, M.L. and Ho, C.-T. (1986). Two novel thiophenes identified from the reaction between cysteine and 2,5-dimethyl-4-hydroxy-3(2H)furanone. *J. Agric. Food Chem.*, **34**, 344–346.

Takken, H.J., van der Linde, L.M., de Valois, P.J., van Dort, H.M. and Boelens, M. (1976). Reaction products of 2,3-dicarbonyl compounds, aldehydes, hydrogen sulfide and ammonia. In *Phenolic, Sulfur and Nitrogen Compounds in Food Flavors*, eds. G. Charalambous and I. Katz. American Chemical Society, Washington, DC, pp. 114–121.

Tonsbeek, C.H.T., Plancken, A.J. and von der Weerdhof, T. (1968). Components contributing to beef flavor: isolation of 4-hydroxy-5-methyl-3(2H)furanone and its 2,5-dimethyl homolog from beef broth. *J. Agric. Food Chem.*, **16**, 1016–1021.

Tonsbeek, C.H.T., Koenders, E.B., van der Zijden, A.S.M. and Losekoot, J.A. (1969). Components contributing to beef flavor: natural precursors of 4-hydroxy-5-methyl-3(2H)furanone in beef broth. *J. Agric. Food Chem.*, **17**, 397–400.

Tressl, R. and Silwar, R. (1981). Investigation of sulfur-containing components in roasted coffee. *J. Agric. Food Chem.*, **29**, 1078–1082.

Tressl, R., Grunewald, H.-G., Silwar, R. and Bahri, D. (1979). Chemical formation of flavor substances. In *Progress in Flavour Research*, eds. D.G. Land and H.E. Nursten. *Applied Science*, London, pp. 197–213.

Tressl, R., Helak, B., Grunewald, H.-G. and Silwar, R. (1983). Formation of flavour components from proline, hydroxyproline and sulphur-containing amino acids. In *Colloque International sur les Aromes Alimentaires*, eds. J. Adda and H. Richard. *Tech. Doc. Lav.*, Paris, pp. 207–230.

Tressl, R., Helak, B., Martin, N. and Kersten, E. (1989). Formation of amino acid specific Maillard products and their contribution to thermally generated aromas. In *Thermal Generation of Aromas*, eds. T.H. Parliment, R.J. McGorrin and C.-T. Ho. American Chemical Society, Washington, DC, pp. 156–171.

van den Ouweland, G.A.M. and Peer, H.G. (1975). Components contributing to beef flavor. Volatile compounds produced by the reaction of 4-hydroxy-5-methyl-3(2H)furanone and its thio analog with hydrogen sulfide. *J. Agric. Food Chem.*, **23**, 501–505.

van den Ouweland, G.A.M., Olsman, H. and Peer, H.G. (1978). Challenges in meat flavor research. In *Agricultural and Food Chemistry: Past, Present and Future*, ed. R. Teranishi. AVI Publishing Co., Westport, CT. pp. 292–314.

van den Ouweland, G.A.M., Demole, E.P. and Enggist, P. (1989). Process meat flavor development and the Maillard reaction. In *Thermal Generation of Aromas*, eds. T.H. Parliment, R.J. McGorrin and C.-T. Ho. American Chemical Society, Washington, DC, pp. 433–441.

van der Linde, L.M., van Dort, J.M., de Valois, P., Boelens, H. and de Rijke, D. (1979). Volatile components from thermally degraded thiamine. In *Progress in Flavour Research*, eds. D.G. Land and H.E. Nursten. Applied Science, London, pp. 219–224.

Van Wassenaar, P.D., Van Den Oord, A.H.A. and Schaapen, W.M.M. (1995). Taste of 'delicious' beefy meaty peptides. Revised. *J. Agric. Food Chem.*, **43**, 2828–2832.

Vernin, G. (1979). Heterocycles in food aromas. I. Structure and organoleptic properties. *Parfum Cosmetiq. Aromes*, **29**, 77–87.

Vernin, G. (1982). *The Chemistry of Heterocyclic Flavouring and Aroma Compounds*. Ellis Horwood, Chichester.

Wang, K., Maga, J.A. and Bechtel, P.J. (1996). Taste properties and synergisms of beefy meaty peptides. *J. Food Sci.*, **61**, 837–839.

Werkhoff, P., Emberger, R., Guntert, M. and Kopsel, M. (1989). Isolation and characterisation of volatile sulfur-containing meat flavor components in model systems. In *Thermal Generation of Aromas*, eds. T.H. Parliment, R.J. McGorrin and C.-T. Ho. American Chemical Society, Washington, DC, pp. 460–478.

Werkhoff. P., Bruning, J., Emberger, R., Guntert, M., Kopsel, M., Kuhn, W. and Surburg, H. (1990). Isolation and characterisation of volatile sulfur-containing meat flavor

components in model systems. *J. Agric. Food Chem.*, **38**, 777–791.
Whitfield, F.B., Mottram, D.S., Brock, S., Puckey, D.J. and Salter, L.J. (1988). Effect of phospholipid on the formation of volatile heterocyclic compounds in heated aqueous solutions of amino acids and ribose. *J. Sci. Food Agric.*, **42**, 261–272.
Wilson, R.A., Vock, M.H., Katz, I. and Shuster, E.J. (174). British Patent 1 364 747.
Yayalan, V. (1990). In search of alternative mechanisms for the Maillard reaction. *Trends in Food Sci. and Technol.*, **1**, 20–22.
Zhang, Y. and Ho, C.-T. (1989). Volatile compounds formed from thermal interaction of 2,4-decadienal with cysteine and glutathione. *J. Agric. Food Chem.*, **37**, 1016–1020.
Zhang, Y., Chien, M. and Ho, C.-T. (1988). Comparison of the volatile compounds obtained from thermal degradation of cysteine and glutathione in water. *J. Agric. Food Chem.*, **36**, 992–996.

4 The flavour of pork

J. CHEN AND C.-T. HO

4.1 Introduction

Due to its excellent nutritional value and unique sensory properties, meat has always made an important contribution to our diet. The meat processing industry is rapidly growing because of the convenience and additional nutritional value of processed meat products.

Flavour is one of the most important factors contributing to the quality of meat products. There has been much research aimed at understanding the chemistry associated with meat flavour, providing a guide for improving the flavour of meat products. So far, however, most research on meat flavour has been devoted to the flavour of beef. The flavour of pork has not received much attention, even though it is one of the most important meat sources in the world. A search of the literature (*Food Science and Technology Abstracts* until December 1996) indicates at least 342 publications related to the flavour of meat, 154 papers on beef flavour, while only 63 on the flavour of pork.

Raw meat possesses little odour and only a mild serum-like taste, which is described as salty, metallic and 'bloody' tasting with a sweet aroma (Wasserman, 1972; Hornstein and Wasserman, 1987). So-called meat flavour is generated primarily through the cooking process, by which various reactions take place and a variety of tastes and aromas are developed. Although over 1000 compounds have so far been identified in meat flavour, the exact contribution of each compound to meat flavour is not yet known. The investigation of key compounds responsible for meat flavour as well as species specific flavour is still very desirable.

4.2 Precursors and flavour compounds in pork

Raw pork meat tissue consists primarily of water, protein, amino acid, nucleotide, sugar, lipid, vitamin and other compounds. These constituents act as important taste compounds, flavour enhancers or aroma precursors. During the cooking process, numerous nonvolatile compounds in pork undergo degradation or reaction with each other to produce hundreds of volatile compounds to form the characteristic pork flavour.

Many compounds in raw pork and their degradation or reaction products during meat processing possess taste properties (MacLeod, 1986; Aristoy and Toldra, 1995). Glucose, fructose, ribose and some free amino acids, such as glycine, alanine, serine, threonine, lysine, proline and hydroxy-proline, exhibit sweet notes. Lactic acid and other organic acids give the sensation of sourness to pork. Certain peptides and amino acids contribute to the bitter taste. Umami taste comes from monosodium glutamate (MSG), 5′-nucleotides (IMP, inosine monophosphate, and GMP, guanosine monophosphate) and some peptides (MacLeod, 1986).

Volatile compounds identified in pork flavour include hydrocarbons, alcohols, carbonyls, carboxylic acids, esters, lactones, ethers, sulphur-containing compounds as well as different classes of heterocyclic compounds, namely furans, pyridines, pyrazines, oxazoles, thiazoles and thiophenes (Table 4.1).

Characteristic pork flavour can not be ascribed to any single chemical compound. Rather, it is a product of the volatile and nonvolatile compounds. It is generally accepted that meaty notes come mainly from such sulphur-containing compounds as 2-methyl-3-furanthiol and bis(2-methyl-3-furyl) disulphide, which are generated from water-soluble precursors in lean meat, while fat and fat-soluble substances would contribute to flavour species difference (van den Ouweland *et al.* 1980; Hornstein and Wasserman 1987; Shahidi, 1988)

On the relationship of meat flavour and its precursors, Hornstein and Wasserman (1987) summarized it well:

> (1) The flavour precursors of lean meats are water soluble. (2) A nonenzymatic reaction between reducing sugars and amino acids may play a major role in the

Table 4.1 Classification of volatiles compounds found in meat (modified from Shahidi *et al.*, 1986)

Compounds	Number of compounds				
	Pork (uncured)	Pork (cured)	Mutton	Chicken	Beef
Hydrocarbons	45	4	26	71	123
Aldehydes	35	29	41	73	66
Ketones	38	12	23	31	59
Alcohols	24	9	11	28	61
Phenols	9	1	3	4	3
Carboxylic acids	5	20	46	9	20
Esters	20	9	5	7	33
Lactones	2	–	14	2	33
Furans	29	5	6	13	40
Pyridines	5	–	16	10	10
Pyrazines	36	–	15	21	48
Other nitrogen compounds	24	3	8	33	37
Sulphur compounds	31	31	12	33	126
Halogenated compounds	4	1	–	6	6
Miscellaneous compounds	7	11	–	6	16
Total	314	135	226	347	681

development of characteristic lean meat flavour. (3) The similarities in composition of the water-soluble fractions of lean beef, pork, and lamb may account for the similarity in flavour of 'fat' free meat from these species. (4) Intact fibrillar and sarcoplasmic protein as such do not contribute to meat flavour. (5) Lipids may contribute to flavour species differences by virtue of their composition and by serving as a reservoir for odouriferous of reactive fat-soluble substances that are characteristic for different animal species.

4.3 Major pathways to generate pork flavour

The reactions involved in pork flavour development are very complex, and they include the thermal degradation of individual components of pork muscle, thermal oxidation of lipids, reactions between amino acids and carbohydrates, and interactions between these various reactions. Several extensive reviews on these reactions have been presented (Shahidi *et al.*, 1986; Whitfield, 1992; Reineccius, 1994; St. Angelo, 1996).

4.3.1 Lipid degradation

Hydrocarbons, aldehydes, ketones, alcohols, fatty acids, esters and long-chain alkyl-substituted heterocyclic compounds in pork volatiles are usually developed from lipid degradation, oxidation, or their further reactions. Lipid-derived volatile compounds dominate the flavour profile of pork cooked at temperatures below 100°C. Table 4.2 lists some of the volatile compounds generated from the degradation of lipids in the cooked pork volatiles.

During cooking, thermal and oxidative degradation of depot triacylglycerols and tissue phospholipids occur simultaneously. In the absence of oxygen, lipids thermally degrade through dehydration, decarboxylation, hydrolysis, dehydrogenation and carbon–carbon cleavage. Through hydrolysis, free fatty acids are released (Nawar, 1969). Other products formed during thermal degradation include saturated and unsaturated hydrocarbons, β-keto acids, methylketones, lactones and esters (Nawar, 1969).

Volatile compounds in cooked pork are also formed by autoxidation. The accepted mechanism of lipid oxidation involves hydroperoxide formation, followed by the production of fragments containing various functional groups. Autoxidation is a free radical chain reaction and is initiated by the extraction of a hydrogen atom from the lipid molecule. For lipids with two or more double bonds in a nonconjugated system, extraction of the hydrogen atoms from the methylene groups between the double bonds is stabilized by delocalization of the free radical over five carbons. Reaction of oxygen with the lipid free radical creates peroxy radicals, and the reaction is then propagated by formation and decomposition of hydroperoxides.

Table 4.2 The volatile lipid oxidation products identified in cooked pork

Compounds	Sample A (mg/kg)	Sample B (relative abundance)	Sample C (ng/100 g)
Aldedydes			
hexanal	12.66	L	245
heptanal			221
3-methylhexanal	0.65		
octanal			52
nonanal		L	209
dodecanal	tr		
tridecanal	0.25		
tetradecanal	0.40		
hexadecanal	0.65		
octadecanal	1.19		
2-hexenal	tr		
2-heptenal	0.34		
(*E*)-2-octenal	0.99		
2-nonenal	0.39		
(*Z*)-4-decenal		S	
2-undecenal	0.39		
2-dodecenal	0.43		
(*E*)-2-tridecenal		S	
(*Z*)-2-tridecenal		L	
(*E*)-2-tetradecenal		S	
(*Z*)-2-tetradecenal		S	
17-octadecenal	tr		
16-octadecenal	8.34		
15-octadecenal	0.70		
9-octadecenal	0.81		
2,4-nonadienal	tr		
(*E,E*)-2,4-decadienal	0.69	S	
(*E,Z*)-2,4-decadienal	0.41	S	
2,4-undecadienal	tr	S	
Ketones			
2-heptanone	0.20		
2-tetradecanone		S	
2-hexadecanone		S	
2,3-octanedione	0.88		
Alcohols			
1-pentanol			6
1-hexanol			47
1-heptanol		M	30
1-octanol			35
(*E*)-2-heptenol		S	
1-octen-3-ol		L	1003
1-nonanol		S	
1-nonen-3-ol	0.75		
Alkylfurans			
2-pentylfuran		M	30

Sample A: Ground pork (250–450 g) was placed in a 2-l beaker. Distilled water was added so as to attain a meat-to-water ratio of 4:1 (w/w), and the contents were heated at 85°C until the meat slurry attained a constant temperature of 73°C and held at that temperature for 10 min (Ramarathnam *et al.*, 1991).
Sample B: 2 kg of pork and 500 ml of water was heated at boiling temperature with refluxing for 7 h (Chou and Wu, 1983).
Sample C: 100 g of minced meat was cooked in Synthene bags for 25 min in a boiling water bath (Mottram, 1985).
L = large; S = small; M = medium; tr = trace.

Hydroperoxides formed during lipid oxidation are typically cleaved to give a lipid oxy-radical and a hydroxy radical.

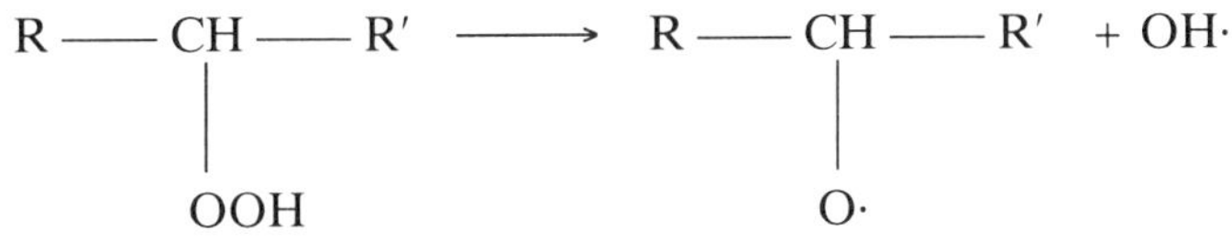

Frankel (1982) has illustrated possible reactions of the resulting lipid oxy-radical as shown below.

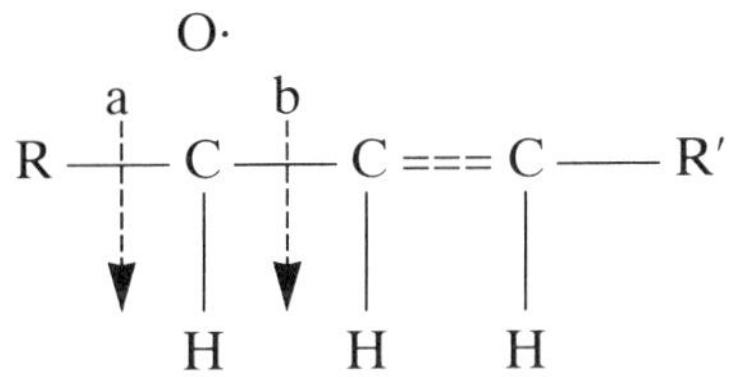

Of the volatiles produced by these reactions, aldehydes are the most significant flavour compounds. Aldehydes can be produced by scission of the lipid molecules on either side of the radical. The products formed by these scission reactions depend on the fatty acids present, the hydroperoxide isomers formed and the stability of the decomposition products. Temperature, time of heating and degree of autoxidation are variables which affect thermal oxidation (Frankel, 1982).

As shown in Table 4.2, aldehydes are the major components identified in volatiles of cooked pork. Octanal, nonanal and 2-undecenal are oxidation products of oleic acid, and hexanal, 2-nonenal and 2,4-decadienal are major volatile oxidation products of linoleic acid. Oleic acid and linoleic acid are the two most abundant unsaturated fatty acids in pork (Schliemann *et al.*, 1987). The unusually high concentration of 1-octen-3-ol reported by Mottram (1985) and Chou and Wu (1983) in cooked pork may be derived from the 12-hydroperoxide of arachidonic acid. Figure 4.1 shows the possible mechanism for its formation. 1-Octen-3-ol has been determined to be the character-impact compound of cooked mushroom. However, the precursor of 1-octen-3-ol in mushroom has been identified as linoleic acid with the unusual 10-hydroperoxide as the intermediate (Whitfield and Last, 1991).

2-Pentylfuran identified in cooked pork is a well-known autoxidation product of linoleic acid and has been known as one of the compounds responsible for the reversion flavour of soybean oil (Ho *et al.*, 1978). Figure 4.2 shows the probable mechanism for its formation. The conjugated diene radical generated from cleavage of 9-hydroxy radical of linoleic acid may react with oxygen to produce vinyl hydroperoxide. The vinyl hydroperoxide will then undergo cyclization via the alkoxy radical to yield 2-pentylfuran (Frankel, 1982).

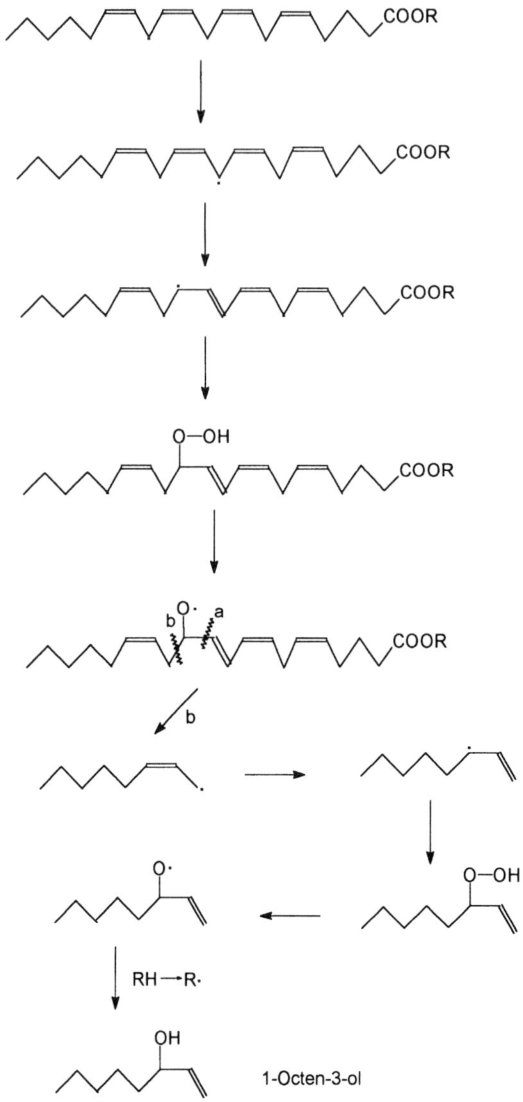

Figure 4.1 Mechanism for formation of 1-octen-3-ol from arachidonic acid.

The concentration of lean and fat tissues in the flavour of cooked meats has been the subject of a number of studies. Hornstein and Crowe (1960, 1963) found that aqueous extracts of beef, pork and lamb have similar aromas when heated, and when heating the fats, they yielded the species characteristic aromas. Fat tissue contains all the normal components of cells, including proteins, amino acids, salts and sugars, as well as lipids.

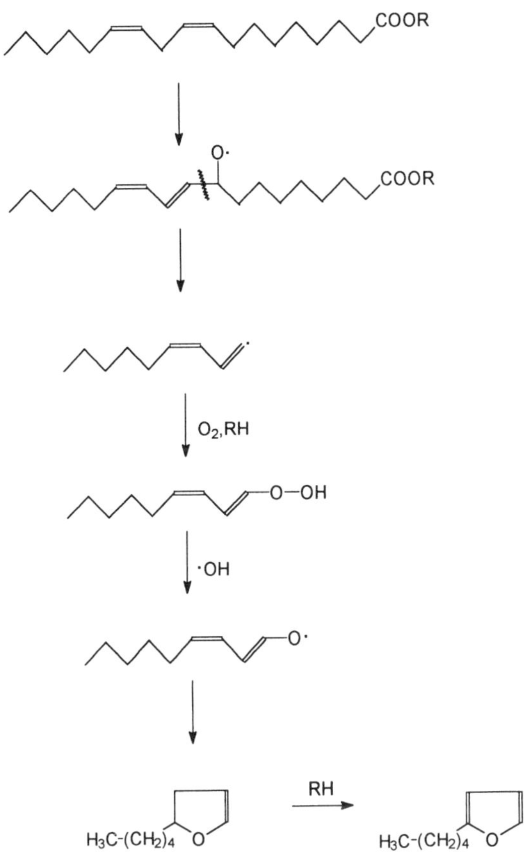

Figure 4.2 Mechanism for the formation of 2-pentylfuran.

Although species-characteristic aromas have been found in heated fatty tissues, lipids alone are not responsible (Mottram *et al.*, 1982).

In one study, Mottram (1979) used triangle taste-tests to differentiate minced pork and beef meat-cakes cooked with and without added fat. The addition of 10% subcutaneous fat (beef or pork) enabled the panel to distinguish the lean meats more easily, no matter which fat was used.

Mottram *et al.* (1982) also reported a comparison of the flavour volatiles from cooked beef and pork meat systems. The addition of minced subcutaneous adipose tissue to the meat-cakes increased the lipid content up to nine-fold. The addition of pork fat to either lean beef or pork resulted in a substantial increase in hexanal but only small changes in most other volatiles. The general lack of a relationship between volatiles and adipose fat level suggested that triacylglycerols of the adipose tissues may not be the major source of volatiles and that the intramuscular triacylglycerols and phospholipids may be important.

4.3.2 Maillard reaction

Many heterocyclic aroma compounds, and some aldehydes and ketones in pork volatiles, are generally derived from the Maillard reaction between reducing sugars and amino acids, peptides or proteins. Maillard reactions ubiquitously occur in food systems. A weak basic condition and water activity of 0.4–0.8 are favourable for such reactions, and high temperature can greatly accelerate the Maillard reaction in pork.

Sugars found in pork include mainly glucose and ribose, which are formed from hydrolysis of glycogen, as well as nucleotides and nucleosides. Nitrogen-containing constituents in pork are free amino acids, peptides, proteins, and nucleotides and nucleosides, among others.

The Maillard reaction generates both volatile aroma compounds and high molecular weight brown-coloured products. It is well documented that the reactions between sulphur-containing amino acids like cysteine and reducing sugars yield meat-like aromas (de Roos, 1992).

The large numbers of heterocyclic compounds reported in the aroma volatiles are associated with roasted, grilled or pressure-cooked meat rather than boiled meat where the temperature does not exceed 100°C (Mottram, 1985). In pressure-cooked pork liver at 163°C, quantitatively over 70% of the total amount of volatiles were furans and pyrazines (Mussinan and Walradt, 1974), while in boiled pork, the volatiles were dominated by aliphatic aldehydes and alcohols with only very small quantities of heterocyclic compounds. In an excellent report, Mottram (1985) compared the effect of cooking conditions on the formation of volatile heterocyclic compounds in pork. Table 4.3 shows the comparison of the volatile heterocyclic compounds in well-done grilled, roasted and boiled pork. Well-done grilled pork contained 66 heterocyclic compounds including pyrazines, thiazoles, thiophenes, furans and pyrroles. Pork cooked under less severe roasting or boiling conditions contained considerably fewer heterocyclic compounds.

The alkylpyrazines accounted for almost 80% of the total headspace volatiles of well-done grilled pork (Mottram, 1985). As a rule, alkylpyrazines produce a roasted nut-like sensory impression (Ohloff and Flament, 1978). The contribution of these pyrazines to the flavour of cooked pork will depend upon their odour threshold values as well as their concentrations and odour characters. Methyl- and dimethylpyrazines have relatively high odour thresholds but much lower values have been observed for pyrazines with increased substitutions (Guadangi *et al.*, 1972); for example, 2-ethyl-3,6-dimethylpyrazine has a threshold of 0.4 parts in 10^9 parts water. It has been well-accepted that alkylpyrazines were formed from a reaction of the degraded nitrogenous substances, NH_3, RNH_2 from proteins, peptides, amino acids and phospholipids, and α-dicarbonyl compounds in food (Shibamoto, 1980).

Bicyclic pyrazines such as 5-methyl-, 2,5-dimethyl- and 3,5-dimethyl-6,7-dihydro-5*H*-cyclopentapyrazine were identified in well-done grilled pork. These cycolopentapyrazines were formed from the cyclotenes, 1,2-dicarbonyl compounds and ammonia (Figure 4.3). Two pentyl-substituted and one butyl-substituted pyrazines were also observed in grilled pork. The formation of higher carbon number-substituted pyrazines is hypothesized to be due to the intervention of lipid-derived aldehydes.

Like alkylpyrazines, alkylthiazoles identified in pork arise from a Maillard-type reaction between amino acids and sugars or other carbonyl compounds. It has been demonstrated that the requirement for the formation of these compounds is severe heating (Table 4.3). However, acetylthiazole was present in large amounts in the boiled pork suggesting that a different mechanism is involved in its production. Four alkylthiazoles,

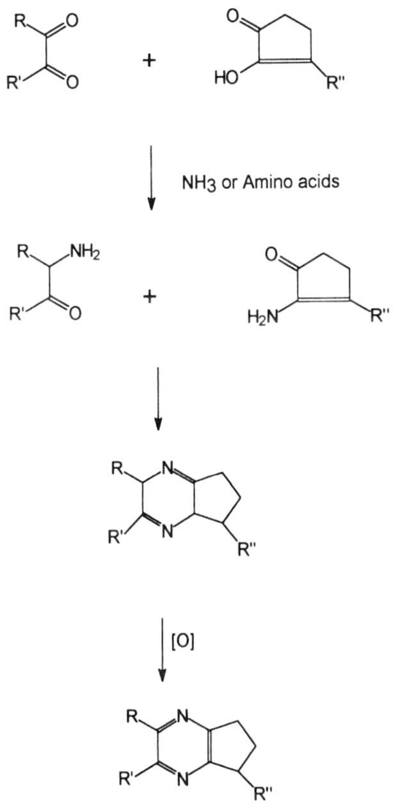

Figure 4.3 Mechanism for the formation of cyclopentapyrazines.

Table 4.3 Major volatile heterocyclic components of pork cooked under different conditions (modified from Mottram, 1985)

Compounds	Grilled (well-done)	Roasted	Boiled
		(ng/100 g)	
Pyrazines			
methylpyrazine	91	3	1
2,5-dimethylpyrazine	2448	4	tr
2,6-dimethylpyrazine	1913	6	3
2,3-dimethylpyrazine	220	2	–
2-ethyl-6-methylpyrazine	545	–	–
2-ethyl-5-methylpyrazine	302	–	–
2-ethyl-3-methylpyrazine	28	–	–
trimethylpyrazine	1952	–	–
2-ethyl-5,6-dimethylpyrazine	30	–	–
2-ethyl-3,6-dimethylpyrazine	1155	–	–
2,5-diethylpyrazine	10	–	–
2-ethyl-3,5-dimethylpyrazine	370	–	–
2-methyl-5-propylpyrazine	10	–	–
tetramethylpyrazine	70	–	–
5,6-diethyl-2-methylpyrazine	45	–	–
3,5-diethyl-2-methylpyrazine	80	–	–
dimethylpropylpyrazine	2	–	–
3,6-diethyl-2-methylpyrazine	10	–	–
butyldimethylpyrazine	10	–	–
triethylpyrazine	10	–	–
acetylpyrazine	7	–	–
pentylmethylpyrazine	15	–	–
5-methyl-6,7-dihydro-5*H*-cyclopentapyrazine	5	–	–
pentyldimethylpyrazine	40	–	–
2,5-dimethyl-6,7-dihydro-5*H*-cyclopentapyrazine	11	–	–
acetylmethylpyrazine	5	–	–
3,5-dimethyl-6,7-dihydro-5*H*-cyclopentapyrazine	2	–	–
Pyridines			
2-methylpyridine	17	–	–
2,6-dimethylpyridine	8	–	–
2-ethylpyridine	10	–	–
4-methylpyridine	8	–	–
2,5-dimethylpyridine	10	–	–
2,4-dimethylpyridine	5	–	–
2,3-dimethylpyridine	8	–	–
4-ethylpyridine	40	–	–
2-pentylpyridine	80	–	–
2-acetylpyridine	5	–	–
Thiazoles			
4-methylthiazole	1	–	–
2,4-dimethylthiazole	3	–	–
2,5-dimethylthiazole	3	–	–
2-ethyl-4-methylthiazole	tr	–	–
4-ethyl-2-methylthiazole	tr	–	–
4,5-dimethylthiazole	10	–	–
trimethylthiazole	8	–	–
5-ethyl-2-methylthiazole	1	–	–
4-ethyl-2,5-dimethylthiazole	1	–	–

Table 4.3 *continued*

Compounds	Grilled (well-done)	Roasted	Boiled
		(ng/100 g)	
4-ethyl-5-methylthiazole	1	–	–
5-ethyl-4-methylthiazole	5	–	–
5-ethyl-2,4-dimethylthiazole	5	–	–
2,5-diethyl-4-methylthiazole	tr	–	–
4-methyl-5-vinylthiazole	tr	–	–
2-acetylthiazole	15	9	54
benzothiazole	tr	tr	2
Thiophenes			
2-ethylthiophene	tr	–	–
2-methyl-5-ethylthiophene	tr	–	–
a thiophene (MW 126)	tr	–	–
dihydro-2-methyl-3(2*H*)-thiophenone	15	–	–
2-acetylthiophene	2	–	–
Furans			
2-pentylfuran	100	824	30
dihydro-2-methyl-3(2*H*)-furanone	24	–	–
2-furanmethanethiol	10	–	–
2-acetylfuran	73	–	–
Oxazole			
2,4,5-trimethyloxazole	2	–	–
Pyrroles			
pyrrole	3	–	–
1-(2-furfuryl)pyrrole	2	–	–

namely 4-methylthiazole, 4,5-dimethylthiazole, 4-methyl-5-ethylthiazole and 4-methyl-5-vinylthiazole, identified in pork may also be the thermal degradation products of thiamine (Figure 4.4) (Ho *et al.*, 1989a).

Several thiophenes were identified in grilled pork. Thiophenes are responsible for the mild sulphurous odour of cooked meat (Shibamoto, 1980). The sulphur atom in a thiophene ring may come either from an amino acid (cysteine or cystine) or from thiamine. Other minor heterocyclic compounds reported in grilled pork, such as furans, pyridine pyrroles, are also well-known Maillard reaction products.

4.3.3 Interaction of Maillard reaction with lipids

The carbonyl compounds generated from lipid degradation can subsequently react with amino acids or Maillard reaction intermediates to form flavour compounds, which contribute to the overall pork aroma. Some long-chain alkyl-substituted heterocyclic compounds have been identified in meat flavour; these may originate from the reaction of aldehydes from lipid degradation with heterocyclic compounds formed from the Maillard reaction. Ho *et al.* (1989b) have studied the contribution of 2,4-decandienal, one of the major degradation products of linoleic acid, to the formation of

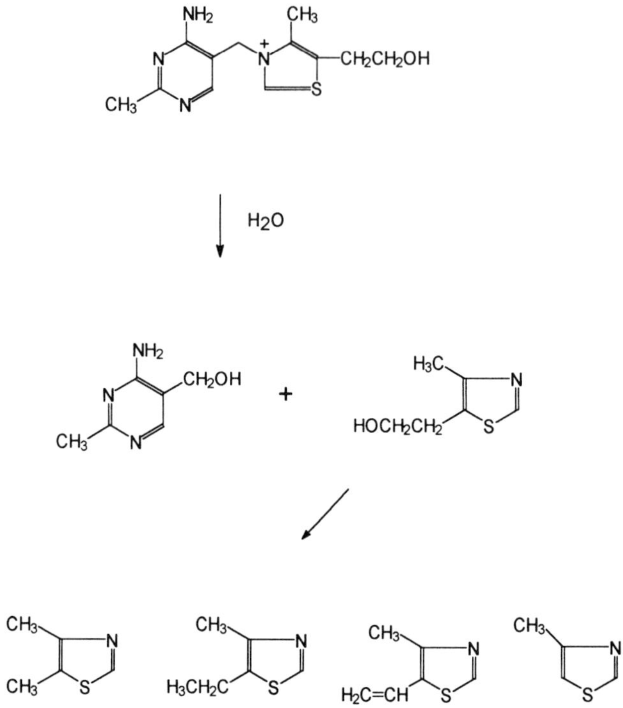

Figure 4.4 Mechanism for the formation of thiazoles from thiamine.

heterocyclic compounds in a model system. A large number of long-chain alkyl-substituted heterocyclic compounds were obtained in the reaction system of 2,4-decadienal and cysteine (Table 4.4). The formation mechanism of 2-pentylpyridine was proposed to be due to reaction of lipid oxidation products through the Maillard reaction (Figure 4.5). Thus, participation of lipid oxidation products in Maillard reaction is the most important reaction in flavour generation during the cooking of pork as their interaction produces its characteristic aroma.

4.3.4 Thiamine degradation

Thiamine degradation generates a series of sulphur-containing meaty flavour compounds; hydrogen sulphide is an important flavour compound itself and can react with furanones to give an intense meaty flavour, and 2-methyl-3-furanthiol and bis-(2-methyl-3-furyl) disulphide are the primary meaty aroma compounds. The proposed pathways of thermal degradation of thiamine may refer to the original paper of Guentert *et al.* (1990) and Werkhoff *et al.* (1990).

Table 4.4 Some heterocyclic compounds identified from the thermal interaction of 2,4-decadienal and cysteine (from Ho *et al.*, 1989)

Compounds	Amount produced (mg/mol)
Furans	
2-butylfuran	12.8
2-pentylfuran	6.4
2-hexylfuran	tr
Thiophenes	
thiophene	3.5
tetrahydrothiophene-3-one	10.5
2-butylthiophene	57.2
2-formyl-3-methylthiophene	29.8
2-pentylthiophene	13.1
2-hexylthiophene	42.0
2-heptylthiophene	1.8
2-formyl-5(or 3)-pentylthiophene	15.6
Thiazoles	
thiazoles	25.6
3-methylisothiazoles	2.0
2-acetylthiazoles	2.2
Cyclic polysulphides	
3,5-dimethyl-1,2,4-trithiolane (isomer)	122.8
3,5-dimethyl-1,2,4-trithiolane (isomer)	18.2
3-methyl-5-pentyl-1,2,4-trithiolane (isomer)	14.3
2,4,6-trimethylperhydro-1,3,5-thiadiazine	828.5
2,4,6-trimethylperhydro-1,3,5-dithiazine	284.2
2,4-dimethyl-6-pentylperhydro-1,3,5-dithiazine	18.9
2-pentyl-4,6-dimethylperhydro-1,3,5-dithiazine	28.7
Pyridine	
2-pentylpyridine	501.5

The intensive study of thiamine degradation and its reaction with other compounds has led to several inventions on meat flavour (Guentert *et al.*, 1990; Werkhoff *et al.*, 1990). There is a higher concentration of thiamine in pork (0.66 mg/kg) than in beef (0.08 mg/kg) (Blitz and Grosch, 1987); it is expected that thiamine-derived compounds contribute to the pork flavour.

4.3.5 Reactions to form polysulphides in roasted pork

Two polysulphides, 3-methyl-2,4,5-trithiahexane and 4,6-dimethyl-2,3,5,7-tetrathiaoctane were reported in the headspace aroma components of roasted pork (Dubs and Joho, 1978; Dubs and Stussi, 1978). It is postulated that they are formed from the reaction of acetaldehyde with methanethiol, hydrogen sulphide and dimethyldisulphide as shown in Figure 4.6.

These polysulphides may play an important role in the flavour of pork. With very low odour threshold values of polysulphides, relatively small amounts could have a significant effect. They may contribute directly to

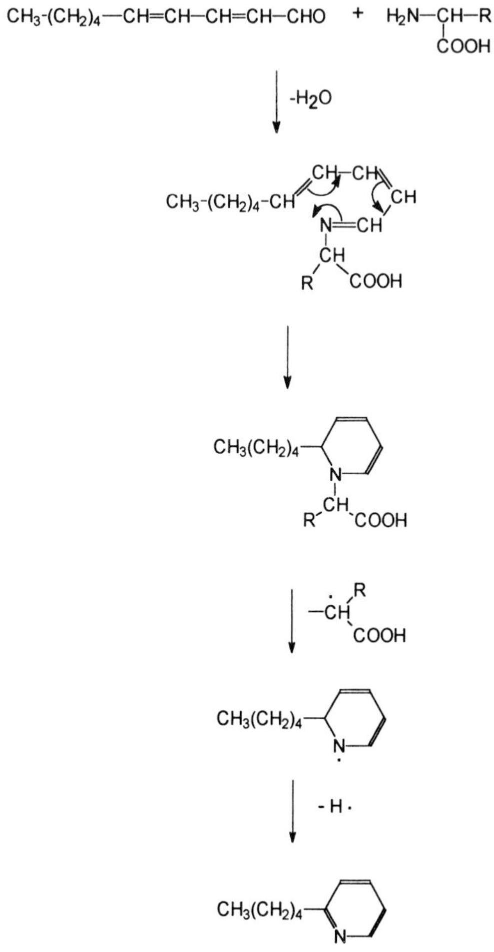

Figure 4.5 Mechanism for the formation of 2-pentylpyridine.

the meaty characteristics, or they may provide an overall sulphurous note that is a part of meat aroma (Mottram, 1991).

4.4 Recently identified pork flavour compounds and their sensory properties

Extensive studies in flavour chemistry of meat volatiles have led to the discovery of many new flavour compounds. In the papers of Werkhoff *et al.* (1992, 1993), they have summarized their new findings with others. In Table 4.5, we have shown their results on the flavour of pork.

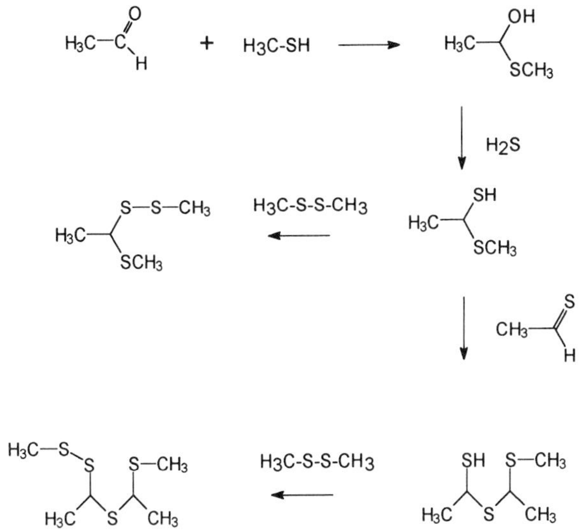

Figure 4.6 Mechanism for the formation of polysulphides in roasted pork.

As mentioned before, furans or thiophenes substituted on position 3 by a thiol, sulphide or disulphide group are considered as key compounds to meat flavour, and long-chain aliphatic aldehydes possess fatty notes.

It is interesting that bicyclic 1,3,5-dithiazines, which were identified in cooked shrimp and dried squid, were also found in the cooked or roasted pork volatiles. As Figure 4.7 shows, the proposed formation mechanism involved Strecker aldehydes from proline, ornithine or lysine with hydrogen sulphide and other aldehydes (Werkhoff *et al.* 1992, 1993).

4.5 Factors affecting pork flavour

Meat flavour development has been shown to be influenced by several antemortem and postmortem factors, as summarized below (Imafidon and Spanier, 1994). The antemortem factors that affect meat flavour are age, species/breed, sex, nutritional status, stress level, fat level and fat composition. Meanwhile, postmortem factors which influence the flavour of meat are manner of slaughter, carcass handling, aging, which increases hydrolytic activity, and levels of products of the activity of endogenous and/or microbial proteinases, lipases and glycosidases. Furthermore, the mode of cooking may also affect the flavour via wet versus dry cooking, convection versus microwave heating, rate of heating, final end-point cooking temperature, and final fat levels and composition. Effects of storage after cooking arise mainly from autoxidation of lipids and off-flavour generation as

Table 4.5 Recently identified flavour compounds in pork volatiles and their sensory properties (modified from Werkhoff *et al.*, 1992, 1993)

Compounds	Sensory impression
Furans and thiophenes	
2-Methyl-3-furanthiol	
Bis-(2-methyl-3-furyl) disulphide	
2-Methyl-3-(methylthio)furan	
2-Methyl-3-(methyldithio)furan	Typical meaty flavour notes, cheese-like, egg-like, yeast-like and onion-like
	Thiamin-like at high concentration, meaty when <1ppb
1-(2-Methyl-3-furylthio)ethanethiol	Typical meat-like, roast beef-like
3-Thiophenethiol	Fatty, oniony, cooked meat flavour note, coffee-like
2-Methyl-3-thiophenethiol	Roasted, cooked meat character, egg-like, fatty, oniony, leek-like
1,2,4-Trithiolanes	
3-Methyl-	Oniony, allium-like, meat-like
3,5-Dimethyl- (two isomers)	Fatty, oniony, nutty, slightly meaty
3-Methyl-5-ethyl- (two isomers)	Herbaceous, oniony, allium-like, leek-like, roasted peanut, Shiitake-like
3-Methyl-5-isopropyl- (two isomers)	
3-Methyl-5-isobutyl-	Oniony, roasted, allium-like, meaty
3-Methyl-5-n-pentyl- (two isomers)	Sulphury, oniony, allium-like, leek-like, meaty
Aliphatic sulphur-containing compounds	
1-Methylthio-1-ethanethiol	Meaty, fatty, oniony, durian-like and tropical fruit-like
1,1-Ethanedithiol	Sulphury, rubbery, cheesy, oniony, slightly meaty, durian-like
2-Methylthiomethyl-2-butenal	Fatty, metallic, intensive potato-like
n-Hexadecanethiol	Meaty, burnt, onion-like
Some aliphatic and cyclic aldehydes	
5-Ethylcyclopentene-1-carbaldehyde	Fatty, slight phenolic note
14-Methylpentadecanal	Fatty, tallowy, train oil-like
14-Methylhexadecanal	Fatty, tallowy, orange-like, tangerine-like
15-Methylhexadecanal	Fatty, tallowy
1,3,5-Dithiazines	
2,4,6-Trimethyl-	Carrot-like, oniony, leek-like, sulphury, roasted, meaty, pyrazine-like, rubbery
2-Ethyl-4,6-dimethyl-	Roasted onion, garlic-like; roasted, roasted peanut, sulphury, green, pea-like, vegetable-like
2,6-Dimethyl-4-ethyl-	Roasted, candy-like; thiazole-like, sulphury, meaty
2,6-Dimethyl-4-isopropyl-	Roasted, roasted peanut, meaty, cocoa-like; sulphury, green, roasted
2-Isopropyl-4,6-dimethyl-	Roasted, onion-like, roasted peanut; nutty, leek-like
2,4-Diethyl-6-methyl-	Garlic-like, liver-sausage-like, roasted, onion-like
2-Propyl-4,6-dimethyl-	Sulphury, roasted, oniony, roasted peanut; roasted, leek-like, raw peanut
2,6-Dimethyl-4-propyl-	Roasted, roasted peanut, egg-like, grilled bacon; green, cabbage-like, garlic-like, mushroom-like
2,4-Dimethyl-4-isobutyl-	Roasted, green, tropical fruit note

Table 4.5 *continued*

Compounds	Sensory impression
2-Isobutyl-4,6-dimethyl-	Roasted, roasted peanut, egg-like, fatty, popcorn-like, sulphury, rubbery, cabbage-like, chives-like
2-*sec*-Butyl-4,6-dimethyl-	Roasted, egg-like, oniony, roasted peanut; oniony
2,6-Dimethyl-4-*sec*-butyl-	
2-Butyl-4,6-dimethyl-	
2-Ethyl-4-isopropyl-6-methyl-	Sweet, sulphury, cocoa-like, chocolate-like, bitter almond
2-Isopropyl-4-ethyl-6-methyl-	Roasted, roasted peanut, onion-like, fatty
2-Ethyl-4-isobutyl-6-methyl-	Leek-like, oniony, garlic-like, sweet
2-Isobutyl-4-ethyl-6-methyl-	Bitter, burnt, roasted
2-Methyl-4-isobutyl-6-ethyl-	Sulphury, rubbery, candy-like, cocoa-like
2,6-Dimethyl-4-pentyl-	
2-Pentyl-4,6-dimethyl-	Sulphury, fatty, green, roasted peanut
2-Methyl-4,6-diisopropyl-	Onion-like, roasted, chocolate-like
2,4-Diisopropyl-6-methyl-	Rubbery, roasted, oniony, roasted peanut
2-Isobutyl-4-isopropyl-6-methyl-	Rubbery, roasted, sulphury, tallowy
2-Hexyl-4,6-dimethyl-	
2,6-Dimethyl-4-hexyl-	
2-Heptyl-4,6-dimethyl-	
2-Ethyl-4-isobutyl-6-isopropyl-	
2-Octyl-4,6-dimethyl-	
2-Butyl-4-pentyl-6-methyl-	
Bicyclic 1,3,5-dithiazines	
2,4-Dimethyltetrahydro-4*H*-pyrrolo[2.1-*d*]-	Fatty, roasted, burnt, oniony, sulphury, roasted peanut, bread-like, coffee-like, roasted onion-like, oniony, metallic
2-Ethyl-4-methyltetrahydro-4*H*-pyrrolo[2.1-*d*]-	Fatty, onion-like, meaty
2,4-Dimethylperhydropyrido[2.1-*d*]-	Sulphury, rubbery, oniony, garlic-like, leek-like, roasted, meaty, match-like, isophorone
2-Isopropyl-4-methyl-tetrahydro-4*H*-pyrrolo[2.1-*d*]-	Fatty, rancid, burnt, roasted, roasted onion-like, leek-like, roasted peanut
2-Isobutyl-4-methyl-tetrahydro-4*H*-pyrrolo[2.1-*d*]-	Rubbery, roasted, fatty, burnt
2,4-Dimethylhexahydropyrido[2.1-*d*]-	Sulphury, rubbery, musty, oniony, and roasted meat notes, garlic-like, leek-like, match-like, thiazole-like, isophorone-like
Other sulphur-containing compounds in pork volatiles	
1-Pentanethiol	
2-Methyl-1-butanethiol	
3-Methyl-1-butanethiol	
Phenylmethanethiol	
2-Furanmethanethiol	
Ethyl isopropyl disulphide	
Dipropyl disulphide	
Diisopropyl disulphide	
Methyl isobutyl disulphide	
Methyl 2-methylbutyl disulphide	
Methyl n-propyl trisulphide	
Dimethyl tetrasulphide	
4-Methyl-2,3,5-trithiahexane	
2-Methyl-4,5-dihydrothiophene	

Table 4.5 *continued*

Compounds	Sensory impression
3-Methylthiophene	
2,5-Dimethylthiophene	
2-Ethylthiophene	
3-Ethylthiophene	
2-Propylthiophene	
3-Propylthiophene	
2-Butylthiophene	
3-Butylthiophene	
3,4-Diethylthiophene	
2-Pentylthiophene	
2-n-Hexylthiophene	
2-n-Heptylthiophene	
3-n-Heptylthiophene	
2-n-Octylthiophene	
Tetrahydrothiophen-3-one	
2-Methyltetrahydrothiophen-3-one	
5-Methyltetrahydrothiophen-3-one	
2,5-Dimethyltetrahydrothiophen-3-one	
2-(1-Hydroxyhexyl)thiophene	
Carbaldehyde	
Benzothiophene	
2-Methylthieno[3.2-*b*]thiophene	
3-Methylthieno[3.2-*b*]thiophene	
Kahweofuran	
3-Methyl-1,2,4-trithiane	
2,4,5-Trimethylthiazine	
2,5-Dimethyl-4-ethylthiazole	
4,5-Dimethyl-2-isopropylthiazole	
2-Methyl-3-thiazoline	
2,4-Dimethyl-3-thiazoline	
2,4,5-Trimethyl-3-thiazoline	

well as decreased hydrophilicity of peptides which enhances the off-flavour perception of meats. Some of the factors that affect pork flavour generation are discussed below.

4.5.1 Composition of pork meat

Many factors influence the composition of flavour precursors in pork muscle, which in turn affect pork aroma.

Diet can have a significant impact on the fatty acid and other fat-soluble substances of porcine subcutaneous fat and lean tissue which affect the flavour of pork (Melton, 1990; Larick *et al.*, 1992). Larick *et al.* (1992) found that cooking meat samples of pigs that were fed a higher linoleic acid content diet had a higher concentration of aldehydes, especially pentanal and hexanal, which indicated increased lipid oxidation.

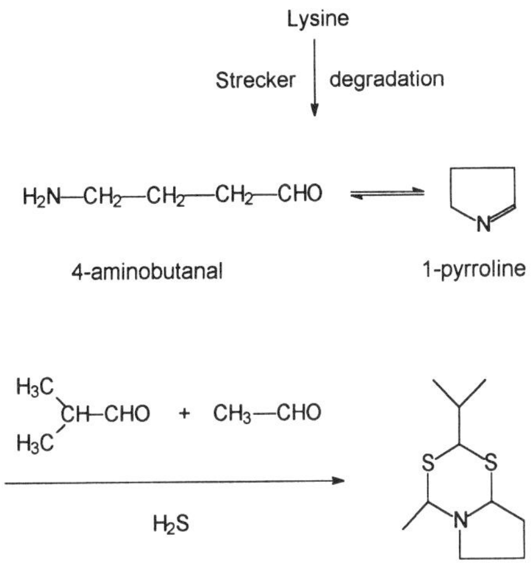

Figure 4.7 Mechanism for the formation of 2,4-dimethylperhydropyrido[2.1-*d*]-1,3,5-diathiazine.

Conditioning can also exert some effect on meat flavour, and a considerable number of studies have investigated it. During the postmortem conditioning, the concentration of sugars and amino acids increases as a result of breakdown of glycogen and the action of the proteolytic enzyme, and free fatty acids and adenosine 5′-triphosphate (ATP) metabolites are released by hydrolysis of phospholipids (Nishimura *et al.*, 1988).

4.5.2 *Additives and processing*

Nitrite and nitrate salts are commonly utilized in meat curing. The effect of nitrite and nitrate on the flavour of pork is well known and is the topic of discussion in another chapter of this book. A quantitative comparison of the flavour of cured and uncured pork indicated that the concentration of carbonyl compounds was higher in uncured pork than in cured meat, where carbonyl compounds were either present in reduced amounts or not detectable (Ramarathnam *et al.*, 1991). For example, hexanal, the major carbonyl compound, was found in the uncured meat at a concentration of 12.66 mg/kg, while only 0.03 mg/kg was present in the cured product, a reduction of 99.8%.

Smoking is also a very important technique in the meat curing process, and it renders a specific flavour to meat products, such as in bacon and sausages. Around 400 compounds were identified in smoke, and 48 acids, 22 alcohols, 131 carbonyls, 22 esters, 46 furans, 16 lactones, 75 phenols and 50 miscellaneous compounds were found among them (Maga, 1988). Phenol compounds contribute significantly to the smoke aroma, and 4-methylguaiacol is the most important due to the quantity found in smoke and its relatively low odour threshold (Wasserman, 1966). Polyphenols can also act as antioxidants in smoked food.

Soy sauce was also found to affect flavour development in cooked pork. Many unsaturated aldehydes including 2,4-decadienal, 2-undecenal, 2-dodecenal and 2-tridecenal identified in the volatiles of cooked pork were not found in the volatiles of pork stewed with soy sauce (Chou and Wu, 1983). It is possible that the very reactive carbonyl compounds such as aldehydes can react with nitrite, nitrate or reactive substances in soy sauce during curing or cooking of pork.

The effect of ingredients on the volatile components generated, other than carbonyl compounds, has also been observed. In the studies of the flavour of soy sauce-stewed pork (Chou and Wu, 1983), several alkyl-substituted trithiolanes such as *syn*- and *anti*-3-methyl-5-ethyl-1,2,4-trithiolanes, *syn*- and *anti*-3-methyl-5-propyl-1,2,4-trithiolanes, *syn*- and *anti*-3-methyl-5-butyl-1,2,4-trithiolanes and *syn*- and *anti*-3-methyl-5-isobutyl-1,2,4-trithiolanes were identified in the headspace volatile components. Neither soy sauce nor cooked pork contained these trithiolanes. Apparently, they were generated when the pork was stewed with soy sauce.

The amount of soy sauce used in the preparation of stewed pork also showed a significant effect on the quantity of some volatile compounds generated in stewed pork. Table 4.6 shows the relative percentage of 3,5-dimethyl-1,2,4-trithiolanes and 2,4,6-trimethylperhydro-1,3,5-dithiazine (thialdine) in stewed pork samples prepared with different amounts of soy sauce. The amount of *syn*- and *anti*-3,5-dimethyl-1,2,4-trithiolanes increased with increasing amounts of soy sauce. Soy sauce may provide the acetaldehyde which is the necessary intermediate for the formation of 3,5-dimethyl-1,2,4-trithiolanes. There is no simple explanation for the lesser amount of thialdine generated in the stewed pork samples cooked with a higher concentration of soy sauce.

The reactivity of aldehydes toward thiols, a class of components in onions, has also been observed. When hexanal was incubated with propanethiol, 1,1-bis(propylthio)-hexane and 1,2-bis(propylthio)-hexane were formed (Kuo and Ho, 1991). Figure 4.8 shows the mechanism for

Table 4.6 Effect of soy sauce on the formation of trithiolanes and thialdine in stewed pork (modified from Chou and Wu, 1983)

Compounds	Relative concentration (%)			
	A	B	C	D
anti-3,5-Dimethyl-1,2,4-trithiolane	0.7	1.8	4.3	5.3
syn-3,5-Dimethyl-1,2,4-trithiolane	1.3	2.2	3.3	5.5
Thialdine	42.9	44.9	41.1	27.2

Cooking conditions: A, pork cooked with 2.5% soy sauce and 27.5% water; B, pork cooked with 5.0% soy sauce and 25.0% water; C, pork cooked with 7.5% soy sauce and 22.5% water; D, pork cooked with 10.0% soy sauce and 20.0% water.

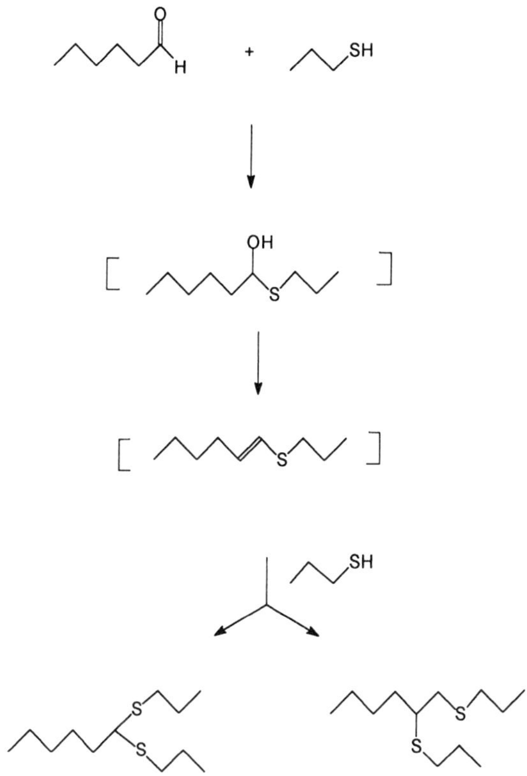

Figure 4.8 Mechanism for the formation of sulphides from the reaction of hexanal with propanethiol.

their formation. It is, therefore, expected that the addition of onion to the pork during heating will definitely modify the flavour profile of cooked pork.

References

Angelo, A.J. St. (1996). Lipid oxidation in foods. *CRC Crit. Rev. Food Sci. Nutr.*, **36**, 175–224.

Aristoy, M.-C. and Toldra, F. (1995). Isolation of flavour peptides from raw pork meat and dry-cured ham. In *Food Flavours: Generation, Analysis and Process Influence*, ed. G. Charalambous. Elsevier, Amsterdam, pp. 1323–1341.

Blitz, H.-D. and Grosch, W. (1987). *Food Chemistry*, Springer-Verlag, Berlin.

Chou, C.C. and Wu, C.M. (1983). The volatile compounds of stewed pork. *Research Report 285*, Food Industry Research and Development Institute, Hsinchu, Taiwan.

de Roos, K.B. (1992). Meat flavour generation from cysteine and sugars. In *Flavour Precursors: Thermal and Enzymatic Conversions*, eds. R. Teranishi, G.R. Takeoka, and M. Guentert. ACS Symp. Ser. 490, American Chemical Society, Washington, DC, pp. 203–216.

Dubs, P. and Joho, M. (1978). Investigation of the head-space of roasted meat. III. Synthesis of 4,6-dimethyl-2,3,5,7-tetrathiaoctane. *Helv. Chim. Acta*, **61**, 2809–2812.

Dubs, P. and Stussi, R. (1978). Investigation of the head-space of roasted meat II. Synthesis of substituted 2,4,5-trithia-hexane. *Helv. Chim. Acta*, **61**, 2351–2359.

Frankel, E.N. (1982). Volatile lipid oxidation products. *Prog. Lipid Res.*, **22**, 1–33.

Gasser, U. and Grosch, W. (1988). Identification of volatile flavour compounds with high aroma values from cooked beef. *Z. Lebensm. Unters. Forsch.*, **186**, 489–494.

Gasser, U. and Grosch, W. (1990). Primary odourants of chicken broth. *Z. Lebensm. Unters. Forsch.*, **190**, 3–8.

Guadagni, D.G., Buttery, R.G., and Turnbaugh, J.G. (1972). Odour thresholds and similarity ratings of some potato chip components. *J. Sci. Food Agric.*, **23**, 1435–1444.

Guentert, M., Bruening, J., Emberger, R., Koepsel, M., Kuhn, W., Thielmann, T., and Werkhoff, P. (1990). Identification and formation of some selected sulphur-containing flavour compounds in various meat model systems. *J. Agric. Food Chem.*, **38**, 2027–2041.

Hartman, G.J., Carlin, J.T., Scheide, J.D. and Ho, C.-T. (1984). Volatile products formed from the thermal degradation of thiamine at high and low moisture levels. *J. Agric. Food Chem.*, **32**, 1015–1018.

Ho, C.-T., Smagula, M.S. and Chang, S.S. (1978). The synthesis of 2-(1-pentenyl)furan and its relationship to the reversion flavour of soybean oil. *J. Am. Oil Chem. Soc.*, **55**, 233–237.

Ho, C.-T., Bruechert, L.J., Kuo, M.C. and Izzo, M.T. (1989a). Formation of volatile compounds from extruded corn-based model systems. In *Thermal Generation of Aromas*, eds. T.H. Parliment, R.J. McGorrin and C.-T. Ho. ACS Symposium Series 409, American Chemical Society, Washington, DC, pp. 504–511.

Ho, C.-T., Bruechert, L.J., Zhang, Y, and Chiu, E-M (1989b) Contribution of lipids to the formation of heterocyclic compounds in model systems. In *Thermal Generation of Aromas,* eds. T.H. Parliment, R.J. McGorrin and C.-T. Ho. ACS Symposium Series 409, American Chemical Society, Washington, DC, pp. 105–113

Hornstein, I. and Crowe, P.F. (1960). Flavour studies on beef and pork. *J. Agric. Food Chem.*, **8**, 494–498.

Hornstein, I. and Crowe, P.F. (1963). Meat flavour: lamb. *J. Agric. Food Chem.*, **11**, 147–149.

Hornstein, I. and Wasserman, A. (1987). Chemistry of meat flavour. In *The Science of Meat and Meat Products*, eds. J.F. Price and B.S. Schweigert. Food and Nutrition Press, Trumbull, CT, pp. 329–348.

Imafidon, G.I. and Spanier, A.M. (1994). Unraveling the secret of meat flavour. *Trends Food Sci. Technol.*, **5**, 315–321.

Kuo, M.C. and Ho, C.-T. (1991). Unpublished results.

Larick, D.K., Turner, B.E., Schoenherr, W.D., Coffey, M.T. and Pilkington, D.H. (1992). Volatile compound content and fatty acid composition of pork as influenced by linoleic acid content of the diet. *J. Anim. Sci.*, **70**, 1397–1403.

MacLeod, G. (1986). The scientific and technological basis of meat flavours. In *Developments in Food Flavours*, eds. G.G. Birch, and M.G. Lindley. Elsevier, London, pp. 191–223.

Maga, J.A. (1988). Flavour chemistry of wood smoke. In *Smoke in Food Processing*. CRC Press, Boca Raton, FL, pp. 49–88.

Melton, S.L. (1990). Effects of feed on flavour of red meat: A review. *J. Anim. Sci.*, **68**, 4421–4435.

Mottram, D.S. (1979). The flavour of cooked meats. *Agricultural Research Council Meat Research Institute (Bristol) Biennial Rep. 1977–79*, pp. 87–88.

Mottram, D.S. (1985). The effect of cooking conditions on the formation of volatile heterocyclic compounds in pork. *J. Sci. Food Agric.*, **36**, 377–382.

Mottram, D.S. (1991). Meat. In *Volatile Compounds in Foods and Beverages,* ed. H. Maarse. Marcel Dekker, New York, pp. 107–177.

Mottram, D.S., Edwards, R.A. and MacFie, H.J.H. (1982). A comparison of the flavour volatiles from cooked beef and pork meat systems. *J. Sci. Food Agric.*, **33**, 934–944.

Mottram, D.S., Croft, S.E. and Patterson, R.L.S. (1984). Volatile components of cured and uncured pork: the role of nitrite and the formation of nitrogen compounds. *J. Sci. Food Agric.*, **35**, 233–239.

Mussinan, C.J. and Walradt, J.P. (1974). Volatile constituents of pressure-cooked pork liver. *J. Agric. Food Chem.*, **22**, 827–831.

Nawar, W.W. (1969). Thermal degradation of lipids: a review. *J. Agric. Food Chem.*, **17**, 18.

Nishimura, T., Rhue, M.R., Okitani, A. and Kato, H. (1988). Compounds contributing to the improvement of meat taste during storage. *Agric. Biol. Chem.*, **52**, 2323–2330.

Ohloff, G. and Flament, I. (1978). Heterocyclic constituents of meat aroma. *Heterocycles*, **11**, 663–695.

Ramarathnam, N., Rubin, L.J. and Diosady, L.L. (1991). Studies on flavour. 1. Qualitative and quantitative differences in uncured and cured pork. *J. Agric. Food Chem.*, **39**, 344–350.

Reineccius, G. (1994) Flavour and aroma chemistry. In *Quality Attributes and Their Measurement in Meat, Poultry and Fish Products*, eds. A.M. Pearson and T.R. Dutson. Blackie Academic & Professional, London, pp. 184–201.

Reineccius, G.A. and Liardon, R. (1985). The use of charcoal traps and microwave desorption for the analysis of headspace volatiles above heated thiamine solutions. In *Topics in Flavour Research*, eds. R.G. Berger, S. Nitz and P. Schreier, Eichhorn. Marzling-Hangenhan, pp. 125–136.

Romarathnam, N., Rubin, L.J. and Diosady, L.L. (1991). Studies of meat flavour. 1. Qualitative and quantitative differences in uncured and cured pork. *J. Agric. Food Chem.*, **39**, 344–350.

Romarathnam, N., Rubin, L.J. and Diosady, L.L. (1993). Studies of meat flavour. 3. A novel method for trapping volatile components from uncured and cured pork. *J. Agric. Food Chem.*, **41**, 933–938.

Schliemann, J., Wolm, G., Schrodter, R. and Ruttloff, H. (1987). Chicken flavour – formation, composition and production. Part 1. Flavour precursors. *Nahrung*, **31**, 47–56.

Shahidi, F. (1988) Flavour of cooked meats. In *Flavour Chemistry Trends and Developments*, eds. R. Teranishi, R.G. Buttery and F. Shahidi. ACS Symposium Series. 388, American Chemical Society, Washington, DC, pp. 188–201.

Shahidi, F., Rubin, L.J. and D'Souza, L.A. (1986). Meat flavour volatiles: A review of the composition, techniques of analysis, and sensory evaluation. *CRC Crit. Rev. Food Sci. Nutr.*, **24**, 141–243.

Shibamoto, T. (1980). Heterocyclic compounds found in cooked meats. *J. Agric. Food Chem.*, **27**, 237–243.

van den Ouweland, G.A.M. and Swaine, R.L. (1980). Investigation of the species specific flavour of meat. *Perfumer Flavourist*, **5**, 15–20.

van den Ouweland, G.A.M., Demole, E.P. and Enggist, P. (1989). Process meat flavour development and the Maillard reaction. In *Thermal Generation of Aroma*, eds. T.H. Parliment, R.J. McGorrin and C.-T. Ho. ACS Symposium Series 409, American Chemical Society, Washington, DC, pp. 431–441.

Wasserman, A.E. (1966). Organoleptic evaluation of three phenols present in wood smoke. *J. Food Sci.*, **31**, 1005–1010.

Wasserman, A.E. (1972). Thermally produced flavour components in the aroma of meat and poultry. *J. Agric. Food Chem.*, **20**, 737–741.

Wasserman, A.E. (1979). Symposium on meat flavour chemical basis of meat flavour: a review. *J. Food Sci.*, **44**, 6–11.

Wasserman, A.E. and Talley, F. (1968). Organoleptic identification of roasted beef, veal, lamb, and pork as affected by fat. *J. Food Sci.*, **33**, 219–223.

Werkhoff, P., Bruening, J., Emberger, R., Guentert, M., Koepsel, M., Kuhn, W. and Surburg, H. (1990). Isolation and characterization of volatile sulphur-containing meat flavour components in model systems. *J. Agric. Food Chem.*, **38**, 777–791.

Werkhoff, P., Guentert, M. and Hopp, R. (1992) Dihydro-1,3,5-dithiazines: Unusual flavour compounds with remarkable organoleptic properties. *Food Rev. Int.*, **8**, 391–442.

Werkhoff, P., Bruening, J., Emberger, R., Guentert, M. and Hopp R. (1993) Flavour chemistry of meat volatiles: New results on flavour components from beef, pork, and chicken. In *Recent Developments in Flavour and Fragrance Chemistry*, eds. R. Hopp and K. Mori. VCH, Weinheim, Basel, pp. 183–213.

Whitfield, F.B. (1992). Volatile from interaction of Maillard reactions and lipids. *CRC Crit. Rev. Food Sci. Nutr.*, **31**, 1–58.

Whitfield, F.B. and Last, J.H. (1991). Vegetables. In *Volatile Compounds in Foods and Beverages*, ed. H. Maarse. Marcel Dekker, New York, pp. 203–281.

5 The flavour of poultry meat

C.-W. CHEN AND C.-T. HO

5.1 Introduction

Poultry meat has become a significant part of the North American diet. Numerous reviews on poultry flavour have been published (Wilson and Katz, 1972; Steverink, 1981; Ramaswamy and Richards, 1982). Thus far, more than 450 components have been characterized in cooked poultry meat (Mottram, 1991). This chapter contains a review of some of the chemical reactions which are responsible for the formation of volatile compounds significant to the aroma of cooked poultry meat.

5.2 Primary odorants of chicken broth

The most important study on chicken flavour in recent years is probably the one published by Gasser and Grosch (1990). Using aroma extract dilution analysis, they identified 16 primary odour compounds in chicken broth. Fourteen of these compounds were structurally identified as 2-methyl-3-furanthiol, 2-furfurylthiol, methional, 2,4,5-trimethylthiazole, nonanal, 2-*trans*-nonenal, 2-formyl-5-methylthiophene, *p*-cresol, 2-*trans*-4-*trans*-nonadienal, 2-*trans*-4-*trans*-decadienal, 2-undecenal, β-ionone, γ-decalactone and γ-dodecalactone. When the primary odorants of chicken were compared with those resulting from the aroma extract dilution analysis of broth from beef (Table 5.1), the major differences were that 2-*trans*-4-*trans*-decadienal (fatty) and γ-dodecalactone (tallowy, fruity) prevailed in the chicken broth, whereas the sulphur compounds, bis-(2-methyl-3-furyl)disulphide (meat-like aroma) and methional (cooked potato-like), predominated in broth prepared from beef.

5.3 Sulphur-containing compounds in chicken flavours

2-Methyl-3-furanthiol identified by Gasser and Grosch (1990) as the most important flavour compound contributing to the meaty perception of chicken broth has been recognized as a character impact compound in the aroma of cooked beef (Gasser and Grosch, 1988) and canned tuna fish (Withycombe and Mussinan, 1988). 2-Methyl-3-furanthiol and its

Table 5.1 Comparison of flavour dilution factors of odourant appearing in broths from chicken and beef.

Compounds	Flavour dilution factor		Odour description
	Chicken	Beef	
2-Methyl-3-furanthiol	1024	512	Meat-like, sweet
bis(2-Methyl-3-furyl)disulphide	<16	2048	Meat-like
2-Furfurylthiol	512	512	Roasted
2,5-Dimethyl-3-furanthiol	256	<16	Meaty
3-Mercapto-2-pentanone	128	32	Sulphurous
Methional	128	512	Cooked potato
2,4,5-Trimethylthiazole	128	<16	Earthy
2-Formyl-5-methylthiophene	64	64	Sulphurous
Phenylacetaldehyde	16	64	Honey-like
2-*trans*-4-*trans*-Decadienal	2048	<16	Fatty
2-*trans*-4-*cis*-Decadienal	128	<16	Fatty, tallowy
2-Undecenal	256	<16	Tallowy, sweet
γ-Dodecalactone	512	<16	Tallowy, fruity
γ-Decalactone	64	<16	Peach-like
Nonanal	64	<16	Tallowy, green
2-*trans*-Nonenal	64	<16	Tallowy, fatty
2-*trans*-4-*trans*-Nonadienal	64	<16	Fatty
β-Ionone	64	<16	Violet-like
p-Cresol	64	<16	Phenolic

oxidative dimer, bis-(2-methyl-3-furyl)disulphide, possessing characteristic meat flavour notes have been found by Evers *et al.* (1976) among the volatile products from heating thiamine hydrochloride with cysteine hydrochloride and hydrolysed vegetable protein. These two compounds were also found in volatile products of thiamine degradation (van der Linde *et al.*, 1979; Hartman *et al.*, 1984; Reineccius and Liardon, 1985) and in heated yeast extract (Ames and MacLeod, 1985). Thiamine has, therefore, been recognized as the precursor for the formation of meaty aroma compounds, 2-methyl-3-furanthiol and bis-(2-methyl-3-furyl)disulphide. However, thiamine is not the sole source of 2-methyl-3-furanthiol. It was found that when ribose or inosine 5′-monophosphate (IMP) react with cysteine or glutathione a significant quantity of 2-methyl-3-furanthiol is formed (Farmer and Mottram, 1990; Grosch *et al.*, 1990; Zhang and Ho, 1991). The mechanisms for the formation of 2-methyl-3-furanthiol and bis-(2-methyl-3-furyl)disulphide are shown in Figure 5.1.

The formation of 2-methyl-3-furanthiol from either ribose or IMP requires interaction with sulphur-containing amino acids, cysteine or cystine, or peptide, glutathione. Cysteine, cystine and glutathione will liberate hydrogen sulphide which is a major reactant for meaty aroma generation. Rapid evolution of hydrogen sulphide occurs from glutathione during the early stages of cooking (Ohloff *et al.*, 1985), but cysteine is the main hydrogen sulphide precursor on prolonged heating (Mecchi *et al.*, 1964).

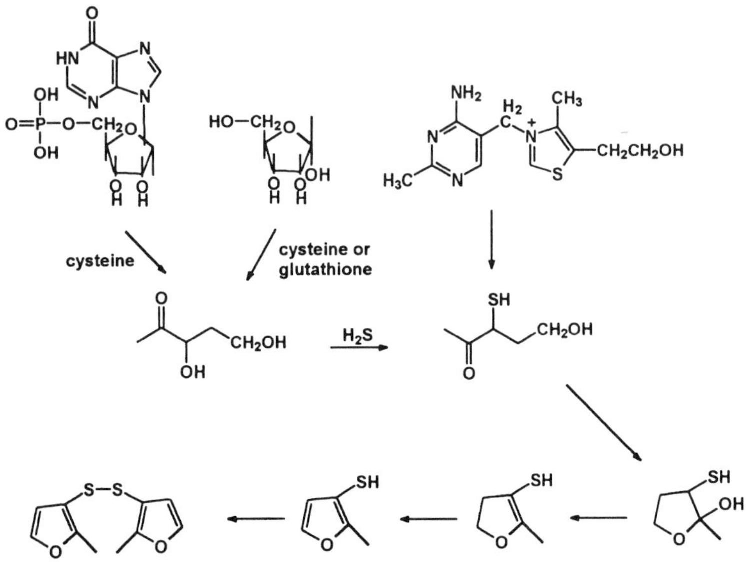

Figure 5.1 Mechanism for the formation of 2-methyl-3-furanthiol and its dimer.

2-Methyl-3-(methylthio)furan and 2-methyl-3-(ethylthio)furan were recently identified as the volatile compounds of chicken (Werkhoff *et al.*, 1993). 2-Methyl-3-(methylthio)furan is thought to be one of the meaty flavour contributors because of its low threshold value of 5 ppt (Werkhoff *et al.*, 1993)

Another compound structurally related to 2-methyl-3-furanthiol and identified as a primary odorant in chicken broth was 2-furfurylthiol (Gasser and Grosch, 1990). This compound possesses a threshold of 5 ppt and aroma qualities such as roasted and sulphury have been recognized as the most important components of roasted coffee (Tressl, 1989). Such compounds are thermally generated from the reaction of furfural and cysteine (Tressl and Silwar, 1981).

Several cyclic sulphur-containing compounds have been identified in chicken flavour. These include alkyl-substituted 1,3,5-dithiazine, 2,4,6-trimethylperhydro-1,3,5-thiadiazine and alkyl-substituted 1,2,4-trithiolanes (Tang *et al.*, 1983; Werkhoff *et al.*, 1992, 1993). The thermal degradation of cysteine and cystine in aqueous solution has been studied by Shu *et al.* (1985 a,b). Two diastereomeric isomers of 3,5-dimethyl-1,2,4-trithiolane were identified as the major degradation products when cysteine or cystine were heated at 160°C. However, when cysteine was degraded at 180°C, representing frying temperature, both 3,5-dimethyl-1,2,4-trithiolane and 2,4,6-trimethylperhydro-1,3,5-dithiazine (thialdine) were identified as major products (Zhang *et al.*, 1988). Degradation of glutathione at 180°C, on the other hand, generated only the 3,5-dimethyl-1,2,4-trithiolanes as the major product (Zhang *et al.*, 1988).

According to the patent literature (Wilson *et al.*, 1974), thialdine is useful as an ingredient of chicken flavour. It is interesting that the organoleptic quality of thialdine and other dithiazines is affected by pH (Kawai, 1991). At a pH of 8.1 and 3.5, thialdine had a medium roasted shrimp flavour. When pH approached the equivalence point at 2.5, a sweet odour was noted. At an acidic pH of 1.6, thialdine had the weak flavour of edible mushrooms (Shiitake).

Although the mechanisms for the formation of dithiazines have been proposed by Boelens *et al.* (1974) and Shu *et al.* (1985a,b,c), the complete understanding of the mechanism was achieved only recently by the extensive experimental study of Kawai (1991). The most probable mechanism for the formation of thiadiazines and dithiazines proposed by Kawai (1991) is shown in Figure 5.2. This proposed mechanism revealed some significant properties observed by Kawai (1991), as follows: (1) the reactions occur under basic conditions; (2) reactions of aldehydes with ammonia prior to the participation of hydrogen sulphide; (3) elimination of ammonia and addition of hydrogen sulphide occur stepwise in the mechanism and (4) the mechanism for the formation of dithiazines is independent of the mechanism for the formation of either trithiolanes or dialkyl polysulphides.

In addition to 2,4,6-trimethylperhydro-1,3,5-dithiazine, other alkyl-substituted dithiazines, including 2-ethyl-4,6-dimethyl, 4-ethyl-2,6-dimethyl, 2-propyl-4,6-dimethyl-, 2-butyl-4,6-dimethyl-, 2-pentyl-4,6-dimethyl- and 4-pentyl-2,6-dimethyl-1,3,5-dithiazines, were also found in chicken flavour (Werkhoff *et al.*, 1992). These alkyl-substituted dithiazines may also be found in the flavour of yeast extracts, beef, pork, roasted peanuts and cocoa (Werkhoff *et al.*, 1992).

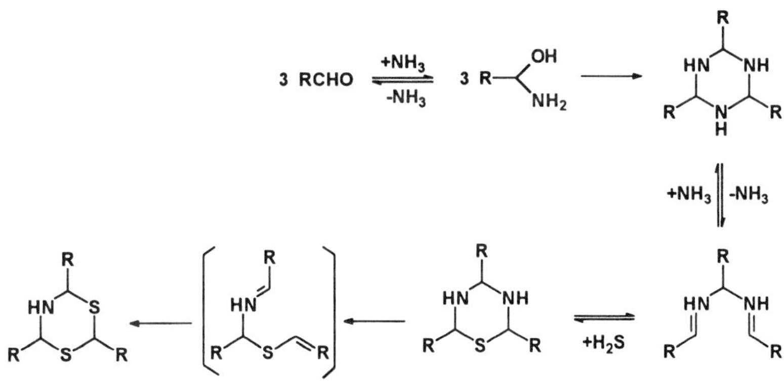

Figure 5.2 Possible mechanism for the formation of thiadiazines and dithiazines (Kawai, 1991).

Interestingly, the higher homologue of 3,5-dimethyl-1,2,4-trithiolane, 3,5-diisobutyl-1,2,4-trithiolane, was identified in the volatiles isolated from fried chicken flavour (Hartman *et al.*, 1984). This trithiolane possesses roasted, roasted-nut, crisp bacon-like and pork rind-like aromas and has been produced in a model system containing isovaleraldehyde, ammonia and hydrogen sulphide (Shu *et al.*, 1981). Figure 5.3 shows the mechanism for the formation of 3,5-diisobutyl-1,2,4-trithiolane as proposed by Shu *et al.* (1981). Isovaleraldehyde arises via Strecker degradation of leucine. Ammonia and hydrogen sulphide may arise via thermal degradation of amino acids and cysteine or cystine. In addition to 3,5-dimethyl-1,2,4-trithiolane and 3,5-diisobutyl-1,2,4-trithiolane, some alkyl-substituted trithiolanes including 3-methyl-5-butyl-, 3-methyl-5-pentyl-, 3-methyl-, 3-methyl-5-ethyl-, 3,5-diethyl-, 3-ethyl-5-propyl- and 3,5-dibutyl-1,2,4-trithiolanes were also reported to be present in chicken flavour (Hwang *et al.*, 1986; Werkhoff *et al.*, 1993). According to Werkhoff *et al.* (1993) the 1,2,4-trithiolanes contribute the overall aroma of meat but do not contribute directly to the characteristic flavour of meat itself.

Some aliphatic sulphur-containing compounds have been reported in chicken flavour (Werkhoff *et al.*, 1993). One aliphatic hemidithioacetal, namely 1-methylthio-1-ethanethiol, which has a meaty, fatty and tropical fruit-like aroma, has been identified in both chicken and pork flavour. 2-Methylthiomethyl-2-butenal and 1-methylthio-3-pentanone were also reported to be present in chicken flavour. 2-Methylthiomethyl-2-butenal which possesses fatty and potato-like flavour notes can be formed by the reaction of methional with acetaldehyde. 1-Methylthio-3-pentanone which has a sweet, creamy and roasted flavour note can be generated by the reaction of methanethiol with α,β-unsaturated ketones (Werkhoff *et al.*, 1993). Other recently identified sulphur-containing compounds in chicken flavour, including thiophenes, alkylthiols and thiozoles, are listed by Werkhoff *et al.* (1993).

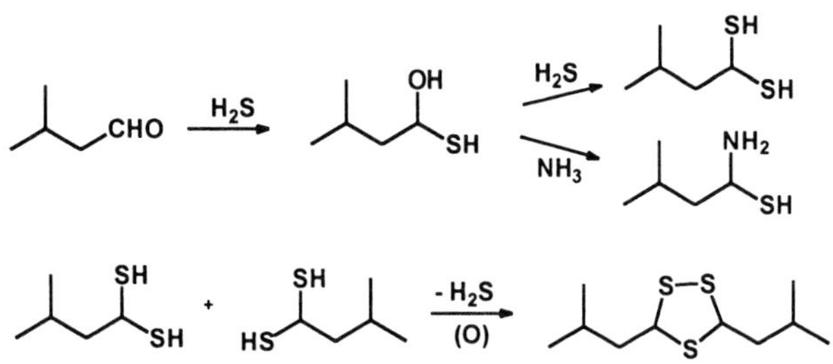

Figure 5.3 Possible mechanism for the formation of 3,5-diisobutyl-1,2,4-trithiolane (Shu *et al.*, 1985c).

5.4 Aldehyde compounds in chicken flavour

The major difference between flavour of chicken broth and broth from beef is the abundance of 2,4-decadienal and γ-dodecalactone in the chicken broth (Table 5.1). Both of them are well-known lipid oxidation products. The role of lipid-derived carbonyl compounds in poultry flavour has been extensively reviewed by Ramaswamy and Richards (1982).

A total of 193 compounds have been reported by Noleau and Toulemonde (1986) in the flavour of roasted chicken. Forty-one of them are lipid-derived aldehydes. When the aroma components of cooked chicken and cooked papain hydrolysates of chicken meat were qualitatively and quantitatively analysed, 23 out of 66 compounds reported were lipid-derived aldehydes (Schroll *et al.*, 1988). Table 5.2 lists the quantitative data of selected aldehydes identified in these two studies. The most abundant aldehydes identified in chicken flavour are hexanal and 2,4-decadienal. In view of the much lower odour threshold of 2,4-decadienal (0.00007 mg/kg) compared to hexanal (0.0045 mg/kg) (van Gemert and Nettenbreijer, 1977), the 2,4-decadienal should be the more important odorant for chicken flavour. Hexanal and 2,4-decadienal are the primary oxidation products of linoleic acid. The autoxidation of linoleic acid generates 9- and 13-hydroperoxides of linoleic acid. Cleavage of 13-hydroperoxide leads to hexanal and the breakdown of 9-hydroperoxide produces 2,4-decadienal (Ho *et al.*, 1989). Subsequent retro-aldolization of 2,4-decadienal will produce 2-octenal and hexanal (Josephson and Lindsay, 1987). 2,4-Decadienal is known to be one of the most important flavour contributors to deep-fat-fried foods (Ho *et al.*, 1987). As shown in Table 5.2, the enzymic hydrolysis of chicken with papain increased the concentration of 2,4-decadienal, as the aroma of cooked meat improved.

In addition to 2,4-decadienal, another important aldehyde structurally related to 2,4-decadienal in chicken flavour is identified as *trans*-4,5-epoxy-*trans*-2-decenal (Schieberle and Grosch, 1991). This compound was found in the flavour of fried chicken (Tang and Ho, 1991), and was also reported in the volatile components of roasted chicken fat as an unidentified compound (Noleau and Toulemonde, 1987). It was characterized as one of the most potent odorants of the crumb flavour of wheat bread. *trans*-4,5-Epoxy-*trans*-2-decenal, having a low odour threshold of approximately 1.5 pg/l (air) (Schieberle and Grosch, 1991), is probably an important flavour contributor of fried chicken. *trans*-4,5-Epoxy-*trans*-2-decenal, an obvious oxidation product of 2,4-decadienal was reported to be one of the major products when pure 2,4-decadienal was incubated at 80°C for three weeks. In the presence of dipropyl disulphide, a possible antioxidant, the formation of *trans*-4,5-epoxy-*trans*-2-decenal was completely inhibited (Kuo, 1991).

Meal from poultry byproducts is a relatively inexpensive protein source and is used in pet foods. The major volatile flavour compounds present

Table 5.2 Selected aldehydes identified in chicken flavour

Aldehyde	Concentration (mg/kg)		
	Roasted[a]	Cooked[b]	Cooked[b] (papain treated)
Butanal	0.133		
Pentanal	0.319		
Hexanal	1.804	25.6	17.2
Heptanal	0.212	2.1	1.5
Octanal	0.422	2.3	1.2
Nonanal	0.467	1.7	1.3
Decanal	0.052	0.3	0.3
Undecanal	0.058		
Dodecanal	0.022		
Tridecanal	0.151		
Tetradecanal	0.125	0.2	0.7
Pentadecanal	0.383		
Hexadecanal	19.788	1.4	9.8
Heptadecanal	0.276	0.1	0.1
Octadecanal	2.664		
trans-2-Butenal	trace		
cis-2-Pentenal	trace		
trans-2-Pentenal	0.085	1.1	0.2
cis-2-Hexenal	trace		
trans-2-Hexenal	0.060	0.3	0.4
trans-2-Heptenal	0.104	1.2	1.5
cis-2-Octenal	0.004		
trans-2-Octenal	0.195	3.7	2.0
trans-2-Nonenal	0.084		
cis-2-Decenal	0.003		
trans-2-Decenal	0.139	1.0	1.2
cis-2-Undecenal	0.002		
trans-2-Undecenal	0.139	0.4	1.1
trans-Dodecenal	0.002	0.3	0.1
trans, cis-2,4-Nonadienal	trace		
trans, trans-2,4-Nonadienal	trace	0.3	0.5
trans, cis-2,4-Decadienal	0.051	1.0	2.7
trans, trans-2,4-Decadienal	0.137	5.2	13.7
trans, trans-2,4-Undecadienal	0.001	0.2	0.2

[a]Noleau and Toulemonde (1986).
[b]Schroll *et al.* (1988).

in the meal have been reported to be hexanal, 3-octen-2-one, 1-pentanol, pentanal, heptanal, octanal, 1-heptanol, 1-octanol and 1-octen-3-ol. All of them are well-known lipid oxidation products (Greenberg, 1981).

In addition to the aliphatic aldehydes, some methyl-branched long-chain aliphatic aldehydes including 14-methylpentadecanal, 14-methylhexadecanal and 15-methylhexadecanal were reported not only in the flavour of beef and pork but also in chicken flavour (Werkhoff *et al.*, 1993). All three compounds possess fatty and tallowy flavour notes. Werkhoff *et al.* (1993) also reported some unsaturated cyclic aldehydes to be present in the chicken flavour. These include 5-ethylcyclopentene-1-carbaldehyde, cyclopentene-1-carbaldehyde, 5-methylcyclopentene-1-carbaldehyde and 5-butylcyclopentene-1-carbaldehyde. The authors suggested that these compounds would not be the chicken flavour contributors because of their high flavour threshold values.

5.5 Heterocyclic compounds in chicken flavour

5.5.1 Pyrazines

Alkylpyrazines have been recognized as important trace flavour components of a large number of cooked, roasted, toasted and deep-fat fried foods (Maga, 1982). As a rule, alkylpyrazines are described as being roasted, nut-like or toasted in flavour character and are highly desirable in foods. Formation pathways for alkylpyrazines have been proposed by numerous researchers (Shibamoto and Bernhard, 1977; Flament, 1981).

Table 5.3 lists pyrazines identified in the flavours of fried chicken and roasted chicken. No pyrazine was identified in the volatiles of chicken broth. This indicates that high temperature and low moisture favour the generation of pyrazines. The formation of higher carbon number substituted pyrazines, such as 2-butylpyrazine and 2-methyl-3-butylpyrazine, is hypothesized to be due to the intervention of aldehydes. The presence of

Table 5.3 Pyrazines identified in chicken flavours

Compound	Fried chicken[a]	Roasted chicken[b]
Pyrazine	+	
2-Methylpyrazine	+	+
2,3-Dimethylpyrazine	+	+
2,5-Dimethylpyrazine	+	
2,6-Dimethylpyrazine	+	+
Trimethylpyrazine	+	+
2-Isopropylpyrazine	+	
2-Methyl-3-ethylpyrazine	+	+
2-Methyl-6(5)-ethylpyrazine	+	+
2-Butylpyrazine	+	
2,3-Dimethyl-5-ethylpyrazine	+	+
2,5-Dimethyl-3-ethylpyrazine	+	
2,6-Dimethyl-3-ethylpyrazine		+
2,6-Diethylpyrazine	+	+
2-Methyl-5,6-diethylpyrazine	+	
2-Methyl-3,5-diethylpyrazine	+	
2-Methyl-3-butylpyrazine	+	
2-Methyl-5-vinylpyrazine	+	
2-Methyl-6-vinylpyrazine	+	
2-Isopropenylpyrazine	+	
6,7-Dihydro-5*H*-cyclopentapyrazine	+	
2-Methyl-6,7-dihydro-5*H*-cyclopentapyrazine	+	

[a]Tang *et al.* (1983).
[b]Noleau and Toulemonde (1986).

aldehydes, originating from lipid degradation, could facilitate addition to the metastable dihydropyrazine compound resulting in longer chain substituted pyrazines (Huang *et al.*, 1987). Figure 5.4 shows the mechanism for the formation of long-chain alkyl-substituted pyrazines.

5.5.2 Pyridines

The occurrence of pyridines in food has been reviewed by Vernin and Vernin (1982). 2-Alkylpyridines were proposed to form from the corresponding unsaturated n-aldehydes with ammonia upon heat treatment (Buttery *et al.*, 1977; Ohnishi and Shibamoto, 1984). Table 5.4 lists pyridines identified in the volatiles of fried chicken and roasted chicken.

2-Pentylpyridine was identified in fried chicken flavour. This compound has a strong fatty and tallow-like odour and was the major product in the volatiles generated from thermal interaction of valine and linoleate (Henderson and Nawar, 1981). Recently, the mechanism for the formation of 2-pentylpyridine in the reaction of 2,4-decadienal with either cysteine or glutathione (γ-glu-cys-gly) in an aqueous solution at high

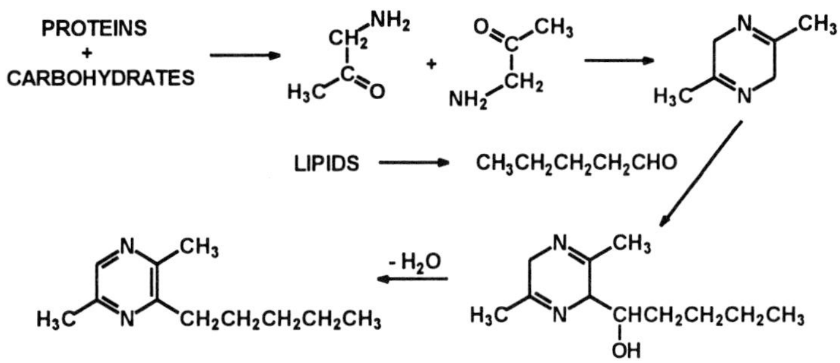

Figure 5.4 Mechanism for the formation of pyrazines.

Table 5.4 Pyridines identified in chicken flavours

Compound	Fried chicken[a]	Roasted chicken[b]
Pyridine	+	+
2-Methylpyridine	+	+
3-Ethylpyridine	+	+
4-Ethylpyridine	+	
2-Methyl-5-ethylpyridine	+	+
2-Ethyl-3-methylpyridine	+	
2-Butylpyridine	+	
2-Pentylpyridine	+	
2-Isobutyl-3,5-dipropylpyridine	+	

[a]Tang *et al.* (1983).
[b]Noleau and Toulemonde (1986).

temperature (180°C) was studied (Zhang and Ho, 1989). The quantities of five major volatile components generated for these two systems are listed in Table 5.5. A greater amount of 2-pentylpyridine and a lesser amount of hexanal were observed in the system of glutathione, suggesting that 2,4-decadienal was involved in forming a Schiff base with the amino group in glutathione directly and thus left less free 2,4-decadienal to involve in autoxidation or retro-aldolization. According to the pathway proposed by Henderson and Nawar (1981), 2-pentylpyridine, which accounted for almost one half of the total volatiles identified in the reaction of glutathione and 2,4-decadienal, can be formed via the Schiff base intermediate from the reaction of 2,4-decadienal with ammonia. However, it is known that the formation of other compounds, dithiazine and thiadiazine, requires the presence of free ammonia (Boelens *et al.*, 1974). Since the absence of dithiazine or thiadiazine formed in the 2,4-decadienal/glutathione indicates that no free ammonia is available, this suggests that free ammonia may not be necessary for the formation of 2-pentylpyridine. It is possible that the amino group from amino acids or peptides condenses directly with the aldehydic group of 2,4-decadienal and is then followed by an electrocyclic reaction and aromatization to form 2-pentylpyridine (see Figure 4.5).

2-Isobutyl-3,5-diisopropylpyridine was identified in fried chicken and has a roasted cocoa-like aroma (Hartman, 1984). Figure 5.5 shows the mechanism for the formation of this compound as proposed by Shu *et al.* (1985a,b,c). It involves the reaction of aldehyde and ammonia at high temperatures and is known as the Chichibabin condensation.

5.5.3 Pyrroles

Pyrrole may have been the first individual heterocyclic compound to be isolated from foods. Table 5.6 lists some pyrroles identified in chicken

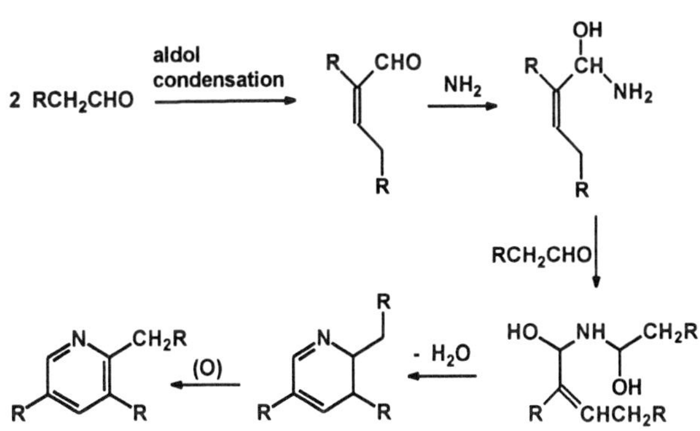

Figure 5.5 Mechanism for the formation of 2-isobutyl-3,5-diisopropylpyridine.

Table 5.5 Quantification of some major flavour components in model systems of 2,4-decadienal/cysteine and 2,4-decadienal/glutathione

Component	System A (mg/mol)	System B (mg/mol)
Total volatiles	2627.3	2511.0
Hexanal	278.6	45.4
3,5-Dimethyl-1,2,4-trithiolane	140.9	368.5
2,4,6-Trimethylperhydro-1,3,5-thiadiazine	828.5	–
2,4,6-Trimethylperhydro-1,3,5-dithiazine	284.2	–
2-Pentylpyridine	501.5	1219.0
Total	2033.7	1632.9
% of total volatiles	77.4	65.0

System A is 2,4-decadienal with cysteine; system B is 2,4-decadienal with glutathione.

flavour volatiles. Some pyrrole derivatives have a pleasant flavour. For example, 2-acetylpyrrole has a caramel-like aroma. Pyrroles have not received as much attention as other heterocyclic flavour compounds such as pyrazines and thiazoles (Shibamoto, 1989).

5.5.4 Thiazoles

Thiazoles are a class of compounds possessing a five-membered ring with sulphur and nitrogen in the 1 and 3 positions, respectively. The potential for thiazole derivatives as flavorants is evident from the work of Stoll *et al.* (1967) who found the strong nut-like odour of a cocoa extract to be due to a trace amount of 4-methyl-5-vinylthiazole. Since then, numerous thiazoles have been identified in food flavours.

The exact origin of thiazoles remains unclear. They can be formed through the thermal degradation of cystine or cysteine (Shu, *et al.*, 1985a,b), or by the interaction of sulphur-containing amino acids and carbonyl

Table 5.6 Pyrroles identified in chicken flavours

Compound	Fried chicken[a]	Roasted chicken[b]
Pyrrole	+	+
N-Methylpyrrole		+
2-Methylpyrrole	+	
2-Ethylpyrrole		+
N-Acetylpyrrole		+
2-Acetylpyrrole	+	
2-Isobutylpyrrole	+	
N-Isobutylpyrrole	+	
N-(2-Butanoyl)pyrrole	+	
N-Furfurylpyrrole		+

[a]Tang *et al.* (1983)
[b]Noleau and Toulemonde (1986)

compounds (Zhang and Ho, 1991; Hartman and Ho, 1984). Thiazoles have been identified as volatile components of thermally degraded thiamine (Hartman *et al.*, 1984).

Table 5.7 lists alkylthiazoles identified in fried and roasted chicken flavours. 2-Pentyl-4-methyl-5-ethylthiazole has a strong paprika pepper flavour and 2-heptyl-4,5-dimethylthiazole has a strong spicy flavour. 2-Octyl-4,5-dimethylthiazole has a sweet fatty aroma (Ho and Jin, 1984). These compounds are probably important contributors to the flavour of fried foods. These and other thiazoles have long-chain alkyl substituents on their rings. The involvement of frying fat or fat decomposition products in the formation of these compounds is suggested.

5.6 Duck and turkey flavour

Reports on duck flavour are limited. The most important study on the volatile compounds of duck is probably the one presented by Wu and Liou (1992). Most of the volatile compounds of duck were lipid oxidation products. The only nitrogen-containing compound found in duck meat was indole. The authors suggested that this compound could be specific for duck meat aroma. Upon roasting, some heteocyclic compounds including pyrazines, pyridines and thiazoles were generated. These results indicate that Maillard reaction and lipid oxidation are important pathways for producing the special roasted duck flavours.

It is worthwhile to mention that there was an unidentified compound (MW = 163) reported in roasted duck (Wu and Liou, 1992). We compared

Table 5.7 Thiazoles identified in chicken flavours

Compound	Fried chicken[a]	Roasted chicken[b]
Thiazole	+	+
2-Methylthiazole	+	+
2,4,5-Trimethylthiazole	+	
2-Methyl-4-ethylthiazole	+	
2-Methyl-5-ethylthiazole		+
2,4-Dimethyl-5-ethylthiazole	+	
2-Isopropyl-4,5-dimethylthiazole	+	
2,5-Dimethyl-4-butylthiazole	+	
2-Isopropyl-4-ethyl-5-methylthiazole	+	
2-Butyl-4,5-dimethylthiazole	+	
2-Butyl-4-methyl-5-ethylthiazole	+	
2-Pentyl-4,5-dimethylthiazole	+	
2-Hexyl-4,5-dimethylthiazole	+	
2-Heptyl-4,5-dimethylthiazole	+	
2-Heptyl-4-ethyl-5-methylthiazole	+	
2-Octyl-4,5-dimethylthiazole	+	

[a]Tang *et al.* (1983).
[b]Noleau and Toulemonde (1986).

the mass spectrum data as reported by Wu and Liou (1992) to those reported by Werkhoff *et al.* (1990) and Hwang *et al.* (1986) as shown in Figure 5.6. This unknown compound could be 2,4,6-trimethylperhydro-1,3,5-dithiazine (thialdine). Thialdine, which possesses carrot-like, oniony, roast and meaty flavour notes, is widespread in foods such as yeast extracts, beef, pork, chicken, grilled liver, roasted peanuts, cocoa, mutton, krill, duck egg and leek (Werkhoff *et al.*, 1992). Because of its flavour characteristics, the large quantity of this compound in roasted duck suggests that thialdine is possibly an important flavour contributor of roasted duck.

Most of the volatile compounds identified from cooked turkey were also lipid degradation products such as aldehydes, hydrocarbons, alcohols and ketones as reported by Wu and Sheldon (1988) and Ramaswamy and

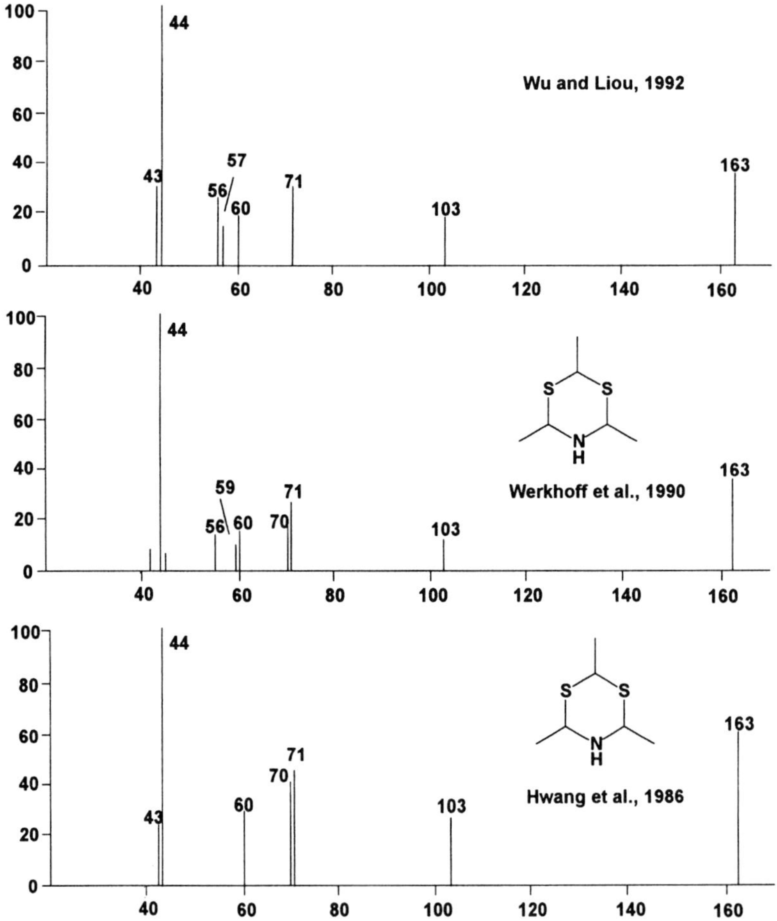

Figure 5.6 Comparison of the mass spectrum of an unidentified compound with those published mass spectra.

Richards (1982). Two sulphur-containing compounds, dimethyl disulphide and dimethyl trisulphide, which can be formed by the Strecker degradation of methionine and cysteine (Schutte, 1976), were found in cooked turkey breast rolls. The possible desirable flavour contributing compound of cooked turkey may be dimethyl disulphide (Wu and Sheldon, 1988).

Some precooked and refrigerated poultry products easily generate 'warmed-over' flavour (WOF) which occurs when these products are reheated (Jones, 1989). It has been reported that turkey meat is more susceptible to WOF than chicken or red meat because of its high concentration of polyunsaturated fatty acids in meat phospholipids. However, presence of tocopherol in chicken meat might affect this trend. Phospholipid concentration in cooked turkey skin was determined by Dimick and MacNeil (1970), who reported that the fractions which were high in phospholipids were highly unstable and produced high concentrations of carbonyl compounds. Volatile compounds formed during lipid deterioration may be directly responsible for the development of WOF. Heptanal and n-nona-3,6-dienal were reported to be well correlated with WOF in cooked turkey (Ruenger *et al.*, 1978).

5.7 Conclusions

Remarkable progress has been made in poultry flavour research in recent years. The use of aroma extract dilution analysis to study the flavour of chicken broth is undoubtly the most outstanding achievement. We now know that the chemical components responsible for the meaty note of chicken meat are not different from that of beef broth. It is the fatty aroma compounds that make the flavour of chicken broth significantly different from that of beef broth. This is very helpful to flavourists interested in compounding savoury flavours of various types. The use of aroma extract dilution analysis or other similar methods such as Charm analysis (Acree *et al.*, 1984) to study other poultry meat products, such as fried chicken or roasted chicken, is highly desirable.

The volatiles resulting from the Maillard reaction and lipid oxidation are obviously the major sources of flavour compounds identified in poultry meat. We begin to see the incorporation of lipid-derived aldehydes into Maillard reaction products, such as long-chain, alkyl-substituted pyrazines, thiazoles and trithiolanes. Recent studies by Apriyantono *et al.* (1997) lend further support to these findings. Other aspects of the interaction between the Maillard reaction and lipid oxidation, such as the antioxidative effect of the Maillard reaction products, need to be explored further.

A recent study by Kawai (1991) on the effect of pH on the formation and sensory quality of perhydrodithiazines is very interesting. The influence of pH on the flavour of chicken broth remains to be determined.

References

Acree, T.E., Bernard, J. and Cunningham, D.G. (1984). The analysis of odour-active volatiles in gas chromatographic effluents. In *Analysis of Volatiles. Method and Application*, ed. P. Schreier. Walter de Gruyter, Berlin, pp. 251–267.

Apriyantono, A., Indrawaty. (1997). Comparison of flavor characteristics of domestic and broiler as affected by different processing methods. In *Food Flavors: Formation, Analysis and Packaging Influences – Proceedings of the Ninth International Flavor Conference*, July 1–4, Island of Limnos, Greece, eds. E.T. Coutis, C.-T. Ho, C.J. Mussinan, T.H. Parliment, F. Shahidi and A. Spanier. Elsevier, Amsterdam, in press.

Ames, J.M. and MacLeod, G. (1985). Volatile components of a yeast extract composition. *J. Food Sci.*, **50**, 125–131.

Boelens, M., van der Linde, L.M., de Valois, P.J., van Dort, H. M. and Takken, H. J. (1974). Organic sulphur compounds from fatty aldehydes, hydrogen sulphide, thiols and ammonia as flavour constituents. *J. Agric. Food Chem.*, **22**, 1071–1076.

Buttery, R.G., Ling, L.C., Teranishi, R. and Mon, T.R. (1977). Roasted lamb fat: basic volatile components. *J. Agric. Food Chem.*, **25**, 1227–1230.

Chang, S. S., Hirai, C., Reddy, B.R., Herz, K.O., Kato, A. and Sipma, G. (1968). Isolation and identification of 2,4,5-trimethyl-3-oxazoline and 3,5-dimethyl-1,2,4-trithiolane in the volatile flavour compounds of boiled beef. *Chem. Ind.*, 1639–1640.

Dimick, P.S. and MacNeil, J.H. (1970). Poultry product quality. 2. Storage time–temperature effects on carbonyl composition of cooked turkey and chicken skin fractions. *J. Food Sci.*, **35**, 186–190.

Dimick, P.S., MacNeil, J.H. and Grunden, L.E. (1972). Poultry product quality. Carbonyl composition and organoleptic evaluation of mechanically deboned poultry meat. *J. Food Sci.*, **37**, 544–546.

Evers, W.J., Heinsohn, H.H. Jr., Mayers, B.J. and Sanderson, A. (1976). Furans substituted at the three position with sulfur. In *Phenolic, Sulfur and Nitrogen Compounds in Food Flavours*, eds. G. Charalambous and I. Katz. American Chemical Society, Washington, DC, pp. 184–193.

Farmer, L.J. and Mottram, D.S. (1990). Recent studies on the formation of meat-like aroma compounds. In *Flavour Science and Technology*, eds. Y. Bessiere and A.F. Thomas. Wiley, Chichester, pp. 113–116.

Flament, I. (1981). Some recent aspects of the chemistry of naturally occurring pyrazines. In *The Quality of Foods and Beverages (Vol. 1)*, eds. G. Charalambous and G. Inglett. Academic Press, New York, pp. 42–48.

Gasser, U. and Grosch, W. (1988). Identification of volatile flavour compounds with high aroma values from cooked beef. *Z. Lebensm. Unters. Forsch.*, **186**, 489–494.

Gasser, U. and Grosch, W. (1990). Primary odourants of chicken broth. *Z. Lebensm. Unters. Forsch.*, **190**, 3–8.

Greenberg, M.J. (1981). Characterization of poultry byproduct meal flavour volatiles. *J. Agric. Food Chem.*, **29**, 831–834.

Grosch, W., Sen, A., Guth, H. and Zeiler-Hilgart, G. (1990). Quantitation of aroma compounds using a stable isotope diluted assay. In *Flavour Science and Technology*, eds. Y. Bessiere and A.F. Thomas. Wiley, Chichester, pp. 191–194.

Hartman, G.J., Carlin, J.T., Scheide, J.D. and Ho, C.-T. (1984). Volatile products formed from the thermal degradation of thiamine at high and low moisture levels. *J. Agric. Food Chem.*, **32**, 1015–1018.

Hartman, G.J. and Ho, C.-T. (1984). Volatile products of the reaction of sulfur-containing amino acids with 2,3-butanedione. *Lebensm. Wiss. u. Technol.* **17**, 171–174.

Hartman, G.J., Carlin, J.T., Hwang, S.S., Bao, Y., Tang, J. and Ho, C.-T. (1984). Identification of 3,5-diisobutyl-1,2,4-trithiolane and 2-isobutyl-3,5-diisopropylpyridine in fried chicken flavour. *J. Food Sci.*, **49**, 1398, 1400.

Henderson, S.K. and Nawar, W.W. (1981). Thermal interaction of linoleic acid and its esters with valine. *J. Am. Oil Chem. Soc.*, **58**, 632–635.

Ho, C.-T. and Jin, Q.Z. (1984). Aroma properties of some alkylthiazoles. *Perfumer Flavourist*, **9**, 15–18.

Ho, C.-T., Carlin, J.T., Huang, T.C., Hwang, L.S. and Hau, L.B. (1987). Flavour development in deep-fat fried foods. In *Flavour Science and Technology*, eds. M. Martens, G.A. Dalen and H. Russwurm Jr. Wiley, Chichester, pp. 35–42.

Ho, C.-T., Zhang, Y., Shi, H. and Tang, J. (1989). Flavour chemistry of Chinese foods. *Food Rev. Int.*, **5**, 253–287.

Huang, T.C., Bruechert, L.J., Hartman, T.G., Rosen, R.T. and Ho, C.-T. (1987). Effect of lipids and carbohydrates on thermal generation of volatiles from commercial zein. *J. Agric. Food Chem.*, **35**, 985–990.

Hwang, S.S., Carlin, J.T., Bao, Y., Hartman, G.J. and Ho, C.-T. (1986). Characterization of volatile compounds generated from the reaction of aldehyde with ammonium sulphide. *J. Agric. Food Chem.*, **34**, 538–542.

Jones, J.M. (1989). Flavour and taint in turkey meat. In *Recent Advances in Turkey Science*, eds. C. Nixey and T.C. Grey. Butterworth, London, pp. 313–330.

Josephson, D.B. and Lindsay, R.C. (1987). Retro-aldol degradation of unsaturated aldehydes: role in the formation of *c*4-heptenal from *t*2, *c*6-nonadienal in fish, oyster and other flavours. *J. Amer. Oil Chem. Soc.*, **64**, 132–138.

Katz, M.A., Dugan, L.R., Jr. and Dawson, L.E. (1966). Fatty acids in neutral lipids and phospholipids from chicken tissues. *J. Food Sci.*, **31**, 717–720.

Kawai, T. (1991). *Thiadiazines and Dithiazines in Volatiles from Heated Seafoods, and Mechanisms of their Formation*, 2nd edn., Shiono Koryo Kaisha, Osaka.

Kuo, M.C. (1991). Formation of volatile polysulphides from Welch onions and scallions. Ph.D. dissertation, Rutgers University, New Brunswick, NJ.

Lee, Y.B., Hargus, G.L., Kirkpatrick, J.A., Berner, D.L. and Forsythe, R.H. (1975). Mechanism of lipid oxidation in mechanically deboned chicken meat. *J. Food Sci.*, **40**, 964–967.

Lillard, D.A. (1987). Oxidative deterioration in meat, poultry, and fish. In *Warmed-Over Flavour of Meat*, eds. A.J. St. Angelo and M.E. Bailey. Academic Press, Orlando, FL, pp. 41–67.

Maga, J.A. (1982). Pyrazines in foods. *CRC Crit. Rev. Food Sci. Nutr.*, **16**, 1–115.

Mecchi, E.P., Pippen, E.L. and Lineweaver, H. (1964). Origin of hydrogen sulphide in heated chicken muscle, *J. Food Sci.*, **29**, 393–399.

Mottram, D.S. (1991). Meat. In *Volatile Compounds in Foods and Beverages*, ed. H. Maarse. Marcel Dekker, New York, pp. 107–177.

Noleau, I. and Toulemonde, B. (1986). Quantitative study of roasted chicken flavour. *Lebensm. Wiss. u. Technol.*, **19**, 122–125.

Noleau, I. and Toulemonde, B. (1987). Volatile components of roasted chicken fat. *Lebensm. Wiss. u. Technol.*, **20**, 37–41.

Ohloff, G., Flament, I. and Pickenhagen, W. (1985). Flavour chemistry. *Food Rev. Int.*, **1**, 99–148.

Ohnishi, S. and Shibamoto, T. (1984). Volatile compounds from heated beef fat and beef fat with glycine. *J. Agric. Food Chem.*, **32**, 987–992.

Pikul, J., Leszcaynski, D.E. and Kummerow, F.A. (1984). Relative role of phospholipids, triacylglycerols, and cholesterol esters on malonaldehyde formation in fat extracted from chicken meat. *J. Food Sci.*, **49**, 704–709.

Ramaswamy, H.S. and Richards, J.F. (1982). Flavour of poultry meat – a review. *Can. Inst. Food Sci. Technol. J.*, **15**, 7–18.

Reineccius, G.A. and Liardon, R. (1985). The use of charcoal traps and microwave desorption for the analysis of headspace volatiles above heated thiamin solutions. In *Topics in Flavour Research*, eds. R.G. Berger, S. Nitz and P. Schreier. Eichhorn, Marzling-Hangenhan, pp. 125–136.

Ruenger, E.L., Reineccius, G.A. and Thompson, D.R. (1978). Flavour compounds related to the warmed-over flavour of turkey. *J. Food Sci.* **43**, 1198–1200.

Schieberle, P. and Grosch, W. (1991). Potent odourants of the wheat bread crumb. *Z. Lebensm. Unters. Forsch.*, **192**, 130–135.

Schroll, W., Nitz, S. and Drawert, F. (1988). Determination of aroma compounds in cooked chicken meat and hydrolysate. *Z. Lebensm. Unters. Forsch.*, **187**, 558–560.

Schutte, L. (1976). Flavor precursors in foodstuffs. In *Phenolic, Sulfur and Nitrogen Compounds in Food Flavors*, eds. G. Charalambous and I. Katz. ACS Symposium Series 26, American Chemical Society, Washington, DC, pp. 96–113.

Shibamoto, T. (1989). Volatile flavour chemicals formed by the Maillard reaction. In *Thermal Generation of Aromas*, eds. T.H. Parliment, R.J. McGorrin and C.-T. Ho. ACS Symposium Series 409, American Chemical Society, Washington, DC, pp. 134–142.

Shibamoto, T. and Bernhard, R.A. (1977). Investigation of pyrazine formation pathways in sugar–ammonia model systems. *J. Agric. Food Chem.*, **25**, 609–614.

Shu, C.K., Mookherjee, B.D. and Vock, M.H. (1981). Flavouring with a mixture of 2,5-dimethyl-3-acetylfuran and 3,5-diisobutyl-1,2,4-trithiolane. US Patent 4 263 331.

Shu, C.K., Hagedorn, M.L., Mookherjee, B.D. and Ho, C.-T. (1985a). pH effect on the volatile components in the thermal degradation of cysteine. *J. Agric. Food Chem.,* **33**, 442–446.

Shu, C.K., Hagedorn, M.L., Mookherjee, B.D. and Ho, C.-T. (1985b). Volatile components of the thermal degradation of cystine in water. *J. Agric. Food Chem.,* **33**, 438–442.

Shu, C.K., Mookherjee, B.D., Bondarovich, H.A. and Hagedorn, M.L. (1985c). Characterization of bacon odour and other flavour components from the reaction of isovaleraldehyde and ammonium sulphide. *J. Agric. Food Chem.,* **33**, 130–132.

Steverink, A.T.G. (1981). Review of the literature on the chemical constituents of chicken meat flavour and suggestions for future research. In *Quality of Poultry Meat, Proceedings of the 5th European Symposium*, eds. R.W.A.W. Mulder, C.W. Scheele and C.H. Veerkamp, pp. 356–371.

Stoll, M., Dietrich, P., Sundte, E. and Winter, M. (1967). Recherches sur les aromes. 15. Sur l'arome du cacao. *Helv. Chim. Acta,* **50**, 2065–2067.

Tang, J., Jin, Q.Z., Shen, G.-H., Ho, C.-T. and Chang, S.S. (1983). Isolation and identification of volatile compounds from fried chicken. *J. Agric. Food Chem.,* **31**, 1287–1292.

Tang, J. and Ho, C.-T. (1991). Unpublished results.

Tressl, R. and Silwar, R. (1981). Investigation of sulfur-containing components in roasted coffee. *J. Agric. Food Chem.,* **29**, 1078–1082.

Tressl, R. (1989). Formation of flavour components in roasted coffee. In *Thermal Generation of Aromas*, eds. T.H. Parliment, R.J. McGorrin and C.-T. Ho. ACS Symposium Series 409, American Chemical Society, Washington, DC, pp. 285–301.

van der Linde, L.M., van Dort, J.M., de Valois, P., Boelens, H. and de Rijke, D. (1979). Volatile components from thermally degraded thiamin. In *Progress in Flavour Research*, eds. Land, D.G. and Nursten, H.E., Applied Science, London, pp. 219–224.

Van Germert, L.J. and Nettenbrijer, A.H. (1977). *Compilation of Odour Threshold Values in Air and Water.* National Institute for Water Supply, Voorburg, The Netherlands.

Vernin, G. and Vernin, G. (1982). Heterocyclic aroma compounds in foods: occurrence and organoleptic properties. In *The Chemistry of Heterocyclic Flavouring and Aroma Compounds*, ed. G. Vernin. Wiley, New York, pp. 72–150.

Werkhoff, P., Brüning, J., Emberger, R., Güntert. M. and Hopp, R. (1993). Flavour chemistry of meat volatiles: new results on flavour components from beef, pork, and chicken. In *Recent Developments in Flavour and Fragrance Chemistry*, eds. R. Hopp and K. Mori. VCH, New York, pp. 183–213.

Werkhoff, P., Brüning, J., Emberger, R., Güntert. M., Köpsel., M. and Kuhn, W. (1990). Studies on volatile sulphur-containing flavour components in yeast extract. In *Chemistry Analysis and Structure (Vol. 4),* eds. S.C. Bhattacharyya, N. Sen and K.L. Sethi. 11th Int. Congr. of Essential Oils, Fragrances and Flavours. Aspect Publishing, London, pp. 215–243.

Werkhoff, P., Güntert, M. and Hopp, R. (1992). Dihydro-1,3,5-dithiazines: unusual flavour compounds with remarkable organoleptic properties. *Food Rev. Int.*, **8**, 391–442.

Wilson, R.A. and Katz, I. (1972). Review of literature on chicken flavour and report of isolation of several new chicken flavour components from aqueous cooked chicken broth. *J. Agric. Food Chem.,* **20**, 741–747.

Wilson, R.A., Vock, M.H., Katz, I. and Shuster, S.J. (1974). British Patent 1 364 747.

Withycombe, D.A. and Mussinan, C.J. (1988). Identification of 2-methyl-3-furanthiol in the steam distillate from canned tuna fish. *J. Food Sci.,* **53**, 658–659.

Wu, C.-M. and Liou S.-E. (1992). Volatile components of water-boiled duck meat and cantonese style roasted duck. *J. Agric. Food Chem.*, **40**, 838–841.

Wu, T.C. and Sheldon, B.W. (1988). Flavour components and factors associated with the development of off-flavours in cooked turkey rolls. *J. Food Sci.*, **53**, 49–54.

Zhang, Y. and Ho, C.-T. (1989). Volatile compounds formed from thermal interaction of 2,4-decadienal with cysteine and glutathione. *J. Agric. Food Chem.,* **37**, 1016–1020.

Zhang, Y. and Ho, C.-T. (1991) Comparison of the volatile compounds formed from the thermal reaction of glucose with cysteine and glutathione. *J. Agric. Food Chem.*, **39**, 760–763.

Zhang, Y., Chien, M. and Ho, C.-T. (1988). Comparison of the volatile compounds obtained from thermal degradation of cysteine and glutathione in water. *J. Agric. Food Chem.,* **36**, 992–996.

6 Sheepmeat odour and flavour

O.A. YOUNG and T.J. BRAGGINS

6.1 Introduction

Sheepmeat (the flesh of *Ovis aries*) is eaten by millions of people all over the world and is probably eaten in every country to some extent. There are no religious or cultural taboos on eating sheepmeat, which contrasts sharply with the taboos that apply to beef (Hindu) and pork (Moslem, Jewish). Nevertheless, many people avoid sheepmeat because they object to its odour (especially during cooking) and/or its flavour. The Chinese even have a special word for the disagreeable cooking odour of sheepmeat, 'soo', meaning sweaty, sour (Wong, 1975). Even in those Western countries that have a greater acceptance of sheepmeat, many dislike it, particularly the meat from mature animals with its apparently stronger odour and flavour. Also, the relatively high melting point of sheep fat contributes to a waxy mouthfeel that is unacceptable to many. On a cool plate the fat tends to harden rapidly, which contrasts with the more oily character of, say, pork fat.

These and other factors, some of which will be alluded to in the following review, conspire to limit the consumption of sheepmeat in countries where consumers are affluent and have a wide choice of meats. The United States and Canada are good examples. In 1987, their (bone-in) annual consumptions were a mere 0.7 and 0.9 kg/person, respectively (Smith and Young, 1991). At the other end of the scale, Mongolians, poor by Western standards, consume about 60 kg/person/year. Sheepmeat consumption is also affected by historical agricultural practices that have led to current dietary traditions. As a domestic animal, sheep were historically suited to arid climates such as are found on either side of the equator. Many Middle Eastern countries, for example Kuwait (35 kg/person/year), Saudi Arabia (21 kg), Libya (18 kg) and Iran (9 kg), have a history of sheepmeat consumption and remain large per capita consumers. They import sheepmeat, indicating a specific demand for it. Evidently their populations like mutton odour and flavour, simply disregard it, or use spices to modify it. All three factors may be important.

This review examines the chemicals thought to be responsible for, and the factors affecting, sheepmeat's characteristic odour and flavour. A particular emphasis will be placed on the characteristic and odorous branched-chain fatty acids (BCFAs) present in sheepfat. Whereas other chemical classes are important in overall odour/flavour, the BCFAs have emerged as clearly distinguishing. At the practical level, the collective research effort will identify ways in which that odour and flavour can be modified. The objective is clear, to sell more sheepmeat to people who currently disdain it because they do not like the way it smells and tastes.

6.2 Assessment of sheepmeat odour and flavour by sensory panels and chemical analysis

Sensory panels assessing sheepmeat odour and flavour must be asked appropriate questions. Crouse (1983) noted that in 24 sensory studies of sheepmeat odour or flavour, 16 used a hedonic scale. Hedonic scales are useful in studies seeking a greater market share for sheepmeat, but an intensity scale is essential for studying 'muttonness' in a scientific way. With hedonic scales, the responses of those who like a strong odour and flavour can cancel out responses of those who dislike it. In both types of assessment, the use of ethnic panels should be considered, although this can be expensive. In New Zealand, Japanese tourists have been used as assessors. However, in hedonic assessment, researchers run more risk of panellists not wanting to offend, giving responses intended to please.

Methods of chemical analysis of sheepmeat cooking volatiles are the same as those for other species. Most current methods rely on capillary or packed-column gas chromatography using various detector types as the means of separating and identifying the components of complex mixtures.

For chromatography, the volatiles must first be trapped in quantities large enough to satisfy the sensitivity requirements of the detector used. Two common trapping methods are headspace sampling, and adsorption on porous polymers, both followed by cryofocusing the volatiles at the start of the column (Suzuki and Bailey, 1985; MacLeod and Ames, 1986a; St. Angelo *et al.*, 1987). To cryofocus the volatiles from porous polymers such as Tenax requires an initial desorption step, often accomplished by heating up to 260°C in a stream of inert gas. However, there have been concerns about conversion of labile compounds during thermal desorption (Lewis and Williams, 1980; Snyder and Mounts, 1990). An alternative is to elute with conventional solvents such as diethylether (Olafsdottir *et al.,* 1985; Buttery *et al.*, 1987; Vercellotti *et al.*, 1992). Solvent elution, however, has the potential to introduce contaminants and there is a risk of volatiles loss during the volume reduction step before chromatography. Supercritical carbon dioxide presents another solvent option. In an experiment that compared the recovery of sheepfat volatiles from Tenax, Braggins (1997) found that thermal desorption in helium was better than elution with either diethyl ether or carbon dioxide. Moreover, thermal desorption did not appear to produce thermal artifacts.

Once cryofocused at the start of the chromatography column, the volatiles are separated by applying a temperature gradient. Sometimes the column effluent is split so that each effluent can be detected in two ways – with a human sniffer (Guadagni *et al.*, 1966) and, typically, a flame ionization detector. Although very labour intensive, odourport sniffing is an extremely powerful technique for identifying cause rather than mere statistical association, especially when refined by odourport dilution (Grosch, 1993; Acree *et al.*, 1984). Odourport sniffing has been applied to several species (MacLeod and Ames, 1986b; Cadwallader *et al.*, 1994) including sheepmeat (Sutherland and Ames, 1995; Braggins, 1996).

Some compounds of interest are present in such low concentrations that a preliminary concentration step is required. This has commonly been required in the case of volatile BCFAs and phenols (see, for example, Wong *et al.*, 1975a; Ha and Lindsay, 1990, 1991c). The usual technique for this is simultaneous distillation and extraction, a laborious procedure. In one study of BCFA concentrations in sheepfat, a concentration step was unnecessary; Young *et al.* (1997a) measured BCFAs after direct injection of conventionally prepared fatty acid methyl esters of whole fat, with quantitation through selective ion monitoring of mass spectra.

Another approach to measuring volatile BCFAs was described by Johnson and Purchas (1997). Long-chain BCFAs such as methylhexadecanoic acid are present in sheepfat at much higher concentrations that their volatile short-chain analogs. Because of their higher concentration they can easily be measured by conventional gas chromatography without the need for prior concentration. Johnson and Purchas found good correlations between concentrations of methylhexadecanoic and 4-methyloctanoic acids in subcutaneous fat of lambs.

6.3 The tissue source of mutton odour and flavour

Even before sheepmeat is cooked, specific odours are evident. Workers handling sheep carcasses develop a distinctive odour on their hands. Subcutaneous fat is the obvious source.

The tissue source of cooked mutton odours, whether from the fat, the lean or both, has received much attention. In a classic paper, Hornstein and Crowe (1963) proposed that lean meat provides the basic meaty flavour common to beef, pork and sheepmeat, whereas the fat is responsible for the species flavour. There is evidence to support this model in sheepmeat. Wasserman and Talley (1968) found that lamb fat was distinctive to the point that whereas the addition of beef fat to veal did not increase panel recognition of veal as beef, addition of lamb fat to veal significantly increased the false identification of veal as lamb. Pearson *et al.* (1973) made water extracts of lean beef and lamb. Panellists smelling the heated extracts could not differentiate between the two species. When the species' fat was added to the respective extract, the panel noticed a difference between the two samples, but still could not pick lamb from beef. However, samples of pure boiled lamb fat were distinguished from equivalent beef fat samples. Wenham (1974) trimmed ewe mutton and beef to nearly zero visible fat, then back-blended mutton and beef fat in various ratios. Although the assessment scale was hedonic, when the added mutton fat content reached 20%, the patties became unacceptable, unlike those with 20% beef fat. Wenham also showed that blends of lean mutton and lean beef could be distinguished (but not correctly identified). Wenham argued that any difference between beef and mutton lean was due to the content of nonvisible fat.

Using a panel selected on ability to distinguish lamb, beef and pork, Brennand and Lindsay (1982) clearly showed that fatty tissues were the most significant source of mutton flavours, but were not so strongly distinguishing for beef and pork flavours.

Echoing Wenham's (1974) comment about nonvisible fat, MacLeod and Seyyedain-Ardebili (1981) concluded in a major review that if fat is an important contributor to species flavour, the intramuscular (nonvisible) fat is sufficient to generate species flavour. Moody (1983) also concluded there is sufficient fat in lean meat to allow the development on cooking of normal (species) flavour. Lean tissue lipids contain a relatively high proportion of unsaturated fatty acids, which are highly susceptible to oxidation. A number of authors have implicated fat oxidation products as contributors to species odour and flavour (see later). However, lipid oxidation products from depot fats will also contribute. Depot fat tissue has cell membranes that contain the susceptible fatty acids, and storage triglycerides of domestic animals contain proportions of unsaturated fatty acids that are capable of being oxidized.

Mutton tallow has a distinct sheepy odour, which can develop in even steam-deodorized samples (Hoffmann and Meijboom, 1968; Brown, 1989). Hoffmann and Meijboom attributed tallow odour to fatty acid oxidation products, several of which they identified. Although tallow odour and mutton cooking odour are probably different, the capacity of mutton tallow to produce odorous compounds with a 'sheep note' clearly implicates tissue fat as a major odour source.

In New Zealand, commercial tallows for domestic soap production are rich in sheepfat. Visitors sometimes report a 'sheepy' note in soaps. All in all, the evidence clearly points to fatty tissue as the principal source of sheepmeat odour and flavour.

6.4 Chemical components involved in sheepmeat odour and flavour

Most of the chemical analyses directed at sheepmeat odour and flavour have centred on trapping volatiles released during cooking. This is a sensible approach as consumers sense odour first, and what they smell before the meal will certainly colour their appreciation of the food. But more importantly, much flavour is perceived as odour, sensed retronasally during mastication and swallowing. Odour and flavour of foods are inextricably linked (Meilgaard *et al.*, 1987).

A given cooking volatile from any species is seldom unique and the common view is that species-characteristic odour will be represented by a blend of components, each having its own detection threshold and each present at different concentrations. Nonetheless, several schools of thought have emerged, focusing on certain classes of compounds as being the most dominant contributors to mutton odour. Such a simplified approach is understandable considering the complexity volatiles chromatograms.

6.4.1. *Fat oxidation products*

Because of the undoubted contribution of fat to cooking odours, several groups have proposed that oxidation products of fat are important contributors to mutton odour. The oxidation products are mainly alkanes, aldehydes, ketones, alcohols and lactones. Hornstein and Crowe (1963) and Jacobson and Koehler (1963) identified carbonyls as important contributors to mutton odour. Caporaso *et al.* (1977) rendered mutton fat at 50°C and judged it to have the characteristic odour even after this minimal heating. The rendered fat was then heated to simulate oven temperatures and steam extracted. The residual fat had no odour. Instead, odour compounds were present in the extract, specifically the neutral fraction rather than the acidic or basic fractions. Subsequent gas chromatographic–mass spectrometric analysis coupled with sensory evaluation allowed the group to shortlist ten aldehydes, three ketones and one lactone as significant contributors to mutton odour. Recent work by Sutherland and Ames (1995) confirmed that these compounds are present in sheepmeat cooking volatiles.

Cross and Ziegler (1965) evaluated volatiles from cured and uncured cooked pork, beef and chicken, and found that the volatiles of cured and uncured chicken and beef, after having been stripped of aldehydes and ketones by passage through a 2,4-dinitrophenylhydrazine solution, possessed an odour very similar to that of cured ham. They suggested that the flavour of cured meat is the basic meat flavour, derived from precursors other than fats, and that the different cooking odours of the various types of cooked meat depend on the types of carbonyl compounds derived from fat oxidation.

Expanding on this hypothesis, Rubin and Shahidi (1988) proposed that meat flavours can be rationally divided into two groups that are relevant here: the fundamental meat flavour obtained by cooking, and the flavour of uncured cooked meat, with species differentiation largely due to different types and concentrations of carbonyls from lipid oxidation.

In the case of sheepmeat, the available data indicate the basic species odour is intrinsic to fat and not its oxidation. Hornstein and Crowe (1963) found that lamb fat developed a mutton aroma when heated either in air or in nitrogen (where lipid oxidation would presumably be limited by a lack of oxygen), but the aromas of beef and pork fats were quite different following heating in a vacuum compared with samples heated in air (Hornstein and Crowe, 1960). Locker and Moore (1977) found that 'bacon' made from sheepmeat has a pronounced mutton flavour in the fat, and Berry *et al.* (1989) found that chopped-and-formed bacon containing 12% lean mutton had a significantly less desirable flavour than commercial bacon or chopped-and-formed bacon containing lean pork.

More recently Reid *et al.* (1993) put the fat oxidation theory to a critical test, where prerigor (initially) lean meat was stored in the presence or total absence of oxygen (air or vacuum) for several days in the cold. It was also cooked in those atmospheres before immediate presentation to a panel experienced in assessing sheepmeat flavours. With lean meat, neither 'mutton odour' intensity nor 'other odour' was affected by the presence of oxygen. Neither were flavours affected. However, mutton fat similarly treated did show significant differences. Mutton odour was more evident in the oxygen-free treatment. In contrast, the oxic treatment scored highest for 'other odour' which Reid *et al.* (1993) attributed to lipid oxidation products (Rubin and Shahidi, 1988). The fact that the oxygen-free treatment scored highest for 'mutton odour' and lowest for 'other odour' indicates that lipid oxidation products do not significantly contribute to the characteristic defining sheepmeat odour/flavour.

In addition to its well-known reactions with haem, nitrite reacts directly with unsaturated bonds in fats (reviewed by Cassens *et al.*, 1979), and as a result possibly contributes to 'cured' flavour by altering the pattern of fat degradation on cooking. However, curing did not reduce mutton odour from fatty tissue (Reid *et al.*, 1993), confirming an observation by Locker and Moore (1977) and reinforcing the view that lipid oxidation is not defining for sheepmeat odour/flavour.

6.4.2 Branched-chain fatty acids

A quite different involvement of fats was described by Wong *et al.* (1975a). They showed that volatile fatty acids from mutton fat included BCFAs with eight to ten carbon atoms that strongly contributed to the characteristic odour of cooked mutton. Compared to the well-known long-chain fatty acids that dominate animal fats, the important BCFAs were present in very low concentrations, and how Wong and colleagues first pinpointed the BCFAs as being important in sheepmeat is unrecorded. The BCFAs are present in goat and sheep fat but largely absent from the fats of other ruminant species (Duncan and Garton, 1978; Ha and Lindsay, 1990).

Two acids, 4-methyloctanoic (especially) and 4-methylnonanoic, were identified by Wong *et al.* (1975a) as being important. Among other experiments, Wong's group showed that lacing a bland mutton sample with 4-methyloctanoic acid significantly increased panel scores for degree of mutton flavour (Wong *et al.*, 1975b). The acid was particularly concentrated in the fat of sheep that panellists described as very muttony. It was absent in beef fat. Brennand and Lindsay (1982) confirmed and extended these observations. Dispersions of 4-methyloctanoic and 4-methylnonanoic acids in water had characteristic sheepmeat odour/flavour (at low odour thresholds), and later work by this research team (Brennand *et al.*, 1989) and another (Karl *et al.*, 1994) clearly characterized 4-methyloctanoic, 4-methylnonanoic and 4-ethyloctanoic acids as 'goaty' and 'muttony' among other related descriptors. Miller *et al.* (1986) showed that concentrations of branched-chain fatty acids were low in intramuscular fats (triglycerides and phospholipids) compared to subcutaneous fats. Since subcutaneous fats are strongly implicated in mutton odour, this between-site difference in concentration supported the Wong's hypothesis. By contrast, Crouse *et al.* (1982) could find no significant correlation between 'lamb flavour' intensity and one of Wong's indicator branched-chain fatty acids, 4-methylnonanoic acid. Crouse *et al.* (1982) even concluded that fatty acids had little effect on variation in lamb flavour intensity. Purchas *et al.* (1986) found that there were no individual fatty acids, or groups of them, which showed consistent relationships with sheepmeat flavour characteristics.

More recently, however, the evidence linking these BCFAs to sheepmeat odour/flavour has been confirmed in several research studies reviewed at various points in this chapter. For example, Ha and Lindsay (1990) examined the distribution of total BCFAs in perirenal fats of red meat species. This tissue was chosen because BCFAs were reported to vary less between animals in fats from that carcass site than for other sites on sheep (Johnson *et al.*, 1977) even though the concentration is lower in perirenal fat. Highly characterizing odours were provided by 4-ethyloctanoic and 4-methyloctanoic acids for goat and sheep fat, but their concentrations were not significant in four other species examined (Table 6.1). A discussion of the metabolic causes of this species difference is described later. The greater part of these BCFAs are present as fatty acid residues in conventional triacylglycerols (Ha and Lindsay, 1990), although a fraction exists as free fatty acids in whole fat (Sutherland and Ames, 1996).

Brennand and Lindsay (1992a) showed that whereas lean lamb contained very low concentrations of 4-methyloctanoic and 4-methylnonanoic acids,

Table 6.1 Concentrations of selected volatile branched chain fatty acids in perirenal fats of several species. Data are μg per g of fat, or not detected (–). Adapted from Ha and Lindsay (1990)

Fatty acid	Goat (male)	Ovine			Bovine		Horse	Pig	Deer
		Ram	Ewe	Lamb	Veal	Beef			
4-Methyloctanoic	26	8	10	0.7	–	–	–	–	0.2
4-Ethyloctanoic	13	12	15	7	–	–	–	–	–
4-Methylnonanoic	–	38	–	–	8	3	37	–	18

subcutaneous fats were relatively rich in these acids. Perirenal and intramuscular fats had lower concentrations confirming the results of Miller *et al.* (1986) and Johnson *et al.* (1977). Brennand and Lindsay further noted that there was more BCFA variation between individual lambs than between adipose sites; some lambs were more 'sheepy' than others. In assessing the potential impact of the BCFAs on odour/flavour, these workers found that 4-methyl- and 4-ethyloctanoic acids were present in concentrations greatly above threshold in all lamb fats tested, and on triacylglycerol hydrolysis, would contribute to characteristic flavour. The contribution of 4-methylnonanoic acid was not so clear.

The same research group also examined the effect of cooking on the BCFAs available for the sheepmeat eater to distinguish the species note (Brennand and Lindsay, 1992b). Although samples were tested from only one sheep (a five-year-old pasture-raised ram), the authors clearly showed that samples (fatty drip, broth, cooking vapours and others), produced by several cooking methods, contained sufficiently high concentrations of free, volatile BCFAs to provide the characterizing note. Hydrolysis of triacylglycerol or other precursors appeared to account for most of the volatile free fatty acids observed. Moreover, the role 4-methylnonanoic acid in sheepmeat odour was largely confirmed in this study.

Lindsay's research group also examined free volatile fatty acids in ruminant cheese. 4-Methyl- and 4-ethyloctanoic acids provided distinguishing flavours to sheep and goat cheeses in contrast to cow's milk cheese (Ha and Lindsay, 1991a,b).

Several other authors have studied the BCFAs and their role in sheepmeat flavour (Sutherland and Ames, 1995, 1996). This work will be discussed in other sections. Of relevance here, however, is the clear statistical relationship between BCFAs and sheepmeat odour (Young *et al.*, 1997b) (Figure 6.1) and the particular importance of 4-methyloctanoic acid (Sutherland and Ames, 1996).

6.4.3 Phenols

In a significant publication, Ha and Lindsay (1991c) proposed that volatile alkylphenols present in fats contributed to sheepmeat odour/flavour more

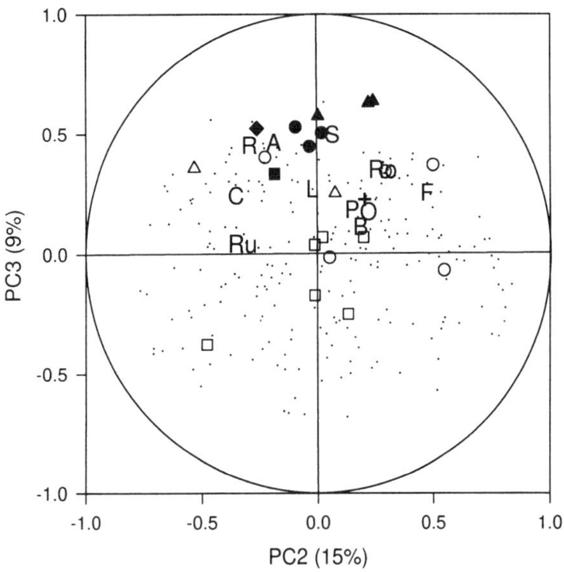

Figure 6.1 Principal component analysis of 244 volatiles from sheep fat; components 2 and 3. Simplified factor loadings were either 4-methyl-substituted or similar fatty acids with about 10 carbon atoms (●), alkyl-substituted fatty acids with fewer carbon atoms (○), long chain lactones (▲), short chain lactones or hydroxyacids (△), 4-methylphenol (■), other phenols (□), benzenethiol (+), 3-methylindole (skatole) (◆), or all other classes, represented by small dots. Supplementary variables are the odour attributes: **F**at, **B**utter, **O**il, **Ro**ast, **P**oultry, **S**heepmeat, **L**iver, **Ru**bber, **A**nimal, **C**abbage, **R**ancid.

than to other species. Alkylphenols involved included the methylphenols, isopropylphenols and others. Although not strictly a phenol, thiophenol (benzenethiol) was found dominant in ram fat, and was thought to be produced during the concentration of phenols by simultaneous distillation and extraction. Ha and Lindsay (1991c) and later Brennand and Lindsay (1992b) proposed that the flavour quality and high potency of thiophenol would contribute unpleasant distinctive burnt sulphury notes, so potentiating unpleasant sheepmeat odours.

Ha and Lindsay (1991c) found that mixtures of phenols and the BCFAs, discussed earlier, yielded a distinct sheepyard mutton odour. In the same study they discussed three possible chemical origins, ruminal fermentation of dietary lignin, dietary diterpenes and dietary tyrosine. Clear mechanisms were outlined for tyrosine fermentations, and the role of amino acids is reconsidered later. Another proposed source is phenolic monomers such as coumaric and ferulic acids that are present in forages (Jung *et al.*, 1983). Given the presence of several of the possible precursor classes in pasture-based diets, Ha and Lindsay (1991c) proposed that higher concentrations of alkylphenols might be encountered in pasture-finished ruminants. This hypothesis has important implications for the huge pasture-finishing ruminant systems of Australasia.

Whatever their origin in ruminant fats, alkylphenols have been detected in a variety of foods (Badings and Neeter, 1980; Heil and Lindsay, 1988), even red wines, where they impart 'animal' and 'stable' off-odours (Chatonnet *et al.*, 1992). They will contribute to cooked meat odour (Brennand and Lindsay, 1992b), even if present as water-soluble conjugates (Kao *et al.*, 1979; Lopez and Lindsay, 1993a) since these are thermally labile (Lopez and Lindsay, 1993b) and evolve as free phenols during cooking.

By contrast, Sutherland and Ames (1995) failed to detect alkylphenols in the cooking volatiles from 12-week-old lambs. Their result was attributed to a failure of the Tenax technique to trap alkylphenols. However, Young *et al.* (1997b) detected several phenols by a similar technique, and 4-methylphenol was correlated with 'animal odour' (Figure 6.1).

Returning briefly to the matter of thiophenol, Ha and Lindsay (1991c) argued that its presence in the ram fat was indicative of a proposed unique sulphur storage system in sheep (see later; Cramer, 1983). Young *et al.* (1997b) detected thiophenol in sheepfat volatiles but it was not statistically related to sheepmeat odour intensity, in marked contrast to other volatiles such as BCFAs (Figure 6.1). The role of thiophenol remains an open question.

6.4.4 *Basic compounds*

Buttery *et al.* (1977) proposed that basic components of volatiles significantly contribute to mutton odour. Of the several pyrazines and pyridines identified, 2-ethyl-3,6-dimethylpyrazine and 2-pentylpyridine were considered likely candidates as specific contributors to mutton odour. These workers proposed that 2-pentylpyridine could be formed from ammonia and a fat oxidation product, deca-2,4-dienal. The former would be formed from the breakdown of amino acids present in sheepfat. Ammonia is a well-known volatile of cooking meats and is reportedly enriched in sheepmeat volatiles (Baines and Mlotkiewicz, 1984). This is perhaps due to the glucose content of sheepmeat, which is lower than that of pork and beef (Macy *et al.*, 1964a). During cooking, the amino acids of sheepmeat are less likely – it is argued – to react with sugars to form Maillard products, and more likely to be thermally degraded to ammonia. Causal or not, 2-pentylpyridine was also identified in sheepfat volatiles by Sutherland and Ames (1995).

Pyrazines and pyridines are two examples of the several classes of heterocyclic volatiles produced from any cooking meat. They include furan derivatives and a variety of *N*- and *S*-heterocyclics. Apart from Buttery *et al.* (1977), no authors have claimed that a given heterocyclic is significantly responsible for mutton odour, but heterocyclics will undoubtedly contribute to the odour. The odour thresholds of many heterocyclic compounds are extremely low (Mussinan *et al.*, 1976).

6.4.5 *Sulphur-containing compounds*

When meat is cooked, sulphur-containing volatiles are generated primarily as a result of degradation of sulphur amino acids. The most dominant sulphur compound in meat volatiles is hydrogen sulphide (Nixon *et al.*, 1979), and, as is examined later, its formation is affected by meat pH. Hydrogen sulphide has its own smell and can act as a precursor for other odorous compounds (Mottram, 1994). Kunsman and Riley (1975) found that on a fresh weight basis, the depot fat tissues of beef and lamb gave off much larger amounts of hydrogen sulphide than the lean tissues, and that lamb samples produced considerably more hydrogen sulphide than beef samples. This result was at the root of a proposal (Cramer, 1983) that an unidentified sulphur store in fat could supply compounds or precursors that would make the cooking odour of lamb different from that of other species. Cramer argued that because wool – a fibre unique to sheep – is rich in cystine, there could be a unique sulphur storage system in sheepfat.

The finding that fat released more hydrogen sulphide than lean on cooking is surprising considering that the protein (and hence amino acid) content of lean tissue is much higher than that of fat. The relative rates of hydrogen sulphide evolution from cooking sheepmeat was reassessed by Reid *et al.* (1992). In contrast to the results of Kunsman and Riley (1975), more hydrogen sulphide evolved from lean meat than from fat.

The literature often states that (lean) lamb has a high cysteine/cystine content (see, for example, Baines and Mlotkiewicz, 1984) and that this is the basic cause of enhanced hydrogen sulphide generation. However, amino acid composition data (Paul and Southgate, 1978; West *et al.*, 1997) for beef and lamb show no significant difference in methionine and half-cystine contents. Clearly though, these data do not show what proportion of each amino acid is free or polymerized in proteins. Differences in distribution could affect cooking odour, as free amino acids are probably more labile. A clue to the compositions of free amino acids is provided by Macy *et al.* (1964b), who found that the free amino acid contents of lean beef, pork and lamb were qualitatively similar.

One sulphur precursor that has been implicated in enhanced hydrogen sulphide formation in sheepmeat is glutathione. Glutathione gives off hydrogen sulphide more rapidly than do sulphur amino acids in proteins (Cramer, 1983), although the latter represent a greater sulphur reserve. Macy *et al.* (1964b) noted that glutathione was present in water extracts of lean lamb but not in those of lean beef and pork. However, as glutathione is an essential part of the free radical scavenging system common to mammalian cells (Munday and Winterbourn, 1989), it will be present in fresh beef and pork.

Schutte (1974) described how thiamine degradation products could play a part in meat odour. Lean lamb contains about 0.14 mg of thiamine per 100 g, double that in beef, 0.07. Pork is higher still at 0.89 mg/100 g (Paul and Southgate, 1978). Although thiamine concentrations are insignificant compared to the total methionine and cysteine concentrations for beef and lamb – around 500 and 250 mg/100 g, respectively – there is general agreement that thiamine degradation products play a very significant part in determining meatiness. But because the concentration of thiamine in lamb is not outstanding, it seems clear that primary degradation products of thiamine are not responsible for mutton odour.

Sutherland and Ames (1995) described a range of sulphur-containing compounds reported for the first time in sheepmeat volatiles. One of these, 4,6-dimethyl-1,3-oxathiane, was probably responsible for a 'stale/wet animal' odour, and was considered the most objectionable of all eluted smells described by odourport sniffers. This compound, for which a formation mechanism was proposed, has not been described before in any food. Its apparent uniqueness and particular association with ram fat (Sutherland and Ames, 1995) marks this compound for special attention in future.

6.5 Factors affecting sheepmeat odour and flavour

6.5.1 *Pre-slaughter factors*

Age. An odour difference between lambs and cull ewes (usually three or more years old) can be noticed during carcass dressing. This is not documented in the scientific literature, but is well known to those who have worked in a sheep slaughterhouse. This observation implies that odorous compounds accumulate with age. Trapping and identifying these volatiles has never been attempted. The raw and cooked odours might be related, and a study of the raw-meat volatiles would not be complicated by the host of compounds present in volatiles released during cooking but not specific to mutton odour.

It is generally believed that as animals grow older their meat becomes more strongly flavoured. Much formal evidence for sheepmeat supports this view (Paul *et al.*, 1964; Batcher *et al.*,1969; Misock *et al.*, 1976; Sutherland and Ames, 1996; Rousset-Akrim *et al.*, 1997), but others have reported the reverse (Weller *et al.*, 1962). In the four early studies, panellists were not questioned specifically on mutton odour or flavour, but any change in disagreeable odour or flavour with age was minimal. It is perhaps significant that these four studies examined animals no older than 16 months. A very pronounced mutton odour and flavour might not be obvious until animals are much older. Nonetheless, Rousset-Akrim *et al.* (1997) reported a clear increase in sheepmeat odour/flavour with age to 6 months, but this was confounded by a possible sex effect (see later).

Of the many biochemical changes that sheep must undergo as they age, three are noted with respect to mutton odour. First, the lean meat of older sheep contains more haem iron, present as myoglobin. New Zealand retail butchers call carcasses from older sheep 'red sheep', reflecting this colour change. During cooking, haem iron and iron released from haem can act as catalysts of lipid oxidation (Rhee, 1988; Johns *et al.*, 1989) with its multitude of downstream effects.

Second, sheep – like humans – become fatter as they age. The fat tends to accumulate in subcutaneous depots rather than intermuscularly or around organs such as kidneys (Broad and Davies, 1980; Butler-Hogg, 1984). Since fats are strongly implicated in mutton odour, it seems reasonable that the increased fattiness alone may be important in odour. This may be especially true when meat is roasted as a whole cut. Precursors such as ammonia and hydrogen sulphide will have more opportunity to interact with lipid oxidation products in a thicker subcutaneous fat layer. The possible importance of the interaction of lean and fat in mutton odour was noted by Cramer *et al.* (1967). As appealing as this theory might be, Field *et al.* (1983) concluded that plane of nutrition, which results in a different amount of fat cover at the same age, has no significant effect on sheepmeat flavour.

Third, as sheep age, the fatty acid composition of their depot fats changes. In general, the fats become more saturated (Barnicoat and Shorland, 1952; Spillane and L'Estrange, 1977; Miller *et al.*, 1986). Although this may relate to animal fatness rather than age (Bensadoun and Reid, 1965), older sheep being generally fatter, the fact remains that a changed fatty composition has the potential to change odour by way of volatile free fatty acids and fatty acid oxidation products. Also, age effects have been observed in the content of branched-chain fatty acids. Early work was conflicting. Bensadoun and Reid (1965) showed increase in high-molecular-weight BCFAs with age, while Spillane and L'Estrange (1977) showed a decrease. The hypothesis of Wong *et al.* (1975a,b) requires that the concentration of lower-molecular-weight BCFAs – specifically around C_9 and C_{10} – must increase with age if it is accepted that older animals have more mutton odour than younger animals. Two recent studies have together confirmed that concentration of characterizing BCFAs increase with age. Young *et al.* (1997b) showed that the fat of 200-day rams was higher in the characterizing BCFAs than that of 100-day lambs. However, in that study, the effects of puberty and age could not be separated. Sutherland and Ames (1996) showed that for both rams and castrates, the subcutaneous fat of 200-day lambs had higher concentrations of (free) causal BCFAs than 80-day lambs. Nonetheless, there is also a clear sex effect, examined later.

Diet. No other factor is so amenable to experimentation as diet. Considering the international impact of the Australian and New Zealand sheep industries, relatively little work has been done on the relationship between diet and odour/flavour, and no findings have been applied commercially in the sense of brand-marketing a specific odour/flavour. This may be due to the following reasons. Historically, their captive markets in Europe required cheap meat, which the two countries could supply economically provided the diet was forage-based, largely ryegrass and clover. Further, the sheep being a dual-purpose animal (meat, wool) meant that at times the meat could be regarded almost as a by-product of the wool industry. There has been little incentive to change diet.

In assessing diet effects it is important to distinguish between 'pastoral' odour/flavour and sheepmeat odour/flavour. The former, defined by Berry *et al.* (1980), is a 'grassy' note in cooked meat from ruminants finished on green-leafed grasses and legumes as opposed to grain-rich diets, whereas the latter odour/flavour is (evidently) dominated by BCFAs. Nonetheless, sheepmeat odour/flavour is affected directly and indirectly by a pastoral diet. In the direct sense, certain pastoral diets can stimulate the formation of BCFAs (Purchas *et al.*, 1986), and this is examined later. Indirectly, undesirable pastoral odour/flavour volatiles deriving from dietary green leaf tissue could simply add to the BCFA flavour, compounding the problem in sensitive markets.

Disregarding these interactions for the moment, the dominance of the legumes, white clover and perennial ryegrass in New Zealand pastures has stimulated research comparing legumes and grasses as dietary alternatives. Cramer *et al.* (1967) found that meat from sheep fed on clover had a significantly more intense flavour and odour than meat from grass-fed sheep. The word 'muttonness' was not used in the sensory work. Several groups of Australasian workers (Czochanska *et al.*, 1970; Shorland *et al.*, 1970; Nicol and Jagusch, 1971; Park *et al.*, 1972, 1975; Purchas *et al.*, 1986) generally agreed with the sensory results of Cramer *et al.*, although only Park *et al.* (1972) clearly quizzed panellists about mutton odour and flavour intensity. In that study, average mutton aroma and flavour values were slightly higher for lucerne-fed (another legume) animals. 'Foreign flavour' and 'foreign aroma' averages were significantly higher for lucerne-fed than for grass-fed animals. There is, moreover, an indication that the foreign odour from legume diets is seasonally related (Park and Minson, 1972; Park *et al.*, 1975), being more dominant in the leafy stage.

Nixon (1981), however, obtained results that contradicted the hypothesis that legumes are responsible for a foreign note. Nixon grazed lambs on ryegrass and each of five legumes. Panellists, selected on their ability to discriminate mutton flavour, were asked to judge samples on 'muttonness' and 'off-flavours'; the latter were defined as not contributing to muttonness. There was no difference in muttonness between any of the samples. Off-flavour was significantly higher in the grass-fed samples than in any of the legume-fed ones.

The biochemical cause of the 'foreign' odour noted in legume-diet experiments is unknown, but there is a clue. Legumes are noted for their high protein content relative to carbohydrate. Rumen bacteria metabolize a fraction of dietary amino acids, and this activity accounts for the high concentrations of 3-methylindole, also called skatole (from tryptophan) (Claus *et al.*, 1994), and phenols (from tyrosine) found in faeces. Since excretion is imperfect, and likewise secretion in urine, it seems likely that rumen fermentation is the origin these compounds in body fat. In a experiment that compared fat volatiles from sheep fed maize or a conventional pastoral diet, Young *et al.* (1997b) found that skatole was far more dominant in the pasture treatment. Since a pasture diet has a relatively higher protein content than a maize diet, protein was probably the cause of this difference in skatole concentration. Moreover, the amino acid catabolites were correlated with 'animal' odour (Figure 6.1), a good example of where a diet effect was superimposed on an intrinsic sheepmeat odour due to BCFAs.

Attributing the 'foreign odour' of sheepmeat raised on legumes to skatole and other amino acid breakdown products is an appealing theory, but must await direct confirmation by an experiment that compares the volatiles derived from grass and legume diets.

Another feeding option for sheep is a high-energy grain diet, and this is common in North American finishing systems. Crouse *et al.* (1978) contrasted alfalfa-rich and maize-rich diets, and found few flavour differences. Kemp *et al.* (1981) contrasted diets ranging from a low-energy bluegrass/clover to high-energy maize-rich, the former being the least acceptable.

Many authors have reported that sheep fats become soft and oily on high-energy diets (see, for example, Field *et al.*, 1978; Hansen and Czochanska, 1978; Busboom *et al.*, 1981; Miller *et al.*, 1986). The main reason for the soft oily condition is the decreased proportions of long-chain saturated fatty acids. Fat melting point is lowered. Another change in fat composition from high-energy diets was described by Garton *et al.* (1972). High-energy grain diets result in increased propionate and butyrate concentrations in the rumen that translate to increased concentrations of odd-chain and branched-chain fatty acids in storage fats. This occurs in sheep and goats but not cattle or deer (Duncan and Garton, 1978). As a class, BCFAs lower fat melting point (Garton *et al.*, 1972) and in their short-chain versions are strongly characteristic of sheepmeat (see earlier).

The metabolic origins of BCFA formation has been explored by Horning *et al.* (1961), Scaife and Garton (1975) and, more recently, Ha and Lindsay (1990). Ha and Lindsay described plausible metabolic pathways to 4-methyl- and 4-ethyloctanoic acids, that hinge on an elevated specificity of sheep and goat fatty acid synthetase enzymes for propionate and butyrate as building blocks. This model can also explain why perirenal fat in sheep has lower concentrations of BCFAs than subcutaneous fat (Brennand and Lindsay, 1992a). In the latter tissue, fats are largely synthesized *in situ* while perirenal fats are diet-derived.

This model implies that any grain-dominated diet is likely to result in increased sheepmeat odour/flavour. However, cereal grains differ in their propensity to generate BCFAs, with wheat possibly being the worst offender (Duncan *et al.*, 1974). Unfortunately no experiments have been performed where different cereal diets have been compared with the primary object of assessing sheepmeat odour/flavour, but scattered studies show how panellists and researchers have variously perceived diet-linked odour/flavour. Wong *et al.* (1975b) found that a barley diet exacerbated sheepmeat flavour compared with a conventional ryegrass/clover diet. Locker (1980) noted that 'bland' was the most common descriptor for barley-fed lamb in the same sensory study that decided that pasture-fed sheepmeat had a 'stronger flavour'. Muttonness was not assessed. In the same experiment a diet of maize silage resulted in a 'porky' character. Young *et al.* (1997b) found that 100-day-old lambs finished on a maize-, wheat- and soy-dominated diet had less sheepmeat odour and flavour than lambs grazed on ryegrass/clover. A diet rich in oats and lupin produced lamb that was less desirable in aroma and flavour, was 'oilier' and 'less meaty' than lamb from lucerne-dominated diet treatments (Hopkins *et al.*, 1995).

The work of Oddy (1980) with wheat-based diets is significant in two respects. First, a pure wheat diet yielded a significantly higher concentration of branched-chain and odd-carbon-number fatty acids than a 70% wheat, 30% lucerne diet. No flavour data were presented, but the work was performed to explain the soft fat and 'terrible smells' of wheat-finished sheepmeat (Oddy and Saville, unpublished observations). Second, Oddy examined the role of cobalt in BCFA formation. Cobalt is essential for vitamin B_{12} formation, and serum propionate is higher in B_{12}-deficient sheep (Marston *et al.*, 1972) suggesting a link to BCFA formation (Garton *et al.*, 1972; Ha and Lindsay, 1990). In Oddy's experiment, cobalt improved live weight gains in lambs fed a pure wheat diet, but had no effect on BCFAs.

While there are sometimes clear differences in BCFA formation between high-energy concentrates diets and lower-energy pastoral diets, one study has shown that some pastoral diets promote BCFA formation more than others. The concentration of 4-methyloctanoic acid in subcutaneous lamb fat was higher when lambs were finished on legumes than on ryegrass (Purchas *et al.*, 1986). These authors also demonstrated how the overall fatty acid profiles in various lamb fat depots were affected by pastoral diet. Although present at lower concentrations than the saturated and monounsaturated fats, the polyunsaturated fats like linoleic and linolenic acid were often affected. Polyunsaturated fatty acids are prone to peroxidation and the downstream generation of markedly odorous compounds that will affect the perception of sheepmeat flavour in the same way that the high concentrations of unsaturated fatty acids in chicken and pork provide characterizing odours, as described in other chapters.

In the diet experiments discussed so far, the fatty acids cross the gut wall after hydrogenation in the rumen. Australian workers (Ford *et al.*, 1975) fed ewes and lambs unsaturated sunflower oil, protected from hydrogenation by suitable encapsulation. The treatment dramatically increased the linoleic acid content of depot fats over that of control animals fed combinations of oats and a legume. Sensory assessment revealed that the aroma and flavour intensities were reduced in the high linoleic acid group and that different aromas and flavours were present. Muttonness was not addressed. Panellists found the meat flavour unacceptable, but in a population accustomed to eating pasture-fed sheepmeat this is perhaps not surprising. Some panellists preferred the high linoleic acid lamb. In a later study by this group (Park *et al.*, 1978a), mature sheep were assessed for mutton odour and flavour after a pasture diet followed by a protected oil supplement. The protected oil supplement reduced mutton odour and flavour. Other work identified some of the diet-specific volatiles derived from high linoleic acid fat that contributed to the altered flavour and reduced muttonness (Park *et al.*, 1978b). Deca-2,4-dienal concentration was an order of magnitude higher than in feed-lot or grazed animals, and an unsaturated 3-dodecalactone was held responsible for a sweet fruity odour. The BCFAs were not examined. Purchas *et al.* (1979) generally confirmed the results of Park and co-workers. Purchas *et al.* (1979) found that meat from sheep fed a protected oil supplement was less acceptable.

Breed and sex. Cramer *et al.* (1970a) found that a fine-woolled breed, Rambouillet, produced meat that was more muttony than two coarse-woolled breeds, Columbia and Hampshire. Later work, which extended the study to the very fine-woolled Merino breed, failed to confirm this (Cramer *et al.*, 1970b). The experiment did show, however, that compared with the other breeds, Merinos had an oily subcutaneous fat, which was reflected in the breed's fatty acid composition. Young *et al.* (1997a) confirmed this. Experiments by others assessing the effects of breed or sire breed (Fox *et al.*, 1962, 1964; Dransfield *et al.*, 1979; Mendenhall and Ercanbrack, 1979; Crouse *et al.*, 1981, 1983; Purchas *et al.*, 1986) generally showed small or no differences.

In Australia and New Zealand it is a common belief that Merino meat is more 'muttony' than that from other breeds. The origin of this belief might be age-related. Merino is primarily a wool breed. So long as a Merino is adequately producing wool it is likely to be held on-farm rather than slaughtered. As a result, Merinos at slaughter are likely to be older and therefore more muttony. Young *et al.* (1993) proposed another reason: Merino lambs were shown to be prone to the high pH condition (at slaughter), which, as discussed later, can adversely affect odour and flavour.

Recently, the BCFAs in the subcutaneous fat of Merino and a common meat/wool breed, the Coopworth, were examined by Young *et al.* (1997a). Castrated lambs were grazed together on ryegrass/clover and slaughtered at 250 days. The concentrations of 4-methyloctanoic and 4-methylnonanoic acids in subcutaneous fat were at least double in the Coopworth group. Therefore, the cause of the stronger odour reported for Merino probably does not lie with these species-defining fatty acids. pH differences remain a possibility as do the higher concentrations of oxidation-prone unsaturated fatty acids in Merino (Young *et al.*, 1997a).

Since ram lambs grow faster than wethers (castrates) (Kirton *et al.*, 1982; Dransfield *et al.*, 1990), there are economic advantages in growing rams. Considerable research has been directed at the effect of sex on flavour differences. Several studies have found ram meat to be significantly less acceptable than ewe meat (Cramer *et al.*, 1970a) or wether meat (Kemp *et al.*, 1972; Misock *et al.*, 1976; Crouse *et al.*, 1981; Field *et al.*, 1984). Misock *et al.* (1976) in particular noted that meat from rams had a stronger and more undesirable odour than meat from wethers. Busboom *et al.* (1981) analysed the fat from rams and wethers used in studies by Crouse *et al.* (1981) and found that the fat of heavy rams, in particular, had more branched-chain fatty acids than that from wethers, and the fat was softer. Vimini *et al.* (1984) obtained similar results. There is some indication (Misock *et al.*, 1976; Crouse *et al.*, 1981) that the heavier the ram, the more intense the flavour, again implicating fatty acids in odour/flavour differences.

More recently, Sutherland and Ames (1996) analysed the free BCFAs of fat from rams and castrates. At 80 days, ram fat had insignificantly higher concentrations of the flavour-significant BCFAs than castrate fat, but by 200 days, levels were much higher in rams (Table 6.2). Thus, not only is there a clear age effect in the accumulation of these (free) fatty acids, as discussed in a previous section, there is a clear male effect. In a parallel study, Sutherland and Ames (1995) found that the higher concentration of BCFAs in ram fat was manifest as higher concentrations in volatiles evolving from cooked fat.

New Zealand is one of the few major sheep producers where the production of entirely males is encouraged (Butler-Hogg and Brown,

Table 6.2 Concentrations of two branched-chain fatty acids present as the free acid in adipose tissue of lambs, castrates and rams, at two ages. Data are means, μg per g tissue. Adapted from Sutherland and Ames (1996)

Fatty acid	Castrates		Rams	
	80 days	200 days	80 days	200 days
4-Methyloctanoic	2.9	3.9	3.8	50.3
4-Methylnonanoic	0.09	0.35	0.35	1.40

1986). Apparently, there have been no adverse market reports regarding the palatability of New Zealand's sheepmeats, which is difficult to reconcile with the clear castrate/ram effects noted above. The answer may lie in the average ages and weights at which lambs are slaughtered in various markets. New Zealand lamb for export is typically four months old and light, whereas lamb production in the USA is geared to a much heavier and older lamb. Entires are not welcome in that market.

Pre-slaughter stress. Pastoral farming and pre-slaughter abattoir practices can contribute to changes in the eating quality of sheepmeat (and other species) by affecting the pH which meat finally attains in rigor, the so-called ultimate pH. Various stresses (e.g. nutritional, human contact, farm dogs, social regrouping, transport) can cause a depletion in muscle glycogen reserves, resulting in high ultimate pH meat. Such meat is darker, is often tougher, and has poorer keeping qualities than meat of normal pH (about 5.5). Surveys by Graafhuis and Devine (1994) revealed that the ultimate pH of New Zealand beef and sheepmeat is variable, with 30% of animals from either species exhibiting pH values greater than 5.8.

Sensory analyses have shown that high pH beef is less flavourful (Dransfield, 1981; Purchas *et al.*, 1986), has more 'off-flavours' (Fjelkner-Modig and Ruderus, 1983) and evokes more negative flavour reactions by panellists than normal pH beef (Dransfield, 1981; Dutson *et al.*, 1981).

Young *et al.* (1993) compared the odour and flavour of meat from Coopworth and Merino lambs grazed on the same pasture. They unexpectedly found that the Merino breed was prone to the high pH condition – recently confirmed by Hopkins *et al.* (1996) – and suggested that pH, rather than breed, might have been the dominant factor affecting odour and flavour. Panellists registered several negative flavour and odour descriptors for the high pH Merino meat. In contrast, Devine *et al.* (1993) found no significant change in aroma and flavour with increasing pH of sheepmeat.

Braggins (1996) probed the chemistry of the effects of sheepmeat pH on odour/flavour using sensory panels and purge-and-trap gas chromatography combined with mass spectrometry or olfactometry. Stress was induced by pre-slaughter adrenaline injections. High pH sheepmeat had a lower 'overall' odour and flavour intensity than meat of normal pH. Sheepmeat flavour was also significantly lower, and desirable odour and flavour notes decreased and undesirable ones increased as pH increased. Braggins found that many aldehydes decreased in concentration with increasing pH, and olfactometry of the chromatographic effluents pinpointed ten odour-significant compounds whose concentration changed with pH. Again, most were aldehydes, indicating that they play a major role in the quality of cooked sheepmeat odour and flavour. Braggins also showed that the flavour effects of high pH developed during cooking rather than in the raw meat.

Proteolysis and lipolysis operate more favourably at lower pH (Buscailhon *et al.*, 1994). These hydrolytic reactions may produce the vital precursors required for generation of odour and flavour compounds produced during cooking. Alternatively, the greater water-holding capacity of high pH meat may influence the release of volatile compounds and affect flavour perception (Lawrie, 1985).

Another possible cause of the differences in odour and flavour intensities at higher meat pH is the evolution of hydrogen sulphide during cooking. Johnson and Vickery (1964) showed increased amounts of hydrogen sulphide was produced from meat of higher pH. This was directly related to the pH of meat. High levels of liberated hydrogen sulphide might influence flavour and odour generation during cooking since hydrogen sulphide is itself reactive.

6.5.2 Post-slaughter factors

Aspects of culinary practice. The studies of Brennand and Lindsay (1992a,b) and more recently Sutherland and Ames (1996) have clearly shown that there are sufficient BCFAs in all fatty tissue of lamb to yield characterizing odours whatever the cooking method. However, it seems reasonable that if lamb with a reduced BCFA content were cooked, odours would be restricted to the immediate area of cooking rather than pervading a wider area. Reductions in BCFA concentrations offer scope for odour/flavour adjustment and this concept applies to all discussion in this section.

An interesting feature of retail meat sales in New Zealand is that quantities of minced (ground) beef outsell those of minced sheepmeat by approximately a factor of ten. Price is not the reason. Mincing probably distributes air through the meat, which might lead, in the case of lamb, to undesirable odours during cooking. Alternatively, because the surface area capable of releasing odours is greater in minced meats, the mince will inevitably release more volatiles, perhaps BCFAs, than will whole-tissue meat.

Sheepmeat is prepared for the table by methods that vary from culture to culture. In Australasia, sheepmeat is commonly dry-roasted as a whole piece. Johnson and Purchas (1997) and Johnson *et al.* (1988) showed that the inner and outer layers of subcutaneous fat differed in their fatty acid profiles. Not only was the outer layer richer in unsaturated fatty acids, but characterizing BCFAs were more concentrated there too, particularly in rams. When sheepmeat is roasted, the outer layer of fat is subjected to the highest temperatures and arguably these high temperatures will liberate excessive quantities of volatile BCFAs. To minimize BCFA liberation, excessive fat trimming would be useful, but this would detract from the traditional appearance of roasted lamb.

In dry roasting, a large quantity of rosemary is often included in the covered roasting pan. It is common anecdotal experience that sheepmeat odour is reduced by this treatment. Two explanations are offered. The antioxidant volatiles of rosemary will permeate the subcutaneous fat and inhibit or modify the oxidation reactions that take place during cooking. However, it is difficult to see how this would alter the liberation of BCFAs. Another explanation is that rosemary volatiles simply mask mutton odour in some way. Some cooks insert slivers of garlic in the roast so that garlic volatiles permeate the tissues during cooking. Although garlic components are antioxidants and may act as such with sheepmeat, the role of garlic in sheepmeat cuisine may go beyond the antioxidant effect (see below).

Specific herbs play an important role in enhancing the enjoyment of sheepmeat in many cultures, a theme explored by Smith and Young (1991). Garlic is the condiment most widely used with sheepmeat, and although garlic is used with many meats, it featured in almost every sheepmeat recipe examined by Smith and Young (1991). Along with onions, this member of the lily family is noted for its complex sulphur chemistry. Alliin (*S*-allylcysteine sulphoxide) comprises about 0.25% of a garlic bulb (Block, 1985). It breaks down to a number of sulphur compounds that could mask volatile sulphur compounds released from cooking sheepmeat. Alternatively, garlic volatiles could react with meat volatiles to produce agreeable odours (Yu *et al.*, 1994). At a practical level, Klettner *et al.* (1989) found that high levels of mutton can be used in processed meat products for the German market if the appropriate seasoning, particularly with garlic, is used.

The markets of the world are awash with processed meat products containing beef and pork where the species is identified as a positive marketing attribute. It is not so with sheepmeat. In Australasia, sheepmeat is used extensively in processed meat products but the inclusion of sheepmeat is never promoted. This deficiency suggests a market niche where the lean and fat of lamb are used in high-value processed meats that are promoted as sheepmeat products. There is scope for curing or smoking, and inclusion of the condiments described by Smith and Young (1991). If unpleasant sheepmeat odours can be inhibited, masked or otherwise changed for the better by the use of unusual herbs and spices, then producers of sheepmeat products should exploit them. For instance, the use of cinnamon, a tropical laurel, is common in Middle Eastern but not in Western sheepmeat cuisines, and presents an opportunity for adding spices. Sheepmeats are traditionally eaten hot, but the future development of novel products might better be applied to products consumed cold. This is because the BCFAs are almost certainly less evident in a more chilled environment.

A novel approach to the use of herbs is the current work in New Zealand by a research company that aims to produce flavour-modified lamb by including substantial quantities of herbs in the diet. Herbs under study include thyme, garlic, coriander and lavender. Lavender is of particular interest. Lamb from the Sisteron region of Provence in southern France is reportedly flavour-distinct because wild lavender, which is common in that region, contributes to lambs' dietary regime. However, to capture benefits from herb supplementation it is clear that meat flavoured in this way would have to be marketed as a branded product.

In contrast to beef and pork, it is uncommon to cure sheepmeats. This may have its origins in the relative hardiness of the species. Slaughter followed by curing and drying of meat presented an alternative to the expensive practice of housing animals over winter, and with a hardy species there may have been less need to slaughter and cure. Curing of sheepmeats for the delicatessen market presents an unexploited marketing opportunity, although as discussed earlier, minimization of oxidation due to curing will not eliminate the fundamental sheepmeat odour/flavour due to BCFAs.

Other methods to modify or reduce mutton odour or flavour. As noted earlier, the lower glucose content of sheepmeat could alter the pattern of amino acid breakdown during cooking, which in turn would affect the pattern of volatiles produced. Hudson and Loxley (1983) studied the effect of pentose sugars on mutton odour and flavour. (Pentose sugars were chosen because a patent [US patent 836 694] claimed that beef flavours could be formed by heating amino acids with them.) An Australian sensory panel chosen on the basis of being able to distinguish species flavour found that xylose-treated mutton (2%) had a significantly different odour and flavour from untreated mutton and untreated beef. The scores indicated that the treated mutton was no more similar to mutton than to beef. Panellists preferred the xylose-mutton to lamb, mutton or beef (Table 6.3). The panellists were not questioned on intensity of mutton odour or flavour.

The patent literature contains several methods claimed to reduce mutton odour. In light of Asians' frequent dislike of mutton, Asian patents are of particular interest. Soaking mutton in a 0.1% solution of either malate, fumarate or succinate has been claimed to be effective in elimi-

Table 6.3 A ranked preference test for xylose-treated mutton, lamb, mutton and beef. Adapted from Hudson and Loxley (1983)

Meat	Frequency of rank score[a]				Weighted mean score[b]
	1	2	3	4	
Xylose-mutton	2	4	8	14	3.21[c]
Lamb	10	8	5	5	2.18[d]
Mutton	8	7	8	5	2.36[d]
Beef	8	9	7	4	2.25[d]

[a]Score 1 is the least preferred, 4 the most.
[b]Means with different superscripts are significantly different ($p<0.05$).

nating mutton odour (Japan patent 80 11 302). Cooking mutton or goat meat in the presence of maltol, ethylmaltol or isomaltol removes the offensive odours, according to Japan patent 80 88 677. Mutton treated with 0.05% to 5% of asparagine, glutamine, alanine or glycine is claimed to lose its mutton odour on cooking (Japan application 70 91 341). Two of the examples in this application also employed the sugars xylose and glucose. Low-molecular-weight dextrins remove the cooking odour of mutton and fish, according to another patent (Japan 80 77 875).

Mechanisms for the above treatments, or for pentose sugars (Hudson and Loxley, 1983), were not proposed by the authors. There are two possibilities. First, the additives are low molecular weight compounds that, as precursors, are likely to substantially alter the pattern of volatiles formation. If true, this would mean that subtle differences between species in the profile of low-molecular-weight precursors are important in specific odour generation. Second, the reaction products of amino acids and sugars heated together (Maillard reaction products) are very good antioxidants (Bailey *et al*., 1987). Maltol is a good example (Sato *et al*., 1973), and xylose is particularly effective as a precursor sugar of these antioxidants (Lingnert and Ericksson, 1981). If the Maillard reaction products derived in the patented treatments reduce mutton odour by reducing oxidation, this would support Rubin and Shahidi's (1988) theory that lipid oxidation products are at the root of all odour/flavour species differences. The extensive work with BCFAs, and the anoxia and curing experiments examined earlier indicate that this is not the case for sheepmeat. Nonetheless, some improvement in sheepmeat flavour by the use of antioxidants is undeniable.

Other patents (New Zealand 159 518; France 2 597 726) call for the use of certain Chinese herbs and 'Herbs de Provence', respectively, also noted by Smith and Young (1991). Again, the herbs may act as antioxidants and/or mask mutton odour. Hayama (Japan 71 30 780) claimed that boiling mutton in water containing ginger, followed by coating the meat with soya, sucrose, rice wine and egg, followed by drying was effective in enhancing the sensory properties. This claim is not doubted. Other combinations of complex additives are similarly likely to be effective, simply by swamping any mutton odour or flavour present.

An Australia patent (596 801) describes a procedure, involving acetic acid, salt and phosphates, and exposure to ultraviolet light, that is claimed to reduce mutton odour. The mechanism is unstated. It is possible that reduced muttonness might stem from reduced hydrogen sulphide evolution due to a more acidic environment (Johnson and Vickery, 1964). In this respect a patent (Japan 61 96 970) calls for browning mutton and then cooking in a 1% acetic acid solution as a means of reducing mutton odour.

World patent 87 03 454 claims that by passing steam at 0.12 atmosphere (45–60°C) through comminuted mutton, the mutton odour of the cooked product approaches that of lamb. The colour data presented for this patent show that the treated raw meat is at risk of turning grey (due to myoglobin denaturation). Interestingly, the patent also extends to the simultaneous use of antioxidants and pentose sugars, methods discussed above.

The Captech process is an advanced controlled-atmosphere packaging system that maintains the microbiological quality of chilled meat by carefully controlling the packaging atmosphere to be extremely low in oxygen and to consist essentially of pure carbon dioxide. Gill (1988) noted that Captech lamb had a 'strong ovine odour' when the packages were opened. He proposed that the carbon dioxide stripped the meat of its characteristic odour, leaving a reduced species flavour after cooking. Carbon dioxide is commonly used in supercritical fluid extraction technology to remove, among other things, odours and flavours from foods, although in the Captech process the supercritical temperature and pressure is not reached.

6.6 Concluding remarks

In seeking to minimize sheepmeat odour/flavour for sensitive markets, it seems the best prospect is to reduce the concentration of BCFAs in sheepfat by diet and/or breeding. Cooking odour is the main objection that sensitive people have to sheepmeat, so any significant reduction in the BCFA levels in fat may be commercially useful in sensitive markets, provided the sheepmeat product can be identified by brand. When sold as a commodity, any sales successes would be eroded by a competitor selling lamb cuts where BCFA concentrations were uncontrolled.

Diet modifications have the advantage that they can be applied more quickly and the scope to reduce pastoral odour/flavour at the same time might be possible. In the longer term, however, breeding for controlled BCFA formation might be possible.

With BCFAs controlled or not, sheepmeat represents a major unexploited resource for added-value branded products to fulfill market demands for novelty and consistency.

References

Acree, T.E., Barnard, J. and Cunningham, D.G. (1984). A procedure for the sensory analysis of gas chromatographic effluents. *Food Chem.*, **14**, 273–285.

Badings, H.T. and Neeter, R. (1980). Recent advances in the study of aroma compounds of milk and dairy products. *Neth. Milk Dairy J.*, **34**, 9–30.

Bailey, M.E., Shin-Lee, S.Y., Dupuy, H.P., St. Angelo, A.J. and Vercellotti, J.R. (1987). In *Warmed-over Flavor of Meat,* eds. A.J. St. Angelo and M.E. Bailey. Academic Press, Orlando, pp. 237–266.

Baines, D.A. and Mlotkiewicz, J.A. (1984). In *Recent Advances in the Chemistry of Meat,* ed. A.J. Bailey. Royal Society Chemistry, Special Publication No. 47.

Barnicoat, C.R. and Shorland, F.B. (1952). New Zealand lamb and mutton. Part II. Chemical composition of edible tissues. *NZ J. Sci. Technol.*, **5**, 16–23.

Batcher, O.M., Brant, A.W. and Kunze, M.S. (1969). Sensory evaluation of lamb and yearling mutton flavours. *J. Food Sci.*, **34**, 272–274.

Bensadoun, A. and Reid, J.T. (1965). Effect of physical form, composition and level of intake of diet on the fatty acid composition of the sheep carcass. *J. Nutr.*, **87**, 239–244.

Berry, B.W., Maga, J.A., Calkins, C.R., Wells, L.H., Carpenter, Z.L. and Cross, H.R. (1980). Flavor profile analysis of cooked beef loin steaks. *J. Food Sci.*, **45**, 1113–1121.

Berry, B.W., Cross, H.R. and Smith, G.C. (1989). Processing, chemical, sensory and physical properties of bacon chopped and formed, made from pork, beef, mutton and chevon. *J. Muscle Foods*, **1**, 45–57.

Block, E. (1985). The chemistry of garlic and onions. *Sci. Am.*, **252**, 94–99.

Braggins, T.J. (1996). Effects of stress-related changes in sheepmeat ultimate pH on cooked odour and flavour. *J. Agric. Food Chem.,* **44**, 2352–2360.

Braggins, T.J. (1997). Ultimate pH and storage related changes to sheepmeat odour and flavour. D.Phil Thesis, Waikato University, Hamilton, New Zealand.

Brennand, C.P. and Lindsay, R.C. (1982). Sensory discrimination of species-related meat flavours. *Lebensm. Wiss. u. Technol.*, **15**, 249–252.

Brennand, C.P. and Lindsay, R.C. (1992a). Distribution of volatile branched-chain fatty acids in various lamb tissues. *Meat Sci.*, **31**, 411–421.

Brennand, C.P. and Lindsay, R.C. (1992b). Influence of cooking on concentration of species-related flavor compounds in mutton. *Lebensm. Wiss. u. Technol.*, **25**, 357–364.

Brennand, C.P., Ha, J.K. and Lindsay, R.C. (1989). Aroma properties and thresholds of some branched-chain and other minor volatile fatty acids occurring in milkfat and meat lipids. *J. Sensory Stud.* **4**, 105–120.

Broad, T.E. and Davies, A.S. (1980). Pre- and postnatal study of carcass growth of sheep. 1. Growth of dissectible fat and its chemical components. *Anim. Prod.*, **31**, 63–71.

Brown, G.I. (1989). New Zealand tallow markets: a survey of opportunities. Meat Industry Research Institute NZ (Hamilton, New Zealand). Publ. No. RM 185.

Busboom, J.R., Miller, G.J., Field, R.A., Crouse, J.D., Riley, M.L., Nelms, G.E. and Ferrell, C.L. (1981). Characteristics of fat from heavy ram and wether lambs. *J. Anim. Sci.*, **52**, 83–92.

Buscailhon, S., Berdague, J.-L., Gandemer, G., Touraille, C. and Monin, G. (1994). Effects of initial pH on compositional changes and sensory traits of French dry-cured hams. *J. Muscle Foods*, **5**, 257–269.

Butler-Hogg, B.W. (1984). The growth of Clun and Southdown sheep: body composition and the positioning of total body fat. *Anim. Prod.*, **39**, 405–411.

Butler-Hogg, B.W. and Brown, A.J. (1986). Muscle weight distribution in lambs: a comparison of entire male and female. *Anim. Prod.,* **42**, 343–348.

Buttery, R.G., Ling, L.C., Teranishi, R. and Mon, T.R. (1977). Roasted lamb fat: basic volatile components. *J. Agric. Food Chem.*, **25**, 1227–1229.

Buttery, R.G, Teranishi, R. and Ling, L.C. (1987). Fresh tomato aroma volatiles: a quantitative study. *J. Agric. Food Chem.*, **35**, 540–544.

Cadwallader, K.R., Baek, H.H., Chung, H.Y. and Moody, M.W. (1994). In *Lipids in Food Flavors*, eds. C.-T. Ho and T.G. Hartman. ACS Symposium Series 558, American Chemical Society, pp. 186–195.

Caporaso, F., Sink, J.D., Dimick, P.S., Mussinan, C.J. and Sanderson, A. (1977). Volatile flavour constituents of ovine adipose tissue. *J. Agric. Food Chem.*, **25**, 1230–1233.

Cassens, R.G., Greaser, M.L., Ito, T. and Lee, M. (1979). Reactions of nitrite in meat. *Food Technol.*, **33**(7), 46–55.

Chatonnet, P., Dubourdieu, D., Boidron, J.-N. and Pons, M. (1992). The origin of ethylphenols in wines. *J. Sci. Food Agric.*, **60**, 165–178.

Claus, R., Weiler, U. and Herzog, A. (1994). Physiological aspects of androstenone and skatole formation in the boar – a review with experimental data. *Meat Sci.*, **38**, 289–305.

Cramer, D.A. (1983). Chemical compounds implicated in lamb flavour. *Food Technol.*, **37**(5), 249–257.

Cramer, D.A., Barton, R.A., Shorland, F.B. and Czochanska, Z. (1967). A comparison of the effects of white clover (*Trifolium repens*) and of perennial ryegrass (*Lolium perenne*) on fat composition and flavour of lamb. *J. Agric. Sci.*, **69**, 367–373.

Cramer, D.A., Pruett, J.B., Kattnig, R.M. and Schwartz, W.C. (1970a). Comparing breeds of sheep. I. Flavor differences. *Proc. Western Sect. Am. Soc. Anim. Sci.*, **21**, 267–269.

Cramer, D.A., Pruett, J.B., Swanson, V.B., Schwartz, W.C., Kattnig, R.M., Phillips, B.R. and Wookey, R.E. (1970b). Comparing breeds of sheep. II Carcass characteristics. *Proc. Western Sect. Am. Soc. Anim. Sci.*, **21**, 270–272.

Cross, C.K. and Ziegler, P. (1965). A comparison of the volatile fractions from cured and uncured meat. *J. Food Sci.*, **30**, 610–614.

Crouse, J.D. (1983). The effects of breed, sex, slaughter weight, and age on lamb flavour. *Food Technol.*, **37**(5), 264–268.

Crouse, J.D., Field, R.A., Chant, G., Jr., Ferrell, C.L., Smith, G.M. and Harrison, V.L. (1978). Effect of dietary energy intake on carcass composition and palatability of different weight carcasses from ewe and ram lambs. *J. Anim. Sci.*, **47**, 1207–1218.

Crouse, J.D., Busboom, J.R., Field, R.A. and Ferrell, C.L. (1981). The effects of breed, diet sex, location and slaughter weight on lamb growth, carcass composition and meat flavour. *J. Anim. Sci.*, **53**, 376–386.

Crouse, J.D., Ferrell, C.L., Field, R.A., Busboom, J.R. and Miller, G.J. (1982). The relationship of fatty acid composition and carcass characteristics to meat flavour in lamb. *J. Food Qual.*, **5**, 203–214.

Crouse, J.D., Ferrell, C.L. and Cross, H.L. (1983). The effects of dietary ingredient, sex and slaughter weight on cooked meat flavour profile of market lamb. *J. Anim. Sci.*, **57**, 1146–1153.

Czochanska, Z., Shorland, F.B., Barton, R.A. and Rae, A.L. (1970). A note on the effect of the length of the resting period before slaughter on the intensity of flavour and odour of lamb. *NZ J. Agric. Res.*, **13**, 662–663.

Devine, C.E., Graafhuis, A.E., Muir, P.D. and Chrystall, B.B. (1993). The effect of growth rate and ultimate pH on meat quality of lambs. *Meat Sci.*, **35**, 63–77.

Dransfield, E. (1981). In *The Problem of Dark-Cutting Beef*, eds., D.E. Hood and P.V. Tarrant. Martinus Nijhoff Publishers, The Hague, pp. 344–358.

Dransfield, E., Nute, G.R., MacDougall, D.B. and Rhodes, D.N. (1979). Effect of sire breed on eating quality of cross-bred lambs. *J. Sci. Food Agric.*, **3**, 805–808.

Dransfield, E., Nute, G.R., Hogg, B.W. and Walters, B.R. (1990). Carcass and eating quality of ram, castrated ram and ewe lambs. *Anim. Prod.*, **50**, 291–299.

Duncan, W.R.H. and Garton, G.A. (1978). Differences in the proportions of branched-chain fatty acids in subcutaneous triacylglycerols of barley-fed ruminants. *Br. J. Nutr.*, **40**, 29–33.

Duncan, W.R.H., Ørskov, E.R. and Garton, G.A. (1974). Effect of different dietary cereals on the occurrence of branched-chain fatty acids in lamb fats. *Proc. Nutr. Soc.*, **33**, 81A–82A.

Dutson, T.R., Savell, J.W. and Smith, C.G. (1981). In *The Problem of Dark-Cutting Beef*, eds. D.E. Hood and P.V. Tarrant. Martinus Nijhoff Publishers, The Hague, pp. 253–268.

Field, R.A., Williams, J.C., Ferrell, C.L., Crouse, J.D. and Kunsman, J.E. (1978). Dietary alteration of palatability and fatty acids in meat from light and heavy weight ram lambs. *J. Anim. Sci.*, **47**, 858–864.

Field, R.A., Williams, J.C. and Miller, G.J. (1983). The effects of diet on lamb flavor. *Food Technol.*, **37**(5), 258–263.

Field, R.A., Williams, J.C., Brewer, M.S., Cross, H.R. and Secrist, J.L. (1984). Influence of sex, NaCl, MSL, nitrite and storage on sensory properties of restructured lamb roasts. *J. Food Qual.*, **7**, 121–129.

Fjelkner-Modig, S. and Ruderus, H. (1983). The influence of exhaustion and electrical stimulation on the meat quality of young bulls: Part 2 – physical and sensory properties. *Meat Sci.*, **8**, 203–220.

Ford, A.L., Park, A.J. and McBride, R.L. (1975). Effect of a protected lipid supplement on flavor properties of sheep meats. *J. Food Sci.*, **40**, 236–239.
Fox, C.W., Eller, R., Sather, L. and McArthur, J.A.B. (1964). Effect of sire and breed on eating qualities from weanling lambs. *J. Anim. Sci.*, **23**, 596.
Fox, C.W., McArthur, J.A.B. and Sather, L. (1962). Effect of sire and breed on flavor scores from weanling lambs. *J. Anim. Sci.*, **21**, 665.
Garton, G.A., Hovell, F.D.D. and Duncan, W.R.H. (1972). Influence of dietary volatile fatty acids on the fatty-acid composition of lamb triglycerides, with special reference to the effect of propionate on the presence of branched-chain components. *Br. J. Nutr.*, **28**, 409–416.
Gill, C.O. (1988). CO_2 packaging – the technical background. In *Proc. 25th Meat Ind. Res. Conf.* (Hamilton, New Zealand), Meat Industry Research Institute NZ Publication No. 855, pp. 181–185.
Gordon, M. (1987). Novel antioxidants. *Food Sci. Technol. Today*, **1**(3), 172–173.
Graafhuis, A.E. and Devine, C.E. (1994). Incidence of high pH beef and lamb II: results of an ultimate pH survey of beef and sheep plants in New Zealand. In *Proceedings of the 28th Meat Ind. Res. Conf.*, Meat Industry Research Institute NZ, Publication No. 942.
Grosch, W. (1993). Detection of potent odorants in foods by aroma extract dilution analysis. *Trends Food Sci. Technol.* **4**(3), 68–73.
Guadagni, D.G., Okano, S., Buttery, R.G. and Burr, H.K. (1966). Correlation of sensory and gas–liquid chromatographic measurements of apple volatiles. *Food Technol.*, **20**, 166–169.
Ha, J.K. and Lindsay, R.C. (1990). Distribution of volatile branched-chain fatty acids in perinephric fats of various red meat species. *Lebensm. Wiss. u. Technol.*, **23**, 433–440.
Ha, J.K. and Lindsay, R.C. (1991a). Contributions of cow, sheep, and goat milks to characterizing branched-chain fatty acid and phenolic flavors in varietal cheeses. *J. Dairy Sci.*, **74**, 3267–3274.
Ha, J.K. and Lindsay, R.C. (1991b). Volatile branched-chain fatty acids and phenolic compounds in aged Italian cheese flavors. *J. Food Sci.*, **56**, 1241–1250.
Ha, J.K. and Lindsay, R.C. (1991c). Volatile alkylphenols and thiophenol in species-related characterizing flavors of red meats. *J. Food Sci.*, **56**, 1197–1202.
Hansen, R.P. and Czochanska, Z. (1978). Fatty acid composition of the subcutaneous fats of lambs fed 71% barley grain, and a comparison with those of lambs grazed on pasture, or fed higher levels of barley grain. *NZ J. Sci.*, **21**, 85–90.
Heil, T.P. and Lindsay, R.C. (1988). Volatile compounds in flavor-tainted fish from the upper Wisconsin river. *J. Environ. Sci. Health*, **B23**, 489–512.
Hoffmann, G. and Meijboom, P.W. (1968). Isolation of two isometric 2,6-nonadienals, and two isomeric 4-heptenals from beef and mutton tallow. *J. Am. Oil Chem. Soc.*, **45**, 468–474.
Hopkins, D.L., Holst, P.J. and Hall, D.J. (1995). Effect of grain or silage supplementation on meat quality attributes of cryptorchid lambs grazing lucerne or annual grass–clover pastures. *Austral. J. Exp. Agric.*, **35**, 461–465.
Hopkins, D.L., Fogarty, N.M. and Menzies, D.J. (1996). Muscle pH of lamb genotypes. *Proc. Aust. Soc. Anim. Prod.*, **21**, 347.
Horning, M.G., Martin, D.B., Karmen, A. and Vagelos, P.R. (1961). Fatty acid synthesis in adipose tissue. II. Enzymatic synthesis of branched chain and odd-numbered fatty acids. *J. Biol. Chem.*, **236**, 669–672.
Hornstein, I. and Crowe, P.F. (1960). Flavor studies on beef and pork. *J. Agric. Food Chem.*, **8**, 494–498.
Hornstein, I. and Crowe, P.F. (1963). Meat flavor: lamb. *J. Agric. Food Chem.*, **11**, 147–149.
Hudson, J.E. and Loxley, R.A. (1983). The effect of pentose sugars on the aroma and flavour of mutton. *Food Technol. Aust.*, **35**(4), 174–175.
Jacobs, J.A., Field, R.A., Botkin, M.P. and Riley, M.L. (1973). Effect of dietary stress on lamb carcass composition and quality. *J. Anim. Sci.*, **36**, 507–510.
Jacobson, M. and Koehler, H.H. (1963). Components of the flavor of lamb. *J. Agric. Food Chem.*, **11**, 336–339.
Johns, A.M., Birkinshaw, L.H. and Ledward, D.A. (1989). Catalysts of lipid oxidation in meat products. *Meat Sci.*, **25**, 209–220.
Johnson, A.R. and Vickery, J.R. (1964). Factors influencing the production of hydrogen sulphide from meat during heating. *J. Sci. Food Agric.*, **15**, 695–701.

Johnson, C.B. and Purchas, R.W. (1998). 4-Methyloctanoic acid (hircinoic acid) and related methyl branched-chain pentadecanoic and heptadecanoic acids in sheep fats. In preparation.

Johnson, C.B., Purchas, R.W. and Birch, E.J. (1988). Fatty acid composition of fats of differing melting points extracted from ram subcutaneous tissue. *Lipids*, **23**, 1049–1052.

Johnson, C.B., Wong, E., Birch, E.J. and Purchas, R.W. (1977). Analysis of 4-methyloctanoic acid and other medium chain-length fatty acid constituents of ovine tissue lipids. *Lipids*, **12**, 340–347.

Jung, H.G., Fahey, G.C. and Garst, J.E. (1983). Simple phenolic monomers of forages and effects of *in vitro* fermentation on cell wall phenolics. *J. Anim. Sci.*, **57**, 1294–1305.

Kao, J., Bridges, J.W. and Faulkner, J.K. (1979). Metabolism of [14C]phenol by sheep, pig and rat. *Xenobiotica*, **9**, 141–147.

Karl, V., Gutser, J., Dietrich, A., Maas, B. and Mosandl, A. (1994). Stereoisomeric flavour compounds LXVIII. 2-, 3-, and 4-alkyl-branched acids, Part 2: Chirospecfic analysis and sensory evaluation. *Chirality*, **6**, 427–434.

Kemp, J.D., Shelley, J.M., Ely, D.G. and Moody, W.G. (1972). Effects of castration and slaughter weight on fatness, cooking losses and palatability of lamb. *J. Food Sci.*, **34**, 560–562.

Kemp, J.D., Mahyuddin, M., Ely, D.G., Fox, J.D. and Moody, W.G. (1981). Effect of feeding systems, slaughter weight and sex on organoleptic properties and fatty acid composition of lamb. *J. Anim. Sci.*, **51**, 321–330.

Kirton, A.H., Clarke, J.N. and Hickey, S.M. (1982). A comparison of the composition and carcass quality of Kelly and Russian castrate, ram, wether and ewe lambs. *Proc. NZ Soc. Anim. Prod*, **42**, 117–118.

Klettner, P., Pollein, H. and Ott, G. (1989). Processing of old sheep in the meat industry. *Fleischwirtsch.*, **69**, 1810–1812.

Kunsman, J.E. and Riley, M.L. (1975). A comparison of hydrogen sulfide evolution from cooked lamb and other meats. *J. Food Sci.*, **40**, 506–508.

Lawrie, R.A. (1985). The eating quality of meat. In *Meat Science*, 4th edn. Pergamon Press, Oxford, pp. 189–207.

Lewis, M.J. and Williams, A.A. (1980). Potential artifacts from using porous polymers for collecting aroma components. *J. Sci. Food Agric.*, **31**, 1017–1026.

Lingnert, H. and Eriksson, C.E. (1981). Antioxidant effect of Maillard reaction products. *Prog. Food Nutr. Sci.*, **5**, 453–466.

Locker, R.H. (1980). In *Developments in Meat Science – 1*, ed. R. Lawrie. Applied Science, London, pp. 181–193.

Locker, R.H. and Moore, V.J. (1977). Lamb ham and bacon. *Food Technol. NZ*, **12**(4), 27–31.

Lopez, V. and Lindsay, R.C. (1993a). Metabolic conjugates as precursors for characterizing flavor compounds in ruminant milks. *J. Agric. Food Chem.*, **41**, 446–454.

Lopez, V. and Lindsay, R.C. (1993b). Thermal stability of some metabolically conjugated potential precursors of flavor components in meat and milk. *J. Food Sci.*, **58**, 292–298.

MacLeod, G. and Ames, J.M. (1986a). Capillary gas chromatography–mass spectrometric analysis of cooked ground beef aroma. *J. Food Sci.*, **51**, 1427–1433.

MacLeod, G. and Ames, J.M. (1986b). The effect of heat on beef aroma: comparisons of chemical composition and sensory properties. *Flavour Fragrance J.*, **1**, 91–104.

MacLeod, G. and Seyyedain-Ardebili, M. (1981). Natural and simulated meat flavours (with particular reference to beef). *CRC Crit. Rev. Food Sci. Nutr.*, **14**, 309–437.

Macy, R.L., Naumann, H.D. and Bailey, M.E. (1964a). Water-soluble flavor and odor precursors of meat. II. Effects of heating on amino nitrogen constituents and carbohydrates in lyophilized diffusates from aqueous extracts of beef, pork, and lamb. *J. Food Sci.*, **29**, 142–148.

Macy, R.L., Naumann, H.D. and Bailey, M.E. (1964b). Water-soluble flavor and odor precursors of meat. I. Qualitative study of certain amino acids, carbohydrates, non-amino acid nitrogen compounds, and phosphoric acid esters of beef, pork, and lamb. *J. Food Sci.*, **29**, 136–141.

Marston, H.R., Allen, S.H. and Smith, R.M. (1972). Production within the rumen and removal from the blood-stream of volatile fatty acids in sheep given a diet deficient in cobalt. *Br. J. Nutr.*, **27**, 147–157.

Meilgaard, M., Civille, G.V. and Carr, B.T. (1987). *Sensory Evaluation Techniques*. CRC Press, Boca Raton, FL.

Mendenhall, V.T. and Ercanbrack, S.K. (1979). Effect of carcass weight, sex and breed on consumer acceptance of lamb. *J. Food Sci.*, **44**, 1063–1066.
Miller, G.J., Field, R.A. and Agboola, H.A. (1986). Lipids in subcutaneous tissues and longissimus muscles of feedlot and grass-fed ewes. *J. Food Qual.*, **9**, 39–47.
Misock, J.P., Campion, D.R., Field, R.A. and Riley, M.L. (1976). Palatability of heavy ram lambs. *J. Anim. Sci.*, **42**, 1440–1444.
Moody, W.G. (1983). Beef flavor - a review. *Food Technol.*, **37**(5), 227–232.
Mottram, D.S. (1994). In *Flavor of Meat and Meat Products*, ed. F. Shahidi. Blackie Academic & Professional, London, pp. 210–230.
Munday, R. and Winterbourn, C.C. (1989). Reduced glutathione in combination with superoxide dismutase as an important biological defence mechanism. *Biochem. Pharmacol.*, **38**, 4349–4352.
Mussinan, C.J., Wilson, R.A., Katz, I., Hruza, A. and Vock, M.H. (1976). In *Phenolic, Sulfur and Nitrogen Compounds in Food Flavours*, eds. G. Charalambous and I. Katz. ACS Symposium Series 26, American Chemical Society, Washington, DC, pp. 133–145.
Nicol, A.M. and Jagusch, K.T. (1971). The effect of different types of pasture on the organoleptic qualities of lambs. *J. Sci. Food Agric.*, **22**, 464–466.
Nixon, L.N. (1981). A comparison of the effects of grass and legume pasture on sheep meat flavour. *NZ J. Agric. Res.*, **24**, 277–279.
Nixon, L.N., Wong, E., Johnson, C.B. and Birch, E. (1979). Nonacidic constituents of volatiles from cooked mutton. *J. Agric. Food Chem.*, **27**, 355–359.
Oddy, V.H. (1980). Cobalt supplementation of whole wheat diets for weaner lambs. *Aust. Soc. Anim. Prod.*, **13**, 221–224.
Olafsdottir, G., Steinke, J.A. and Lindsay, R.C. (1985). Quantitative performance of a simple Tenax–GC adsorption method for use in the analysis of aroma volatiles. *J. Food Sci.*, **50**, 1431–1436.
Park, R.J. and Minson, D.J. (1972). Flavour differences in meat from lambs grazed on tropical legumes. *J. Agric. Sci.*, **79**, 473–478.
Park, R.J., Corbett, J.L. and Furnival, E.P. (1972). Flavour differences in meat from lambs grazed on lucerne (*Medicago sativa*) or phalaris (*Phalaris tuberosa*) pastures. *J. Agric. Sci.*, **78**, 47–52.
Park, R.J, Ford, A., Minson, D.J. and Bacter, R.I. (1975). Lucerne-derived flavour in sheep meat as affected by season and duration of grazing. *J. Agric. Sci.*, **84**, 209–213.
Park, R.J., Ford, A.L. and Ratcliff, D. (1978a). Use of protected lipid supplement to modify the flavor of mutton. *J. Food Sci.*, **43**, 874–881.
Park, R.J., Ford, A.L. and Ratcliff, D. (1978b). A study of 'sweet' flavor in lamb produced by feeding protected sunflower seed. *J. Food Sci.*, **43**, 1363–1367.
Paul, A.A. and Southgate, D.A.T. (1978). In McCance and Widdowson's *The Composition of Foods*. Her Majesty's Stationery Office, London.
Paul, P.C., Torten, J. and Spurlock, G.M. (1964). Eating quality of lamb. *Food Technol.*, **18**(11), 121–124.
Pearson, A.M., Wenham, L.M., Carse, W.A., McLeod, K., Davey, C.L. and Kirton, A.H. (1973). Observations on the contribution of fat and lean to the aroma of cooked beef and lamb. *J. Anim. Sci.*, **36**, 511–515.
Purchas, R.W., O'Brien, L.E. and Pendleton, C.M. (1979). Some effects of nutrition and castration on meat production from male Suffolk cross (Border Leicester–Romney cross) lambs. *NZ J. Agric. Res.*, **22**, 375–383.
Purchas, R.W., Johnson, C.B., Birch, E.J., Winger, R.J., Hagyard, C.J. and Keogh R.G. (1986). *Flavour Studies with Beef and Lamb*. Massey University, Palmerston North, New Zealand.
Raymer, J.H. and Pellizzari, E.D. (1987). Toxic organic compound recoveries from 2,6-diphenyl-p-phenylene oxide porous polymer using supercritical carbon dioxide and thermal desorption methods. *Anal. Chem.*, **59**, 1043–1048.
Reid, D.H., Young, O.A. and Braggins, T.J. (1992). Sulphide evolution from cooking mutton and beef. In *Proceedings of the 38th Int. Congr. Meat Sci. Technol.*, Clermont-Ferrand, France, pp. 579–582.
Reid, D.H., Young, O.A. and Braggins, T.J. (1993). The effects of antioxidative treatments on mutton flavour/odour intensity and species flavour differentiation. *Meat Sci.*, **35**, 171–182.

Rhee, K.S. (1988). Enzymic and nonenzymic catalysis of lipid oxidation in muscle foods. *Food Technol.*, **42**(6), 127–132.

Rousset-Akrim, S., Young, O.A. and Berdagué, J.-L. (1997). Diet and growth effects in panel assessment of sheepmeat odour and flavour. *Meat Sci.*, **45**, 169–181.

Rubin, L.J. and Shahidi, F. (1988). Lipid oxidation and the flavour of meat products. In *Proc. 34th Int. Congr. Meat Sci. Technol.*, Brisbane. Cannon Hill, CSIRO, pp. 295–301.

Sato, K., Hegarty, G.R. and Herring, H.K. (1973). The inhibition of warmed-over flavor in cooked meats. *J. Food Sci.*, **38**, 398–403.

Scaife, J.R. and Garton, G.A. (1975). Methylmalonate as a precursor *in vitro* of branched-chain fatty acids. *Biochem. Soc. Trans.* **3**, 1011–1012.

Schutte, L. (1974). Precursors of sulfur-containing flavor compounds. *CRC Crit. Rev. Food Technol.*, **4**, 457–505.

Shorland, F.B., Czochanska, Z., Moy, M., Barton, R.A. and Rae, A.L. (1970). Influence of pasture species on the flavour, odour and keeping quality of lamb and mutton. *J. Sci. Food Agric.*, **21**, 1–4.

Smith, M. and Young, O.A. (1991). Spicing up sheepmeat. *Food Technol. NZ*, **26**(3), 25–29.

Snyder, J.M. and Mounts, T.L. (1990). Analysis of vegetable oil volatiles by multiple head-space extraction. *J. Am. Oil. Chem. Soc.*, **67**, 800–803.

Spillane, C. and L'Estrange, J.L. (1977). The performance and carcass fat characteristics of lambs fattened on concentrate diets. *Ir. J. Agric. Res.*, **16**, 205–219.

St. Angelo, A.J., Vercellotti, J.R., Legendre, M.G., Vinnett, C.H., Kuan, J.W., James, C. and Dupy, H.P. (1987). Chemical and instrumental analyses of warmed-over flavor of beef. *J. Food Sci.*, **52**, 1163–1168.

Sutherland, M.M. and Ames, J.M. (1995). The effect of castration on the headspace aroma components of cooked lamb. *J. Sci. Food Agric.*, **69**, 403–413.

Sutherland, M.M. and Ames, J.M. (1996). Free fatty acid composition of the adipose tissue of intact and castrated lambs slaughtered at 12 and 30 weeks of age. *J. Agric. Food Chem.*, **44**, 3113–3116.

Suzuki, J. and Bailey, M.E. (1985). Direct sampling capillary GLC analysis of flavor volatiles from ovine fat. *J. Agric. Food Chem.*, **33**, 343–347.

Vercellotti, J.R., Mills, O.E., Bett, K.L. and Sullen, D.L. (1992). Gas chromatographic analyses of lipid oxidation volatiles in foods. In *Lipid Oxidation in Food,* ed. A.J. St. Angelo. ACS Symposium Series, American Chemical Society, Washington, DC, pp. 232–265.

Vimini, R.J., Field, R.A., Crouse, J.D. and Miller, G.J. (1984). Factors affecting melting point of subcutaneous fat from heavy ram and wether lambs. *Int. Goat Sheep Res.*, **2**, 105–113.

Wasserman, A.E. and Talley, F. (1968). Organoleptic identification of roasted beef, veal, lamb and pork as affected by fat. *J. Food Sci.*, **33**, 219–223.

Weller, M., Galgan, M.W. and Jacobson, M. (1962). Flavor and tenderness of lamb as influenced by age. *J. Anim. Sci.*, **21**, 927–929.

Wenham, L.M. (1974). Studies in ewe mutton quality – palatability of beef and mutton patties. *NZ J. Agric. Res.*, **17**, 203–205.

West, J., Quigley, R. and Chrystall, B.B. (1997). *Composition of New Zealand Foods 6, Beef and Lamb*, Institute for Crop and Food Research, Palmerston North, New Zealand, ed. J. Bass. The Conference Committee, Hamilton.

Wong, E. (1975). Mutton flavour. *Food Technol. NZ*, **10**(1), 13–15.

Wong, E., Nixon, L.N. and Johnson, C.B. (1975a). Volatile medium chain fatty acids and mutton flavor. *J. Agric. Food Chem.*, **23**, 495–498.

Wong, E., Johnson, C.B. and Nixon, L.N. (1975b). The contribution of 4-methyloctanoic (hircinoic) acid to mutton and goat meat flavour. *NZ J. Agric. Res.*, **18**, 261–266.

Young, O.A., Reid, D.H. and Scales, G.H. (1993). Effect of breed and ultimate pH on the odour and flavour of sheepmeat. *NZ J. Agric. Res.*, **36**, 363–370.

Young, O.A., Agnew, M.P. and Fraser, T.J. (1997a). Volatile branched chain fatty acids in fat from two sheep breeds. *Proc. 43rd Int. Congr. Meat Sci. Technol.*, Auckland.

Young, O.A., Berdagué, J.-L., Viallon, C., Rousset-Akrim, S. and Theriez, M. (1997b). Fat-borne volatiles and sheepmeat odour. *Meat Sci.*, **45**, 169–181.

Yu, T.-H., Lee, M.-H., Wu, C.-M. and Ho, C.-T. (1994). In *Lipids in Food Flavors*, eds. C.-T. Ho and T.G. Hartman. ACS Symposium Series 558, American Chemical Society, Washington, DC, pp. 61–76.

7 Flavour of fish meat

E. DURNFORD and F. SHAHIDI

7.1 Introduction

The flavour of fish is composed of taste which is comprised of nonvolatile taste-active compounds and odour comprised of volatile compounds. The nonvolatile taste-active compounds are low-molecular-weight extractive components. These compounds are more abundant in the tissues of molluscs and crustaceans than fish which explains why shellfish are considered to be more tasty than finfish. The extractive or nonvolatile taste-active compounds may be divided into two broad groups: nitrogenous compounds, including free amino acids, low-molecular-weight peptides, nucleotides and related compounds, and organic bases; and non-nitrogenous compounds, including organic acids, sugars and inorganic constituents.

The aromas associated with very fresh fish are usually mild, delicate and pleasant (Lindsay, 1990). These aromas are generally described as green, plant-like or melon-like and are provided by various carbonyls and alcohols (Josephson and Lindsay, 1986) along with iodine-like odours in marine fish which are contributed by bromophenols (Boyle *et al.*, 1992). Some species such as salmon have very characteristic fresh odours (Josephson *et al.*, 1991b).

However, the aromas of fish are very perishable and the study of off-odours is therefore important. In the process of deterioration, the very fresh fish odours may be destroyed by microbial and autolytic activity or new compounds produced may mask the very fresh fish aromas. Processing can also have a dramatic impact on the aroma of fish. Besides deterioration, environmental factors can impart off-flavours to fish.

7.2 Very fresh fish aromas

7.2.1 Carbonyls and alcohols

The aroma of very fresh fish may vary considerably among species but most fish have a common sweet and plant-like aroma that is easily recognized and associated with fresh fish. This fresh fish flavour is due to volatile carbonyls and alcohols which are derived from the polyunsaturated fatty acids of fish lipids via specific lipoxygenase activity (Josephson and Lindsay, 1986).

Josephson *et al.* (1984a,b) conducted a cross-species comparison of volatile aroma compounds from freshly harvested freshwater and saltwater fish which revealed a common occurrence of hexanal, 1-octen-3-ol, 1,5-octadien-3-ol, and 2,5-octadien-1-ol. Freshwater fish were also found to contain 1-octen-3-one and 1,5-octadien-3-one. Six of the 12 freshwater species analysed, but none of the saltwater species, contained (E)-2-hexenal, 2-octenal, 2-octen-1-ol, 2,3-octanedione, (E)-2-nonenal, (E,Z)-2,6-nonadienal and 3,6-nonadien-1-ol. Table 7.1 summarizes the volatile carbonyls and alcohols associated with freshly harvested fish aroma.

The nine-carbon compounds (E)-2-nonenal, (E,Z)-2,6-nonadienal and 3,6-nonadien-1-ol are responsible for the cucumber-like, melon-like odour of fresh fish. Berra *et al.* (1982) identified (E,Z)-2,6-nonadienal in Australian grayling (*Prototroctes maraena*) and the flavour isolates of rainbow smelt (*Osmerus mordax*) were described as having a cucumber-like aroma, which was supported by the identification of (E)-2-nonenal, (E,Z)-2,6-nonadienal, 6-nonen-1-ol and 3,6-nonadien-1-ol in this species (Josephson *et al.*, 1984a,b,c).

Six-carbon compounds have also been identified in freshly harvested fish (Josephson and Lindsay, 1986; Josephson *et al.*, 1983a; Suyama *et al.*, 1985) but are not found in all seafoods (Josephson and Lindsay, 1986). Hexanal contributes a distinct coarse, green, plant-like, aldehydic aroma note to immediately harvested finfish, but within minutes is blended with the aromas of the eight- and nine-carbon volatiles (Josephson *et al.*, 1983a).

The five-carbon compound 1-penten-3-ol is also found in all freshwater fish (Josephson and Lindsay, 1986). However, the level of 1-penten-3-ol in fish is lower than its recognition threshold and is therefore not likely

Table 7.1 Volatile carbonyls and alcohols associated with freshly harvested fish aroma

Alcohols	Carbonyls
1-Penten-3-ol[c,d]	(E)-2-Penten-1-al[d]
(Z)-3-Hexen-1-ol[c]	Hexan-1-al[c,d]
1-Octen-3-ol[a–d]	(E)-2-Hexen-1-al[c]
(E)-2-Octen-1-ol[b–d]	(E)-2-Octen-1-al[a–c]
1,5-Octadien-3-ol[a–d]	(E)-2-Nonen-1-al[a–c]
2,5-Octadien-1-ol[a–d]	(E,Z)-2,6-Nonadien-1-al[a–d]
(E)-2-Nonen-1-ol[b]	1-Octen-3-one[c,d]
(Z)-6-Nonen-1-ol[b,c]	2,3-Octadien-1-one[c]
(Z)-3-Nonen-1-ol[b]	1,5-Octadien-3-one[c,d]
(E,Z)-2,6-Nonadien-1-ol[b]	
3,6-Nonadien-1-ol[b,c]	

[a]Hsieh and Kinsella, 1989.
[b]Zhang *et al.*, 1992b,c.
[c]Josephson *et al.*, 1984a,b.
[d]German *et al.*, 1991.

to be an important contributor to the characteristic aroma of freshly harvested fish (Josephson and Lindsay, 1986).

Generally, the volatile carbonyls contribute coarse, heavy aromas whereas the volatile alcohols contribute smoother qualities. Additionally, the carbonyls contribute more to the overall fresh fish-like odours than do their corresponding alcohols because of their lower threshold values (Josephson and Lindsay, 1986).

Several authors have proposed that nicotinamide adenine dinucleotide (NADH)-dependent microsomal oxidase (Shewfelt *et al.*, 1981; Slabyj and Hultin, 1984) and a myeloperoxidase (Kanner and Kinsella, 1983; Kanner *et al.*, 1986) are involved in the formation of the hydroperoxide precursors of the volatile carbonyls and alcohols responsible for fresh fish flavour. However, production of these products without concurrent production of random volatile oxidation products supports the hypothesis of fresh fish flavour volatiles generated via specific lipoxygenase activity.

Figure 7.1 describes the mechanisms involved in the biogeneration of aroma from eicosapentaenoic acid. It is evident that volatile carbonyls and alcohols are important contributors to freshly harvested fish aroma.

The eight-carbon volatile alcohols and carbonyls contribute distinct fresh plant-like aromas, even though these compounds individually exhibit

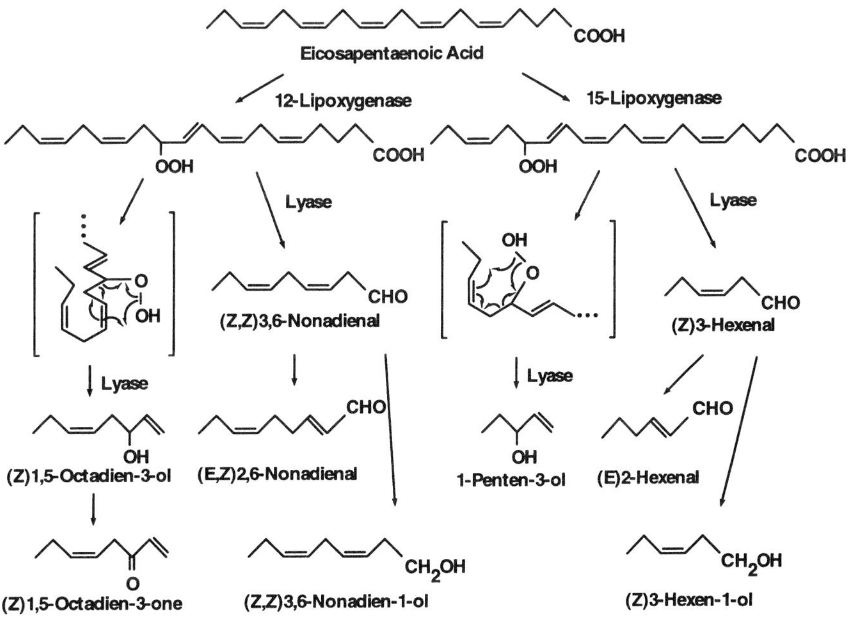

Figure 7.1 Proposed mechanism for enzymatic biogeneration of volatile carbonyls and alcohols important to freshly harvested fish aroma from eicospentaenoic acid (adapted from Josephson and Lindsay, 1986).

Table 7.2 Some volatile aroma compounds and their associated aroma quality

Compound	Description
Hexan-1-al	Green, aldehyde[a]
(E)-2-Nonen-1-al	Cucumber, cardboard[a,c]
(E,Z)-2,6-Nonadien-1-al	Cucumber peel[a,c]
(E,Z)-3,6-Nonadien-1-al	Cucumber, melon rind[a,c]
1-Octen-3-ol	Mushroom[a]
(Z)-2-Octen-1-ol	Fatty, rancid[b]
(Z)-1,5-Octadien-3-ol	Earthy, mushroom[a]
(E,Z)-2,5-Octadien-1-ol	Earthy, mushroom[a]
1-Octen-3-one	Mushroom[a]
(Z)-1,5-Octadien-3-one	Geranium leaves[a]

[a]Josephson *et al.*, 1983.
[b]Pyysalo and Suihko, 1976.
[c]Hirano *et al.*, 1992.

a mushroom and geranium-like aroma. Table 7.2 lists some of the volatile aroma compounds of fresh fish and associated description of individual compounds.

The substrates for the production of the volatile carbonyls and alcohols are the polyunsaturated fatty acids. The lipoxygenase found in fish exhibits the same selectivity towards arachidonic, eicosapentaenoic and docosahexaenoic acids but shows negligible response to linoleic and linolenic acids (Zhang *et al.*, 1992b; Hsieh *et al.*, 1988, 1992c). However, lipoxygenases in plants have specificity for linoleic and linolenic acids (Minamide, 1977; Sessa, 1979; MacLeod and Pikk, 1979).

The specific volatile compounds generated by the lipoxygenase are dependent on the substrate (Zhang *et al.*, 1992b; Hsieh and Kinsella, 1989). For example, if eicosapentaenoic acid or docosahexaenoic acid is the substrate, the compounds 1,5-octadien-3-ol, (E,Z)-2,6-nonadienal, 2,5-octadien-1-ol and 3,6-nonadien-1-ol are the flavour volatiles generated. However, from arachidonic acid, (E)-2-octenal, 1-octen-3-ol, (E)-2-nonenal, (E)-2-octenol and (Z)-3-nonenol are produced (Zhang *et al.*, 1992b).

7.2.2 Sulphur compounds

Volatile sulphur compounds are usually associated with deteriorated seafood. However, there is evidence that sulphur compounds can be produced in fish (Josephson *et al.*, 1986b) and may contribute to aromas that characterize the odours of some fresh seafoods. Dimethyl sulphide is one of the volatile sulphur compounds known to provide a pleasant seashore-like smell in fresh seafoods (Iida, 1988). When dimethyl sulphide is in low concentration (<100 ppb), it gives a pleasant crab-like aroma; however, at higher concentrations, it is perceived as having an offensive odour (Iida, 1988). The formation of methyl mercaptan, dimethyl disulphide and dimethyl sulphide in the flathead (*Calliurichthys doryssus*) at the time of harvest has been reported (Shiomi *et al.*, 1982).

The proposed pathways provided are different from those given by Josephson *et al.* (1984a,b,c) but are consistent with the current data. Further evidence for the proposed pathway is provided by the identification of a 12-lipoxygenase in the skin and gill tissue of trout which can oxidize polyunsaturated fatty acids into position specific hydroperoxides (Hsieh and Kinsella, 1989; German *et al.*, 1986; German and Kinsella, 1985). Lipoxygenase-like activity has also been found in crude extracts from skin and gill of wild and cultured ayu (Zhang *et al.*, 1992a, b) and smelt (Zhang *et al.*, 1992c). The hydroperoxides are decomposed to produce various fragmentation products such as the volatile carbonyls and alcohols which contribute to the aroma of freshly harvested fish (Hsieh *et al.*, 1988).

As shown in Figure 7.1, the 12-lipoxygenase produces hydroperoxides that fragment into the eight- and nine-carbon volatile carbonyls and alcohols. According to the pathways proposed, a 15-lipoxygenase is required for the biogenesis of the five- and six-carbon volatile carbonyls and alcohols (Josephson and Lindsay, 1986). The existence of a 15-lipoxygenase in trout gill was confirmed by German *et al.* (1992)

The 12- and 15-lipoxygenases are not distributed equally amongst fish species (German *et al.*, 1992). For example, trout exhibit very high 12- and virtually undetectable 15-lipoxygenase activities, whereas in carp, although the 12-lipoxygenase is most active, the 15-lipoxygenase is also relatively abundant. In sturgeon, the 15-lipoxygenase is actually the predominant enzyme (German *et al.*, 1992). The existing differences in concentration of different lipoxygenases might account for some of the observed variations in the flavour spectrum of different species.

7.2.3 Bromophenols

Several naturally occurring bromophenols have been reported to be responsible for the iodine-like off-flavour in Australian prawns (Whitfield *et al.*, 1988). However, low concentrations of these bromophenols may be responsible for the desirable brine- or sea-like aromas associated with many saltwater fish (Boyle *et al.*, 1992).

The very potent 2,6-dibromophenol (5×10^{-4} μg/l threshold in water) responsible for the iodine-like off-flavour in prawns (Whitfield *et al.*, 1988) has not been detected in saltwater salmon. However, Pacific salmon harvested from saltwater has been reported to contain several other bromophenols (Table 7.3), with 2,4,6-tribromophenol being the most abundant isomer.

When Pacific salmon migrate from saltwater to brackish water or freshwater environments there is a loss of the high-quality sea-like flavour

Table 7.3 Concentrations of bromophenols (ng/g) in sexually immature Pacific salmon (*Oncorhynchus sp.*) harvested from saltwater and marine fish

Species	2-BP	3-/4-BP	2,4-DBP	2,4,6-TBP
Pink salmon	–	–	–	32.1
Sockeye salmon	1.4	–	0.8	7.4
Chinook salmon	1.6	–	–	33.2
Coho salmon	–	1.0	–	5.1
US pickled herring	1.6	–	2.2	3.7
European brine-cured herring	38.0	–	–	13.5
Icelandic haddock	–	–	2.7	4.5

Source: Adapted from Boyle *et al.* (1992); symbols are BP, bromophenol; DBP, dibromophenol; and TBP, tribromophenol.

(Boyle *et al.*, 1992). Several authors have suggested that this loss of flavour is due to the cessation of feeding and mobilization of muscle lipids and carotenoid pigments into the gonads and skin (Josephson *et al.,* 1991a,b; Kitahara, 1984; Hatano *et al.*, 1983; Ota and Yamada, 1974). However, Boyle *et al.* (1992) found that bromophenols were virtually absent from sexually mature freshwater salmon lacking brine- or sea-like flavours and concluded that the bromophenols were depurated from the saltwater salmon upon cessation of feeding. Bromophenols are virtually absent in freshwater fish (Boyle *et al.*, 1992).

7.2.4 Hydrocarbons

The hydrocarbons (E,Z)- and (E,E)-1,3,5-octatriene have been found in spawning condition salmon and other non-salmonid freshwater fish (Josephson, 1991). The contribution of these unsaturated hydrocarbons to seafood flavour has not been studied, but may be significant because 1,3-octadiene exhibits a mushroom, humus-like aroma (Persson and Juttner, 1983).

7.3 Species-specific characteristic aromas

7.3.1 Canned tuna

Some fish species have distinct characteristic aromas. Canned tuna has an aroma different from other canned fish and is often described as meaty. One of the compounds identified in canned tuna that has an intense beef extract aroma is 2-methyl-3-furanthiol (Withycombe and Mussinan, 1988). 2-Methyl-3-furanthiol along with other similar compounds produces the rich meaty flavour of canned tuna (Withycombe and Mussinan, 1988).

7.3.2 Salmon

Salmon is also recognized as having a distinctive rich flavour that can be described as salmon-like or salmon loaf-like. These distinct flavour compounds are believed to be associated with carotenoid pigments and other lipid substances. The carotenoids either serve as a direct precursor to the salmon loaf-like flavour or they modulate chemical reactions which convert fatty acids or other lipid precursors to the salmon loaf-like compound (Josephson *et al.*, 1991b). The salmon loaf-like aroma compound is considered to be one of the alkyl furanoid-type structures as shown in Figure 7.2.

7.3.3 Sweet smelt

Other fish such as ayu (sweet smelt) also have a characteristic aroma in the raw state. Ayu (*Plecoglossus altivelis*) possesses a sweet smell like the aroma of watermelon or cucumber. The compounds (E,Z)-2,6-nonadienal and (Z)-3-hexenol play an important role in the characteristic odour of wild ayu (Suyama *et al.*, 1985).

7.4 Derived or process-induced flavours

7.4.1 Canning

Many flavours in fish are developed during heat processing. As previously discussed, the compound 2-methyl-3-furanthiol gives canned tuna its characteristic meaty flavour (Withycombe and Mussinan, 1988). This compound is produced by the thermally mediated reaction between ribose and cysteine (Lindsay, 1990). Tuna, containing an abundant supply of ribose from the degradation of ribonucleotides, produces 2-methyl-3-furanthiol when heated during the canning process (Lindsay, 1990).

7.4.2 Dried and salted fish

In a study on the odour of dried and salted marine products, volatile bases had no correlation with odour whereas there was a close relationship between volatile fatty acid concentration and odour. However, volatile

Figure 7.2 Alkyl furanoid-type structure of salmon loaf-like aroma compound (adapted from Josephson *et al.*, 1991b).

carbonyl compounds are considered to be the most important contributor to the odour of dried and salted fish (Nakamura *et al.*, 1982). A later study of several kinds of dried and smoked fish identified 142 volatile compounds (Sakakibara *et al.*, 1990a). Katsuobushi (dried bonito) is a traditional flavour enhancer regularly used in Japanese cuisine and is known to have superior flavour qualities. The final product is produced through boiling, sun-drying, smoking and moulding which develop a complex flavour composition. One study identified 237 volatile compounds in katsuobushi (Sakakibara *et al.*, 1990b).

7.4.3 *Smoked fish*

Smoking is a common processing method for various fish species. The principal reason for smoking fish is flavour improvement and several studies have investigated the flavour of various smoked fish (Kasahara and Nishibori, 1979a,b, 1982). A study of the volatile components of smoked salmon identified 16 phenols, 17 acids, an ester, an alcohol and three hydrocarbons. Of these, phenols such as guaiacol, 4-methylguaiacol, 4-ethylguaiacol, 2,6-dimethoxyphenol and 4-methyl-2,6-dimethoxyphenol were considered most important in the aroma of smoked salmon (Kasahara and Nishibori, 1979a).

7.4.4 *Pickled fish*

Another common processing method for fish is pickling. Pickled fish refers to fish that are treated with salt brine and acidified. A study on the influence of the pickling process on the volatile flavour compounds of fish found that fresh fish volatile carbonyls and alcohols are extracted into the brine during processing. Therefore, only traces of carbonyls and modest amounts of alcohols remain in the pickled fish and it is the alcohols that are more influential in the flavour of pickled fish (Josephson *et al.*, 1987). Hence, the flavour of pickled fish is less pronounced in fishiness character than either unprocessed fresh fish or abused, oxidized fish (Josephson *et al.*, 1987). High-quality pickled fish have a mild, but distinct fresh fish-like flavour, similar to fresh, green-plant-like flavours (Josephson *et al.*, 1987).

7.4.5 *Fermented fish*

Fermented fish sauce is a processed product that is common in South-East Asia. This product is prepared by mixing small, uneviscerated fish with salt, at a ratio of three to one, and fermented. During production of fish sauce, the flavour developed is usually described as cheesy, ammoniacal and meaty (Beddows *et al.*, 1976). The cheesy odour is produced by volatile fatty acids while the ammoniacal odour is due to ammonia and amines

(Dougan and Howard, 1975). The meaty aroma is complex and varies with the origin of the sauce (Beddows *et al.*, 1976). Dougan and Howard (1975) identified acetic, propionic, n-butyric and isobutyric, and isovaleric acids as specific volatile acids contributing to the aroma of fermented fish sauce with acetic and n-butyric acids being the two major components. Sanceda *et al.* (1983) reported that propionic and n-butyric acids were the major acids and identified additional volatile acids with two to ten carbon atoms, both n-acids and iso-acids. Sanceda *et al.* (1983) concluded that the volatile acids appear to be responsible for the cheesy aroma and rancid odour of fermented fish sauce. McIver *et al.* (1982) reported that the neutral fraction of a fish sauce extract which possessed a meaty aroma contained three lactones as the main components along with alcohols, heterocyclic compounds and benzaldehyde.

Fish enzymes, microorganisms and fat oxidation have all been considered as possible contributors to the development of fermented fish sauce aroma (Beddows *et al.*, 1976). In a study on the use of enzymes on the hydrolysis of mackerel and the investigation of fermented fish aroma, Beddows *et al.* (1976) concluded that bacteria play an important role in the development of the cheesy aroma of fermented fish sauce obtained from mackerel.

Anchovies are usually eaten salted and cured (ripened). The maturing or ripening process is thought to have a significant impact on the final product. As anchovies ripen, their contents of 2,4-heptadienal and (E,Z)-3,5-octadien-2-one increase (Triqui and Reineccius, 1995a). The anchovy flavour is due to both the enzymatically generated C_8 alcohols and ketones along with (E,Z)-2,6-nonadienal, which contribute plant- and cucumber-like aromas, and autoxidatively derived C_7–C_{10} conjugated aldehydes, which impart fatty and fried fat-like aromas to products (Triqui and Reineccius, 1995b). More recently, Cha *et al.* (1997) reported 98 volatile compounds in salt-fermented anchovy with 1-octen-3-one, (Z)-4-heptenal, (E,Z)-2,6-nonadienal, 3-methylbutanal, 3-(methylthio)propanal, ethyl 2-methylbutanoate and ethyl 3-methylbutanoate being the most potent odorants.

7.4.6 Cooking

The characteristic aroma of fish which develops upon cooking is due to a mixture of low-molecular-weight aldehydes and browning reaction products (Shibamoto and Horiuchi, 1997). The browning-flavour compounds are produced from the interactions of carbonyls and amines (Shibamoto, 1983). In the absence of amines, nC_1–C_{10} saturated aldehydes and some branched aldehydes are detected in the headspace of heated fish oil. However, with TMAO added as an amine precursor, the compounds N,N-dimethylformamide and N-methylpyrrole are formed which make a strong contribution to the cooked flavour of fish oil (Shibamoto and Horiuchi, 1997).

Some fish in the process of cooking generate distinct cooked odours. In Japanese conger, sardine and pale chub the compounds 2-phenylethanol, 1-penten-3-ol and dimethyl sulphide were specific components which contributed to the roasted aroma of these fish, respectively (Kasahara and Nishibori, 1985). Sardines, upon cooking, generally produce a strong undesirable odour. Various aldehydes, alcohols, hydrocarbons and fatty acids are believed to contribute to the strong odour produced when sardine is cooked (Koizumi *et al.*, 1979; Nakamura *et al.*, 1980b).

7.5 Deteriorated fish odours

Compounds causing off-flavours in fish and their control have been extensively reviewed (Obata *et al.*, 1950; Wyatt and Day, 1963; Meijboom and Stroink, 1972; McGill *et al.*, 1974; Ke *et al.*, 1975; Kikuchi *et al.*, 1976; Reineccius, 1979; Ross and Love, 1979; Josephson *et al.*, 1983b; Hsieh *et al.*, 1989; Kasahra *et al.*, 1989, 1990; Kawai, 1990; Haard, 1992). These compounds may arise from the environment or through deterioration. Environmentally derived odour compounds will be discussed in a later section.

7.5.1 Trimethylamine and related compounds

Historically, trimethylamine (TMA) and dimethylamine (DMA) have been associated with the odour of deteriorating fish (Hebard *et al.*, 1982; Regenstein *et al.*, 1982). Trimethylamine is produced by the reduction of trimethylamine oxide (TMAO) during microbial spoilage. Dimethylamine and formaldehyde are produced from the breakdown of TMAO by enzymes in the muscles of various fish species (Hebard *et al.*, 1982; Regenstein *et al.*, 1982).

Trimethylamine oxide is used for osmoregulation in fish in saltwater environments (Love, 1970) and is therefore absent from freshwater species (Hebard *et al.*, 1982). In marine species, elasmobranchs have the highest levels of TMAO followed by gadoids, pelagics and flatfish (Herbard *et al.*, 1982).

Trimethylamine oxide has no odour whereas TMA is a very potent odour compound described as 'old-fishy' or fish house-like (Lindsay, 1991) with a very low threshold (300–600 ppb) (Amoore *et al.*, 1975; Kikuchi *et al.*, 1976; Ikeda, 1979). Trimethylamine itself, however, is not responsible for the fishy odour as it smells like ammonia in its purified form (Hebard *et al.*, 1982). Trimethylamine reacts with fat in the fish tissue to produce the fishy odour (Davies and Gill, 1936). Dimethylamine exhibits an ammoniacal aroma and is less fishy than TMA (Lindsay, 1990).

In unfrozen fish, TMA production is much greater than the production of DMA (Castell *et al.*, 1973). In gadoid fish, the formation of DMA precedes that of TMA (Amano and Yamada, 1964) and in frozen gadoids that are kept at high subfreezing temperatures, DMA and formaldehyde are produced enzymatically, but TMA formation is prevented due to inhibition of microbial growth (Castell *et al.*, 1973).

Many studies have positively correlated levels of TMA and sensory scores of unfrozen marine fish (Dussault, 1957; Spencer, 1962; Farber, 1965; Magno-Orejana *et al.*, 1971; Sen Gupta *et al.*, 1972) and thus TMA is often used as a spoilage index for unfrozen fish. However, production of DMA may serve as a better measure of deterioration in frozen gadoids (Castell *et al.*, 1970).

7.5.2 Autoxidation

Autoxidizing fish lipids have long been linked to the production of fishy flavours in both chill-stored and frozen fish. Similar compounds are formed in fish and fish oil as a result of autoxidation of polyunsaturated fatty acids (Josephson *et al.*, 1984c, 1986b; Karahadian and Lindsay, 1989a,b).

The oxidized aromas in autoxidized fish oils vary from just perceptible to extremely unpleasant fish oil-like odours. Initially, aromas described as green or cucumber-like (Karahadian and Lindsay, 1989a) arise, but as oxidation progresses odours described as fishy, cod liver oil-like, or burnt are developed (Lindsay, 1990).

The initial aromas are due to the production of short-chain saturated and unsaturated aldehydes and include hexanal and (E)-2-hexenal. The most important contributors to the fishy and cod liver oil-like aromas are (E,Z,Z)-2,4,7-decatrienal and (E,E,Z)-2,4,7-decatrienal (Meijboom and Stroink, 1972: Karahadian and Lindsay, 1989a). Ke *et al.* (1975) reported the presence of 2,4,7-decatrienals in autoxidized mackerel oil.

The 2,4,7-decatrienals are derived from autoxidation of long-chain polyunsaturated ω-3 fatty acids which are abundant in fish lipids. The compound 2,4,7-decatrienal and other aldehydes are produced via β-scission of alkoxy radicals generated by the homolytic cleavage of each isomer of the hydroperoxides (Fujimoto, 1989) (Figure 7.3).

It is proposed that the H-donating character of tocopherol-type compounds causes a preferential formation of *cis-trans* rather than *trans-trans* monohydroperoxide that provides the direct precursors of the 2,4,7-decatrienals. A stepwise mechanism for the formation of *trans,trans*-hydroperoxide compared to *trans,cis*-hydroperoxide is provided in Figure 7.4.

7.5.3 (Z)-4-Heptenal

Deteriorated aromas that develop in cod and related species have also been associated with the compound (Z)-4-heptenal (McGill *et al.*, 1974,

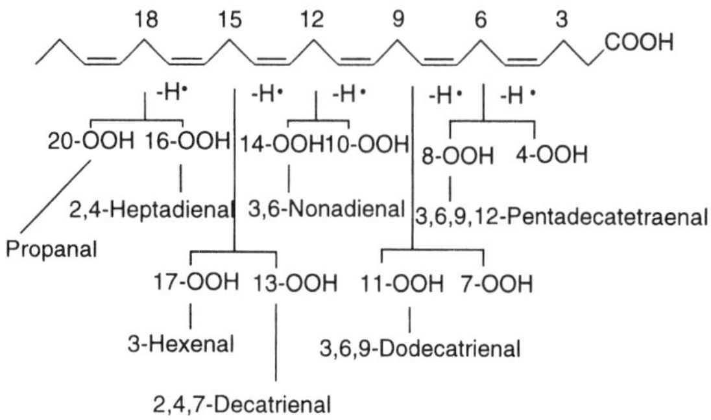

Figure 7.3 Formation of 2,4,7-decatrienal and other aldehydes from autoxidation of docosa hexaenoic acid.

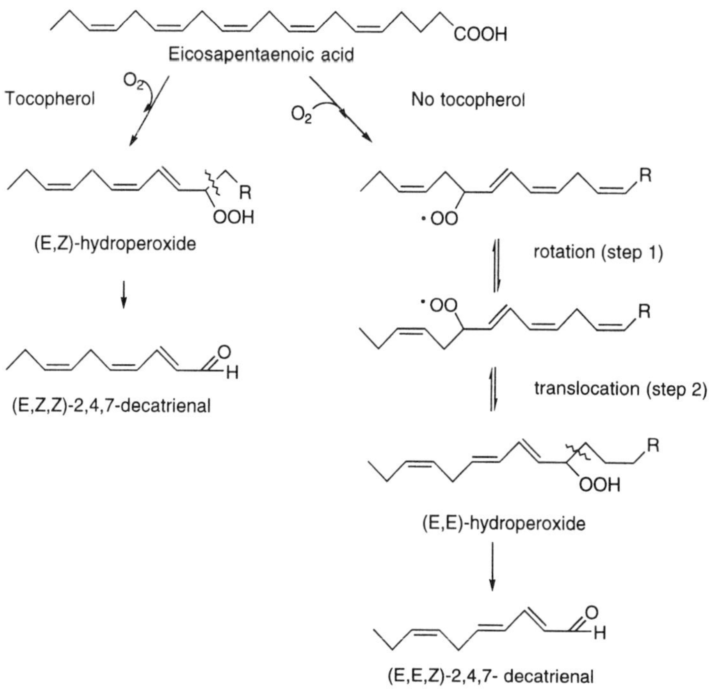

Figure 7.4 Formation of isomeric hydroperoxides of polyunsaturated fatty acids (adapted from Karahadian and Lindsay, 1989b).

1977; Hardy *et al.*, 1979). This compound does not contribute distinct fishy-type flavours but rather it potentiates the stale, burnt/fishy, cod liver oil-like flavour contributed by the 2,4,7-decatrienals (Karahadian and Lindsay, 1989a). At low concentrations in water (Z)-4-heptenal exhibits a cardboardy character while at higher concentration the aroma is more putty-, paint- or linseed oil-like (Lindsay, 1990). The aroma of (Z)-4-heptenal has also been described as cold boiled potato-like and is believed to be responsible for much of the aroma of boiling potatoes (Josephson and Lindsay, 1987a).

(Z)-4-Heptenal is produced by the water-mediated retro-aldol condensation of (E,Z)-2,6-nonadienal and the proposed mechanism is shown in Figure 7.5 (Josephson and Lindsay, 1987b).

The production of (Z)-4-heptenal is accelerated with increased temperatures and at high pH (Josephson and Lindsay, 1987b) and is therefore commonly found in cooked, stored seafoods (McGill *et al.*, 1974, 1977).

7.5.4 Volatile acids

During the storage of fish, various volatile acids are formed. Kikuchi *et al.* (1976) have reported that formic, acetic, propionic, n- and isobutyric, and n- and isovaleric acids are formed in fish flesh during storage. A study

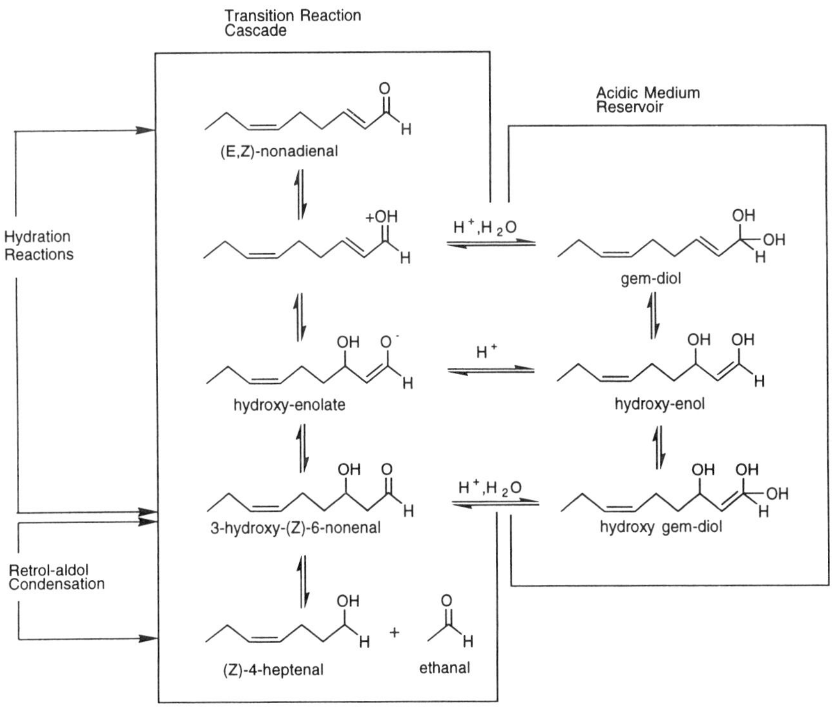

Figure 7.5 Proposed mechanism for the formation of (Z)-4-heptenal from (E,Z)-2,6-nonadienal via alpha/beta double bond hydration and retro-aldol condensation (adapted from Josephson and Lindsay, 1987b).

on oxidized sardine oil found that propionic acid followed by acetic acid were dominant (Table 7.4).

Although the concentrations of butyric and valeric acid are much lower, their lower odour thresholds make them more important than other acids (Kikuchi *et al.*, 1976). The short-chain volatile acids give very intense and objectionable sweaty odours and are considered important markers for flavour quality of fish oil (Hsieh *et al.*, 1989). However, Karahadian and Lindsay (1989b) concluded that short-chain fatty acids found in oxidizing fish were of insignificant concentrations to contribute characterizing burnt/fishy flavours and aromas.

7.5.5 Other compounds

Karahadian and Lindsay (1989a) identified a compound in fish oils described as fish bowl-like or minnow bucket-like when eluted via packed-column gas chromatography. This compound was identified as 5,8,11-tetradecatrien-2-one.

During the advanced stages of spoilage of chill-stored cod, off-odours described as 'sulphidy', 'hydrogen sulphide', 'stale cabbage' and 'mercaptan-like' develop (Herbert *et al.*, 1975). Herbert *et al.* (1975) reported that hydrogen sulphide, methyl mercaptan and dimethyl sulphide are responsible for the 'sulphidy' off-flavours associated with chill-stored cod in the advanced stages of spoilage. These volatile sulphides result from the microbial degradation of free cysteine and methionine in the fish muscle (Herbert and Shewen, 1975).

In oxidized cod liver oil, nonadienal, (E)-2-hexenal and 1,5-octadien-3-one have been identified as contributing to a green aroma and the distinctly fishy odour of fish oils (Karahadian and Lindsay, 1989a). In oxidized sardine oil 1,5-octadien-3-hydroperoxide has also been identified as a green aroma compound (Wada and Lindsay, 1992).

Table 7.4 Contents of volatile fatty acids in oxidized sardine oil and their corresponding odour thresholds

	Content (ppm)[a]	Odour threshold (ppm)[b]
Acetic acid	959	34.2
Propionic acid	1270	32.8
n-Butyric acid	17.4	3
Isobutyric acid	8.4	9.2
n-Valeric acid	13.5	1.1
Isovaleric acid	13.3	1.7
n-Caproic acid	500	7.5
Isocaproic acid	54.5	–

[a]Nakamura *et al.*, 1980.
[b]Kikuchi et *al.*, 1976.

In a study of changes in the odorants of boiled trout (*Salmo fario*) during storage it was found that the concentrations of (Z)-3-hexenal and (Z,Z)-3,6-nonadienal increased to levels which contributed strongly to the fatty, fishy off-flavour of boiled trout (Milo and Grosch, 1993). A later study on boiled cod and trout reported that an increase of acetaldehyde, propionaldehyde, butane-2,3-dione, pentane-2,3-dione and C_6, C_8 and C_9 carbonyl compounds in trout and trimethylamine, butane-2,3-dione, methylpropanal, and 2- and 3-methylbutanal in cod contributed to the development of off-flavours (Milo and Grosch, 1995). In subsequent studies on boiled salmon and cod, the off-flavours in cod were attributed to an increase in the concentration of 3-methylbutanal while in salmon the off-flavours were due to an increase in the concentrations of (Z)-3-hexenal, 2,6-nonadienal and (Z,Z)-3,6-nonadienal (Milo and Grosch, 1996, 1997).

Several studies have investigated the control of fish deterioration odours. Josephson *et al.* (1983) reported that addition of sodium bisulphite (100–500 ppm) to water extracts of slime from fresh and oxidized whitefish (*Coregonus clupeaformis*) suppressed fishy aromas. The sodium bisulphite reduces the fishy aroma by reacting with aldehydes and many ketones that contribute to fresh and oxidized fish aromas to form nonvolatile adducts with bisulphite (Josephson *et al.*, 1983b) The general scheme for the reaction of aldehydes and unhindered ketones with bisulphite ion is shown below (Scheme 7.1).

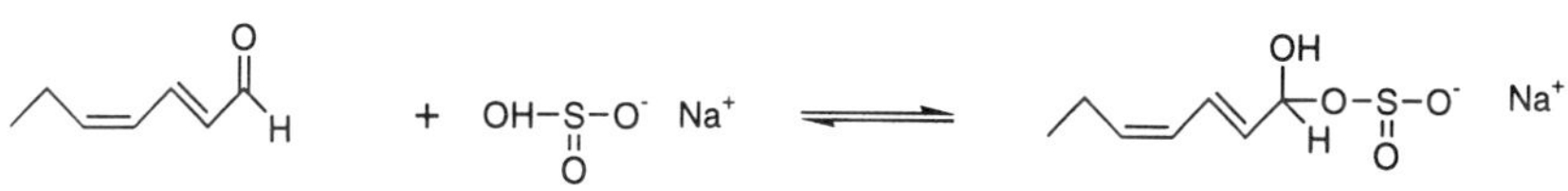

Scheme 7.1

Trimethylamine has historically been associated with off-flavours in fish (Hebard *et al.*, 1982; Regenstein *et al.*, 1982). It has been reported that *dl*-[3-amino-3-carboxypropyl] dimethyl sulphonium chloride (also known as vitamin U chloride) can suppress the fishy odours of spoiling fish (Kawai *et al.*, 1990). It was proposed that the compound reacts with trimethylamine to reduce the fishy odour, as given below.

$$(CH_3)_2S^+(Cl^-)CH_2CH_2CH(NH_2)COOH + (CH_3)_3N$$
$$\rightarrow CH_3SCH_2CH_2CH(NH_2)COOH + (CH_3)_4N^+Cl^-$$

Efforts have also been made to improve sardine odour by adding soy sauce flavouring or Mirin flavourings (Kasahara *et al.*, 1989, 1990) and the suppression of the odour of roasted sardine has been achieved using lemon juice (Kasahara and Nishibori, 1992).

7.6 Environmentally derived flavours

7.6.1 Muddy off-flavours

Many flavours present in fish are due to environmental factors. Musty, muddy, earthy and mouldy are common off-flavours in wild (Kuusi and Suihko, 1983) and farmed fish populations (Yurkowski and Tabachek, 1980) that are caused by environmental factors. Geosmin (Lovell *et al.*, 1986) and 2-methylisoborneol (Martin *et al.*, 1987) are the two primary chemical compounds responsible for the musty or earthy flavours. Two other compounds identified as 2-methylenebornane and 2-methyl-2-bornene, which are dehydration products of 2-methylisoborneol, were also believed to cause the musty off-flavour in catfish (Martin *et al.*, 1988). However, Mills *et al.* (1993) reported that these dehydration products were not responsible for off-flavours in catfish as they do not have discernible odours and are often present in catfish free of earthy or musty off-flavours.

The chemical structures of geosmin ((E)-1,10-dimethyl-(E)-9-decalol) (Gerber, 1968) and 2-methylisoborneol (1,2,7,7-tetramethyl-exo-bicyclo-heptan-2-ol) (Schrader and Blevins, 1993) are shown in Figure 7.6(a) and (b), respectively.

Geosmin and 2-methylisoborneol are secondary metabolites produced by various actinomycetes (Medsker *et al.*, 1968; Gerber, 1967, 1968, 1969) and cyanobacteria (Schrader and Blevins, 1993; Matsumoto and Tsuchiya, 1988; Negoro *et al.*, 1988; Tabachek and Yurkowski, 1976; Martin *et al.*, 1991). It has been shown that geosmin can be absorbed through the gills, skin, small intestine and stomach of trout with absorption being most rapid through the gills and slowest through the stomach (From and Horlyck, 1984). Johnsen *et al.* (1996) reported that temperature is an important factor in the rate of absorption and depuration of 2-methylisoborneol in catfish.

Geosmin has been identified as the compound causing the earthy off-flavour in channel catfish (Lovell and Sackey, 1973) whereas 2-methylisoborneol is the compound that causes the musty off-flavour in channel catfish (Martin *et al.*, 1987, 1988). Other cultured species such as bream

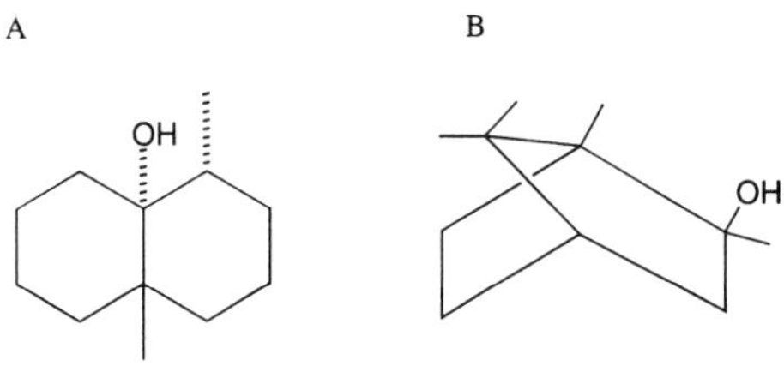

Figure 7.6 Chemical structure of geosmin (A) and 2-methylisoborneol (B) (adapted from Schrader and Blevins, 1993).

(Persson, 1979), trout (Yurkowski and Tabachek, 1974) and shrimp (Lovell and Broce, 1985) and various wild commercial species such as walleye, northern pike, cisco and lake whitefish (Yurkowski and Tabachek, 1980) have also been tainted with these earthy or musty off-flavour compounds.

7.6.2 'Blackberry' off-flavour

Another off-flavour problem associated with environmental conditions is referred to as the 'blackberry' problem in cod from the Labrador area of Canada. This condition results when cod consume invertebrates locally called 'blackberry' making the fillets smell unpleasant. Sipos and Ackman (1964) associated the off-flavour with dimethyl sulphide. The authors proposed that the dimethyl sulphide smell originated in algae and was passed to the invertebrates and then on to the fish when they consumed the invertebrates.

7.6.3 Environmental pollutants

The flavour of fish may also become tainted due to environmental pollutants. In a study by Lindsay and Heil (1992) fish harvested from the Upper Wisconsin River had a pronounced chemical, petroleum, phenolic and sulphurous flavour. They identified several alkylphenols and thiophenol as the compounds responsible for this flavour taint. The alkylphenols (2-isopropyl-, 3-isopropyl-, 4-isopropyl-, 2,4-diisopropyl-, 2,5-diisopropyl-, 2,6-diisopropyl-, 3,5-diisopropyl-, 5-methyl-2-isopropyl- and 2-methyl-5-isopropyl-) and thiophenol were reported to be the principal contributors to the flavour tainting (Lindsay and Heil, 1992).

Lindsay and Heil (1992) concluded that thiophenol entered the river through discharges from paper mills. However, these authors proposed that alkylphenols were formed in the environment from precursors, such as diterpenes or phenolic glycoside conjugates, which are produced by paper pulping activities.

Petroleum substances can have serious flavour tainting effects on exposed fish. Toluene and benzene have been identified as substances that cause offensive flavours in fish (Ogata and Miyake, 1973). The odour and flavour characteristics of salmonids have been shown to be affected by contamination with crude oil especially when dispersants are used (Martinsen *et al.*, 1992). Ackman *et al.* (1996) reported that a portion of the water-soluble fraction of crude petroleum oil, which is rich in methyl- and alkyl-substituted monoaromatic and low-molecular-weight polyaromatic hydrocarbons and has a strong petroleum flavour, is dissolved into adipocyte cells of farmed salmon. These substances are retained much longer in the adipocyte cells than in intercellular fluids where they are rapidly depurated.

7.7 Nonvolatile nitrogenous compounds

The previous sections have dealt with various volatile compounds that contribute to the odour rather than the taste of fish. The taste of fish is dependent on its extractive components. The extractive components are defined as water-soluble, low-molecular-weight compounds and they are divided into two broad groups: nitrogenous compounds and non-nitrogenous compounds (Fuke and Konosu, 1991). The nitrogenous extractive components include free amino acids, low-molecular-weight peptides, nucleotides and related compounds, urea and quateranary ammonium compounds (Konosu and Yamaguchi, 1982; Haard *et al.*, 1994). The distribution of nonprotein nitrogenous compounds in a teleost and elasmobranch is provided in Table 7.5.

7.7.1 Free amino acids and related compounds

The free amino acid content of fish is relatively low when compared to shellfish. However, some authors have reported that certain free amino acids can occur in fish muscle at high enough concentrations to contribute to fish flavour, independent of other constituents. There are reports that glycine contributes to the sweetness of fish (Amano and Bito, 1951) and histidine contributes to the 'meaty' character of some seafoods (Simidu *et al.*, 1953). Others argue that the individual free amino acids in fish such as gadoids are at levels below their flavour thresholds and are therefore unlikely to be important contributors to flavour.

The most notable feature of free amino acid contents in fish is the high content of histidine in makerel and tuna and a high taurine content in white-fleshed fish (Konosu and Yamaguchi, 1982). A study of wild and cultured red sea bream reported that wild fish had a higher content of many of the free amino acids, except for histidine, than their cultured counterparts (Morishata *et al.*, 1989). However, other studies on the extractive components of wild and cultured fish, including red sea bream (Konosu and Watanabe, 1976), yellowtail (Endo *et al.*, 1974) and ayu

Table 7.5 Distribution (%) of non-protein nitrogen compounds in a teleost (mackerel) and elasmobranch (shark)

Class of compounds	Mackerel	Shark
Free amino acids	25	5
Peptides	5	5
Nucleotides	10	5
Creatine and creatinine	35	10
TMAO	15	20
Urea	–	55
Ammonia and amides	10	–

Source: Adapted from Finne (1992).

(Suyama *et al.*, 1970), have concluded that the composition of extractive components, including free amino acids, of wild and cultured fish is similar.

Taurine, a major constituent of white-fleshed fish, has been reported to be slightly bitter (Jones, 1967). More important than the individual contributions of free amino acids to flavour is the mutual enhancement of flavour by the free amino acid fraction and nucleotides. An inter-relationship between glutamic acid and adenosine 5′-monophosphate has been demonstrated to provide the meaty character of some fish and between mononucleotides and glycine, alanine, glutamic acid and methionine in sea urchin gonads (Jones, 1967).

Carnosine and anersine are two of a limited number of peptides that have been identified in the extracts of fish and their structures are shown in Figure 7.7.

Carnosine is abundant in eel and skipjack while anserine is abundant in tuna, skipjack and some species of shark (Konosu and Yamaguchi, 1982). Several studies on beef and pork indicate that carnosine may be a naturally occuring antioxidant that would have an impact on flavour development (Chan *et al.*, 1993; Decker and Crum, 1993; Decker *et al.*, 1995)

Peptides and free amino acids are important contributors to the flavour of fish sauce. Raksakulthai and Haard (1992) reported that the typical flavour of fish sauce was correlated with large peptides and free amino acids. Major amino acid residues in peptides were aspartic acid, serine, glutamic acid and leucine (Raksakulthai and Haard, 1992).

7.7.2 *Nucleotides and related compounds*

Nucleotides and related compounds are important because of their palatable taste (umami)-producing factors. Umami taste of seafood has been reviewed (Komata, 1990; Fuke, 1994). In the muscle of live fish, adenosine triphosphate (ATP) predominates but shortly after death it is enzymatically degraded according to the pathway shown in Figure 7.8.

In this degradation pathway, the reaction inosine monophosphate (IMP) → inosine is slow and IMP usually accumulates in fresh fish muscle (Konosu and Yamaguchi, 1982). IMP is a desirable flavour enhancer in fish extracts (Murata and Sakaguchi, 1989). IMP is at highest concentration

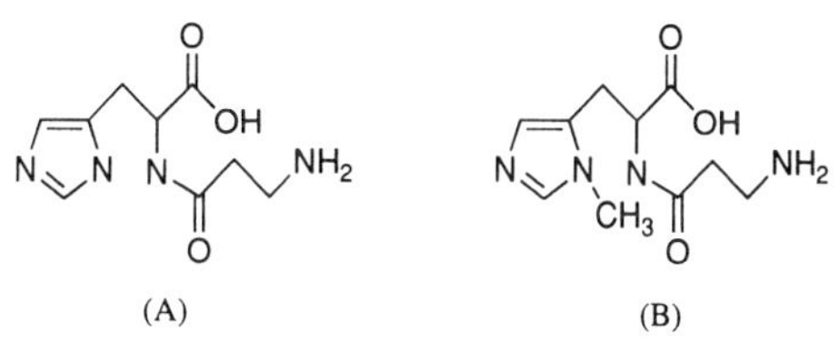

(A) (B)

Figure 7.7 Structures of carnosine (β-alanylhistidine) (A) and anserine (β-alanyl-l-methyl histidine) (B) (adapted from Konosu and Yamaguchi, 1982).

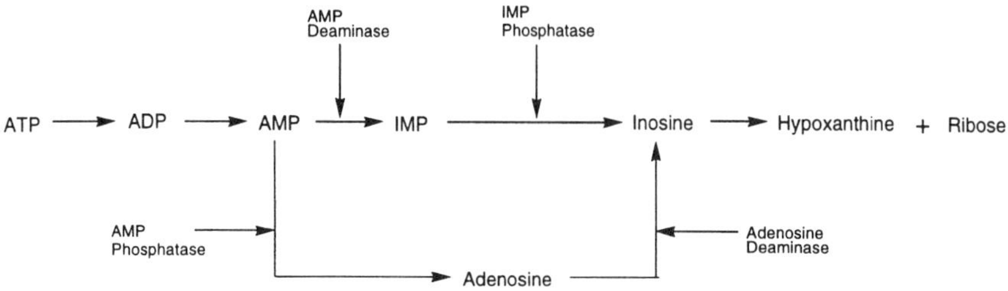

Figure 7.8 The postmortem enzymatic degradation of ATP (adapted from Komata, 1990), where ATP = adenosine triphosphate; ADP = adenosine diphosphate; AMP = adenosine monophosphate; and IMP = inosine monophosphate.

within one to two days postmortem and as its concentration decreases the flavour of the fish becomes less acceptable (Fletcher and Statham, 1988a,b). In contrast to the desirable sweet or salty taste contributed by IMP, hypoxanthine, the end-product of nucleotide degradation contributes a bitter taste to muscle foods (Bremner *et al.*, 1988). Measurement of nucleotides and calculation of K values (Saito *et al.*, 1959; Karube *et al.*, 1984) can be used to determine the freshness of fish (Greene and Bernatt-Byrne, 1990) and other seafoods (Shahidi *et al.*, 1994).

Creatine (Figure 7.9(A)), a guanidino compound is often found in high amounts in fish while creatinine, a dehydration product of creatine (Figure 7.9(B)), is usually found in much smaller quantities.

7.7.3 *Urea and quaternary ammonium compounds*

Urea is present in small quantities in tissues of all fish. Marine elasmobranchs, however, contain relatively high amounts (1–2.5%) of urea for osmoregulation (Haard *et al.*, 1994). Urea has no flavour, but it is readily decomposed to ammonia and carbon dioxide. Bacterial urease catalyses this reaction and the pungent odour of the resulting ammonia may contribute to unacceptable quality of fish (Finne, 1992).

Trimethylamine oxide, a quateranary ammonium compound, commonly found in marine teleosts and elasmobranchs, has no odour or taste. However, the breakdown products of TMAO have very potent odours which contribute to fish spoilage (see section 7.5).

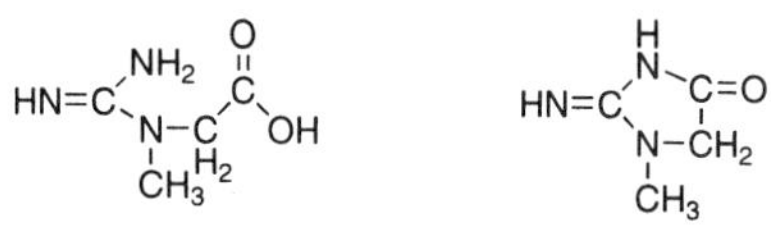

(A) (B)

Figure 7.9 Structures of creatine (A) and creatinine (B) (adapted from Konosu and Yamaguchi, 1982).

7.8 Nonvolatile non-nitrogenous compounds

Very few studies have investigated nonvolatile non-nitrogenous components in fish when compared to the studies of the nitrogenous compounds. Some of the acids that have been found in fish extracts include acetic, propionic, lactic, pyruvic, succinic and oxalic acids. Lactic acid, which is produced through glycolysis, is the main acid and can reach high concentrations in active fish such as tuna and skipjack (Konosu and Yamaguchi, 1982).

7.8.1 Sugars and related compounds

Most fish contain some free glucose and ribose whereas fructose occurs in some species of fish (Jones, 1958). When plaice are stored on ice, their ribose content increases (Ehira and Uchiyama, 1967). In addition to free sugars, various sugar phosphates and inositol, a sugar alcohol, are also known to occur in fish (Konosu and Yamaguchi, 1982). Konosu and Yamaguchi (1982) concluded that the content of free sugars and their derivatives in fish muscle are fairly low, so it is unlikely that they contribute to fish flavour. However, another study found that the sugar phosphates at levels equal to maximum concentrations in cod tasted 'sweetish-salty' according to a taste panel (Jones, 1961).

7.8.2 Inorganic salts

Very little work has been done on the flavour of inorganic salts. However, the cations Na^+ and K^+ and the anions Cl^- and PO_4^{3-} are believed to be important contributors to fish flavour (Fuke and Konosu, 1991).

7.9 Summary

The area of fish flavours, especially volatiles, has been the subject of renewed studies. As a result, compounds such as various C_8- and C_9-carbonyls and alcohols have been elucidated as being responsible for the characteristic sweet, plant-like aroma in freshly harvested fish. More recently, the efforts have been concentrated on identifying the pathways for the formation of aroma-active compounds. Many of the data reported to date have been supportive of the hypothesis that such products are produced from polyunsaturated fatty acids via lipoxygenase-mediated oxidation.

Fresh fish flavours are very delicate and can be easily destroyed or masked by deteriorative odours. Fresh fish flavours are very unstable under abusive conditions, upon which undesirable flavour compounds associated with deterioration are produced.

Processing of fish also affects its flavour. Sometimes processing results in the production of desirable aromas while undesirable flavours may be produced under different conditions. Similarly, environmental factors can positively or negatively affect flavour. For example, saltwater fish have a different flavour to freshwater fish and this is most likely due to the presence of bromophenols in the marine environment. These compounds give a flavour that is indicative of high-quality saltwater fish. Examples of off-flavours contributed by environmental factors include muddy off-flavour in catfish, sulfurous off-odour in cod feeding on 'blackberry', and off-flavour contributed by man-made pollution.

The taste of fish is also important. However, fish contain much less taste-active compounds than shellfish and are therefore considered to be much less tasteful. As a result, much of the research into taste-active compounds of seafood have focused on shellfish.

References

Ackman, R.G., Heras, H. and Zhou, S. (1996). Salmon lipid storage sites and their role in contamination with water-soluble petroleum materials. *J. Food Lipids*, **3**, 161–170.

Amano, K. and Bito, M. (1951). Consequence of free amino acids generated from decomposing fish muscle. *Bull. Jpn. Soc. Sci. Fish.*, **16**, 10–16.

Amano, K. and Yamado, K. (1964). A biological formation of formaldehyde in the muscle tissue of gadoid fish. *Bull. Jpn. Soc. Sci. Fish.*, **30**, 430–435.

Amoore, J.E., Forrester, L.J. and Buttery, R.G. (1975). Specific anosmia to 1-pyrroline: The spermous primary odor. *J. Chem. Ecol.*, **1**, 299–302.

Beddows, C.G., Ismail, M. and Steinkraus, K.H. (1976). The use of bromelain in the hydrolysis of mackerel and the investigation of fermented fish aromas. *J. Food Technol.*, **11**, 379–388.

Berra, T.M., Smith, J.F. and Morrison, M. (1982). Probable identification of the cucumber odor of the Australian grayling (*Prototroctes maraena*). *Trans. Am. Fish. Soc.*, **111**, 78–82.

Boyle, J.L., Lindsay, R.C. and Stuiber, D.A. (1992). Bromophenol distribution in salmon and selected seafoods of fresh- and saltwater origin. *J. Food Sci.* **57**, 918–922.

Bremner, H.A.M., Olley, J., Statham, J.A. and Vail, A.M. (1988). Nucleotide catabolism: Influence on the storage life of tropical species of fish from the North West Shelf of Australia. *J. Food Sci.*, **53**, 6–11.

Castell, C.H., Neal, W. and Smith, B. (1970). Formation of dimethylamine in stored frozen sea fish. *J. Fish. Res. Bd. Can.*, **27**, 1685–1690.

Castell, C.H., Neal, W.B. and Dale, J. (1973). Comparison of changes in trimethylamine, dimethylamine and extractable protein in iced and frozen gadoid fillets. *J. Fish. Res. Bd. Can.*, **30**, 1246–1248.

Cha, Y.J., Lee, G.H. and Cadwallader, K.R. (1997). Aroma-active compounds in salt-fermented anchovy. In *Flavor and Lipid Chemistry of Seafoods*, eds. F. Shahidi and K.R. Cadwallader. American Chemical Society, Washington, DC, pp. 131–147.

Chan, K.M., Decker, E.A. and Means, W.J. (1993). Extraction and activity of carnosine, a naturally occurring antioxidant in beef muscle. *J. Food Sci.*, **58**, 1–4.

Davies, W.L. and Gill, E. (1936). Investigations on fishy flavor. *J. Soc. Chem. Ind.*, **55**, 141–146.

Decker, E.A. and Crum, A.D. (1993). Antioxidant activity of carnosine in cooked ground pork. *Meat Sci.*, **34**, 245–253.

Decker, E.A., Chan, W.K.M., Livisay, S.A., Butterfield, D.A. and Faustman, C. (1995). Interactions between carnosine and the different redox states of myoglobin. *J. Food Sci.*, **60**, 1201–1204.

Dougan, J. and Howard, G.E. (1975). Some flavoring constituents of fermented fish sauces. *J. Sci. Food Agric.*, **26**, 887–894.

Dussault, H.P. (1957). TMA test for evaluating the quality of rosefish fillets. *Fish Res. Bd. Can. Progr. Rep. Atl. Coast Stn. 67.*

Ehira, S. and Uchiyama, H. (1967). Change in the ribose content of plaice muscle during ice storage. *Bull. Jpn. Soc. Sci. Fish.*, **33**, 126–140.

Endo, K., Kishimoto, Y. and Shimizu, Y. (1974). Seasonal variations in chemical constituents of yellowtail muscle. II. Nitrogenous extracts. *Bull. Jpn. Soc. Sci. Fish.*, **40**, 67–72.

Farber, L. (1965). Freshness tests. In *Fish as Food*, ed. G. Borgstrom. Academic Press, New York, NY, pp. 65–126.

Finne, G. (1992). Non-protein nitrogen compounds in fish and shellfish. In *Advances in Seafood Biochemistry, Composition and Quality*, eds. G.J. Flick and R.E. Martin. Technomic Publishing, Lancaster, PA, pp. 393–401.

Fletcher, G.C. and Statham, J.A. (1988a). Deterioration of sterile chill-stored and trumpeter fish (*Latridopsis forsteri*). *J. Food Sci.*, **53**, 1336–1339.

Fletcher, G.C. and Statham, J.A. (1988b). Shelf-life of sterile yellow-eyed mullet (*Aldrichetta forsteri*). *J. Food Sci.*, **53**, 1030–1035.

From, J. and Horlyck, V. (1984). Sites of uptake of geosmin, a cause of earthy-flavor in rainbow trout (*Salmo gairdneri*). *Can. J. Fish. Aquat. Sci.*, **41**, 1224–1226.

Fujimoto, K. (1989). Flavor chemistry of fish oils. In *Flavor Chemistry of Lipid Foods*, eds. D.B. Min and J.H. Smouse. American Oil Chemists' Society, Champaign, IL, pp. 190–195.

Fuke, S. (1994). Taste-active components of seafoods with special reference to umami substances. In *Seafoods: Chemistry, Processing Technology and Quality*, eds. F. Shahidi and J.R. Botta. Blackie Academic & Professional, Glasgow, pp. 115–139.

Fuke, S. and Konosu, S. (1991). Taste-active components in some foods: A review of Japanese research. *Phys. Behav.*, **49**, 863–868.

Gerber, N.N. (1967). Geosmin, an earthy smelling substance isolated from actinomycetes. *Biotechnol. Bioeng.*, **9**, 321–327.

Gerber, N.N. (1968). Geosmin from microorganisms is *trans*-1,10-dimethyl-*trans*-9-decalol. *Tetrahedron Lett.*, **26**, 2971–2974.

Gerber, N.N. (1969). A volatile metabolite of actinomycetes, 2-methylisoborbeol. *J. Antibiot.*, **22**, 508–509.

German J.B. and Kinsella, J.E. (1985). Lipid oxidation in fish tissue. Enzymatic initiation via lipoxygenase. *J. Agric. Food Chem.*, **33**, 680–683.

German, J.B., Bruckner, G. and Kinsella, J.E. (1986). Lipoxygenases in trout gill tissue affecting on arachidonic, eicosapentaenoic, docosahexaenoic acids. *Biochim. Biophys. Acta*, **879**, 378–387.

German, J.B., Berger, R.G. and Drawert F. (1991). Generation of fresh fish flavour: Rainbow trout (*Salmo gairdner*) gill homogenate as a model system. *Chem. Mikrobiol. Technol. Lebensm.*, **13**, 19–24.

German, J.B., Zhang, H. and Berger, R. (1992). Role of lipoxygenases in lipid oxidation in foods. In *Lipid Oxidation in Foods*, ed. A.J. St. Angelo. American Oil Chemical Society, Washington, DC, pp. 74–92.

Greene, D.H. and Bernatt-Byrne, E.I. (1990). Adenosine triphosphate catabolites as flavor compounds and freshness indicators in Pacific cod (*Gadus macrocephalus*) and pollock (*Theragra chalcogramma*). *J. Food Sci.*, **55**, 257–258.

Haard, N.F. (1992). Biochemical reactions in fish muscle during frozen storage. In *Seafood Science and Technology*, ed. E.G. Bligh. Blackwell Scientific, Cambridge, MA, pp. 176–209.

Haard, N.F., Simpson, B.K. and Sun Pan, B. (1994). Sarcoplasmic proteins and other nitrogenous compounds. In *Seafood Proteins*, eds. Z.E. Sikorski, B. Sun Pan and F. Shahidi. Chapman & Hall, London, pp. 13–39.

Hardy, R., McGill, A.S. and Gunstone, F.D. (1979). Lipid autoxidative changes in cold stored cod (*Gadus morhua*). *J. Sci. Food Agric.*, **30**, 999.

Hatano, M., Takama, K., Kojima, H. and Zama, K. (1983). Proximate composition of fall chum salmon. *Bull. Jpn. Soc. Sci. Fish.*, **49**, 213–218.

Hebard, C.E., Flick, G.J. and Martin, R.E. (1982). Occurrence and significance of trimethylamine oxide and its derivatives in fish and shellfish. In *Chemistry and Biochemistry of Marine Food Products*, eds. R.E. Martin, G.E. Flick, C.E. Hebard and D.R. Ward. AVI Publishing Co., Westport, CT, pp. 149–305.

Herbert, R.A. and Shewan, J.M. (1975). Precursors of the volatile sulfides in spoiling North Sea cod (*Gadus morhua*). *J. Sci. Food Agric.*, **26**, 1195–1202.

Herbert, R.A., Ellis, J.R. and Shewan, J.M. (1975). Isolation and identification of the volatile sulfides produced during chill-storage of North Sea cod (*Gadus morhua*). *J. Sci. Food Agric.* **26**, 1187–1194.

Hirano, T., Zhang, C., Morishita, A., Suzuki, T. and Shirai, T. (1992). Identification of volatile components in ayu and its feeds. *Nippon Suisan Gakkaishi*, **58**, 547–557.

Hsieh, R.J., German, J.B. and Kinsella, J.E. (1988). Lipoxygenase in fish tissue: Some properties of the 12-lipoxygenases from trout gill. *J. Agric. Food Chem.*, **36**, 680-685.

Hsieh, R.J. and Kinsella, J.E. (1989). Lipoxygenase generation of specific volatile flavor carbonyl compounds in fish tissues. *J. Agric. Food Chem.*, **37**, 279–286.

Hsieh, T.C.Y., Williams, S.S., Vejaphan, W. and Meyers, S.P. (1989). Characterizationof volatile components of menhaden fish (*Brevoortia tyrannus*) oil. *J. Am. Oil Chem. Soc.*, **66**, 114–117.

Iida, H. (1988). Studies on the accumulation of dimethyl-β-propiothetin and the formation of dimethyl sulfide in aquatic organisms. *Bull. Tokai Reg. Fish. Res. Lab.*, **124**, 35–63.

Ikeda, S. (1979). Other organic components and inorganic components. In *Advances in Fish Science and Technology*, ed. J.J. Connell, (Jubilee Conf. Torry Res. Station, Aberdeen, Scotland). Fishing News Books, Surrey, England, pp. 226–232.

Johnsen, P.B., Lloyd, S.W., Vinyard, B.T. and Dionigi, C.P. (1996). Effects of temperature on the uptake and depuration of 2-methylisoborneol (MIB) in channel catfish *Ictalurus punctatus*. *J. World Aqua. Soc.*, **27**, 15–20.

Jones, N.R. (1958). Free sugars in chilled-stored, trawled cod (*Gadus callarias*) muscle. *J. Sci. Food Agric.*, **9**, 672–677.

Jones, N.R. (1961). Fish flavors. *Proc. Flavor Chem. Symp.*, Campbell Soup Co., Camden, NJ, pp. 61–81.

Jones, N.R. (1967). Fish flavors. In *Chemistry and Physiology of Flavors*, eds. H.W. Schultz, E.A. Day and L.A. Libbey. AVI Publishing Co. Inc., Westport, CT, pp. 267–295.

Josephson, D.B. (1991). Seafood. In *Volatile Compounds in Foods and Beverages*, ed. H. Maarse. Marcel Dekker, New York, pp. 179–202.

Josephson, D.B. and Lindsay, R.C. (1986). Enzyme generation of volatile aroma compounds from fresh fish. In *Biogeneration of Aromas*, eds. T.H. Parliment and R. Croteau. American Chemical Society, Washington, DC, pp. 201–226.

Josephson, D.B. and Lindsay, R.C. (1987a). c4-Heptenal: An influential volatile compound in boiled potato flavor. *J. Food Sci.*, **52**, 328–331.

Josephson, D.B. and Lindsay, R.C. (1987b). Retro-aldol degradations of unsaturated aldehydes: Role in the formation of the c4-heptenal from t2,c6-nonadienal in fish, oyster and other flavors. *J. Am. Oil Chem. Soc.*, **64**, 132–138.

Josephson, D.B., Lindsay, R.C. and Stuiber, D.A. (1983a). Identification of compounds characterizing the aroma of fresh whitefish (*Coregonus clupeaformis*). *J. Agric. Food Chem.*, **31**, 326–330.

Josephson, D.B., Lindsay, R.C. and Stuiber, D.A. (1983b). Bisulfite suppression of fish aromas. *J. Food Sci.* **48**, 1064–1067.

Josephson, D.B., Lindsay, R.C. and Stuiber, D.A. (1984a). Variations in the occurrences of enzymatically derived volatile aroma compounds in salt- and freshwater fish. *J. Agric. Food Chem.*, **32**, 1344–1347.

Josephson, D. B., Lindsay, R.C. and Stuiber, D.A. (1984b). Biogenisis of lipid-derived volatile aroma compounds in the emerald shiner (*Notropis atherinoides*). *J. Agric. Food Chem.*, **32**, 1347–1350.

Josephson D.B., Lindsay, R.C. and Stuiber, D.A. (1984c). Identification of volatile aroma compounds from oxidized frozen whitefish. *Can. Inst. Food Sci. Technol.*, **17**, 178–181.

Josephson, D., Lindsay, R. and Olafsdotter G. (1986). Measurement of volatile aroma constituents as a means of following sensory deterioration of fresh fish and fishery products. In *Seafood Quality Determination*, eds. D.E. Kramer and J. Liston, Elsevier, New York, pp. 27–47.

Josephson, D.B., Lindsay, R.C. and Stuiber, D.A. (1987). Influence of processing on the volatile compounds characterizing the flavor of pickled fish. *J. Food Sci.*, **52**, 10–14.

Josephson, D.B., Lindsay, R.C. and Stuiber, D.A. (1991a). Influence of maturity on the volatile aroma compounds from fresh Pacific and Great Lakes salmon. *J. Food Sci.* **56**, 1576–1579.

Josephson, D.B., Lindsay, R.C. and Stuiber, D.A. (1991b). Volatile carotenoid related oxidation compounds contributing to cooked salmon flavor. *Lebensm. Wiss. u. Technol.*, **24**, 424–432.

Kanner, J. and Kinsella, J. (1983). Lipid deterioration initiated by phagocytic cells in muscle foods: β-carotene destruction by a myeloperoxidase–hydrogen peroxide–halide system. *J. Agric. Food Chem.* **31**, 370–376.

Kanner, J., Harel, S. and Hazan, B. (1986). Muscle membranal lipid peroxidation by an 'iron redox cycle' system: initiation by oxy radicals and site specific mechanisms. *J. Agric. Food Chem.*, **34**, 506–510.

Karahadian, C. and Lindsay, R.C. (1989a) Evaluation of compounds characterizing fishy flavors in fish oils. *J. Am. Oil Chem. Soc.*, **66**, 953–960.

Karahadian, C. and Lindsay, R.C. (1989b). Role of oxidative processes in the formation of and stability of fish flavors. In *Flavor Chemistry Trends and Developments*, eds. R. Teranishi, R.G. Buttery and F. Shahidi. American Chemical Society, Washington, DC, pp. 60–65.

Karube, I., Matsuoka, H., Suzuki, S., Watanabe, E. and Toyama, K. (1984). Determination of fish freshness with an enzyme sensor system. *J. Agric. Food Chem.*, **32**, 314–319.

Kasahara, K. and Nishibori, K. (1979a). Volatile components of smoked salmon – I. *Bull. Jpn. Soc. Sci. Fish.*, **45**, 1543–1545.

Kasahara, K. and Nishibori, K. (1979b). Volatile components of smoked salmon – II: The relationship between flavors of smoked salmon and smoke-tar. *Bull. Jpn. Soc. Sci. Fish.*, **45**, 1547–1549.

Kasahara, K. and Nishibori, K. (1982). Volatile carbonyls and acids of smoked sardine. *Bull. Jpn. Soc. Sci. Fish.*, **48**, 691–695.

Kasahara, K. and Nishibori, K. (1985). Volatile components of roasted fishes. *Bull. Jpn. Soc. Sci. Fish.*, **51**, 489–492.

Kasahara, K. and Nishibori, K. (1992). Suppressing effect of lemon juice on the odor of roasted sardine. *Nippon Suisan Gakkaishi*, **58**, 529–531.

Kasahara, K., Itaya, M. and Nishibori, K. (1989). Effect of mirin flavoring on the improvement of sardine odor. *Nippon Suisan Gakkaishi*, **55**, 715–719.

Kasahara, K., Itaya, M. and Nishibori, K. (1990). Effect of soysauce flavoring on improvement of sardine odor in 'mirin-boshi'. *Nippon Suisan Gakkaishi*, **56**, 619–623.

Kawai, T., Ishida, Y. and Higashida, T. (1990). Elimination of unpleasant odors of putrefied seafoods. *Nippon Suisan Gakkaishi*, **56**, 1013.

Ke, P.J., Ackman, R.G. and Linke, B.A. (1975). Autoxidation of polyunsaturated fatty compounds in mackerel oil: Formation of 2,4,7-decatrienals. *J. Am. Oil Chem. Soc.*, **52**, 349–353.

Kikuchi, T., Wada, S. and Suzuki, H. (1976). Significance of volatile bases and volatile acids in the development of off-flavor of fish meat. *Eiyo to Shokuryo*, **29**, 147–152.

Kitahara, T. (1984) Behavior of carotenoids in the chum salmon (*Oncorhynchus keta*) during anadromous migration. *Comp. Biochem. Physiol.*, **76B**, 97–101.

Koizumi, C., Kieu-Thu, C. and Nonaka, J. (1979). Undesirable odor of cooked sardine meat. *Bull. Jpn. Soc. Sci. Fish.*, **45**, 1307–1312.

Komata, Y. (1990). Umami taste of seafoods. *Food Rev. Int.*, **6**, 457–487.

Konosu, S. and Watanabe, K. (1976). Comparison of nitrogenous extractives of cultured and wild red sea bream. *Bull. Jpn. Soc. Sci. Fish.*, **39**, 645–648.

Konosu, S. and Yamaguchi, K. (1982). The flavor components in fish and shellfish. In *Chemistry and Biochemistry of Marine Food Products*, eds. R.E. Martin, G.E. Flick, C.E. Hebard and D.R. Ward. AVI Publishing Co., Westport, CT, pp. 367–404.

Kuusi, T. and Suihko, M. (1983). Occurrence of various off-flavors in fish in Finland from 1969 to 1981. *Wat. Sci. Technol.*, **15**, 47–58.

Lindsay, R.C. (1990). Fish flavors. *Food Rev. Int.*, **6**, 437–455.

Lindsay, R.C. (1991). Chemical basis of the quality of seafood flavors and aromas. *MTS J.*, **25**, 16–22.

Lindsay, R.C. and Heil, T.P. (1992), Flavor tainting of fish in the Upper Wisconsin River caused by alkyl- and thiophenols. *Wat. Sci. Technol.*, **25**, 35–40.

Love, R.M. (1970). *The Chemical Biology of Fishes*. Academic Press, New York.

Lovell, R.T. and Broce, D. (1985). Cause of musty flavor in pond-cultured penaeid shrimp. *Aquaculture*, **50**, 169–174.

Lovell, R.T. and Sackey, L.A. (1973). Absorption by channel catfish of earthy-musty flavor compounds synthesised by cultures of blue-green algae. *Trans. Am. Fish. Soc.*, **102**, 774–777.
Lovell, R.T., Lelana, J.Y., Boyd, C.E. and Armstrong, M.S. (1986). Geosmin and musty muddy flavors in pond raised channel catfish *Ictalurus punctatus*. *Trans. Am. Fish. Soc.*, **115**, 485–489.
MacLeod, A.J. and Pikk, H.E. (1979). Formation of (E)-hex-2-enal and (Z)-hex-3-en-1-ol by fresh leaves of brassica. *J. Agric. Food Chem.*, **27**, 469–475.
Magno-Oregana, F., Juliano, R.O. and Banasihan, E.T. (1971). Trimethylamine and volatile reducing substances in frigate mackerel (*Auxis thazard lacepeda*). *Philip. J. Sci.*, **100**, 209–226.
Martin, J.F., McCoy, C.P., Greenleaf, W. and Bennet, L. (1987). Analysis of 2-methylisoborneol in water, mud, and channel catfish *Ictalurus punctatus* from commercial culture ponds in Mississippi. *Can. J. Fish. Aquat. Sci.*, **44**, 909–912.
Martin, J.F., McCoy, C.P., Tucker, C.S. and Bennet, C.W. (1988). 2-Methylisoborneol implicated as cause of off-flavor in channel catfish, *Ictalurus punctatus*, (Rafinesque), from commercial culture ponds in Mississippi. *Aquacult. Fish. Manage.*, **19**, 151–157.
Martin, J.F., Izaguirre, G. and Waterstrat, P. (1991). A planktonic *Oscillatoria* species from Mississippi catfish ponds that produces the off-flavor compound 2-methylisoborneol. *Wat. Res.*, **25**, 1447–1451.
Martinsen, C, Lauby, B., Nevissi, A. and Brannon, E. (1992). The influence of crude oil and dispersant on the sensory characteristics of steelhead (*Oncorhynchus mykiss*) in marine waters. *J. Aquat. Food Prod. Technol.*, **1**, 37–51.
Matsumoto, A. and Tsuchiya, Y. (1988). Earthy-musty odor producing cyanophytes isolated from five water areas in Tokyo. *Wat. Sci. Technol.*, **20**, 179–183.
McGill, A.S., Hardy, R., Burt, J.R. and Gunstone, F.D. (1974). Hept-cis-4-enal and its contribution to the off flavor in cold stored cod. *J. Sci. Food Agric.*, **25**, 1477–1481.
McGill, A.S., Hardy, R. and Gunstone, F.D. (1977). Further analysis of the volatile components of frozen cold stored cod and the influence of these on flavor. *J. Sci. Food Agric.*, **28**, 200–204.
McIver, R.C., Brooks, R.I. and Reineccius, G.A. (1982). Flavor of fermented fish sauce. *J. Agric. Food Chem.*, **30**, 1017–1020.
Medsker, L.L., Jenkins, P. and Thomas, J.F. (1968). An earthy-smelling compound associated with blue-green algae and actinomycetes. *Environ. Sci. Technol.*, **2**, 461–464.
Meijboom, P.W. and Stroink, J.B.A. (1972). 2-trans, 4-cis, 7-cis-decatrienal, the fishy off-flavor occurring in strongly autoxidized oils containing linolenic acid or ω 3, 6, 9 etc., fatty acids. *J. Am. Oil Chem. Soc.*, **49**, 555–558.
Mills, O.E., Chung, S. and Johnsen, P.B. (1993). Dehydration products of 2-methylisoborneol are not responsible for off-flavor in the channel catfish. *J. Agric. Food Chem.*, **41**, 1690–1692.
Milo, C. and Grosch, W. (1993). Changes in the odorants of boiled trout (*Salmo fario*) as affected by the storage of the raw material. *J. Agric. Food Chem.*, **41**, 2076–2081.
Milo, C. and Grosch, W. (1995). Detection of odor defects in boiled cod and trout by gas chromatography–olfactometry of headspace samples. *J. Agric. Food Chem.*, **43**, 459–462.
Milo, C. and Grosch, W. (1996). Changes in the odorants of boiled salmon and cod as affected by the storage of the raw material. *J. Agric. Food Chem.*, **44**, 2366–2371.
Milo, C. and Grosch, W. (1997). Potent odorants in boiled cod as affected by the storage of raw material. In *Flavor and Lipid Chemistry of Seafoods*, eds. F. Shahidi and K.R. Cadwallader. American Chemical Society, Washington, DC, pp. 110–119.
Minamide, T. (1977). Lipids and physiological role of lipoxygenase in fruits and vegetables. *Nippon Shokuhin Kogyo Gakkaishi*, **24**, 186–199.
Morishata, T., Uno, K., Araki, T. and Takahashi, T. (1989). Comparison of the amounts of extractive nitrogenous constituents in the meats of cultured red sea bream of different localities and culture methods, and those of wild fish. *Nippon Suisan Gakkaishi*, **55**, 1565–1573.
Murata, M. and Sakaguchi, M. (1989). The effects of phosphatase treatment of yellowtail muscle extracts and subsequent addition of IMP on flavor intensity. *Nippon Suisan Gakkaishi*, **55**, 1599–1603.
Nakamura, K., Iida, H. and Tokunaga, T. (1980a). Separation and identification of odor in oxidized sardine oil. *Bull. Jpn. Soc. Sci. Fish.*, **46**, 355–360.

Nakamura, K., Iida, H., Tokunaga, T. and Miwa, K. (1980b). Volatile flavor compounds of grilled red-fleshed fish volatile carbonyl compounds and volatile fatty acids. *Bull. Jpn. Soc. Sci. Fish.*, **46**, 221–224.

Negaro, T., Ando, M. and Ichiwaka, N. (1988). Blue-green algae in Lake Biwa which produce earthy-musty odors. *Wat. Sci. Technol.*, **20**, 117–123.

Obata, Y., Yamanishi, T. and Ishida, M. (1950). Chemical studies on the substance of fish smell. 2. Pyridine group compounds as the substance concerned with fishy smell. *Bull. Jpn. Soc. Sci. Fish.*, **15**, 551.

Ogata, M. and Miyake, Y. (1973). Identification of substances in petroleum causing objectionable odors in fish. *Wat. Res.*, **7**, 1493–1504.

Ota, T. and Yamada, M. (1974). Lipids in Masu salmon. II. Seasonal variations in the lipids of Masu salmon parr during the life in freshwater. *Bull. Jpn. Soc. Sci. Fish.*, **40**, 699–706.

Persson, P.E. (1979). The source of muddy odor in bream (*Abramis brama*) from the Porvoo Sea area (Gulf of Finland). *J. Fish. Res. Bd. Can.*, **36**, 883–890.

Persson, P.E. and Juttner, F. (1983). Threshold odor conditions of odorous algal metabolites occurring in lake water. *Aqua. Fenn.*, **13**, 3–7.

Pyysalo, H. and Suihko, M. (1976). Odour characterization and threshold values of some volatile compounds in fresh mushrooms. *Lebensm. Wiss u. Technol.,*, **9**, 371–373.

Raksakulthai, N. and Haard, N.F. (1992). Correlation between the concentration of peptides and amino acids and the flavor of fish sauce. *ASEAN Food J.*, **7**, 86–90.

Regenstein, J.M., Schlosser, M.A., Samson, A. and Fey, M. (1982). Chemical changes in trimethylamine oxide during fresh and frozen storage of fish. In *Chemistry and Biochemistry of Marine Food Products*, eds. R.E. Martin, G.E. Flick, C.E. Hebard and D.R. Ward. AVI Publishing Co., Westport, CT, pp. 149–305.

Reneccius, G.A. (1979). Off-flavors in meat and fish – a review. *J. Food Sci.*, **44**, 12–21.

Ross, D.A. and Love, R.M. (1979). Decrease in the cold store flavor developed by frozen fillets of starved cod (*Gadus morhua* L.). *J. Food Technol.*, **14**, 115–122.

Saito, T., Arai, K. and Matsuyoshi, M. (1959). A new method for estimating the freshness of fish. *Bull. Jpn. Soc. Sci. Fish.*, **24**, 749–752.

Sakakibara, H., Ide, J., Yanai, T., Yajima, I. and Hayashi, K. (1990a). Volatile flavor compounds of some kinds of dried and smoked fish. *Agric. Biol. Chem.*, **54**, 9–16.

Sakakibara, H., Hosokawa, M., Yajima, I. and Hayashi, K. (1990b). Flavor constituents of dried bonito (katsuobushi). *Food Rev. Int.*, **6**, 553–572.

Sanceda, N.G., Kurata, T. and Arakawa, N. (1983). Volatile compounds in the acidic fraction of the steam distillate of Philippine fish sauce (Patis). *Philip. Agric.*, **66**, 176–182.

Schrader, K.K. and Blevins, W.T. (1993). Geosmin-producing species of *Streptomyces* and *Lyngba* from aquaculture ponds. *Can. J. Microbiol.* **39**, 834–840.

Sen Gupta, P., Mondal, A. and Mitra, S.N. (1972). Separation and quantitative estimation of dimethylamine and trimethylamine in fish by paper chromatography. *J. Inst. Chem.*, **44**, 49–50.

Sessa, D.J. (1979). Biochemical aspects of lipid derived flavor in legumes. *J. Agric. Food Chem.*, **27**, 234–239.

Shahidi, F., Chong, X. and Dunajski, E. (1994). Freshness quality of harp seal (*Phoca groenlandica*) meat. *J. Agric. Food Chem.*, **42**, 868–872.

Shewfelt, R. L., MacDonald, R.E. and Hultin, H.O. (1981). Effect of phospholipid hydrolysis on lipid oxidation in flounder muscle microsomes. *J. Food Sci.*, **46**, 1297–1301.

Shibamoto, T. (1983). Heterocyclic compounds in browning and browning/nitrite model systems: Occurrence, formation mechanisms, flavor characteristics, and mutagenic activity. In *Instrumental Analysis of Foods: Recent Progress*, eds. G. Charalambous and G. Inglett. Academic Press, New York, pp. 229–278.

Shibamoto, T. and Horuichi, M. (1997). Role of aldehydes in cooked fish flavors. In *Flavor and Lipid Chemistry of Seafoods*, eds. F. Shahidi and K.R. Cadwallader. American Chemical Society, Washington, DC, pp. 110–119.

Shiomi, K., Noguchi, A., Yamanaka, H., Kikuchi, T. and Iida, H. (1982). Volatile sulfur compounds responsible for an offensive odor of the flat head (*Calliurichthys doryssus*). *Comp. Biochem. Physiol.*, **71B**, 29–31.

Simidu, W., Hibiki, S., Sabata, S. and Takeda, K. (1953). Studies on muscle of aquatic animals. XVI. Distribution of extractive nitrogen in muscle of several kinds of gastropod. *Bull. Jpn. Soc. Sci. Fish.*, **19**, 871–881.

Sipos, J.C. and Ackman, R.G. (1964). Association of dimethyl sulphide with the 'Blackberry' problem in cod from the Labrador area. *J. Fish. Res. Bd. Can.*, **21**, 423–425.

Slabyj, B.M. and Hultin, H.O. (1984). Oxidation of a lipid emulsion by a peroxidizing microsomal fraction from herring muscle. *J. Food Sci.*, **49**, 1392–1396.

Spencer, R. (1962). Odor and the assessment of fish freshness. *Soap Perfum. Cosmet.*, **35**, 795–799.

Suyama, M., Hirano, T. and Yamazaki, S. (1985). Odor of ayu and its volatile components. *Bull. Jpn. Soc. Sci. Fish.*, **51**, 287–294.

Suyama, M., Hirano, T., Okada, N. and Shibuya, T. (1970). Quality of wild and cultured ayu – I. On the proximate composition, free amino acids and related compounds. *Bull. Jpn. Soc. Sci. Fish.*, **43**, 535–540.

Tabachek, J.L. and Yurkowski, M. (1976). Isolation and identification of blue-green algae producing muddy odor metabolites geosmin and 2-methylisoborneol in saline lakes in Manitoba. *J. Fish. Res. Bd. Can.*, **33**, 25–35.

Triqui, R. and Reineccius, G.A. (1995a). Changes in flavor profiles with ripening of anchovy (*Engraulis encrasicholus* L.). *J. Agric. Food Chem.*, **43**, 1883–1889.

Triqui, R. and Reineccius, G.A. (1995b). Flavor development in the ripening of anchovy (*Engraulis encrasicholus* L.). *J. Agric. Food Chem.*, **43**, 453–458.

Wada, S. and Lindsay, R.C. (1992). Occurrence of the green aroma compound, 1,5-octadien-3-hydroperoxide, in stored sardine oil. *Nippon Suisan Gakkaishi*, **58**, 1195.

Whitfield, F.B., Last, J.H., Shaw, K.J. and Tindale, C.R. (1988). 2,6-Dibromophenol, the cause of the iodoform-like off-flavor in some Australian crustacea. *J. Sci. Food Agric.*, **46**, 29.

Withycombe, D.A. and Mussinan, C.J. (1988). Identification of 2-methyl-3-furanthiol in the steam distillate from canned tuna fish. *J. Food Sci.*, **53**, 658.

Wyatt, C.J. and Day, E.A. (1963). Autoxidation of fish oils. 2. Changes in the carbonyl distribution of autoxidizing salmon oils. *J. Food Sci.*, **28**, 305–309.

Yurkowski, M. and Tabachek, J.L. (1974). Identification, analysis, and removal of geosmin from muddy-flavored trout. *J. Fish. Res. Bd. Can.*, **31**, 1851–1858.

Yurkowski, M. and Tabachek, J.L. (1980). Geosmin and 2-methylisoborneol implicated as a cause of muddy odor and flavor in commercial fish from Cedar Lake, Manitoba. *Can. J. Fish. Aquat. Sci.*, **37**, 1449–1450.

Zhang, C., Shirai, T., Suzuki, T. and Hirano, T. (1992a). Lipoxygenase-like activity and formation of characteristic aroma compounds from wild and cultured ayu. *Nippon Suisan Gakkaishi*, **58**, 959–964.

Zhang, C., Shirai, T., Suzuki, T. and Hirano, T. (1992b). Enzymatically generated specific volatile compounds in ayu tissues. *Nippon Suisan Gakkaishi*, **58**, 559–565.

Zhang, C., Hirano, T., Suzuki, T., Shirai, T. and Morishita, A. (1992c). Volatile compounds and their generation in smelt. *Nippon Suisan Gakkaishi*, **58**, 773–779.

8 Flavour of shellfish

S. SPURVEY, B.S. PAN AND F. SHAHIDI

8.1 Introduction

Flavour is defined as sensations arising from the integration of signals produced as a consequence of smell and taste (Laing and Jinks, 1996). The flavour sensation of shellfish usually begins with its visual assessment, but the main components of it include volatile compounds (aroma), non-volatile components (taste) and compounds that are perceived in the mouth as mouthfeel (Taylor and Linforth, 1996).

The volatile components of shellfish are regarded as the most important determinant for their flavour quality. The contribution of volatiles to flavour is dependent upon their recognition threshold values and concentrations. The flavour of most fresh shellfish can be described as sweet, distinctly plant-like, often accompanied by metallic and very slight to pronounced fishy attributes (Josephson, 1991). This fresh flavour aroma is generated by unsaturated alcohols and aldehydes of a chain length of less than ten carbon atoms. Alkylpyrazines and sulphur-containing compounds are important contributors to the cooked aroma of crustaceans.

The nonvolatile components of shellfish are water-soluble low-molecular-weight components which are regarded as the principal flavour producers. The most important nonvolatile components are free amino acids, nucleotides, inorganic salts and quaternary ammonium bases which produce the individual tastes associated with each shellfish. 5′-Ribonucleotides, adenosine monophosphate (AMP), inosine monophosphate (IMP) and guanosine monophosphate (GMP), combined with monosodium glutamate (MSG) produce the characteristic umami taste of foods. Umami substances are widely distributed in shellfish such as crab, scallop, shrimp and lobster and contribute to the taste of shellfish in cooperation with other substances such as free amino acids and inorganic ions (Komata, 1990).

Several review papers have reported the flavour of shellfish. However, none has considered both the volatile and nonvolatile components involved in producing flavour. This overview provides information on compounds that contribute both to the aroma and taste of shellfish.

8.2 Volatile components in shellfish

Important volatile components of shellfish may be divided into several classes of compounds. Among these, alcohols, aldehydes, ketones, furans, nitrogen-containing compounds, sulphur-containing compounds, hydrocarbons, esters and phenols are the most important.

8.2.1 Alcohols

Alcohols may be formed by the decomposition of secondary hydroperoxides of fatty acids (Tanchotikul and Hsieh, 1989), action of lipoxygenase on fatty acids (Suzuki *et al.*, 1990), oxidative decomposition of fat or reduction of a carbonyl to an alcohol (Pan and Kuo, 1994). Alcohols are generally minor contributors to food flavour because of their high thresholds unless they are present at high concentrations or are unsaturated (Heath and Reineccius, 1986). Only trace amounts of unsaturated alcohols, derived from unsaturated fatty acids by the action of lipoxygenase, were present in raw scallop (Suzuki *et al.*, 1990).

Alcohols mostly possess fragrant, planty, rancid and earthy odours (Cadwallader *et al.*, 1995). Dodecanol possesses a floral-like odour and is important in the volatiles of cooked shrimp (Mandeville *et al.*, 1992), snow crab (Cha *et al.*, 1993), blue crab (Chung and Cadwallader, 1993) and rangia clams (Tanchotikul and Hsieh, 1991). The most noticeable compound detected in rancid sardine oil, namely 1-penten-3-ol (Nakamura *et al.*,1980; Peterson and Chang, 1982), was identified in oyster, snow crab, blue crab, roasted shrimp, steamed rangia clams and crayfish waste. This compound was suggested to be formed through autoxidation of unsaturated fatty acids (Sakakibara *et al.*, 1988). The presence of 2-butoxyethanol, a compound identified in crayfish tail meat (Vejaphan *et al.*, 1988), has been reported in beef products (Peterson and Chang, 1982). Gas chromatographic aroma perception of 2-butoxyethanol gave a spicy and woody note, hence it could be an important flavour component in boiled crayfish tail meat.

The eight-carbon alcohols including 1-octen-3-ol, 1,5-octadien-3-ol and 2,5-octadien-1-ol contribute a heavy, plant-like aroma (Josephson *et al.*, 1985; Lindsay, 1990). These compounds were found in raw fish (Hsieh *et al.*, 1989; Phleger *et al.*, 1978; Josephson *et al.*, 1984a) and oyster (Josephson *et al.*, 1985). The compound 1-octen-3-ol, a degradation product of linoleic acid hydroperoxides (Baldings, 1970; Josephson *et al.*, 1983; Wurzenberger and Grosch, 1984), possesses a mushroom-like odour. It has been identified as one of the major volatile alcohols in oyster, rangia clams, crab, prawns, crayfish and sand-lobster (Vejaphan *et al.*, 1988; Kim *et al.*, 1994; Cha *et al.*, 1992; Tanchotikul and Hsieh, 1991). Josephson *et al.* (1984a) found that 1-octen-3-ol is one of the volatile components widely distributed in fresh and saltwater fish. The nine-carbon volatile alcohols contribute a distinct melon-like and cucumber-like aroma to seafoods. For example, 3,6-nonadien-1-ol contributes a distinct melon-like and cucumber-like aroma to oysters (Lindsay, 1990; Josephson *et al.*, 1985). Josephson *et al.* (1984a) suggested that the formation of eight- and nine-carbon volatile alcohols in fresh fish results from the action of lipoxygenase on unsaturated fatty acids. Alcohols identified in different species of shellfish are listed in Table 8.1.

8.2.2 Aldehydes

The odour thresholds of aldehydes are generally lower than those of alcohols (Hsieh *et al.*, 1989). Therefore, aldehydes have a great potential for over-riding the flavour effect of many other substances, even when present in trace amounts. Long-chain aldehydes, which apparently do not have detectable aromas, may act as precursors to other important aroma compounds, such as heterocyclic compounds (Cha *et al.*, 1993). As an example, formaldehyde, acetaldehyde and malonaldehyde react with ammonia, hydrogen sulphide and glucose to produce heterocyclic compounds (Pan and Kuo, 1994; Baek and Cadwallader, 1996).

Alkanals, alkenals and alkadienals are degradation products of linoleate and linolenate hydroperoxides (Grosch, 1982; Josephson *et al.*, 1983, 1984a,b; Selka and Frankel, 1987; Josephson and Lindsay, 1987; Karahadian and Lindsay, 1989). Oxidation of polyunsaturated acids results in the formation of various aldehydes, such as octanal and nonadienal. These aldehydes play an important role in food products and are responsible for a wide range of oxidized flavours. The saturated straight-chain aldehydes normally have unpleasant, sharp, pungent and irritating odours with oily and waxy topnotes (Heath and Reineccius, 1986). According to Tanchotikul and Hsieh (1991), these aliphatic aldehydes are major components of steamed clam flavour. Hsieh *et al.* (1989) described the aroma of 3-methylbutanal in crayfish tail meat and pasteurized crab meat and found them to possess a green plant-like note. The aroma characteristics of this aldehyde have also been reported as being green, fruity, nutty, cheesy and sweat, depending on the concentration (Fors, 1983). Alkadienals such as 2,6-nonadienal are derived from the degradation of omega-3-polyunsaturated fatty acids. This compound was found in the volatile components of the hepatopancreas of crayfish (Tanchotikul and Hsieh, 1989; Cha *et al.*, 1992), steamed rangia clams (Tanchotikul and Hsieh, 1991), spiny lobster tail meat (Cadwallader *et al.*, 1995), pickled fish and fresh oyster (Josephson *et al.*, 1985, 1987). The compound (*E*,*Z*)-2,6-nonadienal, which has a cucumber-like aroma, may be degraded to (*Z*)-4-heptenal via retro-aldol condensation in an aqueous medium (Josephson and Lindsay, 1987). (*Z*)-4-Heptenal, the compound responsible for cold-stored odour of cod (Josephson and Lindsay, 1987; Lindsay, 1990), was also found in the flavour

Table 8.1 Alcohols in shellfish volatiles

Compounds	Occurrence	References
Ethanol	Oyster, raw/fermented/cooked shrimp, dried krill shell powder, cooked corbicula, boiled scallops	1–11, 22
2-butoxy-	Cooked crayfish, steamed clams	12, 13, 23
phenyl-	Raw and fermented shrimp	3, 10, 13
2-dimethylamino-	Roasted squid, cooked clams	35, 36
2-Propanol	Cooked crab meat	23
2-methyl-	Raw and fermented shrimp	3, 6, 9, 10
3-methylthio-	Cooked snow and blue crabs	25, 26, 27
2,3-dichloro-	Spray dried krill, spray dried shrimp powder, shrimp hydrolysate	14
1-butoxy-	Crayfish waste	28
Butanol	Cooked crayfish, raw/fermented/cooked shrimp, steamed clams, boiled/pasteurized crab meat	3, 5, 6, 10, 12, 15, 23, 25, 29–31, 32
3-methyl-1-	Cooked crayfish, raw/fermented/cooked shrimp, raw/cooked crab meat	3–7, 10, 12, 13, 16, 23, 31
3-methyl-2-	Roasted shrimp	17
Pentanol	Cooked crayfish, oyster, cooked/roasted shrimp, steamed clams, raw/cooked crab meat	1, 12, 13, 17, 23, 29, 31
3-Pentanol	Raw and boiled scallop	22
1-Penten-3-ol	Oyster, raw/cooked/fermented/roasted shrimp, dried krill shell powder, crayfish waste, steamed clams, boiled scallop, cooked crab	1, 2–8, 10, 12, 13, 17–20, 22, 25, 28–31
2-Penten-1-ol	Boiled crayfish, boiled crab, steamed clams, raw/boiled scallop	22, 23, 25, 28–30, 33
Hexanol	Cooked crayfish, pasteurized crab meat, steamed clams	12, 13, 23, 25, 28–31
2-ethyl-	Steamed clams, crayfish waste	23, 29
2-Hexanol	Boiled/pasteurized crab meat, crayfish waste	13, 32
3-Hexanol	Boiled/pasteurized crab meat, crayfish waste	13, 32
1-Hexen-3-ol	Raw scallop, crayfish waste	13, 22
2-Hexen-1-ol	Crayfish hydrolysate	33
3-Hexen-1-ol	Oyster	1
Heptanol	Cooked crayfish, raw/fermented shrimp, raw/cooked crab meat, steamed clams	6, 12, 13, 23, 25, 28, 29, 31
2-methyl-	Cooked/roasted shrimp	17
3-methyl-	Raw/fermented shrimp	6, 10
6-methyl-	Fermented shrimp	3
2-Heptanol	Crayfish waste	13
Octanol	Raw/cooked crab meat, boiled crayfish, steamed clams	12, 23, 29, 31
1-Octen-3-ol	Cooked crayfish, oyster, prawn, sand-lobster, cooked shrimp, dried krill shell powder, raw/cooked scallop, steamed clams, raw/cooked crab meat	1, 2, 8, 10, 12, 13, 22, 23, 25, 28-31, 33
2-Octen-1-ol	Oyster	1
2,5-Octadien-1-ol	Oyster	1
Octa-1,5-dien-3-ol	Oyster, prawns, sand-lobster	21
Nonanol	Steamed clams, boiled crayfish, crayfish waste, raw/cooked crab meat	13, 23, 29, 31
2-Nonanol	Boiled crayfish, crayfish waste	12, 29
(*E*)-6-Nonen-1-ol	Crayfish waste	29

Table 8.1 *continued*

Compounds	Occurrence	References
3,6-Nonadien-1-ol	Oyster	1
Decanol	Cooked crayfish, boiled scallops	12, 22, 23
Undecanol	Cooked crayfish	12, 23
2-Undecanol	Boiled crab meat	31
Dodecanol	Cooked crab, boiled shrimp waste, steamed clams	25, 29, 31, 34
2-methyl-	Boiled shrimp waste	34
1,14-tetradecanediol	Boiled shrimp waste	34
Hexadecanol	Raw/processed crab meat, boiled shrimp waste, steamed clams	29, 31, 34
Heptadecanol	Boiled shrimp waste	34
Cyclopentanol	Oyster	1
Cyclohexanol		
2-(dimethylamino)-	Boiled shrimp waste	34
Benzyl alcohol	Fermented/roasted shrimp, cooked corbicula, cooked crab meat	3, 6, 10, 11, 17, 25, 30

References: (1) Josephson *et al.*, 1985; (2) Kubota *et al.*, 1982; (3) Choi and Kato, 1984a; (4) Choi *et al.*, 1983; (5) Choi and Kato, 1984b; (6) He and Kato, 1985; (7) Choi and Kobayashi, 1983; (8) Kubota *et al.*, 1980a; (9) Shye *et al.*, 1988; (10) Choi and Kato, 1983; (11) Kubota *et al.*, 1991; (12) Vejaphan *et al.*, 1988; (13) Tanchotikul and Hsieh, 1989; (14) Kubota *et al.*, 1985; (15) Suyama *et al.*, 1985; (16) Soliman *et al.*, 1983; (17) Kubota *et al.*, 1986; (18) Shye *et al.*, 1987; (19) Kubota *et al.*, 1980b; (20) Josephson and Lindsay, 1987; (21) Whitfield *et al.*, 1982; (22) Suzuki *et al.*, 1990; (23) Hsieh *et al.*, 1989; (24) Kawai *et al.*, 1990; (25) Cha *et al.*, 1993; (26) Chung and Cadwallader, 1994; (27) Chung *et al.*, 1995; (28) Cha *et al.*, 1992; (29) Tanchotikul and Hsieh, 1991; (30) Kim *et al.*, 1994; (31) Chung and Cadwallader, 1993; (32) Matiella and Hsieh, 1990; (33) Baek and Cadwallader, 1996; (34) Mandeville *et al.*, 1992; (35) Kawai *et al.*, 1991; and (36) Kawai *et al.*, 1990.

volatiles of oyster, raw and cooked shrimp (Kubota *et al.*, 1986), cooked crab meat (Chung and Cadwallader, 1993), crayfish waste (Tanchotikul and Hsieh, 1989; Cha *et al.*, 1992), steamed clams (Tanchotikul and Hsieh, 1991) and spiny lobster tail meat (Cadwallader *et al.*, 1995). The aroma of (*Z*)-4-heptenal was described as being cardboardy, painty, linseed oil-like or cream-like. However, its presence in freshly cooked crab meat might be desirable. McGill *et al.* (1974) described the aroma of (*Z*)-4-heptenal in crayfish as being similar to that of boiled potatoes and its threshold was determined to be 0.04 ppb. Benzaldehyde, probably derived from Strecker degradation of amino acids, has been identified as the major monocarbonyl compound in roasted peanuts (Mason *et al.*, 1967). Benzaldehyde has been perceived as having a pleasant almond, nutty and fruity aroma and is an important flavour volatile of crayfish tail meat and pasteurized crab meat (Hsieh *et al.*, 1989).

Long-chain aldehydes do not contribute to the flavour of shellfish in spite of their high concentration because of their high boiling points and low volatility. Aldehydes identified in different species of shellfish are listed in Table 8.2.

Table 8.2 Aldehydes in shellfish flavour

Compounds	Occurrence	References
Acetaldehyde	Oyster, raw/fermented shrimp, boiled scallop	3, 20, 22
phenyl-	Spray dried shrimp powder, shrimp hydrolysate, cooked corbicula	11, 14
Propanal		
2-methyl-	Roasted dried squid, cooked crab meat	25, 35
3-(methylthio)-	Spiny lobster tail meat, cooked crayfish tail meat, cooked crab meat	23, 26, 27, 37
Butanal	Raw/fermented shrimp, steamed clams	3, 29
2-methyl-	Cooked shrimp, cooked crayfish tail meat, cooked crab tail meat, cooked crab meat, roasted dried squid	9, 10, 13, 25, 30, 35
3-methyl-	Cooked crayfish, cooked shrimp, spray dried shrimp powder, roasted clam, roast dried squid, boiled/pasteurized crab meat, steamed clams	2, 5–7, 10, 12–14, 16, 17, 23, 25, 28–30, 32, 35, 36
2-phenyl-	Spray dried shrimp powder	14
2-Butenal	Crayfish waste, crab processing by-products, cooked crab meat	25, 28, 30, 31, 33
2-methyl-	Cooked shrimp, crayfish waste, steamed clams, cooked crab meat	5, 6, 17, 13, 23, 25, 29, 30, 33
Pentanal	Oyster, raw and fermented shrimp, dried shell krill powder, cooked corbicula, boiled and pasteurized crab meat, crab processing by-products, crayfish waste, steamed clams	3, 6, 8, 11, 13, 23, 29, 30, 32
2-Pentenal	Cooked crab meat, crayfish waste, crab processing by-products, steamed clams	23, 25, 28–30
2-methyl-	Oyster, crab processing by-products	1, 31
Hexanal	Cooked crayfish, cooked shrimp, dried krill shell powder, cooked spiny lobster tail meat, crayfish waste, boiled/pasteurized crab meat, steamed clams	2, 8, 12, 13, 16, 19, 23, 25, 28–30, 32, 33, 37
2-Hexenal	Cooked shrimp, crayfish waste, steamed clams, crab processing by-products, cooked crab meat	19, 23, 25, 29–30
5-methyl-2-phenyl-	Spray dried shrimp powder	14
2,4-Hexadienal	Steamed clams, crab processing by-product, cooked crab meat, crayfish waste, boiled scallop	22, 25, 28–30
Heptanal	Cooked crayfish, cooked shrimp, crayfish waste, cooked spiny lobster tail meat, crab processing by-products, steamed clams, boiled scallop	2, 12, 13, 22, 23, 28–31, 33, 37
2-Heptenal	Crayfish waste, crab processing by-products	23, 30
4-Heptenal	Oyster, raw/fermented/cooked shrimp, dried krill shell powder, crayfish waste, raw crab meat, steamed clams, crab processing by-products	2, 3, 5, 6, 8, 13, 19, 20, 28–31, 33, 37
2*E*,4*E*-Heptadienal	Oyster, cooked crayfish, crab processing by-products, crayfish waste, raw/cooked crab meat, steamed clams, boiled scallops	1, 12, 13, 22, 23, 25, 28-31, 33
2*E*,4*Z*-Heptadienal	Oyster	1
Octanal	Cooked crayfish, cooked shrimp, crayfish waste, steamed clams, raw crab meat	2, 12, 13, 16, 23, 28, 29, 31
2-Octenal	Crayfish waste, crab processing by-products, steamed clams	28–30, 33

Table 8.2 *continued*

Compounds	Occurrence	References
2,4-Octadienal	Steamed clams, crab processing by-products, crayfish waste	29, 30, 33
Nonanal	Cooked crayfish, oyster, cooked shrimp, crayfish waste, raw crab meat, steamed clams, crab processing by-product, boiled scallop	1, 2, 12, 13, 16, 22, 23, 28, 29, 30, 31
2-Nonenal	Cooked crayfish, crayfish waste, crab processing by-products	12, 13, 23, 28, 30, 31, 33
2,6-Nonadienal	Oyster, crayfish waste, crab processing by-products	13, 23, 28, 30, 33
Decanal	Cooked shrimp, pasteurized crab meat, crayfish waste, steamed clams, boiled scallops	2, 12, 22, 23, 28, 29
2,4-Decadienal	Steamed clams, crayfish waste, raw crab meat	23, 28, 29, 31
Benzaldehyde	Cooked crayfish, oyster, raw fermented and cooked shrimp, dried krill shell powder, spray dried shrimp, shrimp hydrolysate, cooked corbicula, crayfish waste, crab processing by-product, raw/cooked crab meat, steamed clams, boiled scallops	1, 2–8, 11, 12–14, 17, 19, 22, 23, 25, 28–31, 33, 37
2,5-dimethyl-	Crayfish waste	28
Retanal	Cooked shrimp	19

References: (1)–(36) as in footnotes to Table 8.1; (37) Cadwallader *et al.*, 1995.

8.2.3 *Ketones*

Ketones may be produced by thermal oxidation/degradation of polyunsaturated fatty acids (Josephson and Lindsay, 1986; Matiella and Hsieh, 1990; Cha *et al.*, 1992), amino acid degradation (Chung and Cadwallader, 1994; Pan and Kuo, 1994) or microbial oxidation (Pan and Kuo, 1994). Some of the most intense odorants in this group are 2,3-butanedione, 3,5-octadien-2-one, 1-octen-3-one and 2,3-pentanedione. The Maillard reaction product, 2,3-butanedione, is a characteristic volatile in cooked foods (Hodge, 1967). It has been identified in the volatiles of cooked crayfish, shrimp, fermented shrimp, cooked shrimp, cooked spiny lobster tails, cooked crab claws and cooked beef (Chang and Peterson, 1977; Choi and Kato, 1983, 1984b; He and Kato, 1985; Vejephan *et al.*, 1988; Tanchotikul and Hsieh, 1989; Chung and Cadwallader, 1994; Cadwallader *et al.*, 1995). The unsaturated ketones, 3,5-octadien-2-one and 3-penten-2-one, have been reported as lipid oxidation products in fish (Josephson *et al.*, 1984a; Karahadian and Lindsay, 1989), krill (Kubota *et al.*, 1982), roasted shrimp (Kubota *et al.*, 1986) and cooked snow crab (Cha *et al.*, 1993). The vinyl ketone, 1,5- octadien-3-one was found in the volatile components of fresh oyster (Josephson *et al.*, 1985), fish (Josephson *et al.*, 1983, 1984a) and fish oils (Karahadian and Lindsay, 1989).

Ketones contribute to the sweet floral and fruity flavours of crustacea (Cha *et al.*, 1992). The methyl ketones (C_3–C_{17}) found in shellfish, such as 2-heptanone and 2-pentanone, were probably formed from beta-oxidation of their carbon chain followed by decarboxylation (Tanchotikul and Hsieh, 1989; Lindsay, 1990). These compounds have a distinct green and fruity aroma and give a more floral note as the chain length increases (Tanchotikul and Hsieh, 1991). Vinyl ketones are products of lipid oxidation formed during heating and have a heavy, green geranium leaf-like aroma (Karahadian and Lindsay, 1989). The alkanediones, 2,3-butanediones and 2,3-pentanedione, impart an intense buttery and desirable aroma to foods (Vejaphan *et al.*, 1988; Tanchotikul and Hsieh, 1991). The diketones might be important aroma components of crayfish tail meat and provide a desirable balance of the meaty and buttery notes (Hsieh *et al.*, 1989). Acetophenone imparts a sweet rose floral odour (Hsieh *et al.*, 1989). Table 8.3 lists ketones found in shellfish.

8.2.4 Furans and other oxygen-containing cyclic compounds

Furans are found in flavour components of boiled and pasteurized crab meat, spray-dried shrimp powder, shrimp hydrolysate, roasted dried squid, boiled crayfish and boiled crab (Table 8.4). The compound 2-ethylfuran with an odour threshold of 8.0 ppm in water (Maga, 1979) has been identified in pasteurized crab meat (Matiella and Hsieh, 1990). This furan has been reported to have a warm sweet aroma when diluted (Fors, 1983). However, it has been reported to have a burnt, sweet, bitter and coconut flavour in coffee, red meat and poultry (Maga, 1979; Ho *et al.*, 1983). Hsieh *et al.* (1989) found that 2-pentylfuran contributes negatively to the flavour quality of crayfish and crab meats. It has also been reported as having an off-flavour effect in certain fats and oils and imparting a beany and grassy odour to stored soyabean oil (Krishnamurthy *et al.*, 1967). However, Vejaphan *et al.* (1988) claimed that this compound did not contribute to the flavour of crayfish regardless of its contribution to the flavour of meat and soyabean oil. Furans such as 2-acetylfuran have a powerful, balsamic, sweet flavour and a tobacco-like odour, while furfural possess a sweet, caramel- and bread-like odour (Maga, 1979). Furfural and 2-acetylfuran are derived from sugar–amino acid reactions (Maga, 1979).

Lactones (Table 8.4) have been identified in roasted shrimp and boiled scallops (Kubota *et al.*, 1985, 1986; Suzuki *et al.*, 1990). Lactones are odorous compounds and may contribute either desirable or undesirable notes to meat flavour at different concentrations (Wasserman, 1972). Some lactones have been reported as having a pleasant flavour, ranging from nutty–fatty to coconut and peach-like (Peterson and Chang, 1982) and have been found in many heat-treated foods (Liebich *et al.*, 1972; Nixon

Table 8.3 Ketones in shellfish volatiles

Compounds	Occurrence	References
2-Butanone	Cooked/roasted shrimp, roasted clams, roasted dried squid, cooked clams, boiled scallop	17, 22, 35, 36
3-hydroxy-	Raw/fermented shrimp, cooked corbicula, roasted clam, cooked clam, cooked snow crab, steamed clams, fresh crab meat, crab by-products	3, 5, 6, 10, 11, 25, 29, 31, 36
3-methyl-	Steamed clams, boiled crayfish tail meat	13, 29
2,3-Butanedione	Cooked crayfish, raw/fermented/cooked shrimp, fresh/cooked crab, crab by-products, crayfish waste, boiled crayfish tail meat, cooked spiny lobster tail meat, boiled scallop	3, 6, 10, 12, 13, 22, 23, 25, 26, 27, 28, 31, 33, 37
2-Pentanone	Cooked shrimp, crayfish by-products, boiled and pasteurized crab meat, cooked krill	2, 13, 17–19, 23, 25, 32, 33
3-methyl-	Crayfish waste	13
4-methyl-	Crab by-products, steamed clams, boiled/pasteurized crab meat	25, 29, 31, 32
4-hydroxy-4-methyl-	Steamed clams	36
3-Penten-2-one	Crayfish waste, raw crab meat	28, 31, 33
4-methyl-	Roasted shrimp, steamed clams	17, 29
4-Penten-2-one		
4-methyl-	Oyster, steamed clams, cooked crab meat	1, 25, 29
1-Penten-3-one	Roasted shrimp	17
2-Penten-4-one	Roasted shrimp	17
2,3-Pentanedione	Oyster, crayfish tail meat, boiled scallop, crayfish waste, raw crab meat, crab by-products, steamed clams	1, 12, 22, 23, 28, 29, 31, 33
2-Hexanone	Cooked crayfish, crab by-products, boiled and pasteurized crab meat, crayfish by-products	12, 13, 31–33
3-Hexanone	Crab by-products, boiled/pasteurized crab meat	23, 31–33
3-Hexen-2-one		
5-methyl-	Roasted shrimp	17
3-butyl-	Roasted shrimp	17
2-Heptanone	Crayfish waste, crab by-products, steamed clams	13, 28–30, 33
5-Hepten-2-one		
6-methyl-	Cooked crayfish, cooked/roasted shrimp, dried krill shell powder, boiled crayfish tail meat and hepatopancreas, pasteurized crab meat, cooked krill, crab by-products, raw crab meat	2, 3, 6–8, 10–13, 16, 17, 23, 25, 28, 30, 31, 33
Octanone	Steamed clams	29
2-Octanone	Cooked crayfish, cooked/roasted shrimp, dried krill shell powder, cooked corbicula, raw/pasteurized crab meat, cooked krill, crayfish waste, crab by-products	2, 8, 11–13, 17, 23, 28, 30, 31, 33, 37
1-Octen-3-one	Oyster, lobster tail meat	1, 37
3-Octen-2-one	Crayfish waste	13, 28
3,5-Octadien-2-one	Cooked shrimp, dried krill shell powder, crayfish waste, crab by-products	1, 3, 5–8, 18, 28, 30, 37

Table 8.3 *continued*

Compounds	Occurrence	References
3*E*,5*E*-Octadien-2-one	Oyster, spray dried shrimp powder, crab by-products, cooked crab, cooked krill, roasted/boiled shrimp	2, 14, 17, 25, 30
1,5-Octadien-3-one	Oyster	29
2,3-Octanedione	Crayfish waste	1
Nonanone	Steamed clams	34
2-Nonanone	Cooked shrimp, dried krill shell powder, crayfish waste, crab by-products, cooked krill, boiled scallop, boiled crayfish tail meat, raw crab meat	2, 8, 12, 13, 17, 19, 22, 23, 28, 30, 33, 37
3-Nonanone	Cooked crayfish waste	12, 13
3-Nonen-2-one	Cooked crab by-products, crayfish waste	13, 30
5-Nonen-2-one	Cooked crayfish waste	12, 23
2-Decanone	Cooked crayfish, cooked/roasted shrimp, dried krill shell powder, cooked corbicula, crab by-products, cooked krill	2, 4, 7, 8, 11–13, 17, 19, 23, 28, 31, 33, 37
3-Decen-2-one		
3-methyl-	Steamed clams	29
Undecanone	Steamed clams, crab by-products	29, 30
2-Undecanone	Crab by-products, boiled crayfish tail meat, cooked krill, crayfish waste	2, 12, 13, 23, 28, 30, 31
3-Undecanone	Cooked crayfish	12, 23
Undecatrien-2-one	Cooked shrimp	5
5,9-Undecadien-2-one		
6*E*,10*E*-dimethyl-	Steamed clams	29
2-Dodecanone	Crayfish waste, crab by-products	28, 30
Tridecanone	Steamed clams, crab by-products	29, 30
2-Tridecanone	Crayfish waste, raw crab meat, crab by-products	13, 28, 31
Tetradecanone	Crab by-products	30
5*E*,8*Z*,11*Z*-Tetradecatrien-2-one	Cooked shrimp	17, 38, 39
5*Z*,8*Z*,11*Z*-Tetradecatrien-2-one	Cooked shrimp	4, 6, 7, 10, 38, 39
Pentadecanone	Crab by-products, steamed clams	29, 30
2-Pentadecanone	Cooked shrimp	2
6,10,14-trimethyl-	Cooked crayfish waste	5, 6, 10
Cyclopentanone		
1-methyl-	Cooked shrimp	12, 25, 29
2-Cyclopenten-1-one		
2-methyl-	Crayfish waste	28
2,3-dimethyl-	Cooked crab, cooked and roasted shrimp, cooked crab meat	17, 25, 28, 32
1,3-Cyclopentane-dione	Shrimp waste	34
Cyclohexanone	Steamed clams, crayfish tail meat, cooked crab meat	23, 25, 29, 31
2-Cyclohexen-1-one	Steamed clams, crayfish by-products, raw crab meat	29, 31, 33
3-methyl-	Steamed clams	29
3,5,5-trimethyl-	Crayfish waste	28

Table 8.3 *continued*

Compounds	Occurrence	References
Cyclohexen-1,4-dione		
2,6,6-trimethyl-	Roasted shrimp	17
Cyclooctanone		
2-methyl-	Boiled shrimp waste	28
Acetophenone	Cooked crayfish, cooked shrimp, raw/cooked crab meat, crab by-products, crayfish waste, boiled scallops, cooked krill, roasted shrimp	2, 12, 13, 17, 22, 23, 25, 31
2-Thienyl ketone		
1,5-dimethyl-	Cooked shrimp	2

References: (1)–(36), as in footnotes to Table 8.1; (37) as in footnotes to Table 8.2; (38) Kubota and Kobayashi, 1988; (39) Kobayashi *et al.*, 1989.

et al., 1979; Kubota *et al.*, 1986). Some lactones with deisrable aromas found in crayfish waste are γ-octalactone and 2,3-dimethylcyclopent-2-en-1-one (Tanchotikul and Hsieh, 1989; Cha *et al.*, 1992).

8.2.5 Pyrazines and other nitrogen-containing compounds

Pyrazines may be formed via Maillard and pyrolysis reactions through Strecker degradation in heat processed foods (Whitfield, 1992). Additional precursors for the formation of pyrazines could be produced by proteolysis during heating and via treatment with proteolytic enzyme prior to heating (Maga, 1982). Pyrazines, with their large quantities and low flavour thresholds, are generally considered important volatile components in many thermally processed foods since they contribute nutty, roasted and toasted characteristics to products (Maga and Sizer, 1973; Maga, 1982; Shibamoto, 1989).

More than 100 different pyrazines have so far been identified in various food products. Maga (1983) has reviewed the distribution and occurrence of pyrazines in foods. Pyrazines (Table 8.5) contribute greatly to the flavour of boiled crayfish (Vejaphan *et al.*, 1988; Cha *et al.*, 1992), cooked crab (Cha *et al.*, 1993; Chung and Cadwallader, 1993; Kim *et al.*, 1994), raw and fermented shrimp (Choi *et al.*, 1983), roasted shrimp (Kubota *et al.*, 1986), cooked clams (Tanchotikul and Hsieh, 1991) and cooked krill (Kubota *et al.*, 1982). Alkylpyrazines, including methylpyrazines, 2,5- and 2,6-dimethylpyrazines, were identified as having a roasted, nutty/meaty aromas in boiled crayfish tail meat and hepatopancreas (Vejaphan *et al*, 1988; Hsieh *et al.*, 1989), and cooked rangia clams (Tanchotikul and Hsieh, 1991). Alkylpyrazines may be formed by the involvement of lipid oxidation products in Maillard reaction (Huang *et al.*, 1987). Since the formation of alkylpyrazines increases by heating of food at or above 100°C (Koehler and Odell, 1970; Shibamoto and Bernhard, 1977), cooking

Table 8.4 Furans and other oxygen-containing cyclic compounds in shellfish volatiles

Compounds	Occurrence	References
Furan		
2-acetyl-	Spray dried shrimp powder, shrimp hydrolysate, roasted clam	14, 16, 36
2-acetyl-5-methyl-	Spray dried shrimp powder, shrimp hydrolysate, roasted dried squid	14, 23
2-ethyl-	Pasteurized/boiled crab meat, boiled crayfish	23, 25, 28, 30, 32
2-pentyl-	Cooked crayfish, raw, boiled/pasteurized crab meat, steamed clams, crayfish waste	12, 13, 14, 23, 25, 28–32, 36
octyl-	Crayfish waste	30
Furfuryl alcohol	Fermented/cooked shrimp, dried krill shell powder, spray dried shrimp powder, shrimp hydrolysate, roasted dried squid, cooked clams, raw/fermented krill, crayfish waste, cooked crab meat	2, 4–8, 10, 12, 14, 16, 19, 25, 28, 29, 35
2-Furfural	Spray dried shrimp powder, shrimp hydrolysate, pasteurized/boiled crab meat, crayfish waste	13, 14, 16, 23, 25
5-methyl-	Roasted clam, cooked crab meat	25, 36
2-Furanone		
5-methyl-	Cooked shrimp waste	34
Lactones		
4-pentanolide	Spray dried shrimp powder, shrimp hydrolysate	14
4-hexanolide	Spray dried shrimp powder, shrimp hydrolysate	14
2-hexen-4-olide	Roasted shrimp	17
4-heptanolide	Spray dried shrimp powder, shrimp hydrolysate	14
4-octanolide	Spray dried shrimp powder, shrimp hydrolysate	14
4-nonanolide	Roasted shrimp	17
5-nonanolide	Roasted shrimp, spray dried shrimp powder, shrimp hydrolysate	14, 17
γ-butyrolactone	Boiled scallop	22
Maltol	Spray dried powder, roasted clam	14, 19, 24
Dioxane	Cooked shrimp	3–6, 10

References: (1)–(36) as in footnotes to Table 8.1.

of crayfish tail meat at higher temperatures may enhance the formation of these compounds and accentuate their nutty flavours. However, a decrease in the content of pyrazines during roasting of dried squid occurred. Thus, pyrazines in proteinaceous food volatiles contribute to boiled rather than roasted odours (Kawai *et al.*, 1991). Meanwhile, acetylpyrazines contribute a popcorn-like odour to cooked lobster tail meat (Cadwallader *et al.*, 1995), boiled scallops (Suzuki *et al.*, 1990) and cooked crab meat (Cha *et al.*, 1993; Chung and Cadwallader, 1993) (Table 8.5).

1*H*-Pyrrole does not have the strong nutty aroma of alkylpyrazines but it may impart a sweet and slightly burnt character to boiled crayfish. Shigematsu *et al.* (1972) reported that alkylpyrroles possess undesirable odours at high concentrations, but had a sweet, burnt flavour in high dilutions.

Table 8.5 Pyrazines in shellfish volatiles

Compounds	Occurrence	References
Pyrazine	Cooked shrimp, raw/boiled crab meat, boiled scallop, boiled crayfish, cooked krill	2, 4, 7, 10, 22, 28, 31, 33
2-methyl-	Cooked crayfish, raw/fermented/cooked and roasted shrimp, dried krill shell powder, spray dried shrimp powder, shrimp hydrolysate, roasted clam, roasted dried squid, cooked crab	2–4, 7, 9–12, 14, 17, 19, 23, 25, 28, 29, 30, 33, 35–37, 40, 41
2,3-dimethyl-	Raw/fermented/cooked/roasted shrimp, dried krill shell powder, spray dried shrimp powder, shrimp hydrolysate, roasted clam, roasted dried squid, raw/cooked crab, cooked crayfish, cooked clams, boiled scallop	2–4, 7–10, 12, 14, 17, 22, 25, 28–31, 33, 35–37, 40, 41
2,5-dimethyl-	Cooked crayfish, raw/fermented/cooked/ roasted shrimp, spray dried shrimp powder, shrimp hydrolysate, roasted clam, roasted dried squid, cooked krill, raw/cooked crab meat	2-4, 7, 9, 10, 12–14, 17–19, 25, 28–31, 33, 35–37, 41
2,6-dimethyl-	Cooked crayfish, raw/fermented/cooked/ roasted shrimp, dried krill shell powder, spray dried shrimp, shrimp hydrolysate, cooked corbicula, raw/ cooked crab meat, cooked krill, cooked clams, boiled scallop	2–4, 7, 8, 10, 12–14, 17, 19, 22, 23, 25, 28–31, 33, 35–37, 41
2,3,5-trimethyl-	Raw/fermented/cooked/roasted shrimp, dried krill shell powder, spray dried shrimp powder, shrimp hydrolysate, roasted clam, roasted dried squid, boiled scallop, lobster tail meat, cooked krill, raw/cooked crab meat	2–4, 7–10, 13, 14, 17, 19, 22, 23, 28, 31, 35-37, 40
2,3,5,6-tetramethyl-	Raw/fermented/cooked/roasted shrimp, dried krill shell powder, roasted dried squid, raw/cooked crab meat, cooked crayfish	2–4, 8, 9, 17–19, 35, 37–39
2-methyl-3,5-diethyl-	Raw/fermented shrimp, boiled crayfish	3, 13, 17
2,5-dimethyl-3-ethyl-	Roasted shrimp, dried krill shell powder, roasted dried squid, boiled clams, boiled crayfish hepatopancreas	8, 17, 18, 23, 28, 35, 36
2,6-dimethyl-3-ethyl-	Roasted shrimp, roasted clam, roasted dried squid	17, 35, 36
2,5-dimethyl-3-propyl-	Roasted shrimp	9, 17, 18
2-ethyl-	Roasted shrimp, spray dried shrimp powder, shrimp hydrolysate, roasted clam, crayfish hepatopancreas, raw/ cooked crab meat, cooked crayfish	14, 17, 23, 25, 28, 30–31, 33, 36, 39
2-ethyl-3-methyl-	Fermented/cooked shrimp	3, 40
2-ethyl-5-methyl-	Cooked/roasted shrimp, spray dried shrimp powder, cooked crayfish waste, cooked crab	2, 7, 10, 14, 17, 18, 25, 28, 40
2-ethyl-6-methyl-	Cooked/roasted shrimp, spray dried shrimp powder, cooked crayfish waste	2, 7, 17, 17, 28

Table 8.5 *continued*

Compounds	Occurrence	References
2-ethyl-3,5-dimethyl-	Raw/fermented/cooked shrimp, spray dried shrimp powder, shrimp hydrolysate, cooked crayfish waste, cooked crab meat, cooked krill	2, 7, 13, 14, 17, 19, 25, 28, 30, 33, 40
2-ethyl-3,6-dimethyl-	Raw/roasted/fermented shrimp, cooked krill, cooked crayfish waste, cooked crab meat	2, 4, 7, 13, 17, 25, 30, 33, 37, 40, 41
2-ethyl-5,6-dimethyl-	Cooked shrimp	40
2,6-diethyl-3-methyl-	Roasted shrimp	9, 17, 18

References: (1)–(36) as in footnotes to Table 8.1; (37) as in footnotes to Table 8.2; (38)–(39) as in footnotes to Table 8.3; (40) Kubota *et al.*, 1981; (41) Kubota *et al.*, 1988; (42) Ferretti and Flanagan, 1971; (43) Ferretti *et al.*, 1970.

2-Acetylpyrrole, a major component of foodstuffs, has been reported as having a caramel-like odour (Heath and Reineccuis, 1986). Its presence in the volatile components of roasted shrimp (Kubota *et al.*, 1985, 1986), spray-dried shrimp powder (Kubota *et al.*, 1985, 1986) and roasted clams (Kawai *et al.*, 1990) has been documented.

At high concentrations, pyridines (Table 8.6) impart an unpleasant and pungent odour to foods; however, at low concentrations they typically have pleasant aroma notes. Pyridines have been known to have both a positive and a negative impact on the flavour of cooked foods (Shibamoto, 1989). For example, 2-pentylpyridine (Ho and Carlin, 1989) and 5-ethyl-2-methylpyridine had an offensive fatty or green odour while 2-methyl-pyridine yielded a popcorn or astringent odour (Vernin, 1984) and 3-methyl-4-ethylpyridine was reported to have a sweet and hazelnut-like odour (Kawai *et al.*, 1990). Alkylpyridines have been known to impart an unpleasant flavour in roasted lamb fat (Buttery *et al.*, 1977). Various pyridines, identified in roasted clam and squid, might be formed by the interaction of 1,5-dicarbonyl compounds with ammonia or amino acids (Mandeville *et al.*, 1992). They may also arise from thermal degradation of sulphur-containing amino acids (Kato *et al.*, 1973), alone or in the presence of glucose (Kato *et al.*, 1969).

Pyrrolidine with an alkaline, raw egg-like aroma has been identified in the volatiles of blue crab (Chung and Cadwallader, 1994), roasted dried squid (Kawai *et al.*, 1991), roasted coffee (Tressl *et al.*, 1981), whisky (Viro, 1984) and alfalfa (Srinivas, 1988). Pyrrolidine has also been detected in seawater (Yang *et al.*, 1993) and can be formed biologically by bacteria such as *Clostridium perfringens* under low sugar conditions (Allison and MacFarlane, 1989; Griffith and Hammond, 1989). Maillard reaction of proline with monosaccharides may also lead to the formation of pyrrolidines (Tressl *et al.*, 1985b). Since proline is a major free amino acid in the mantle of fresh squid (Suyama and Kobayashi, 1980), its role in

Table 8.6 Pyrroles and pyridines and cyclic nitrogen-containing compounds in shellfish volatiles

Compounds	Occurrence	References
Pyridine	Raw/fermented/cooked/roasted shrimp, cooked corbicula, roasted clam, roasted dried squid, cooked crayfish waste, cooked crab meat, cooked krill, boiled scallop, cooked clams	2–7, 10, 11, 17, 22, 25, 28, 30, 35, 36, 42
2-methyl-	Cooked shrimp, roasted clam, roasted dried squid, cooked krill, cooked crab meat, cooked crayfish waste	2, 4, 7–9, 17, 25, 28, 35, 36, 42
3-methyl-	Cooked shrimp, roasted clam, roasted dried squid, cooked crab meat, cooked krill	2, 25, 35, 36
2,3-dimethyl-	Roasted clam, roasted dried squid	35, 36
2,4-dimethyl-	Roasted clam, roasted dried squid	35, 36
3,4-dimethyl-	Cooked shrimp	9, 17
3,5-dimethyl-	Roasted dried squid	35
2-ethyl-	Cooked crab meat, cooked crayfish waste	28, 30
3-ethyl-	Roasted clam, roasted dried squid	35, 36
5-ethyl-2-methyl-	Cooked shrimp, cooked crab meat	4, 9, 30
2-phenyl-	Roasted dried squid	35
4-hydroxy-	Cooked crab meat, cooked crayfish waste	25, 28
Pyrrole	Roasted clam, roasted dried squid, cooked crayfish waste, pasteurized crab meat	13, 30, 32, 35, 36
1*H*-	Cooked crayfish waste, raw/cooked crab meat, cooked clams	12, 23, 25, 28, 29, 31
1-methyl-	Raw/fermented/cooked shrimp, roasted dried squid, cooked crab tissue, cooked crayfish hepatopancreas	3, 5, 9, 28, 30, 35
2,5-dimethyl-	Roasted dried squid, roasted clams	35, 36
2-acetyl-	Roasted shrimp, spray dried shrimp powder, roasted clam, boiled scallops	14, 17, 22, 36
Pyrroline	Cooked corbicula, roasted dried squid	11, 35
2-acetyl-	Cooked crayfish waste, cooked lobster tail meat, cooked crab meat	26, 27, 33, 37
Pyrrolidine	Cooked crab meat	26, 27
1-methyl-	Roasted dried squid	35
1-ethyl-	Roasted dried squid	35
1-acetyl-	Roasted dried squid, roasted clams	35, 36
1-propyl-	Roasted dried squid	35
1-butyl-	Roasted dried squid	35
2-Pyrrolidinone		
N-methyl-	Roasted shrimp	17
Quinazoline		
2,4-dimethyl-	Cooked shrimp	4, 9

References: (1)–(36) as in footnotes to Table 8.1; (37) as in footnotes to Table 8.2; (42) as in footnotes to Table 8.5.

producing pyrrolidines can not be ignored (Tressl *et al.*, 1985a,b; Helak *et al.*, 1989; Kawai *et al.*, 1991).

2-Acetyl-1-pyrroline has been identified in the volatiles of cooked lobster tail meat (Cadwallader *et al.*, 1995) and cooked crab meat (Chung and Cadwallader, 1994) as well as aromatic rice (Buttery *et al.*, 1983) and wheat bread crust (Schieberle and Grosch, 1987). Tressl *et al.* (1985b) reported that pyrrolines were Maillard reaction products of proline and monosaccharides. Recently, Schieberle (1990) demonstrated that 2-acetyl-1-pyrroline was formed from the reaction of 2-oxypropanal with either proline or ornithine. This compound was reported as having a nutty/popcorn-like aroma (Chung *et al.*, 1995; Cadwallader *et al.*, 1995).

Trimethylamine (TMA) has been associated with off-flavour in fish and shrimp due to its low threshold in water (Pan and Kuo, 1994), and was also recognized as being an important contributor to the overall boiled crab-type aromas (Josephson and Lindsay, 1986). TMA is mainly derived from trimethylamine oxide by microbial reduction (Hebard *et al.*, 1982). *N,N*-Dimethyl-2-phenylethylamine is a main component of cooked Antarctic krill odour (Maga, 1978; Pan and Kou, 1994). This compound occurs naturally in krill and may originate from phytoplanktons (Kubota *et al.*, 1980b).

Imines are another class of nitrogen-containing compounds identified in roasted dried squid and are known to contribute a plant-like green note. They are formed by the reaction of aldehydes with primary amines (Flath *et al.*, 1989).

Table 8.7 Volatile amines and other nitrogen-containing compounds in shellfish

Compounds	Occurrence	References
Amine		
trimethyl-	Raw/fermented shrimp, raw/cooked crab meat, boiled scallop	3, 22, 26, 27, 31
N,N-(2-methylpropylidene)-2-(methyl-1-propenyl)-	Roasted dried squid	35
N,N-(2-methylpropylidene)-3-methylbutyl-	Roasted dried squid	35
N,N-(3-methylbutylidene)-3-methylbutyl-	Roasted dried squid	35
N,N-dimethyl-2-phenylethyl-	Raw/fermented/cooked shrimp	4, 10
Cyanamide		
N,N-dimethyl-	Roasted shrimp	17
Isovaleramide	Roasted shrimp	17
Indole	Cooked crayfish, raw/fermented/cooked shrimp, cooked corbicula	3, 4, 6, 11, 12, 17
1*H*-	Raw/cooked crab meat	31
Urea		
tetramethyl-	Roasted shrimp	17

References: (1)–(36) as in footnotes to Table 8.1.

8.2.6 Sulphur-containing compounds

Straight-chain and heterocyclic sulphur-containing compounds have been identified in the volatiles of a variety of shellfish such as krill, shrimp and crab. They are formed during processing or storage of foods and contribute to both desirable and undesirable aromas.

Two straight-chain sulphur-containing compounds, dimethyl disulphide and dimethyl trisulphide, have been found in most thermally processed crustacea such as prawn, crab meat, oyster, crayfish and shrimp (Table 2.8). They usually affect the overall food aroma because of their low threshold values (Buttery *et al.*, 1976). Tanchotikul and Hsieh (1989) described dimethyl trisulphide in crayfish waste as having a green vegetable-like aroma. The aroma of dimethyl disulphide has been characterized as onion- or cabbage-like (Fors *et al.*, 1983; Vejaphan *et al.*, 1988) or as spoiled sulphurous and bad egg-like (MacLeod and Cave, 1976). These compounds have been reported to give a spoiled and cooked cabbage odour to crayfish meat (Vejaphan *et al.*, 1988) and an onion off-flavour to prawn (Whitfield *et al.*, 1981). Dimethyl disulphide may be an oxidation product of methanethiol or a bacterial degradation product of methionine (Deck *et al.*, 1973; Christensen *et al.*, 1981). Alasalvar *et al.* (1997) examined the aroma compounds present in raw, cooked and frozen mackerel that were stored for 5 days at $15 \pm 2°C$. Mackerel samples that had been cooked in steam for 15 min and then stored, showed a small increase in dimethyl trisulphide which imparted a slight sulphurous aroma. The frozen mackerel samples accumulated large amounts of dimethyl disulphide which conveyed onion or cabbage-like odours. Dimethyl sulphide, with a petroleum-like odour, was found in frozen raw krill and is thought to originate from dimethyl-β-propiothetin by enzymatic action (Tokunaga *et al.*, 1977).

Heterocyclic sulphur-containing compounds play important roles in generating meaty aromas in a variety of meat products (Shahidi *et al.*, 1986; St. Angelo *et al.*, 1987) and were considered important volatile components of marine crustaceans (Kubota *et al.*, 1980c, 1986; Choi *et al.*, 1983). 2-Methylthiophene, a thermal degradation product of sulphur-containing amino acids (Vernin and Parkanyi, 1982), was identified in boiled crayfish tail meat (Vejaphan *et al.*, 1988) and pasteurized crab meat (Hsieh *et al.*, 1989), where it imparted an onion- or gasoline-like odour (Fors, 1983). Alkylthiophenes and thiophene aldehyde were derived from cysteine and cystine degradation (Schutte, 1974; Shankaranarayana *et al.*, 1974) or from the pyrolysis of amino acids in the presence of xylose (Mussinan and Katz, 1973) or from the interaction of glucose degradation products with hydrogen sulphide (Shibamoto and Russell, 1977). Alkylthiophenes has a sesame-like or popcorn-like aroma, depending on the temperature (Schutte, 1974).

3,5-Dimethyl-1,2,4-trithiolane was found in the volatiles of clams (Tanchotikul and Hsieh, 1991), cooked beef (Brinkman *et al.*, 1972), Antarctic krill (Kubota *et al.*, 1980b), crayfish hepatopancreas (Hsieh *et al.*, 1989) and cooked shrimp (Kawai *et al.*, 1990). Trithiolanes may be produced from the reaction of aldehydes, ammonia and hydrogen sulphide, which are normally generated from the thermal degradation of sugars, lipids and amino acids, in cooked foods (Boelens *et al.*, 1974; Kawai *et al.*, 1985). These compounds impart an onion-like odour in boiled krill.

Thiazoles such as 2,4-dimethyl- and 2,4,5-trimethylthiazole which have been identified in cooked krill (Kubota *et al.*, 1982) and beef (Mussinan and Katz, 1973) are important in generating meaty flavours in marine crustacea (Kubota *et al.*, 1986). Alkylthiazoles have been reported to have green, nutty, roasted, vegetable or meaty notes. Trimethylthiazoles have a cocoa and nutty character (Pan and Kou, 1994). 2-Acetylthiazole has a nutty, cereal (Ho and Jin, 1985), cracker- or popcorn-like aroma (Teranishi and Buttery, 1985). Thiazoles may be formed through the thermal degradation of cysteine or cystine (Shu *et al.*, 1985) or by interaction of sulphur-containing amino acids with carbonyl compounds (Hartman and Ho, 1984).

Dithiazines contribute important flavour aromas to heated foods (Pan and Kou, 1994). Various alkyl-substituted 1,3,5-dithiazines were identified in cooked shrimp, roasted dried squid, cooked krill and steamed clams (Table 8.8). These compounds were also found in model reactions, such as heat degradation of sulphur-containing amino acids (Mussinan and Katz, 1973), from a mixture of acetaldehyde, ammonia and hydrogen sulphide (Kawai *et al.*, 1985, 1986) or from the interaction of acetaldehyde with propionaldehyde and ammonium sulphide (Shu *et al.*, 1985). Thialdine (2,4,6-trimethyldihydro-1,3,5-dithiazine) is an important component in the volatiles of cooked krill (Kubota *et al.*, 1982), boiled shrimp (Kubota *et al.*, 1986; Choi *et al.*, 1983), steamed clams (Tanchotikul and Hsieh, 1991) and roasted dried squid (Kawai *et al.*, 1991). Thialdine had a medium-roast shrimp aroma (Kawai *et al.*, 1985).

Dimethyl- and methylethyl-1,3-dithiines and ethyl-, propyl- and butyl-substituted homologues of dihydrodithiazines have more recently been identified volatiles in shrimp (Kubota *et al.*, 1989). The concentration of these compounds was not as high as that of thialdine, but their low threshold values make them more important flavour compounds in shrimp. Each compound had a characteristic odour of fuel gas or Allium plant. Sulphur-containing compounds identified in shellfish are listed in Table 8.8.

8.2.7 Hydrocarbons

Various alkanes (C_8–C_{19}) have been identified in shellfish volatiles. For instance, n-nonane, n-decane and n-undecane were found in the volatiles of boiled and pasteurized crab meat (Matiella and Hsieh, 1990; Cha *et al.*,

Table 8.8 Sulphur-containing compounds in shellfish volatiles

Compounds	Occurrence	References
Sulphide		
dimethyl-	Shrimp, oyster hydrolysate, raw/boiled scallop	22, 44, 45
Disulphide		
carbon-	Cooked shrimp, raw/cooked crab meat	19, 31
dimethyl-	Cooked crayfish, oyster, raw/fermented/ pasteurized and boiled shrimp, spray dried shrimp powder, cooked crab by-products, cooked crab meat, boiled clams, boiled scallops, oyster hydrolysate	1, 4–7, 10, 12–14, 17, 18, 22, 25, 28–32, 33, 44
Trisulphide		
dimethyl-	Cooked crayfish, oyster, raw/fermented shrimp, boiled scallop, raw/cooked crab, oyster hydrolysate	1, 4–7, 10, 12, 13, 17, 18, 22, 25, 28, 31, 33, 44
methyl propyl-	Cooked shrimp	17
Sulphoxide		
dimethyl-	Oyster, roasted shrimp	1, 17
Methanethiol	Cooked shrimp, boiled scallop	17, 22
Thioethanol		
2-methyl-	Roasted shrimp	17
Thiopropanol		
3-methyl-	Roasted shrimp	17
Methional	Raw/fermented/cooked shrimp, boiled scallop	3, 5, 6, 22
2-(Methylthio)-methyl-2-butenal-	Spray dried shrimp powder, cooked crayfish waste	13, 14
Thiopropanal		
3-methyl-	Fermented/cooked shrimp, cooked crayfish waste	14, 23
Thiobutanal		
3-methyl-	Cooked shrimp	2, 5, 6
Thiazole	Cooked shrimp, cooked crayfish, cooked crab	5, 6, 10, 23, 25
2,4-dimethyl-	Cooked shrimp	2, 51
4,5-dimethyl-	Cooked shrimp	2, 51
2,3,5-trimethyl-	Cooked/fermented shrimp	4, 7
2,4,5-trimethyl-	Cooked shrimp, cooked krill	4, 37, 43, 51
2-acetyl-	Cooked shrimp, cooked corbicula, cooked krill, cooked crayfish waste, cooked lobster tail meat, cooked clams, raw/cooked crab meat	4–7, 9–11, 17, 18, 28, 29, 31, 37
benzo-	Cooked crayfish waste, cooked crab by-products, roasted dried squid	25, 28, 30, 31, 35, 51
2-ethyl-4,5-dimethyl-	Cooked shrimp	2, 51
2-isopropyl-4,5-dimethyl-	Cooked shrimp	2, 51
1,2,4-Trithiolane		
3,5-dimethyl-	Cooked shrimp, roasted clam	36, 47, 51
(Z,Z)-3,5-dimethyl-	Raw/fermented/cooked shrimp, cooked corbicula, cooked crayfish, cooked crab meat, steamed clams	4, 7, 11, 17, 28–30, 37
(E,E)-3,5-dimethyl-	Raw/fermented/cooked shrimp, cooked corbicula, steamed clams, cooked crayfish, cooked crab meat	4, 7, 11, 17, 28–30, 37
3-ethyl-5-methyl-	Cooked shrimp	17, 24, 46, 49, 51
3,5-diethyl-	Cooked shrimp	18, 24, 43, 46, 49, 51

Table 8.8 *continued*

Compounds	Occurrence	References
Thiophene		
2-methyl-	Cooked crayfish, steamed clams, raw/pasteurized/cooked crab meat	12, 23, 29, 31, 32
2-acetyl-	Cooked krill, cooked shrimp	2, 51
2-*N*-octyl-	Raw/fermented shrimp	3
carbaldehyde-2-	Raw/fermented/cooked shrimp cooked crab by-products, cooked crayfish, steamed clams	2, 28–30
carbaldehyde-3-	Cooked shrimp, cooked crab by-product, cooked crayfish, steamed clams	2, 28–30
carbaldehyde-5-methyl-2-	Cooked shrimp	2
1,3,5-Dihydrodithiazine		
4,6-dimethyl-2-propyl-	Cooked shrimp	8
4-ethyl-2,6-dimethyl-4,5,6-dihydro-	Cooked shrimp	2, 43, 46, 51
dihydro-2,4,6-trimethyl-4*H*-	Cooked shrimp, roasted clam	4, 7, 36
dihydro-2-ethyl-4,6-dimethyl-4*H*-	Cooked shrimp	4, 51
5,6-dihydro-2,4,6-trimethyl-4*H*-	Cooked shrimp, steamed clams, roasted dried squid	2, 4, 17, 29, 35, 37, 49, 51
5,6-dihydro-2,4-dimethyl-6-isopropyl-4*H*-	Dried/roasted squid	35, 50
5,6-dihydro-ethyl-2,6-dimethyl-4*H*-	Cooked shrimp	2, 37, 42, 43, 49
5,6-dihydro-2-ethyl-4,6-dimethyl-4*H*-	Cooked shrimp, roasted dried squid	2, 35, 37, 38, 49
5,6-dihydro-4,6-dimethyl-2-isopropyl-4*H*-	Dried/roasted dried squid	35, 50
5,6-dihydro-2,4,6-triethyl-4*H*-	Cooked shrimp	17
pyrrolidino[1,2-*e*]-4*H*-2,4-dimethyl-	Cooked shrimp, cooked corbicula, roasted dried squid	11, 35, 39
pyrrolidino[1,2-*e*]-2-butyl-4-methyl-4*H*-	Roasted dried squid	35
3,4,5,6-tetrahydro-2,4,6-trimethyl-2*H*-	Roasted clam, roasted dried squid	35, 36
1,3-Dithiine		
2,6-dimethyl-	Cooked shrimp	2, 17, 37, 40, 50
2-ethyl-6-methyl-	Cooked shrimp	6
6-ethyl-2-methyl-	Cooked shrimp	2

References: (1)–(36) as in footnotes to Table 8.1; (37) as in footnotes to Table 8.2; (38)–(39) as in footnotes to Table 8.3; (40)–(43) as in footnotes to Table 8.5; (44) Cha, 1995; (45) Ho and Jin, 1985; (46) Shu *et al.*, 1985; (47) Shankaranarayana *et al.*, 1974; (48) Schutte, 1974; (49) Kawai *et al.*, 1985; (50) Kubota *et al.*, 1980c; (51) Kubota *et al.*, 1989.

1993) and boiled crayfish (Vejaphan *et al.*, 1988). Alkanes are reported to contribute very little to the overall flavour of foods due to their high aroma thresholds (Grosh, 1982). However, there may be notable exceptions, especially when branched-chain alkanes are concerned. The compound 2,4,10,14-tetramethylpentadecane was reported to contribute a green, sweet aroma to crayfish processing waste (Tanchotikul and Hsieh, 1989). Hydrocarbons may be formed via lipid autoxidation processes through alkyl radicals or from decomposition of carotenoids (Pippen *et al.*, 1969). Alkylbenzenes and naphthalene have been identified in cooked crayfish, cooked crab and cooked shrimp (Table 2.8). Hsieh *et al.* (1989) reported that these compounds might be transferred to crayfish from environmental pollutants. Two chlorinated benzenes identified in crayfish, namely 1,2- and 1,3-dichlorobenzenes, were probably degradation products of various pesticides (Vejaphan *et al.*, 1988; Reineccius, 1991). Hydrocarbons identified in shellfish are listed in Table 8.9.

8.2.8 Phenols

Phenols have been identified in raw and cooked crab meat (Chung and Cadwallader, 1993), cooked crayfish (Vejaphan *et al.*, 1988; Tanchotikul and Hsieh, 1989) and raw, fermented and cooked shrimp (Choi *et al.*, 1983; Choi and Kobayashi, 1983; Choi and Kato, 1984a). The content of phenols in the by-products of shellfish was high compared to other aromatic compounds. The aromas associated with phenols are woody, smoky and burnt (Kubota *et al.*, 1980b). However, at high concentrations, phenols impart a medicinal odour to crayfish waste (Tanchotikul and Hsieh, 1989). The formation of simple phenols occurs via two pathways, namely decarboxylation of phenolic carboxylic acids and thermal degradation of lignin. Phenols found in shellfish are listed in Table 8.10.

8.2.9 Esters

Esters have been identified in the volatiles of boiled crayfish waste (Tanchotikul and Hsieh, 1989) and boiled shrimp waste (Mandeville *et al.*, 1992). Esters are believed to be products of esterification of carboxylic acids and alcohols (Peterson and Chang, 1982) previously formed from fermentation or lipid metabolism. A large number of esters have been identified in the volatiles of fresh, roasted and fermented foods. In general, esters give a sweet fruity aroma to foods. Esters present in shellfish volatiles are listed in Table 8.11.

Table 8.9 Hydrocarbons in shellfish volatiles

Compounds	Occurrence	References
Hexane	Boiled scallops	22
Heptane	Boiled scallops	22
2,2,4,6,6-penta-methyl-	Raw/cooked crab meat, crayfish waste	13, 31
Octane	Cooked crayfish waste, cooked crab meat	12, 13, 30
2-Octene	Cooked crayfish waste, cooked crab meat	13, 30
Octadiene	Boiled scallop	22
Octatriene		
1,3,5-	Oyster	1
3,7-dimethyl-1,3,6-	Cooked shrimp	4
3-Octyne	Cooked shrimp	7
Nonane	Cooked crayfish, cooked/pasteurized crab meat, boiled scallop	12, 22, 25, 31, 32
Decane	Cooked crayfish, roasted shrimp, pasteurized/cooked crab meat	12, 17, 31, 32
Undecane	Cooked crayfish, dried krill shell powder, boiled scallop, boiled/pasteurized crab meat	12, 22, 25, 28, 30, 32
Dodecane	Cooked shrimp, raw/cooked crab meat, boiled scallop, cooked crayfish	22, 25, 30, 31, 37
Tridecane	Cooked crayfish, cooked shrimp, raw/cooked crayfish	12, 17, 19, 31, 37
Tetradecane	Cooked crayfish, dried krill shell powder, raw/cooked crab meat	8, 12, 17, 31
Pentadecane	Cooked/roasted shrimp, dried shell krill powder, cooked crayfish, raw/cooked crab meat, raw oyster, oyster hydrolysate, cooked krill	2, 8, 12, 17, 31, 44
2,6,10,14-tetramethyl-	Cooked crayfish, cooked shrimp, raw/cooked crab meat	2, 3, 6, 7, 13, 19, 28, 31, 37
Hexadecane	Cooked shrimp, dried krill shell powder, cooked crayfish, oyster hydrolysate	2, 8, 13, 44
Heptadecane	Cooked crayfish, dried krill shell powder, steamed clams, raw/cooked crab meat, oyster hydrolysate	8, 12, 25, 28, 30, 31, 36, 44
Octadecane	Cooked crayfish, raw/cooked crab meat	12, 25, 31
Nonadecane	Cooked crayfish, raw crab meat	12
Heneicosane	Cooked crab meat, cooked crayfish	12, 25, 31
Benzene	Raw and fermented shrimp, cooked crayfish, pasteurized/cooked crab meat	22, 31
1,2-dichloro-	Cooked crayfish	10, 25, 30, 32
1,3-dichloro-	Cooked crayfish, raw/cooked crab meat	12, 23
1,4-dichloro-	Boiled/pasteurized crab meat, cooked crayfish	12, 23, 31
methyl-	Raw/fermented/cooked shrimp, raw/cooked crab meat	23, 25, 30, 32
1,2-dimethyl-	Boiled/pasteurized crab meat, cooked crayfish	13, 23, 25, 30–32
1,3-dimethyl-	Cooked crayfish, cooked shrimp, boiled/pasteurized crab meat	12, 13, 17, 23, 25, 30, 32
1,4-dimethyl-	Raw/boiled/pasteurized crab meat, cooked crayfish	13, 23, 25, 30–32
1,2,3-trimethyl-	Cooked crayfish, boiled/pasteurized crab meat	3, 5, 6, 10, 13, 31
1,2,4-trimethyl-	Cooked crayfish, boiled/pasteurized crab meat	12, 32
1,3,5-trimethyl-	Cooked crayfish, boiled/pasteurized crab meat	12, 25, 32

Table 8.9 *continued*

Compounds	Occurrence	Reference
ethyl-	Cooked crayfish, raw/pasteurized/ cooked crab meat	12, 13, 23, 25, 30–32
1-methyl-2-ethyl-	Cooked shrimp	6
1-ethyl-3-methyl-	Raw/cooked crab meat	31
1,4-diethyl-	Cooked crayfish	12, 23
propyl-	Cooked crayfish, boiled/pasteurized crab meat	12, 13, 23, 25, 32
cyclopropyl-	Cooked shrimp	12
Limonene	Cooked crayfish, cooked/roasted shrimp, dried krill shell powder	2, 8, 12, 17, 19, 37
Naphthalene	Cooked crayfish, raw/cooked crab meat	12, 23, 30, 32
1-methyl-	Cooked crayfish, cooked crab meat	12, 23, 30
2-methyl-	Cooked crayfish	12, 13, 23

References: (1)–(36) as in footnotes to Table 8.1; (37) as in footnotes to Table 8.2; (44) as in footnotes to Table 8.8.

Table 8.10 Phenols in shellfish volatiles

Compounds	Occurrence	References
Phenol	Cooked crayfish, raw/fermented/cooked shrimp, cooked corbicula, roasted dried squid, cooked clams, raw, pasteurized/cooked crab meat	2, 3, 4, 6, 7, 10–12, 13, 23, 25, 30, 31, 33, 36
2-methyl-	Smoked octopus, smoked cuttlefish, cooked crayfish, cooked crab meat	13, 25, 52
3-methyl-	Smoked octopus, smoked cuttlefish	52
4-methyl-	Smoked octopus, smoked cuttlefish, raw/ cooked crab meat, cooked crayfish, roasted dried squid	13, 31, 35, 52
2,3-dimethyl-	Smoked octopus, smoked cuttlefish, cooked crab meat	30, 52
2,4-dimethyl-	Smoked octopus, smoked cuttlefish	52
2,6-dimethyl-	Smoked octopus, smoked cuttlefish	52
ethyl-	Cooked crayfish, smoked octopus	13, 52
Ionol	Dried krill shell powder	4, 7

References: (1)–(36) as in footnotes to Table 8.1; (52) Kasahara and Nishibori, 1983.

Table 8.11 Esters in shellfish volatiles

Compounds	Occurrence	References
Ethyl acetate	Raw and fermented shrimp, raw oyster, oyster hydrolysate, cooked crayfish	3, 9, 13, 44
Methyl tetradecanoate	Roasted shrimp, roasted dried squid	17, 35
Methyl pentadecanoate	Roasted dried squid	35
Methyl hexadecanoate	Roasted dried squid, roasted shrimp, roasted clams	17, 35, 36
Methyl octadecanoate	Roasted clams, roasted dried squid	35, 36

References: (1)–(36) as in footnotes to Table 8.1; (44) as in footnotes to Table 8.8.

8.3 Nonvolatile flavour components in shellfish

The nonvolatile of shellfish are referred to as extractive components, and are defined as being water-soluble, low-molecular-weight compounds. They are classified into nitrogenous (free amino acids, nucleotides, organic bases and related compounds, etc.) and non-nitrogenous compounds (sugars, organic acids and inorganic compounds) with the exception of vitamins, minerals and pigments (Konosu and Yamaguchi, 1982).

8.3.1 Nitrogenous compounds

Free amino acids. The content of the total free amino acids in shellfish is higher than that in fish, but remains lower than that in crustacea. Higgins and Munday (1968) stated that a striking feature of the free amino acid pool of crustacea is the presence of large amounts of arginine, glycine and proline along with lesser amounts of alanine, glutamic acid and taurine (Figure 8.1). Camien *et al.* (1951) showed that the high concentrations of free amino acids, particularly glycine, proline, arginine, glutamic acid and alanine, were found intracellularly in muscle and that their concentrations were higher in marine species than in their freshwater counterparts. They suggested that certain free amino acids are probably important for the regulation of osmotic pressure in crustacean muscle.

Crabs were found to contain high amounts of glycine and arginine and high, but somewhat lower (compared to those for glycine and arginine), amounts of proline and taurine (Konosu, 1979). The four amino acids comprised 69–80% of the total free amino acids in crab and were the most important taste-active components present (Fuke, 1994). A distinct difference between crabs and prawns is the remarkably high amounts of glycine in prawns, reaching more than 1% of the fresh muscle in most species, whereas crabs accumulate more taurine (Konosu and Yamaguchi, 1982).

Prawns and lobsters are characterized by the presence of high amounts of free glycine in their muscle tissues which is thought to make an important contribution to taste. Hujita (1972) observed that the amount of free glycine in the muscle of crustacea paralleled their palatability. Moreover, they suggested that the other three sweet amino acids (alanine, serine and proline) might contribute somewhat to the acceptability of these species to some extent.

Other seafoods such as molluscs are rich in taurine, proline, glycine, alanine and arginine but levels of these amino acids varies from species to species. The content of glycine varied from 1455 mg (/100 g raw muscle) in scallops to 10 mg in squids (Konosu and Yamaguchi, 1982). Further-

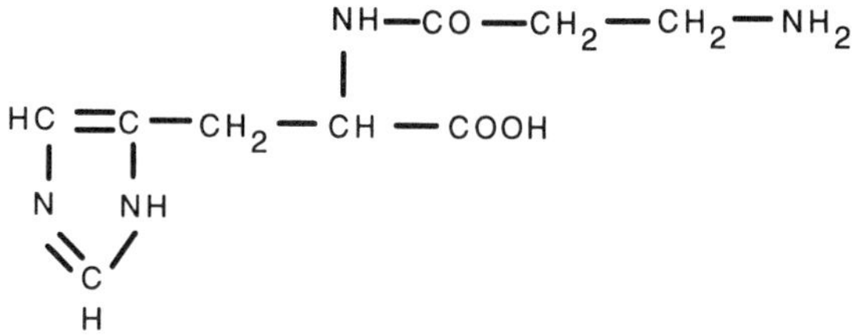

Carnosine (β-alanylhistidine)

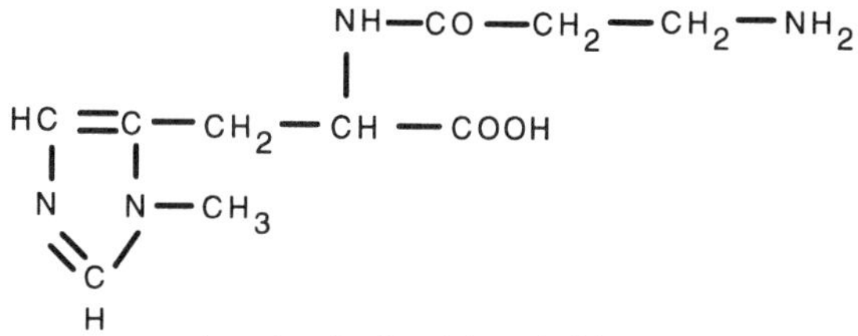

Anserine (β-alanyl-1-methylhistidine)

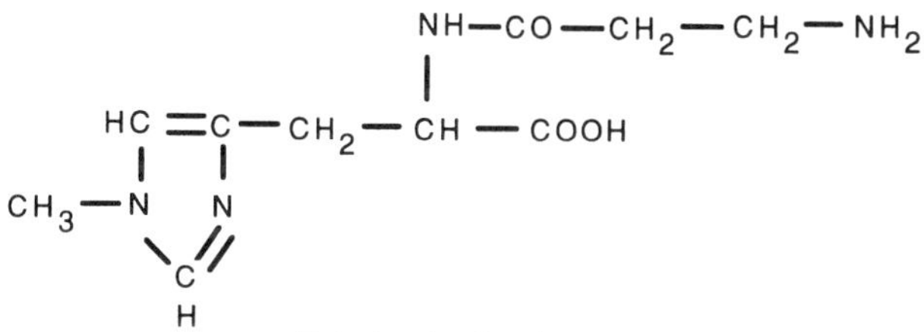

Balenine (β-alanyl-3-methylhistidine)

Figure 8.1 Peptides in seafoods (adapted from Konosu and Yamaguchi, 1982).

more, Fuke (1994) has reported that squids contain more proline while shellfish contain more glutamic acid. Table 8.12 shows the free amino acids present in some seafoods.

Peptides. A number of seafoods contain dipeptides such as carnosine, anserine and balenine. Carnosine is abundant in eels, anserine in tuna and balenine in baleen whales. Small amounts of carnosine, anserine and balenine are found in the muscles of crustaceans and molluscs. Lukton and Olcott (1958) demonstrated the presence of carnosine in the muscles of prawn, squid and crab and anserine in oyster. However, Suyama *et al.* (1970) analysed prawns, squids, crabs and clams, but the presence of anserine and carnosine was only confirmed in blue crab. Konosu *et al.*

Table 8.12 Contents of free amino acids in selected seafoods (mg/100 g)

Amino acids	Abalone	Scallop	Prawn	Snow crab
Tau	946	176	150	243
Asp	9	TR	TR	10
Thr	82	38	13	14
Ser	95	6	133	14
Gln + Asn	ND	ND	ND	TR
Glu	109	99	34	19
Pro	83	36	203	327
Gly	174	613	1222	623
Ala	98	82	43	187
Cys	ND	3	TR	ND
Val	37	10	17	30
Met	13	12	12	19
Ile	18	3	9	29
Leu	24	ND	13	30
Tyr	57	2	20	19
Phe	26	4	7	17
Trp	20	ND	TR	10
His	23	10	16	8
Lys	76	7	52	25
Arg	299	935	902	579

TR, trace; ND, not detected.
Adapted from Kato *et al.*, 1989.

(1978) analysed five species of crab and found trace amounts of carnosine in Alaskan king crab only. The structure of these dipeptides is shown in Figure 8.1.

Nucleotides and related compounds. Nucleotides are important in producing a palatable taste known as umami. Umami was first defined as the characteristic taste elicited by glutamates, and has since also been associated with monosodium glutamate (MSG) and with the disodium salts of 5′-nucleotides, such as inosine monophosphate (IMP), guanosine monophosphate (GMP) and adenosine monophosphate (AMP) (Fuke and Ueda, 1996). More than 90% of nucleotides in the muscles of fish and shellfish are purine derivatives with small amounts of uracil and cytosine being present (Seki, 1971).

In live muscles of animals, adenosine triphosphate (ATP) predominates, but after death it is enzymatically degraded to hypoxanthine and ribose through the pathway shown in Figure 8.2.

During enzymatic breakdown of ATP in fish and crustaceans, 5′-adenylic acid is converted to 5′-inosinic acid by adenylic acid deaminase. During the degradation of ATP, fresh fish accumulate IMP due to the slow conversion of IMP to inosine whereas crustaceans accumulate AMP due to the low activity of AMP deaminase (Konosu and Yamaguchi, 1982). On the other hand, molluscs convert 5′-adenylic acid to adenosine by

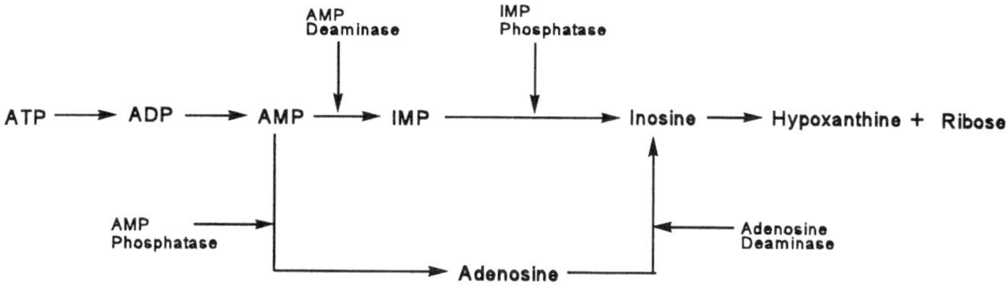

Figure 8.2 Enzymatic breakdown of ATP, where ATP = adenosine 5′-triphosphate, ADP = adenosine 5′-diphosphate, AMP = adenosine 5′-monophosphate and IMP = inosine 5′-monophosphate (adapted from Komata, 1990).

adenylic acid dephosphorylase due to the absence of adenylic acid deaminase. Adenosine, formed from ATP, is eventually converted to inosine by adenosine deaminase (Komato, 1990).

Nucleotides identified by Hayashi *et al.* (1978) in the leg meat extracts of boiled crabs included AMP as the main component, about half as much CMP as AMP and small amounts of IMP, GMP, adenosine diphosphate (ADP) and uridine 5′-monophosphate (UMP). AMP has been identified as being an important contributor to abalone, scallop and short neck clam flavour, while CMP has been reported to be a prominent component of sea urchins (Fuke and Konosu, 1991). Table 8.13 shows the distribution of nucleotides in the extracts of boiled crab legs.

Table 8.13 Nucleotides in the extracts of boiled crab legs (mg/100 g of tissue)

Species	CMP	AMP	GMP	UMP	IMP	ADP
Snow crab						
male	6	32	4	ND	5	7
female	6	20	1	ND	7	1
Blue crab						
male	59	63	1	ND	2	2
female	39	46	1	ND	TR	TR
Horsehair crab						
male	10	82	TR	ND	ND	3
female	22	87	ND	ND	ND	2
King crab						
male	21	70	ND	ND	1	3
female	33	65	ND	ND	ND	2
Alaska king crab						
male	25	46	1	ND	1	3
female	13	53	1	ND	5	2

Abbreviations used: CMP, cytidine 5′-monophosphate; AMP, adenosine 5′-monophosphate; GMP, guanosine 5′-monophosphate; UMP, uridine 5′-monophosphate; IMP, inosine 5′-monophosphate; ADP, adenosine 5′-diphosphate; TR, trace; ND, not detected.
Adapted from Hayashi *et al.*, 1978.

Quaternary ammonium salts. Trimethylamine oxide (TMAO) and betaines are the most common nitrogenous bases found in the muscle of shellfish. TMAO is abundant in squids and constitutes up to 27.2% of its total extractive nitrogen (Fuke, 1994). However, the content of TMAO varies widely among individual squids and a range of 100–1000 mg per individual has been reported (Endo *et al.*, 1962). TMAO was also present at a level of 70–350 mg in Pacific crabs (Konosu and Yamaguchi, 1982). Fuke and Konosu (1991) found comparably high amounts of TMAO in snow crab but only small amounts in clam, abalone and scallop (Table 8.14). Trimethylamine (TMA) and dimethylamine (DMA) have been associated with the deteriorated odour and flavour quality of fish. TMA is very low in fresh muscle but increases with the postmortem bacterial reduction of TMAO (Konosu and Yamaguchi, 1982). DMA is produced by the breakdown of TMAO by enzymes in the muscle of fish (Hebard *et al.*, 1982).

Betaines are a main nonprotein nitrogenous constituents in the muscle of crustaceans and molluscs (Hayashi and Konosu, 1977). The main compound is glycine betaine which is present at levels between 400 and 900 mg in hard clams, octopus, abalone, squid, oysters, mussels and prawns (Konosu and Kasai, 1961; Shiau *et al.*, 1994). Glycine betaine, unlike TMAO, is relatively uniform among species as was demonstrated by Endo *et al.* (1962). They found the content of glycine betaine among six species of squid to be between 74 and 105 mg. Next to glycine betaine, hormarine has been widely distributed among marine invertebrates (Hayashi *et al.*, 1978). Table 8.15 shows the amount of glycine betaine present in shellfish.

Table 8.14 Trimethylamine oxide (TMAO) in the muscles of shellfish

Species	TMAO (mg/100 g)
Crustaceans	
Blue crab	65
Horsehair crab	140
Prawn	213
Squilla	128
Molluscs	
Octopus	134
Squid	217–1045
Fan mussel	72
Scallop	52
Abalone	3

Adapted from Konosu and Yamaguchi, 1982.

Table 8.15 Glycine betaine in the muscles of shellfish

Species	Glycine betaine (mg/100 g)
Crustaceans	
Blue crab	646
Horsehair crab	711
Snow crab	357
King crab	476
Alaska king crab	417
Prawn	251–961
Krill	106
Molluscs	
Octopus	1434
Squid	619–928
Short-necked clam	679
Hard clam	727
Oyster	805
Scallop	211
Fan-mussel	964
Abalone	668

Adapted from Konosu and Yamaguchi, 1982.

8.3.2 *Non-nitrogenous compounds*

Organic acids. Organic acids such as acetic, propionic, oxalic, succinic, pyruvic and lactic acids have been identified in crustaceans (Hayashi *et al.*, 1979). Lactic acid, which is produced by glycerolysis, is the main acid present in shellfish (Table 8.16). Hayashi *et al.* (1979) found the content of lactic acid in five species of crab to vary between 30 and 200 mg. Succinic acid was also found but its level was far lower than that of lactic acid. Traces of acetic, propionic, and oxalic acids were also identified in crab meats. Succinic acid was recognized to be the main component in short-necked clams and it is claimed to be one of the taste-producing substances in clams (Takagi and Simidu, 1962).

Table 8.16 Organic acids in the muscles of shellfish

Species	Propionic	Acetic	Pyruvic	Succinic	Lactic	Malic	Citric
Snow crab	TR	TR	NR	9	100	NR	NR
Alaska king crab	TR	TR	NR	4	130	NR	NR
Prawn	4–7	4–5	7	6–27	232	NR	NR
Short-necked clam	63	5	1	274	4	15	22
Hard clam	5	9	8	80	26	58	7
Oyster	32	26	9	59	52	6	10
Scallop	NR	NR	NR	14	2	NR	NR

TR, trace; NR, not reported.
Adapted from Konosu and Yamguchi, 1982.

Sugars. The content of free sugars in fish and shellfish muscles is fairly low, so it is unlikely that they contribute to flavour. Glucose and ribose are the main free sugars present in the muscles of fish and shellfish (Table 8.17). Hayashi *et al.* (1979) examined the amount of free sugars in five species of crab. They found that glucose was the most abundant monosaccharide in all species examined. Ribose, which is distributed widely in marine products, was found at relatively low levels in crab. Small amounts of fructose, arabinose and inositol were detected but no oligosaccharides such as sucrose or maltose were identified. The occurrence of galactose in squids and prawns has been reported. Mandeville *et al.* (1992) examined the amount of free sugars present in shrimp waste. The most abundant sugar was ribose followed by fructose. Mannose, xylose and glucose were present in small amounts.

Inorganic salts. Sidwell *et al.* (1977) reviewed the minerals present in over 160 species of fish, crustaceans and molluscs. However, they were concerned with the nutritional aspects of the minerals rather than their contribution to flavour. Hayashi *et al.* (1979) examined six inorganic ions in five species of crab. They found that sodium and potassium ions comprised the major part of the cations ranging from 72 to 422 mg/100 g of tissue. As for the anions, the content of chloride ion was approximately 600 mg/100 g of tissue while phosphate ion was present at 100–200 mg/100 g of tissue in most samples. It has been shown that these ions make an important contribution to flavour (Hayashi *et al.*, 1981). See Table 8.18.

8.4 Summary

Perception of the flavour of shellfish is a complex process that involves both senses of taste and smell. Generally, substances that stimulate the

Table 8.17 Free sugars present in seafoods (mg/100 g raw muscle)

Free sugars	Shrimp waste	Snow crab (legs)	Clam	Alaska king crab (legs)	Sea-urchin
Ribose	378.0[a]	2.0[b]	NR	TR[b]	193.0[c]
Glucose	5.2[a]	12.0[b]	3.0[c]	61[b]	NR
Fructose	102.4[a]	14.0[b]	NR	TR[b]	NR
Xylose	6.6[a]	NR	NR	TR[b]	NR
Mannose	19.0[a]	NR	3.0[c]	TR[b]	NR

TR, trace; NR, not reported.

[a]Adapted from Mandeville *et al.*, 1992.
[b]Adapted from Hayashi *et al.*, 1979.
[c]Adapted from Fuke and Konosu, 1991.

Table 8.18 Inorganic constitutents in the muscle of seafoods (mg/100 g raw muscle)

Species	Na^+	K^+	Ca^{2+}	Mg^{2+}	Cl^-	PO_4^{3-}
Scallop	73[a]	218[a]	TR[a]	TR[a]	95[a]	213[a]
Clam	378[a]	273[a]	52[a]	40[a]	452[a]	72[a]
Snow crab	191[b]	197[b]	TR[b]	TR[b]	336[b]	217[b]
Alaska king crab	336[b]	277[b]	TR[b]	TR[b]	584[b]	169[b]

TR, trace.

[a]Adapted from Fuke and Konosu, 1991.
[b]Adapted from Konosu and Yamaguchi, 1982.

sense of taste are nonvolatile components, whereas substances that stimulate the sense of smell are volatile.

Volatile components of shellfish are considered as being the most important determinant for their flavour quality. Aldehydes, ketones, nitrogen- and sulphur-containing compounds are considered the most important contributors to the odour of shellfish. Short-chain aldehydes contribute a fresh or raw shellfish flavour which has been described as having a green and plant-like aroma. Ketones contribute to the sweet floral and fruity flavour of raw crustaceans. However, vinyl ketones are products of lipid oxidation formed during heating and have a green-leafy aroma. Alkylpyrazines and sulphur-containing compounds contribute to the cooked odour of shellfish. Pyrazines produce nutty, roasted and toasted aromas in foods, while sulphur-containing compounds have been characterized as onion- or cabbage-like or as spoiled sulphurous and bad egg-like.

The specific taste of each food relies on extractive components. The extractive components are referred to as nonvolatile components and are defined as water-soluble, low-molecular-weight compounds. The most important nonvolatile components are amino acids and nucleotides. Nucleotides are important in producing the palatable taste known as umami. Umami substances contribute to the taste of shellfish in cooperation with other substances such as free amino acids and inorganic ions (Komata, 1990). Shellfish contain appreciable amounts of arginine, taurine, glycine and proline (Higgins and Munday, 1968). These free amino acids contribute to the sweet taste of raw crustaceans and to the formation of flavour compounds during heating. Nonvolatile components of shellfish are regarded as principal flavour producers.

Great advances in the understanding of fish and shellfish flavours have been made in recent years. Not only important compounds have been identified, but also great strides in understanding the biological and chemical mechanisms of formation of flavour compounds have been undertaken. However, other important compounds and character-impact volatiles remain to be identified.

References

Alasalvar, C., Quantick, P.C. and Grigor, J.M. (1997). Aroma compounds of fresh and stored mackerel (*Scomber scombrus*). In *Flavour and Lipid Chemistry of Seafoods*, eds. F. Shahidi and K.R. Cadwallader. ACS Symposium Series 674, American Chemical Society, Washington, DC, pp. 39–54.

Allison, C. and MacFarlane, G.T. (1989). Influence of pH, nutrient availability and growth rate on amine production by *Bacteriodes fragilis* and *Clostridium perfringens*. *Appl. Environ. Microbiol.*, **55**, 2894–2898.

Baek, H.H. and Cadwallader, K.R. (1996). Volatile compounds in flavor concentrates produced from crayfish-processing by-products with and without protease treatment. *J. Agric. Food Chem.*, **44**, 3262–3267.

Baldings, H.T. (1970). Cold storage defects in butter and their relation to the autoxidation of unsaturated fatty acids. *Neth. Milk Dairy J.*, **24**, 149–153.

Boelens, M., van der Linde, L.M., de Valois, P.J., van Dort, H.M. and Takken, H.J. (1974). Organic sulphur compounds from fatty aldehydes, hydrogen sulphide, thiols and ammonia as flavor constituents. *J. Agric. Food Chem.*, **22**, 1071–1074.

Brinkman, H.W., Copier, J.J, deLeuw, M. and Tjan, S.B. (1972). Components contributing to beef flavor. Analysis of the headspace volatiles of beef broth. *J. Agric. Food Chem.*, **20**, 177–181.

Buttery, R.G., Guadagni, D.G., Ling, L.C., Seifert, R.M. and Lipton, W. (1976). Additional volatile components of cabbage, broccoli and cauliflower. *J. Agric. Food Chem.*, **24**, 829–832.

Buttery, R.G., Ling, L.C. and Teranishi, R. (1977). Roasted lamb fat: basic volatile components. *J. Agric. Food Chem.*, **25**, 1222–1229.

Buttery, R.G., Ling, L.C., Juliano, B.O. and Turnbaugh, J.G. (1983). Cooked rice aroma and 2-acetyl-1-pyrroline. *J. Agric. Food Chem.*, **31**, 823–826

Cadwallader, K.R., Tan, Q., Chen, F. and Meyers, S.P. (1995). Evaluation of the aroma of cooked spiny lobster tail meat by aroma extract dilution analysis. *J. Agric. Food Chem.*, **43**, 2432–2437.

Camian, M.N., Sarlet, H., Duchateau, G. and Florkin, M. (1951) Non-protein amino acids in muscle and blood of marine and fresh water crustacea. *J. Biol. Chem.*, **193**, 881–885.

Cha, Y.J. (1995). Volatile compounds in oyster hydrolysate produced by commercial protease. *J. Korean Soc. Food Nutr.*, **24**, 420–426.

Cha, Y.J., Baek, H.H., and Hsieh, T.C.-Y. (1992). Volatile components in flavor concentrates from crayfish processing waste. *J. Sci. Food Agric.*, **58**, 239–248.

Cha, Y.J., Cadwallader, K.R. and Baek, H.H. (1993). Volatile flavor components in snow crab cooker effluent and effluent concentrate. *J. Food Sci.*, **58**, 525–530.

Chang, S.S and Peterson, R.J. (1977). Recent developments in the flavor of meat. *J. Food Sci.*, **42**, 298–305.

Choi, S.H. and Kato, H. (1983). Volatile components of fermented *Euphausia pacifica* Hansen changes during fermentation and after cooking. *Nippon Nogeikagaku Kaishi*, **57**, 1121–1125.

Choi, S.H. and Kato, H. (1984a). Volatile components of *Sergia lucens* and its fermented product. *Agric. Biol. Chem.*, **48**, 1479–1486.

Choi, S.H. and Kato, H. (1984b). Changes in cooked odor of Antarctic krill during frozen storage. *Agric. Biol. Chem.*, **48**, 545–547.

Choi, S.H. and Kobayashi, A. (1983). Cooked odor of commercial Shoikara made from Myside shrimps (odor of fermented small shrimp part II). *Nippon Shokuhin Kogyo Gakkaishi*, **30**, 404–408.

Choi, S.H., Kobayashi, A. and Yamanishi, T. (1983). Odor of cooked small shrimp *Acetes japonicus* kishinouye: Difference between raw material and fermented product. *Agric. Biol. Chem.*, **47**, 337–342.

Christensen, B.W. Kjaer, A. and Madsen, J.O. (1981). Volatile sulphur compounds and other headspace constituents of north sea fish oils. *J. Am. Oil Chem. Soc.*, **58**, 1053–1056.

Chung, H.Y. and Cadwallader, K.R. (1993). Volatile components in blue crab meat and processing by-product. *J Food Sci.*, **6**, 1203–1207.

Chung, H.Y. and Cadwallader, K.R. (1994). Aroma extract dilution analysis of blue crab claw meat volatiles. *J. Agric. Food Chem.*, **42**, 2867–2870.

Chung, H.Y., Chen, F. and Cadwallader, K.R. (1995). Cooked blue crab meat aroma compared with lump meat. *J. Food Sci.*, **60**, 289–291.

Deck, R.E., Pokorny, J. and Chang, S.S. (1973). Isolation and identification of volatile compounds from potato chips. *J. Food Sci.*, **38**, 345–349.

Endo, K., Hujita, M. and Simidu, W. (1962). Studies on the muscle of aquatic animals – XXX. Free amino acids, trimethylamine oxide and betaine in squids. *Bull. Jpn. Soc. Sci. Fish.*, **28**, 833–836.

Ferretti, A. and Flanagan, V.P. (1971). The lactose–casein (Maillard) browning system: volatile components. *J. Agric. Food Chem.*, **38**, 245–249.

Ferretti, A., Flanagan, V.P. and Ruth, J.M. (1970). Nonenzymatic browning in a lactose–casein model system. *J. Agric. Food Chem.*, **18**, 13–18.

Flath, R.A., Matsumoto, K.E., Binder, R.G., Cunningham, R.T. and Mon. R.T. (1989). Effects of pH on the volatiles of hydrolyzed protein insect baits. *J Agric. Food Chem.*, **37**, 814–819.

Fors, S. (1983). Sensory properties of volatile Maillard reaction products and related compounds. In *Maillard Reaction in Food and Nutrition*, eds. G.R. Waller and M.S. Feather. ACS Symposium Series 215, American Chemical Society, Washington, DC, pp. 185–286.

Fuke, S. (1994). Taste-active components of seafoods with special reference to umami substances. In *Seafoods: Chemistry, Processing, Technology and Quality*, eds. F. Shahidi and J.R. Botta. Blackie Academic and Professional, Glasgow, pp. 115–139.

Fuke, S. and Konosu, S. (1991). Taste-active components in some foods: A review of Japanese research. *Physiology and Behavior,* **49**, 863–868.

Fuke, S. and Udea, Y. (1996). Interactions between umami and other flavor characteristics. *Trends Food Sci. Technol.*, **7**, 407–411.

Grosch, W. (1982). Lipid degradation products and flavour. In *Food Flavours: Part A. Introduction*, eds. I.D. Morton and A.J. Macleod. Elsevier Scientific, New York, pp. 325–398.

Griffith, R. and Hammond, E.G. (1989). Generation of Swiss cheese flavor components by the reaction of amino acids with carbonyl compounds. *J. Dairy Sci.*, **72**, 664–673.

Hartman, G.J. and Ho, C.T. (1984). Volatile products of the reaction of sulphur-containing amino acids with 2,3-butanedione. *Lebens. Wiss. u. Technol.* **17**, 171–174.

Hayashi, T. and Konosu, S. (1977). Quaternary ammonium bases in adductor muscle of fan-mussel. *Bull. Jpn. Soc. Sci. Fish.*, **43**, 343–348.

Hayashi, T., Yamaguchi, K. and Konosu, S. (1978). Studies on flavor components in boiled crabs. II. Nucleotides and organic bases in the extracts. *Bull. Jpn. Soc. Sci. Fish.*, **44**, 1357–1362.

Hayashi, T., Asakawa, A., Yamaguchi, K. and Konosu, S. (1979). Studies on flavor components in boiled crabs. III. Sugars, organic acids and minerals in the extracts. *Bull. Jpn. Soc. Sci. Fish.*, **45**, 1325–1329.

Hayashi, T., Yamaguchi, K. and Konosu, S. (1981). Sensory analysis of taste-active components in the extract of boiled snow crab meat. *J. Food Sci.*, **46**, 479–483.

He, C.S. and Kato, H. (1985). Flavor of fermented product of Antarctic krill prepared by modified method (odor of fermented small shrimp part VI). *Nippon Shokuhin Kogyo Gakkaishi*, **32**, 274–280.

Heath, H.B. and Reineccius, G. (1986). Flavor and its study. In *Flavour Chemistry and Technology*, eds. H.B. Heath and G. Reineccius. AVI Publishing Co., Westport, CT, pp. 71–111.

Hebard, C.E., Flick, G.J. and Martin, R.E. (1982). Occurrence and significance of trimethylamine oxide and its derivatives in fish and shellfish. In *Chemistry and Biochemistry of Marine Food Products*, eds. R.E. Martin, G.J. Flick, C.E. Hebard and D.R. Ward. AVI Publishing Co., Westport, CT, pp. 149–304.

Helak, B., Kerstem, E., Spengler, K., Tressl, R. and Rewicki, D. (1989). Formation of furrylpyrrolidines and -piperidines on heating L-proline with reducing sugars and furancarboxaldehydes. *J. Agric. Food Chem.*, **37**, 405–410.

Higgins, A.E. and Munday, K.A. 1968. In *Advances in Comparative Physiology and Biochemistry*, ed. O. Lowenstein. Academic Press, New York, pp. 293–315.

Ho, C.T. and Carlin, J.T. (1989). Formation and aroma characteristics of heterocyclic compounds in foods. In *Flavour Chemistry: Trends and Developments*, eds. R. Teranishi,

R.G. Buttery and F. Shahidi. ACS Symposium Series 388, American Chemical Society, Washington, DC, pp. 92–104.

Ho, C.T. and Jin, Q.Z. (1985). Aroma properties of some alkylthiazoles. *Perfumer Flavorist*, **9**, 15–18.

Ho, C.T., Lee, K.N. and Jin, Q.Z. (1983). Isolation and identification of volatile flavor compounds in fried bacon. *J. Agric. Food Chem.*, **31**, 336–342.

Hodge, J.E. (1967). Origin of flavour in foods. Nonenzymic browning reactions. In *Chemistry and Physiology of Flavour*, eds. H.E. Schultz, E.A. Day and L.M. Libbey. AVI Publishing Co., Westport, CT, pp. 465–491.

Hsieh, T.C.-Y., Vejaphan, W., Williams, S.S. and Matiella, J.E. (1989). Volatile flavour components in thermally processed Louisiana red swamp crayfish and blue crab. In *Thermal Generation of Aromas*, eds. T.H. Parliment, R.J. McGorrin and C.T. Ho. ACS Symposium Series 409, American Chemical Society, Washington, DC, pp. 386–395.

Huang, T.C., Bruechert, L.J, Hartman, T.G., Rosen, R.T. and Ho, C.T. (1987). Effects of lipids and carbohydrates on thermal generation of volatiles from commercial zein. *J. Agric. Food Chem.*, **35**, 985–988.

Hujita, M. (1972). Studies on the muscle of aquatic animals. XXXXVI. Free amino acids, trimethylamine oxide and betaine in shrimp muscle. *Mem. Fac. Agric. Kinki Univ. Kinki. Daigaku Nogakubu Kiyo*, **5**, 61–67.

Josephson, D.B. (1991). Seafoods. In *Volatile Compounds in Foods and Beverages*, ed. H. Maarse. Marcel Dekker, New York, pp. 179–202.

Josephson, D.B. and Lindsay, R.C. (1986). Enzymic generation of volatile aroma compounds from fresh fish. In *Biogeneration of Aromas*, eds. T.H. Parliment and R. Croteau. ACS Symposium Series 317, American Chemical Society, Washington, DC, pp. 201–21.

Josephson, D.B. and Lindsay, R.C. (1987). Retro-aldol degradations of unsaturated aldehydes: role in the formation of 4-heptenal from t2,c6-nonadienal in fish, oyster and other flavors. *J. Am. Oil Chem. Soc.*, **64**, 132–138.

Josephson, D.B., Lindsay, R.C. and Stuiber, D.A. (1983). Identification of compounds characterizing the aroma of fresh white fish (*Coregonus clupeaformus*). *J. Agric. Food Chem.*, **31**, 326–330

Josephson, D.B., Lindsay, R.C. and Stuiber, D.A. (1984a). Variations in the occurrences of enzymatically derived volatile aroma compounds in salt and fresh water fish. *J. Agric. Food Chem.*, **31**, 1344–1347.

Josephson, D.B., Lindsay, R.C. and Stuiber, D.A. (1984b). Biogenesis of lipid-derived volatile aroma compounds in the Emerald shiner (*Notropis atherinoides*). *J. Agric. Food Chem.*, **32**, 1347–1352.

Josephson, D.B., Lindsay, R.C. and Stuiber, D.A. (1985). Volatile compounds characterizing the aroma of fresh Atlantic and Pacific oysters. *J. Food Sci.*, **50**, 5–9.

Josephson, D.B., Lindsay, R.C. and Stuiber, D.A. (1987). Enzymic hydroperoxide initiated effects in fresh fish. *J. Food Sci.*, **52**, 596–600.

Karahadian, C. and Lindsay, R.C. (1989). Evaluation of compounds contributing characterizing fishy flavors in fish oils. *J. Amer. Oil Chem. Soc.*, **66**, 953–960.

Kasahara, K. and Nishibori, K. (1983). Volatile components of smoked cuttlefish and octopus. *Bull. Jap. Soc. Sci. Fish.*, **49**, 913–918.

Kato, H., Rhue, M.R. and Nishimura, T. (1989). Role of free amino acids and peptides in food taste. In *Flavour Chemistry Trends and Developments*, eds. R. Teranishi, R.G. Buttery, and F. Shahidi. ACS Symposium Series 388, American Chemical Society, Washington, DC, pp. 158–174.

Kato, S., Fujimaki, M. and Kurata, T. (1969). Pyrolysis of sulphur-containing amino acids. *Agric. Biol. Chem.*, **33**, 1144–1151.

Kato, S., Kurata, T. and Fujimaki, M. (1973). Volatile compounds produced by the reaction of L-cysteine or L-cystine with carbonyl compounds. *Agric. Biol. Chem.*, **37**, 539–544.

Kawai, T., Irie, M. and Sakaguchi, M. (1985). Degradation of 2,4,6-trialkyltetrahydro-1,3,5-thiadiazines during storage. *J. Agric. Food Chem.*, **33**, 393–397.

Kawai, T., Irie, M. and Sakaguchi, M. (1986). Heterocyclic compounds: A mechanism for the formation from aldehydes, ammonia and hydrogen sulphide. *Koryo*, **151**, 53–60.

Kawai, T., Ishida, Y., Kakiuchi, H., Ikeda, N. and Tsuneya, T. (1990). Volatile compounds of boiled and roasted short-necked clam *Tapes philippinarium. Nippon Suisan Gakkaishi*, **56**, 795–802.

Kawai, T., Ishida, Y., Kakiuchi, H., Ikeda, N. Higashida, T. and Nakamura, S. (1991). Flavor components of dried squid. *J. Agric. Food Chem.*, **39**, 770–777.
Kim, H.R., Baek, S.P., Cadwallader, K.R. and Godber, J.S. (1994). Crayfish hepatopancreatic extract improves flavor extractability from a crab processing by-product. *J. Food Sci.*, **59**, 91–96.
Kobayashi, A., Kubota, K., Iwamoto, M. and Tamura, H. (1989). Syntheses and sensory characterization of 5,8,11-tetradeca-trien-2-one isomers. *J. Agric. Food Chem.*, **37**, 151–154.
Koehler, P.E. and Odell, G.V. (1970). Factors affecting the formation of pyrazine compounds in sugar–amine reactions. *J. Agric. Food Chem.*, **18**, 895–898.
Komtata, Y. (1990). Umami taste of seafoods. *Food Rev. Int.*, **6**, 457–487.
Konosu, S. (1979). The taste of fish and shellfish. In *Food Taste Chemistry*, ed. J.C. Boudreau. ACS Symposium Series 115, American Chemical Society, Washington, DC, pp. 185–203.
Konosu, S. and Kasai, E. (1961). Muscle extracts of aquatic animals – III on the method for determination of betaine and its content of the muscle of some marine animals. *Bull. Jpn. Soc. Sci. Fish.*, **27**, 194–198.
Konosu, S. and Yamaguchi, K. (1982). The flavour components in fish and shellfish. In *Chemistry and Biochemistry of Marine Food Products*, eds. R.E. Martin, G.J. Flick, C.E. Hebard and D.K. Ward. AVI Publishing Co., Westport, CT, pp. 367–403.
Konosu, S., Yamaguchi, K. and Hayashi, T. (1978). Studies on flavor components in boiled crabs – I. Amino acids and related compounds in extracts. *Bull. Jpn. Soc. Sci. Fish.*, **44**, 505–510.
Krishnamurthy, R.G., Smouse, S.T., Mookherjee, B.D., Reddy, R.D. and Chang, S.S. (1967). Identification of 2-pentyl furan in fats and oils and its relationship to the reversion flavor of soyabean oil. *J. Food Sci.*, **32**, 372–377.
Kubota, K. and Kobayashi, A. (1988). Identification of unknown methyl ketones in volatile flavor components from cooked small shrimp. *J. Agric. Food Chem.*, **36**, 121–123.
Kubota, K., Yokoyama, K., Yamanishi, T. and Akatsuka, S. (1980a). Odor of dried shell powder of Antarctic krills and liquid seasoning of the hydrolysate. *Nippon Nogeikagaku Kaishi*, **54**, 727–731.
Kubota, K., Matsufusa, K., Yamanishi, T. and Yanase, M. (1980b). Cooked odor of Antarctic krills. *Nippon Nogeikagaku Kaishi*, **54**, 1–5.
Kubota, K., Kobayashi, A. and Yamanishi, T. (1980c). Some sulphur containing compounds in cooked odor concentrate from boiled Antarctic krills (*Euphausia superba dana*). *Agric. Biol. Chem.*, **44**, 2677–2682.
Kubota, K., Kobayashi, A. and Yamanishi, T. (1981). Effect of precooking on the odor of Antarctic krills (cooked odor of Antarctic krills, part V). *Nippon Shokuhin Kogyo Gakkaishi*, **28**, 457–460.
Kubota, K., Kobayashi, A. and Yamanishi, T. (1982). Basic and neutral compounds in the cooked odor from Antarctic krill. *Agric. Biol. Chem.*, **46**, 2835–2839.
Kubota, K., Takasaki, S., Kobayashi, A. and Akatsuka, S. (1985). Changes of aroma constituents of krill hydrolyzate during spray drying process. *Nippon Shokuhin Kogyo Gakkaishi*, **32**, 765–768.
Kubota, K., Shuimaya, H. and Kobayashi, A. (1986). Volatile components of roasted shrimp. *Agric. Biol. Chem.*, **50**, 2867–2873.
Kubota, K., Watanabe, K. and Kobayashi, A. (1988). Novel dithiazine in volatile components from cooked Sakuraebi (*Sergia lucens Hansen*). *Agric Biol. Chem.*, **52**, 1537–1540.
Kubota, K., Uchida, C., Kurosawa, K., Komuro, A. and Kobayashi, A. (1989). Identification and formation of characteristic volatile compounds from cooked shrimp. In *Thermal Generation of Aromas*, eds. T.H. Parliment, R.J. McGorrin and C.T. Ho. ACS Symposium Series 409, American Chemical Society, Washington, DC, pp. 376–385.
Kubota, K., Nakamoto, A., Moriguchi, M., Kobayashi, A. and Ishii, H. (1991). Formation of pyrrolidino[1,2-e-]-2,4-dimethyl-1,3,5-dithiazine in the volatiles of boiled short necked clam, and corbicula. *J. Agric. Food Chem.*, **39**, 1127–1130.
Laing, D.G and Jinks, A. (1996). Flavor perception mechanisms. *Trends Food Sci. Technol.*, **7**, 387–389.
Liebich, H.M., Douglas, D.R., Zlatkis, A., Muggler-Chavan, F. and Donzel, A. (1972). Volatile components in roast beef. *J. Agric. Food Chem.*, **20**, 96–99.
Lindsay, R.C. (1990). Fish flavors. *Food Rev. Int.*, **6**, 437–455.
Lukton, A. and Olcott, H.S. (1958). Content of free imidazole compounds in the muscle tissue of aquatic animals. *Food Res.*, **23**, 611–615.

MacLeod, A.J. and Cave, S.J. (1976). Volatile flavor components of eggs. *J. Sci. Food Agric.*, **27**, 799–806.

Maga, J.A. (1978). Simple phenol and phenolic components in food flavor. *CRC Crit. Rev. Food Sci. Nutr.*, **16**, 323–372.

Maga, J.A. (1979). Furans in food. *CRC Crit. Rev. Food Sci. Nutr.*, **17**, 355–400.

Maga, J.A. (1982). Pyrazines in foods: An update. *CRC Crit. Rev. Food Sci. Nutr.*, **20**, 1–48.

Maga, J.A. (1983). Heterocyclic flavoring and aroma compounds A. Pyrazines in foods: An update (I). *Koryo*, **140**, 69–86.

Maga, J.A. and Sizer, C.E. (1973). Pyrazines in foods: A review. *J. Agric. Food Chem.*, **21**, 22–30.

Mandeville, S., Yaylayan, V. and Simpson, B. (1992a). Proximate analysis, isolation and identification of amino acids and sugars from raw and cooked commerical shrimp waste. *Food Biotechnol.*, **6**, 51–64.

Mandeville, S., Yaylayan, V. and Simpson, B. (1992b). GC/MS analysis of flavor-active compounds in commercial shrimp waste. *J. Agric. Food Chem.*, **40**, 1275–1279.

Mason, M.E., Johnson, B. and Hamming, M.C. (1967). Volatile components of roasted peanuts. The major monocarbonyls and some noncarbonyl components. *J. Agric. Food Chem.*, **15**, 66–70.

Matiella, J.E. and Hsieh, C.Y. (1990). Analysis of crabmeat volatile compounds. *J. Food Sci.*, **55**, 962–966.

McGill, A.S., Hardy, R. and Burt, R.J. (1974). Hept-*cis*-4-enal and its contribution to the off-flavor in cold stored cod. *J. Sci. Food Agric.*, **25**, 1477–1489.

Mussinan, C.J. and Katz, I. (1973). Isolation and identification of some sulphur chemicals present in two model systems approximating cooked meat. *J. Agric. Food Chem.*, **21**, 43–45.

Nakamura, K., Iida, H. and Tokunaga, T. (1980). Separation and identification of odor in oxidized sardine oil. *Bull. Jpn. Soc. Sci. Fish.*, **46**, 355–360.

Nixon, L.N., Wong, E., Johnson, C.B. and Birch, E.J. (1979). Nonacidic constituents of volatiles from cooked mutton. *J. Agric. Food Chem.*, **27**, 355–359.

Pan, B.S. and Kou, J.M. 1994. Flavour of shellfish and kamaboko flavourants. *In Seafoods: Chemistry, Processing, Technology and Quality*, eds. F. Shahidi and J.R. Botta. Blackie Academic and Professional, New York, pp. 85–114.

Peterson, R.J. and Chang, S.S. (1982). Identification of volatile flavor compounds of fresh, frozen beef stew and a comparison of these with those of canned beef stew. *J. Food Sci.*, **47**, 1444–1448.

Phleger, C.F., Holtz, R.B., Grines, P.W., Leighton, D.L. and Jacobson, F. (1978). Chemical and sensory analysis of the purple-hinge rock scallop. *J. Food Sci.*, **43**, 1793–1796.

Pippen, E.L., Mecchi, M.P. and Nonaka, N. (1969). Origin and nature of aroma in fat of cooked poultry. *J. Food Sci.*, **34**, 436–438.

Reineccius, G. (1991). Off flavor in foods. *CRC Crit. Rev. Food Sci. Nutr.*, **29**, 381–402.

Sakakibara, H., Yanai, T., Yajima, I. and Hayashi, K. (1988). Changes in volatile flavor compounds of powdered dried bonito (Katsuobushi) during storage. *Agric. Biol. Chem.*, **52**, 2731–2739.

Schieberle, P. (1990). The role of free amino acids present in yeast as precursors of the odorants 2-acetyl-1-pyrroline and 2-acetyltetrahydropyridine in wheat bread crust. *Z. Lebensm. Unters. Forsch.*, **191**, 206–209.

Schieberle, P. and Grosch, W. (1987). Evaluation of the flavor of wheat and rye bread crusts by aroma extract dilution analysis. *Z. Lebensm. Unters. Forsch.*, **184**, 111–113.

Schutte, L. (1974). Precursors of sulphur-containing flavor compounds. *CRC Crit. Rev. Food Technol.*, **14**, 457–505.

Seki, N. (1971). Nucleotides in aquatic animals and seaweeds. *Bull. Jpn. Soc. Sci. Fish.*, **37**, 777–783.

Selka, E. and Frankel, E.N. (1987). Dynamic headspace capillary gas chromatographic analysis of soyabean oil volatiles. *J. Am. Oil Chem. Soc.*, **64**, 749–753.

Shahidi, F., Rubin, L.J. and D'Souza, L. (1986). Meat flavor volatiles: a review of the composition, techniques of analysis, and sensory evaluation. *CRC Crit. Rev. Food Sci. Nutr.*, **24**, 141–243.

Shankaranarayana, M.I., Raghavan, B., Abraham, K.O. and Natarajan, C.P. (1974). Volatile sulphur compounds in food flavors. *CRC Crit. Rev. Food Technol.*, **14**, 395–435.

Shiau, C.Y., Chai, T.J. and Chiou, T.K. (1994). Extractive nitrogenous components of oysters (*Crassostrea virginica*) and their released liquors. *J. Fish. Soc. Taiwan*, **21**, 281–291.

Shibamato, T. (1989). Volatile flavour chemicals formed by the Maillard reaction. In *Thermal Generation of Aromas*, eds. T.H. Parliment, R.J. McGorrin and C.T. Ho. ACS Symposium Series 409, American Chemical Society, Washington, DC, pp. 134–142.

Shibamoto, T. and Bernhard, R.A. (1977). Investigation of pyrazine formation pathways in glucose–ammonia model system. *Agric. Biol. Chem.*, **41**, 143–153.

Shibamoto, T. and Russell, G.F. (1977). A study of volatiles isolated from a D-glucose–hydrogen sulfide–ammonia model system. *J. Agric. Food Chem.*, **25**, 109–112.

Shibamoto, T., Akiyama, T., Sakaguchi, M., Enomoto, Y. and Masuda, H. (1985). A study of pyrazine formation. *J. Agric. Food Chem.*, **27**, 1027–1031.

Shigematsu, H., Kurata, T., Kato, H. and Fujimaki, M. (1972). Volatile compounds formed on roasting DL-alanine with D-glucose. *Agric. Biol. Chem.*, **36**, 1631–1637.

Shu, C.K., Mookherjee, B.D., Bondarovich, H.A. and Hagedorn, M.L. (1985). Characterization of bacon odor and other flavor components from the reaction of isovaleraldehyde and ammonium sulfide. *J. Agric. Food Chem.*, **33**, 130–132.

Shye, S.C., Pan, B.S. and Wu, C.M. (1987). Preliminary identification of flavor components extracted from shrimp (*Parapenaues fissurus*). *J. Chinese Agric. Chem. Soc.*, **25**, 169–176.

Shye, S.C., Pan, B.S. and Wu, C.M. (1988). Changes in polar lipids extracted from shrimp (*Parapenaues fissurus*) during accelerated storage. *J. Food Sci.*, **15**, 244–253.

Sidwell, V.D., Buzzell, D.H., Foncannon, P.R. and Smith, A.L. (1977). Composition of edible portion of raw (fresh or frozen) crustaceans, finfish, mollusks. II. Macroelements: sodium, potassium, chlorine, calcium, phosphorus and magnesium. *Marine Fish Rev.*, pp. 1–11.

Simidu, W. and Hujita, M. 1954. Studies on the muscle of aquatic animals – XXI. On glycine content in extracts of shrimp with special reference to their taste. *Bull. Jpn. Soc. Sci. Fish.*, **20**, 720–722.

Soliman, M.M., Fahmy, A.A., Sawy, A.A.E. and Osman, F. (1983). Effect of salting treatment on the flavor of oyster, donax and sea urchin. *Agric. Biol. Chem.*, **47**, 1655–1659.

Srinivas, S.R. (1988). Volatile constituents in alfalfa and red clover extracts. *Dev. Food Sci.*, **18**, 343–353.

St. Angelo, A.J., Vercellotti, J.R., Legendre, M.G., Vinnett, C.H., Kuan, J.W., James, C. Jr, and Dupuy, H.P. (1987). Chemical and instrumental analysis of warmed-over flavor in beef. *J. Food Sci.*, **52**, 1163–1167.

Suyama, M. and Kobayashi, H. (1980). Free amino acids and quaternary ammonium bases in mantle muscle of squids. *Bull. Jpn. Soc. Sci. Fish.*, **46**, 1261–1264.

Suyama, M., Suzuki, T., Maruyama, M. and Saito, K. (1970). Determination of carnosine, anserine, and balenine in the muscle of animals. *Bull. Jpn. Soc. Sci. Fish.*, **36**, 1048–1053.

Suyama, M., Hirano, T. and Yamazaki, S. (1985). Odor of ayu and its volatile components. *Bull. Jap. Soc. Sci. Fish.*, **51**, 287–294.

Suzuki, J., Ichimura, N. and Etoh, T. (1990). Volatile components of boiled scallop. *Food Rev. Int.*, **6**, 537–552.

Takagi, I. and Simidu, W. (1962). Muscle of aquatic animals. XXXI. Constituents and extractive nitrogen in a few species of shellfish. *Bull. Jpn. Soc. Sci. Fish.*, **28**, 1192–1198.

Tanchotikul, U. and Hsieh, T.C.-Y. (1989). Volatile flavor components in crayfish waste. *J. Food Sci.*, **54**, 1515–1520.

Tanchotikul, U. and Hsieh, T.C.-Y. (1991). Analysis of volatile flavor components in steamed rangia clam by dynamic headspace sampling and simultaneous distillation and extraction. *J. Food Sci.*, **56**, 327–331.

Taylor, A.J. and Linforth, R.S.T. (1996). Flavor release in the mouth. *Trends Food Sci. Technol.*, **7**, 444–447.

Teranishi, R. and Buttery, R.G. (1985). Chemical changes in flavour compounds during processing. In *Chemical Changes in Food During Processing*, eds. T. Richardson and J.W. Finley. AVI Publishing Co., Westport, CT, pp. 327–346.

Tokunaga, T., Iida, H. and Nakamura, K. (1977). Formation of dimethyl sulfide in Antarctic krill (*Euphausia superba*). *Bull. Jap. Soc. Sci. Fish.*, **43**, 1209–1217.

Tressl, R., Grunewald, G.R., Kamperschroer, H. and Silwar, R. (1981). Formation of pyrroles and aroma contributing sulphur components in malt and roasted coffee. *Prog. Food Nutr. Sci.*, **5**, 71–79.

Tressl, R., Helak, B., Koppler, H. and Rewicki, D. (1985a). Formation of 2-(1-pyrrolidinyl)-2-cyclopentennones and cyclopent(b)azepin-8(1H)-ones as proline specific Maillard products. *J. Agric. Food Chem.*, **33**, 1132–1137.

Tressl, R., Helak, B., Rewicki, D. and Kamperchroer, H. (1985b). Formation of pyrrolidines and piperidines on heating L-proline with reducing sugars. *J. Agric. Food Chem.*, **33**, 924–928.

Vejaphan, W., Hsieh, T.C.Y. and Williams, S.S. (1988). Volatile flavor components from boiled crayfish (*Procambarus clarkii*) tail meat. *J. Food Sci.*, **53**, 1666–1670.

Vernin, G. (1984). Heterocyclic flavoring and aroma compounds B. Pyridine and derivatives. *Koryo*, **144**, 47–62.

Vernin, G. and Parkanyi, C. (1982). Mechanisms of formation of heterocyclic compounds in Maillard and pyrolysis reactions. In *The Chemistry of Heterocyclic Flavoring and Aroma Compound*, ed. G. Vernin. Halsted Press, New York, pp. 151–209.

Viro, M.N. (1984). Heterocyclic aroma compounds in whiskey. *Found. Biotechnol. Ind. Ferment. Res.*, **3**, 227–233.

Wasserman, A.E. (1972). Thermally produced flavor components in the aroma of meat and poultry. *J. Agric. Food Chem.*, **20**, 737–741.

Whitfield, F.B. (1992). Volatiles from interactions of Maillard reactions and lipids. *CRC Crit. Rev. Food Sci. Nutr.*, **31**, 1–58.

Whitfield, F.B., Freeman, D.J. and Bannister, P.A. (1981). Dimethyl trisulphide: an important off-flavor component in the royal red prawn (*Hymenopenaeus sibogal*). *Chem. Ind. (London)*, **October**, 692–695.

Whitfield, F.B., Freeman, D.J., Last, J.H., Bannister, P.A. and Kennett, B.H. (1982). Oct-1-en-3-ol and (5z)-octa-1,5-dien-3-ol, compounds important in the flavor of prawns and sand-lobsters. *Aust. J. Chem.*, **35**, 373–383.

Wurzenberger, M. and Grosch, W. (1984). Stereochemistry of the cleavage of the 10-hydroperoxide isomer linoleic acid to 1-octen-3-ol by a hydroperoxide lyase from mushrooms (*Psalliota bispora*). *Biochem. Biophys. Acta*, **795**, 163–165.

Yang, X.H., Lee, C. and Scranton, M.I. (1993). Determination of nanomolar concentrations of individual dissolved low molecular weight amines and organic acids in seawater. *Anal. Chem.*, **65**, 572–576.

9 Umami flavour of meat

J.A. MAGA

9.1 Introduction

Many compounds have been shown to be present in the flavour fraction of food. However, the flavour chemist has to decide which compound or series of compounds are the major contributors to specific food flavours. This has been a long and difficult task, and as a result, the flavours of foods such as meats are still not completely understood.

Even more complex is the situation where certain compounds have been shown to intensify, modify or mask the flavours of certain foods. The fact that a specific compound or combination of compounds, when intentionally added or formed in foods by biological or thermal pathways, has the ability to change the perceived flavour properties of certain foods is a research area that is fascinating.

Over the years, various nomenclatures have been proposed for compounds that have the ability to modify flavour perception. These include terms such as flavour potentiators, flavour enhancers and umami. Currently, the scientific community appears to be adopting the name umami, defined as the taste of monosodium glutamate (MSG) and 5′-nucleotides such as 5′-inosinate (IMP) and 5′-guanylate (GMP).

The major objectives of this chapter are to define and discuss the properties of umami in model system studies, and to review the formation, identification, quantitation and stability together with the sensory significance in beef, pork, chicken, turkey and lamb. The influence of meat aging/processing and compound synergism is also discussed, in this attempt to update the role of umami in meat flavour chemistry.

In addition, the identification of certain naturally occurring peptides having umami-like properties has opened another interesting research area. Also, the incorporation of yeast extracts or hydrolysed proteins from both animal and plant sources has resulted in the flavour intensification/modification of various foods, including processed meats. This latter approach has specific application with the introduction of low-fat meat products, which can be typically described as lacking charactristic meat flavour. The intentional addition of these materials, which normally contain significant amounts of flavour enhancers, has resulted in increased acceptance of low-fat meat products.

9.2 Definitions

Umami can be defined as the taste properties resulting from the natural occurrence or intentional addition of compounds such as monosodium glutamate (MSG) and certain 5′-nucleotides such as 5′-inosine monophosphate (IMP) and 5′-guanosine monophosphate (GMP). Other researchers have used terms such as 'savoury', 'beefy' and 'brothy' to describe the same taste sensations. These nucleotides have also been referred to respectively as inosinic and guanylic acids, 5′-inosine and 5′-guanylic acids, inosine 5′- and guanosine 5′-phosphates and disodium 5′-inosinate or disodium 5′-guanylate. IMP and GMP have also been marketed under the trade name Ribotide®.

Compounds of these types are especially interesting in that they have the ability to modify taste, even though they do not possess characteristic flavours of their own, especially at the low concentrations at which they affect food flavour. In using the above definition of flavour alteration without taste contribution, one could also consider that sodium chloride, if used as subthreshold levels, also possesses umami properties. This is an area that deserves research attention.

9.3 Historical background

Many cultures throughout the world have long used ingredients or food preparation techniques that result in the presence of umami compounds that intensify certain food flavours. Experience has taught cooks what it took scientists many years to discover.

It was not until the early 1900s that a specific compound that was proven to be responsible for an umami sensation was isolated. Ikeda (1909) was able to identify the compound monosodium glutamate (MSG) in the naturally occurring form in an extract from dried kombu or sea tangle, a type of seaweed. The importance of his discovery was soon evident because the commercial production of MSG for the intentional addition to foods began shortly thereafter. Apparently, Ikeda was the first to propose the name 'umami', which means 'deliciousness' in Japanese, for the taste sensation associated with MSG. Today MSG is produced in many countries and is consumed internationally.

A few years later, another food common to Oriental cuisine, dried bonito tuna, was the source for the identification of another umami compound, namely inosine monophosphate (IMP). It was reported in the initial study (Kodama, 1913) that the compound in question was the histidine salt of 5′-inosinic acid. However, it was later concluded that histidine was not a significant contributor to umami. In contrast to MSG, the commercialization of IMP was not begun until the early 1960s.

In the early 1960s another compound, guanosine monophosphate (GMP), was identified from another natural source, the Shiitake black mushroom (Nakajima *et al.*, 1961; Shimazono, 1964). It is quite interesting to note that the three commercially available umami compounds were first identified from natural sources. In fact, it has been postulated (Hashimoto, 1965) that all types of marine products possess umami compounds due to the high amounts of glutamic acid and nucleotides that they contain.

More recently, other umami compounds, including ibotenic and tricholomic acids, have been identified as naturally occurring compounds in other types of Japanese mushrooms (Takemoto and Nakajima, 1964; Takemoto *et al.*, 1964). Currently these umami compounds are not commercially available due to the effectiveness and ready availability of MSG, IMP and GMP. In the meantime, over 80 years of research on umami compounds has accumulated and their commercialization now represents an international industry worth several hundred million dollars.

9.4 Structural considerations

To date, umami compounds have been found to structurally contain either L-amino acids containing five carbon atoms or a purine ribonucleotide 5′-monophosphate having an oxy group in the 6-position. These structures are representative of MSG, IMP and GMP as shown in Figure 9.1.

In commercial practice the amino acid based compound is in the monosodium salt form whereas the nucleotides are in the disodium form. In addition, MSG can also be obtained in the potassium, ammonium or calcium forms for utilization in products where low sodium levels are desired.

The isomeric structure of umami compounds can also dramatically influence their taste properties. In the case of MSG, the D-form, which is not

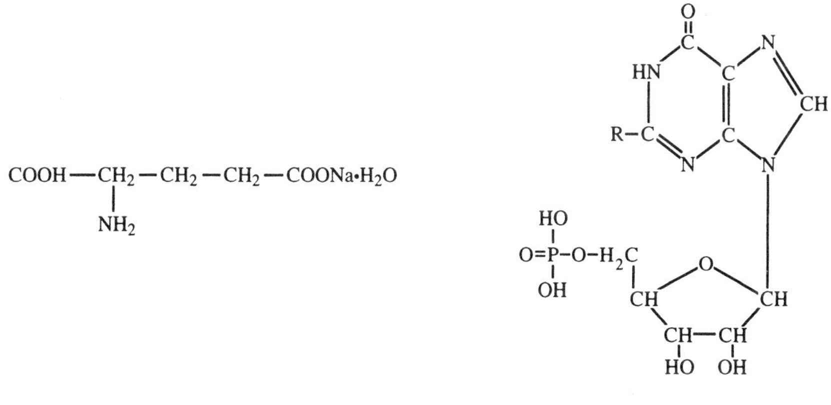

Figure 9.1 Umami structures.

naturally occurring, has no umami properties, whereas the L-form, which is naturally occurring, does. The nucleotides IMP and GMP can occur as the 2′-, 3′- or 5′- forms but only the 5′-form is taste active.

In the case of MSG, it is interesting to note that L-glutamine, which has the same basic structure as MSG, has no umami taste. The umami properties of other structurally related amino acids have been investigated (Akabori, 1939) and it was concluded that the L-forms of α-amino dicarboxylic cids with four to seven carbons also have umami properties similar to those of L-glutamic acid. The flavour properties of various substituted nucleotides have also been reported (Yamazaki *et al.*, 1968a,b) and as expected, some structures do result in compounds possessing umami properties.

9.5 Stability

When in solution, MSG acts as an ampholyte and as such can exist in a number of ionic forms, in equilibrium, which are pH dependent. At various pHs forms such as glutamic acid hydrochloride, free glutamic acid, neutral MSG and basic disodium glutamate can exist. The influence of pH on these forms in summarized in Table 9.1, and as can be seen, the neutral MSG form is predominant over most of the pH scale, except in very acid conditions. The neutral MSG form in turn possesses the most potent umami sensation.

The thermal stability of IMP and GMP has been shown to be also pH dependent. As seen in Table 9.2, acidic conditions decrease compound stability and IMP appears to be slightly more stable than GMP. Their thermal degradation has been shown to follow first order kinetics with the major degradation products being nucleosides and phosphoric acid, which would suggest degradation via hydrolysis of the phosphoric ester bond (Matoba *et al.*, 1988).

Table 9.1 Ionic form distribution of MSG as influenced by pH

pH	Distribution (%)			
	R_1	R_2	R_3	R_4
8.0	0	0	96.9	3.1
7.0	0	0	99.8	0
6.0	0	1.8	98.2	0
5.5	0	5.3	94.7	0
5.0	0	15.1	84.9	0
4.5	0	36.0	64.0	0
4.0	0.9	63.1	36.0	0
3.5	4.0	80.9	15.1	0
3.0	13.4	81.3	5.3	0

Table 9.2 Stability of GMP and IMP to temperature and pH

Compound	pH	Temperature (°C)	Half-life (h)
GMP	4.0	100	6.4
	7.0	100	8.2
	9.0	100	38.5
IMP	4.0	100	8.7
	7.0	100	13.1
	9.0	100	46.2

A 10°C increase in temperature lowered all half-lives by one-third.
From Matoba *et al.* (1988).

Table 9.3 Aqueous hydrolysis rates for nucleotides at 121°C and various pH values

pH	Compound	Half-life (min)
3.0	IMP	32.4
	GMP	31.5
4.0	IMP	45.3
	GMP	62.4
5.0	IMP	45.9
	GMP	57.8
6.0	IMP	63.0
	GMP	41.0

From Shaoul and Sporns (1987).

Nucleotide stability at temperatures normally used in retorting also presents problems, especially for larger-size cans where extended retorting times are required. The data summarized in Table 9.3 show that under highly acidic conditions half of either IMP or GMP can be lost in approximately 30 min. In contrast, it has been estimated that at pH 5 and at 23°C the half-life for IMP and GMP is 36 and 19 years, respectively (Shaoul and Sporns, 1987).

9.6 Synergism

One of the most fascinating aspects of umami compounds that is far from understood is their ability to act synergistically when used in combination with each other. This concept has been extensively reviewed (Maga, 1983) and thus will only be briefly touched upon. As shown in Table 9.4, if MSG is assigned a umami intensity of 1.0, the addition of an equal amount of GMP increases relative flavour intensity 30 times. Even addition of as little as 1% GMP to MSG significantly increases flavour intensity. Similar effects are noted for combinations of IMP and MSG.

Table 9.4 Synergistic flavour intensity of MSG–GMP

Ratio of MSG:GMP	Relative flavour intensity
1:0	1.0
1:1	30.0
10:1	18.8
20:1	12.5
50:1	6.4
100:1	5.5

From Ribotide® Product Data Sheet No. 3.

9.7 Taste properties

The individual taste thresholds for various umami compounds, alone and in combination with each other, are summarized in Table 9.5. It can be seen that when used alone, GMP has a threshold an order of magnitude lower than that for MSG and IMP. The role of synergism as described above is also evident in Table 9.5 when all three compounds are utilized in equal proportions. Therefore, if naturally present in combination or intentionally added in combination, a relatively small total amount is required to elicit a umami response.

Relative to the basic tastes of sweet, salt, sour and bitter, the taste threshold of MSG is well within their range (Table 9.6). This is quite interesting in light of the fact that there are many studies attempting to relate the taste of MSG to the other basic tastes. As seen in Figure 9.2, various portions of the MSG molecule have been proposed to possess the potential to elicit other tastes. This approach has led Birch (1987) to propose that MSG not only possesss a umami taste but also a salty one (Table 9.7). In addition, he proposes that aspartic and glutamic acids also have a umami taste in combination with an acidic sensation.

Yamaguchi (1987) has attempted to locate the relative position of umami taste to the other basic tastes as well as to that of various foods, utilizing multi-dimensional plotting. As seen in Figure 9.3, umami falls outside the regions occupied by sweet, salt, sour and bitter tastes but is closely associated with tastes derived from various marine and meat prod-

Table 9.5 Umami compound taste thresholds

Compound	Taste threshold (%)
MSG	0.012
IMP	0.014
GMP	0.0035
IMP + GMP	0.0063
IMP + GMP + MSG	0.000031

From Maga (1983).

Table 9.6 Detection thresholds in water of various taste substances

Compound	Threshold (%)
Sucrose	0.086
Sodium chloride	0.0037
Tartaric acid	0.00094
Quinine sulphate	0.000049
MSG	0.012

From Yamaguchi (1987).

Table 9.7 Compounds related to MSG and having umami taste

Compound	Tastes
MSG	Umami and salty
Aspartic acid	Acidic and umami
Glutamic acid	Acidic and umami

From Birch (1987).

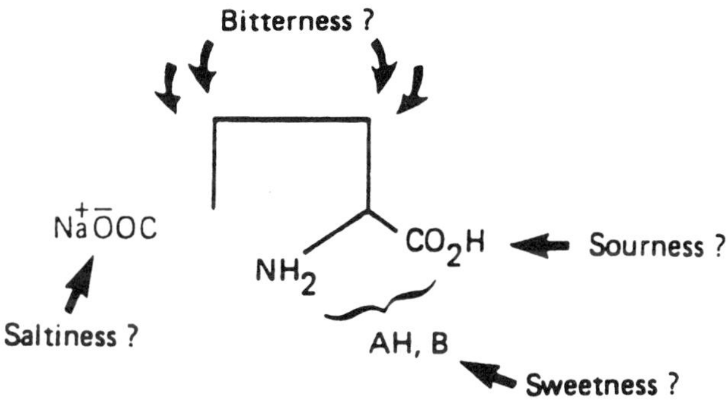

Figure 9.2 Proposed MSG taste properties (Birch, 1987).

ucts. These data can be interpreted in at least two different ways. First, the data clearly demonstrate that umami is neither associated with, nor is the result of, the four acknowledged basic tastes, from which one could conclude that it indeed is a separate and distinctive taste. On the other hand, one also conclude that umami is a specialized form of taste associated with meat or marine-based foods.

9.8 Food occurrence

Umami compounds and their precursors are present in a wide variety of foods. This can perhaps be best appreciated by viewing the information

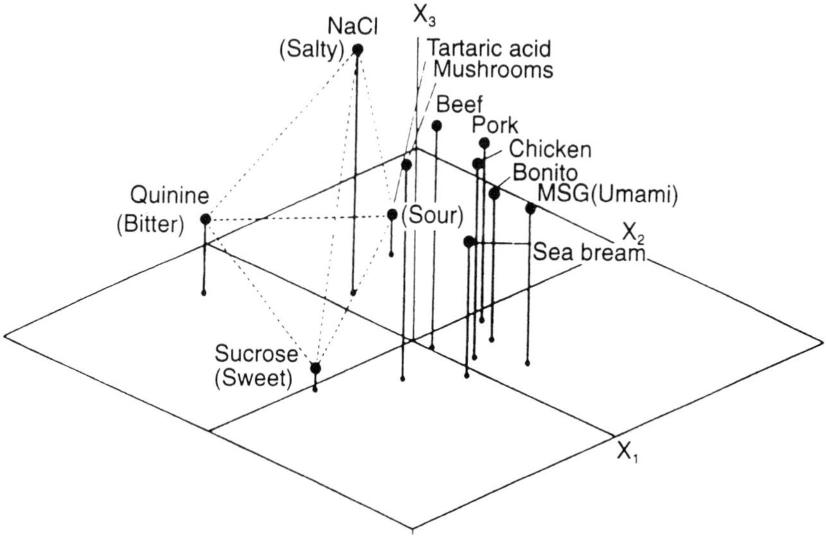

Figure 9.3 Relationship of umami taste to other tastes and foods (Yamaguchi, 1987).

summarized in Table 9.8. These foods include raw and processed, fermented and unfermented, as well as plant and animal sources. Therefore, those individuals who may be concerned with the intentional addition of umami compounds to various foods would be hard pressed to identify a varied diet that did not contain naturally occurring sources of the compounds in question.

If one looks at the free glutamate levels that are present in various foods (Table 9.9), it can quickly be appreciated that significant amounts of glutamate may be present, especially in cheeses that are aged for long periods.

Glutamic acid is usually the major amino acid in protein, and as seen in Table 9.10, it represents approximately 20% of all amino acids in various animal protein sources. When these proteins are processed or intentionally hydrolysed in the manufacture of hydrolysed protein, glutamic acid is freed, which in turn can result in the formation of MSG.

Table 9.8 Foods containing umami compounds

Beverages (beer, tea, wine)
Fruits (various)
Marine products (seaweed, fish, clams, crab, oysters, shrimp)
Meats (beef, chicken, lamb, pork)
Milk products (milk, cheese)
Plant proteins (barley, coconut, corn, cottonseed, peanut, soybean, wheat)
Vegetables (various)

From Maga (1983).

Table 9.9 Free glutamate levels in various foods

Food	Glutamate level (mg/100 g)
Tomato	140
Tomato juice	260
Grape juice	258
Broccoli	176
Parmesan cheese	1 200
Gruyère cheese	1 050

From Giacometti (1979).

Table 9.10 Glutamate levels in various animal proteins

Protein	Glutamate level (g/100 g)
α-Casein (milk)	22.5
β-Lactoglobulin (milk)	20.0
Actin (muscle)	14.8
Myosin (muscle)	21.0
Albumin (egg white)	16.5

From Giacometti (1979).

If one was to rank the importance of umami compounds to the flavour properties of various foods, as shown in Table 9.11, it becomes quite apparent that they make major contributions to meat flavours. Therefore, the remainder of this review will be devoted to the role of umami compounds in meat flavour.

9.9 Umami compounds and meat flavour

Numerous early studies have demonstrated that the intentional addition of MSG and nucleotides significantly modifies meat flavour perception. For example, Girardot and Peryam (1954) showed that the addition of MSG to a variety of processed meats resulted in products that were preferred over the same products to which MSG was not added (Table 9.12). This was especially true in the case of chicken. Of the products they evaluated, meat loaf had the least improvement in flavour perception.

When using added IMP, other researchers (Kurtzman and Sjostrom, 1964) concluded that canned chicken-containing noodle soup was not flavour-enhanced (Table 9.13). Based on IMP thermal stability data presented earlier, perhaps most or all of the added IMP was degraded. However, other products evaluated, including canned beef noodle soup, did show improvement with IMP addition. For some unexplained reason, no improvement was found with canned beef hash.

Table 9.11 Major taste-active compounds in foods

Compound	Meat	Vegetables	Fruit	Roots	Seeds
Inorganic ions	× ×	×	×	×	×
Amino acids	× × ×	× ×	×	× ×	× ×
Peptides and proteins	× ××	× ×		×	××
Histidine dipeptides	× × ×				
Nucleotides	× × ×	×		××	× ×
Amines	× ×	×			
Sugars			× × ×	× ×	××
Phenols (simple)		× ×	× ×		×
Hydroxy compounds		×	× ×		
Polyphenolic compounds	× ×	× ×			×
Carbonyl compounds		× ×	× × ×		×
Esters			× ×		
Sulphur compounds	× ×	× ×			×
Acids		× ×	× × ×		
Furans		× ×	× ×	×	×
Lactones		× ×	× × ×		
N,*S*-heterocycles	×	× × ×	×		×

From Grill and Flynn (1987).
× of minor importance; × × somewhat important;
× × × very important.

Table 9.12 Preference or processed meats containing MSG

Product	% Preferring MSG sample
Beef stew	61
Beef and gravy	61
Boned chicken	75
Hamburger	65
Meat loaf	53
Pork and gravy	66

From Giradot and Peryam (1954).

Table 9.13 Influence of IMP addition to the flavour of various meat products

Product	Effect
Beef bouillon	Enhanced
Beef noodle soup, canned	Enhanced
Chicken noodle soup, canned	Undecided
Canned luncheon meats	Enhanced
Corned beef hash	Not enhanced
Ham	Enhanced

From Kurtzman and Sjostrom (1964).

Bauer (1983) clearly demonstrated that the aging of meat clearly influences resulting glutamic acid content. He aged beef and pork for 4 and 7 days, and as seen in Table 9.14, in the case of both products, the longer-aged meat had significantly higher levels of glutamic acid. Apparently, microbial activity during aging resulted in partial protein hydrolysis thereby liberating free glutamic acid. However, as seen in Table 9.15, the levels of glutamate present in a range of processed meats are quite variable. The relatively low levels found in some of these products could perhaps be attributed to the degree of heat processing or low pH.

The free glutamate content in a variety of cooked meats has been summarized in Table 9.16, and as can be seen, most of the values are relatively low except for the two poultry products and the free-run juice from beef rib roast. It is also interesting to note that cooking beef rib roast to well done compared to medium rare resulted in approximately an 80% reduction in the amount of free glutamate present. Other products that had low values included lamb and mutton as well as boiled frankfurters. In the case of the frankfurters, the low value could be attributed to one of two reasons; either their high fat content had a dilution effect on the actual amount of meat present, or the free glutamate was lost to the cooking water.

The nucleotide levels naturally found in beef and chicken are summarized in Table 9.17. These data indicate that beef actually has higher levels of IMP than does chicken whereas the levels of GMP are the same.

Table 9.14 Meat age versus glutamic acid content

Product	Age (days)	Glutamic acid content (mg/100 g)
Beef	4	9.3
	7	21.1
Pork	4	11.4
	7	16.0

From Bauer (1983).

Table 9.15 Glutamate content of processed meats

Product	Glutamate content (%)
Dry sausages	0.03–0.25
Frankfurters	0.02–0.15
Liver sausages	0.01–0.29
Hams	0.01–0.11
Pickled tongue	0.07–0.17

From Gerhardt and Schulz (1985).

Table 9.16 Free glutamate content of various cooked meats

Product	Glutamate content (%)
Beef, tenderloin	0.042
Beef, shank	0.014
Beef, standing rib, medium rare	0.057
Juice from above	0.088
Beef, standing rib, well done	0.013
Bologna	0.004
Chicken	0.055
Duck	0.064
Frankfurters, boiled	0.001
Lamb	0.003
Mutton	0.008
Pork, loin	0.029

From Maga (1983).

Table 9.17 Nucleotide composition (%) of various meats

Nucleotide	Beef	Chicken
IMP	0.106–0.443	0.075–0.122
GMP	0.002	0.002
AMP	0.007	0.007–0.013
CMP	0.001	0.002
UMP	0.001	0.003

From Maga (1983).

Interestingly, chicken has somewhat higher levels of other nucleotides (AMP, CMP, UMP), which have limited umami properties, than beef.

Kato and Nishimura (1987) measured the IMP levels in beef, pork and chicken that had been aged for varying lengths of time, and as seen in Table 9.18, chicken had the highest IMP level even though it had only been aged 2 days. Pork had the next highest level while beef, although aged the longest, had the lowest amount of IMP. When a panel was asked to judge whether samples had a more intense umami taste before or after again, no difference was found for beef, which had the lowest IMP level, while storage significantly influenced umami taste intensity for pork and chicken (Table 9.19). Interestingly, cooking stored beef, pork and chicken

Table 9.18 Effect of meat storage on IMP levels

Meat and storage days	IMP level (μmol/g meat)
Beef (12)	3.2
Pork (6)	6.7
Chicken (2)	7.2

From Kato and Nishimura (1987).

Table 9.19 Effect of meat storage on umami taste intensity

Meat and storage days	Number of samples judged to have a more intense umami taste		Significance
	Before storage	After storage	
Beef (12)	12	4	NS
Pork (6)	2	14	S
Chicken (2)	8	23	S

NS, not significant; S, significant.
From Kato and Nishimura (1987).

(Table 9.20) resulted in levels of IMP that were 50% higher in all three products, as compared to their noncooked but aged counterparts.

Most researchers have exclusively investigated the influence of umami compounds on taste perception but one should also question if these compounds can influence odour perception. A limited number of articles has appeared addressing this issue. Maga and Lorenz (1972) prepared beef stock samples to which they added either no umami compounds, or a total of 0.05% of only MSG, IMP or GMP, or a combination of IMP, GMP and MSG. They sampled the headspace above the various stocks and measured total peak areas using gas chromatography. The stock sample which had no added compounds served as the control. As seen in Table 9.21, the addition of MSG alone as well as combinations of umami compounds significantly increased peak area ratios thereby indicating that these compounds can also influence aroma properties.

Table 9.20 Effect of storage on IMP levels in cooked meats

Meat and storage days	IMP level (μmol/g meat)
Beef (12)	4.1
Pork (6)	10.5
Chicken (2)	10.9

From Kato and Nishimura (1987).

Table 9.21 Beef broth GLC headspace peak area ratios versus umami additions

Peak area ratio[a]	Additive
1.66	0.05% MSG
2.30	IMP + GMP (0.05% total)
2.35	IMP + GMP + MSG (0.05% total)

[a] Control versus compound addition.
From Maga and Lorenz (1972).

Yamaguchi and Kimizuka (1979) performed a flavour profile analysis on cooked hamburgers to which 1% MSG had been added as compared to a no additive control. As seen in Table 9.22, they observed a slight increase in the 'meaty' and 'acceptability' descriptors associated with the aroma portion of the profile of hamburgers containing added MSG. It is also interesting to note that added MSG also intensified many of the other sensory properties.

Yamaguchi (1987) has conducted extensive research on umami compounds in meat stocks. For example, she found that the major nucleotide in chicken stock was IMP (Table 9.23). Interestingly, no GMP was detected in this study. In addition to IMP, relatively high levels of glutamic acid were found. In comparing beef and chicken stocks, she reported (Table 9.24) significant differences between minimum detectable levels for MSG and IMP. Twice the amount of MSG was required before it was detected in chicken stock as compared to beef stock, whereas only one-third the amount of IMP required for beef stock needed to be added to chicken stock before it was detected.

Table 9.22 Flavour profile of cooked hamburger with 1% MSG

Descriptor	Score
Aroma	
Whole aroma	0
Meaty	0.2
Beefy	0
Acceptability	0.4
Basic taste	
Whole taste	0.6
Salty	0.3
Sweet	0.1
Sour	0.2
Bitter	0
Flavour characteristic	
Continuity	0.5
Mouthfulness	0.6
Impact	0.5
Mildness	0.5
Thickness	0.4
Other flavours	
Spicy	0.2
Oily	0
Meaty	0.2
Beefy	0.3
Overall preference	
Palatability	0.6

0, Same as control (no MSG); 1, slight increase; 2, marked increase.
From Yamaguchi and Kimizuka (1979).

Table 9.23 Umami content (mg/100 ml) in chicken stock

Compound	Amount
AMP	2.26
IMP	5.84
GMP	0
ATP	ND
Glutamic acid	15.00

From Yamaguchi (1987).
ND = not detected.

Table 9.24 Detection levels of umami compounds added to stocks

Stock	Detection level (%)	
	MSG	IMP
Beef	0.00625	0.025
Chicken	0.0125	0.00625

From Yamaguchi (1987).

Maga (1987) attempted to evaluate the role of MSG, IMP and GMP on the taste intensities of various purified meat proteins including beef, pork, lamb, chicken and turkey as compared to the same meat proteins containing no added compounds. From Table 9.25 it can be seen that with no additions, beef and pork had the highest intensity ratings while lamb had the lowest. In all cases, the addition of any of the three compounds increased taste intensity although the increase for lamb was minimal. In looking at percentage increases over the control (Table 9.26) it is apparent that the most affected meat protein was chicken. Also, not all additives were equally effective. Therefore, these data would indicate that the functionality and perhaps mechanism of interaction between umami compounds and meat proteins vary with protein source.

Table 9.25 Taste intensities (0–100) of various 1% meat proteins with added umami compounds

Meat	No added umami	Added umami compound		
		MSG (0.015%)	IMP (0.010%)	GMP (0.004%)
Beef	64	83	80	89
Pork	60	68	70	76
Lamb	41	46	45	48
Chicken	53	84	86	90
Turkey	49	61	60	65

From Maga (1987).

Table 9.26 Percentage increase in taste intensity caused by umami addition to meats

Meat	Added umami compound		
	MSG	IMP	GMP
Beef	30	25	39
Pork	13	17	27
Lamb	12	10	17
Chicken	58	62	70
Turkey	24	22	33

From Maga (1987).

9.10 Yeast extracts

The fact that yeast is generally high in ribonucleic acid (RNA), and that it is a readily available by-product from various food processing operations, has led to the widespread use of yeast extracts or powders to modify or intensify meat flavour. As as result, 5′-nucleotides as well as MSG can be derived from the nucleotide associated with yeast RNA. Yeast extracts in turn can be used in conjunction with added MSG to produce an extremely effective umami effect.

Basically, yeast extract is a concentrated form of soluble material obtained from yeast cells that have undergone hydrolysis by various techniques, including autolysis, plasmolysis or acid hydrolysis. By controlling temperature and pH, autolysis is an autodegradation process that minimizes the loss of inherent hydrolytic enzymes such as proteases, nucleases and carbohydrases. Plasmolysis is usually brought about by the addition of salt, while acid addition is used to induce acid hydrolysis. Based on the type and source of yeast, along with the type of hydrolysis employed, a wide range of umami types and concentrations can result. In addition, most yeast extracts have a wide array of precursors such as amino acids, sugars and B-vitamins that can react during subsequent heat treatment when added to meat systems to produce additional flavour compounds.

9.11 Hydrolysed proteins

Historically the use of products such as soy sauce as a flavour stimulant is well documented. In actuality, soy sauce represents a hydrolysed protein obtained by either natural fermentation or via chemical acid hydrolysis. Both processes result in a product high in MSG.

Through the past 30 years, various plant protein materials, such as wheat gluten, corn gluten, defatted legume flours (soy, peanut) and defatted cottonseed flour have served as starting materials for the manufacture of what is commonly known as hydrolysed vegetable protein (HVP). Hydrolysis can be either by enzymatic addition or by acid/base addition. Acid-treated products are normally neutralized with sodium hydroxide, thereby resulting in the formation of sodium chloride, up to levels of 45%, which in turn can add to umami flavour properties and interactions via synergism. Depending on the protein source, as well as processing conditions, typical HVP can contain up to 16% MSG. The utilization of hydrolysed meat by-products also can serve as a source of umami compounds, as well as numerous meat-like flavours.

9.12 Peptides

Throughout the flavour literature, numerous reports have appeared where peptides of varying structure and length possess unique taste properties including sweet, salty, sour, bitter and umami/beefy (Otagiri *et al.*, 1985; Asao *et al.*, 1987; Mojarro-Guerra *et al.*, 1991; Izzo and Ho, 1992).

Relative to beef, the octapeptide (H-Lys-Gly-Asp-Glu-Glu-Ser-Leu-Ala-OH) was first isolated by Yamasaki and Maekawa (1978) from papain-treated beef. They reported that a panel of three people described the isolated peptide as having a 'delicious' taste. In a later publication, Yamasaki and Maekawa (1980) reported on the synthesis of the octapeptide in question and described the taste of a 5% solution as meat-like. They also synthesized a series of structurally similar peptides and reported that the meat-like character was not evident. Similar structures primarily possessed sour, bitter and astringent taste properties.

A similar approach was utilized by Tamura *et al.* (1989) to synthesize the originally described 'delicious' peptide as well as structurally related compounds. They concluded that the octapeptide in question possessed both sour and umami properties and had a taste threshold of 1.4 mM. Based on the taste properties of similar peptides, they concluded that the taste observed with the octapeptide in question was due to the interaction between the basic and acidic fractions of the peptide.

Later, Spanier (1992) and Spanier *et al.* (1992) reported that the octapeptide was naturally present in beef that had not undergone enzymatic treatment and proposed the name beefy meaty peptide (BMP). Earlier, Spanier and Edwards (1987) were able to isolate two polypeptide fractions from cooked beef. One group was composed of hydrophilic peptides, which can lead to sweet taste, while the other group represented hydrophobic peptides, which are normally associated with sour and bitter tastes. Also, Cagan (1984) reported that a portion of the BMP structure is similar to that of protein-based monellin, which has a strong sweet taste.

More recently, Spanier *et al.* (1995) reported on the taste properties of BMP as compared to MSG. In a beef-flavoured gravy system, flavour enhancement by BMP was found to be more pronounced than with MSG, especially at sub-threshold levels. Both BMP and MSG were reported to have their optimum effect at their threshold levels (1.41 mM/0.16% for BMP and 1.56 mM/0.26% for MSG). They also reported that the addition of MSG at its taste threshold increased the salty note in a beef gravy, whereas the addition of BMP at its taste threshold did not enhance perceived saltiness.

Recently, Wang *et al.* (1996) reported on the taste response of BMP as influenced by pH and in conjunction with salt and MSG additions. They determined BMP taste thresholds at pH 3.5, 6.5 and 9.5. From their data they concluded that BMP taste threshold was not influenced by pH but taste descriptors used by the panel did change with pH. For example, at pH 3.5, BMP taste was described as sour, whereas as pH 6.5, the predominant taste was umami, while at 9.5, sweet and sour notes were also noted, along with umami.

Wang *et al.* (1996) clearly demonstrated that BMP displays synergism with added salt and MSG. For example, the taste threshold for BMP decreased approximately 3.5-fold when either salt or MSG was present at their individual taste thresholds. When both salt and MSG were added to BMP, the threshold for BMP decreased 9-fold. The major taste descriptor reported for combinations of BMP, salt and MSG was umami.

Wang *et al.* (1966) added 2 mM of synthesized BMP to a beef extract prepared from a heated ground beef and water mixture as compared to the same beef extract without added BMP. Triangle test results demostrated that the panel could statistically distinguish between the two variables. Sensory descriptors used by the panel to describe the taste of added BMP to beef extract included meaty, salty, savoury, sweet and umami.

Since BMP is protein-based, a practical question is compound stability, especially during heating. Wang *et al.* (1995) subjected a synthesized 0.35 mM solution of BMP to heating at 71°C for 15 s (pasteurization) and 121°C for 20 min (sterilization) to determine thermal compound stability. Unheated and heated solutions were analysed using both HPLC and mass spectrometry. From their data, they concluded that BMP was over 97% stable to both pasteurization and sterilization conditions. From these data, it can be concluded that either naturally occurring or intentionally added BMP to meat products would be stable to subsequent heat processing.

9.13 Conclusions

The literature clearly demonstrates that umami compounds have the ability to alter the taste and possibly aroma properties of a wide range of foods independent of the four basic tastes. Umami compounds react synergistically, therefore reducing the total amount required to elicit a response.

They and their precursors occur naturally in a wide range of foods and their effectiveness can be improved by their intentional addition to most foods. Acidic conditions and high processing temperatures minimize the effectiveness of 5′-nucleotide-based umami compounds. Umami compounds are very effective in contributing to meat flavour, especially chicken, and exhibit minimum effectiveness with lamb.

References

Akabori, S. (1939). Relationships between the taste of amino acids and their structures. *J. Japan Biochem. Soc.*, **14**, 185–197.

Asao, M., Iwamura, H., Akamatsu, M. and Fujita, T. (1987). Quantitative structure–activity relationships of the bitter thresholds of amino acids, peptides, and their derivatives. *J. Med. Chem.*, **30**, 1873–1879.

Bauer, F. (1983). Free glutamic acid in meats products. *Ernahrung*, **7**, 688.

Birch, G.G. (1987). Structure, chirality, and solution properties of glutamates in relation to taste. In *Umami: A Basic Taste*, eds. Y. Kawamura and M.R. Kare. Marcel Dekker, New York, pp. 173–184.

Cagan, R.H. (1984). Peptide interaction with taste receptors: Overlap in taste receptor specificity. *Experienta*, **40**, 843–844.

Gerhardt, U. and Schulz, W. (1985). The presence of free glutamic acid in foods, with special reference to meat products. *Fleischwirtshaft*, **65**, 1483–1486.

Giacometti, T. (1979). Free and bound glutamate in natural products. In *Glutamic Acid: Advances in Biochemistry and Physiology*, eds. L.J. Filer, S. Garattini, M.R. Kare, W.A. Reynolds and R.J. Wurtman. Raven Press, New York, pp. 25–34.

Girardot, N.F. and Peryam, D.R. (1954). MSG's power to perk up foods. *Food Eng.*, **26**(12), 71–74.

Grill, H.J. and Flynn, F.W. (1987). Behavioral analysis of oral stimulating effects of amino acid and glutamate compounds on the rat. In *Umami: A Basic Taste*, eds. Y. Kawamura and M.R. Kare. Marcel Dekker, New York, pp. 461–480.

Hashimoto, Y. (1965). Taste-producing subtances in marine products. In *The Technology of Fish Utilization*, ed. R. Kreuzery. Elsevier, London, pp. 57–65.

Ikeda, K. (1909). On a new seasoning. *J. Tokyo Chem.*, **30**, 820–826.

Izzo, H.V. and Ho, C.-T. (1992). Peptide-specific Maillard reaction products: A new pathway for flavor chemistry. *Trends Food Sci. Technol.*, **3**, 253–257.

Kato, H. and Nishimura, T. (1987). Taste components and conditioning of beef, pork, and chicken. In *Umami: A Basic Taste*, eds. Y. Kawamura, and M.R. Kare. Marcel Dekker, New York, pp. 289–306.

Kodama, S. (1913). On a procedure for separating inosinic acid. *J. Tokyo Chem.*, **34**, 751–755.

Kurtzman, C.H. and Sjostrom, L.B. (1964). The flavor-modifying properties of disodium inosinate. *Food Technol.*, **18**, 221–225.

Maga, J.A. (1983). Flavor potentiators. *CRC Crit. Rev. Food Sci. Nutr.*, **18**, 231–312.

Maga, J.A. (1987). Organoleptic properties of umami substances. In *Umami: A Basic Taste*, eds. Y. Kawamura, and M.R. Kare. Marcel Dekker, New York, pp. 255–269.

Maga, J.A. and Lorenz, K. (1972). The effect of flavor enhancers on direct headspace gas–liquid chromatography profiles of beef broth. *J. Food Sci.*, **37**, 963–964.

Matoba, T., Kuchiba, M., Kimura, M. and Hasegawa, K. (1988). Thermal degradation of flavor enhancers, inosine 5′-monophosphate, and guanosine 5′-monophosphate in aqueous solution. *J. Food Sci.*, **53**, 1156–1170.

Mojarro-Guerra, S.H., Amado, R., Arrigoni, E. and Solms, J. (1991). Isolation of low-molecular-weight taste peptide from Vacherin Mont d'Or cheese. *J. Food Sci.*, **56**, 943–947.

Nakajima, N., Ishikawa, K., Kamada, M. and Fujita, E. (1961). Food chemical studies on 5′-ribonucleotides. I. On the 5′-ribonucleotides in foods. Determination of the 5′-ribonucleotides in various stocks by ion exchange chromatography. *J. Agric. Chem. Soc. Japan*, **35**, 797–804.

Otagiri, K., Nosho, Y., Shinoda, I., Fukui, H. and Okai, H. (1985). Studies on a model of bitter peptides including arginine, proline, and phenylalanine residues. I. Bitter taste of di- and tripeptides, and bitterness increase of the model peptides by extension of the peptide chain. *Agric. Biol. Chem.*, **49**, 1019–1026.

Ribotide® Product Data Sheet No. 3. The synergistic effect between disodium 5′-inosinate, disodium 5′-guanylate and monosodium glutamate. Takeda Chemical Industries, Ltd., Tokyo, Japan.

Shaoul, O. and Sporns, P. (1987). Hydrolytic stability at intermediate pHs of the common purine nucleotides in food, inosine-5′-monophosphate, guanosine-5′-monophosphate and adenosine-5-monophosphate. *J. Food Sci.*, **52**, 810–812.

Shimazono, H. (1964). Distribution of 5′-ribonucleotides in foods and their application to foods. *Food Techol.*, **18**, 294–298.

Spanier, A.M. (1992). Current approaches to the study of meat flavor quality. In *Food Science and Human Nutrition*, ed. G. Charalambous. Elsevier, New York, pp. 695–709.

Spanier, A.M. and Edwards, J.V. (1987). Chromatographic isolation of presumptive flavor principles from red meat. *J. Liq. Chromatogr.*, **10**, 2745–2758.

Spanier, A.M., Miller, J.A. and Bland, J.M. (1992). Lipid oxidation: Effect on meat proteins. In *Lipid Oxidation in Foods*, ed. A.J. St. Angelo. ACS Books, Columbus, OH, pp. 104–119.

Spanier, A.M., Bland, J.M., Miller, J.A., Glinka, J., Wasz, W. and Duggins, T. (1995). BMP: A flavor enhancing peptide found naturally in beef. Its chemical synthesis, descriptive sensory analysis, and some factors affecting its usefulness. In *Food Flavors: Generation, Analysis, and Process Influence*, ed. G. Charalambous. Elsevier, New York, pp. 1365–1378.

Takemoto, T. and Nakajima, T. (1964). Studies on the constituents of indigenous fungi. I. Isolation of the flycidal constituent from *Tricholoma muscarium. J. Pharm. Soc. Japan*, **84**, 1183–1185.

Takemoto, T., Yokobe, T. and Nakajima, T. (1964). Studies on the constituents of indigenous fungi. II. Isolation of the flycidal constituent from *Amanita strobiliformis. J. Pharm. Soc. Japan*, **84**, 1186–1189.

Tamura, M., Nakasuka, T., Tada, M., Kawasaki, Y., Kikuchi, E. and Okai, H. (1989). The relationship between taste and primary structure of 'delicious peptide' (Lys-Gly-Asp-Glu-Glu-Ser-Leu-Ala) from beef soup. *Agric. Biol. Chem.*, **53**, 319–325.

Wang, K., Maga, J.A. and Bechtel, P.J. (1995). Stability of beefy meaty peptide to pasteurization and sterilization temperatures. *Lebensm. Wiss. u. Technol.*, **28**, 539–542.

Wang, K., Maga, J.A. and Bechtel, P.J. (1996). Taste properties and synergism of beefy meaty peptide. *J. Food Sci.*, **61**, 837–839.

Yamaguchi, S. (1987). Fundamental properties of umami in human taste sensation. In *Umami: A Basic Taste*, eds. Y. Kawamura and M.R. Kare. Marcel Dekker, New York, pp. 41–73.

Yamaguchi, S. and Kimizuka, A. (1979). Psychometric studies on the taste of monosodium glutamate. In *Glutamic Acid: Advances in Biochemistry and Physiology*, eds. L.J. Filer, S. Garattini, M.R. Kare, W.A. Reynolds and R.J. Wurtman. Raven Press, New York, pp. 35–54.

Yamasaki, Y. and Maekawa, K. (1978). A peptide with a delicious taste. *Agric. Biol. Chem.*, **42**, 1761–1765.

Yamasaki, Y. and Maekawa, K. (1980). Synthesis of a peptide with a delicious taste. *Agric. Biol. Chem.*, **44**, 93–97.

Yamazaki, A., Kumashiro, I. and Takenishi, T. (1968a). Synthesis of 2-alkyl-thioinosine 5′-phosphate and N′-methylated guanosine 5′-phosphates. *Chem. Pharm. Bull.*, **16**, 338–343.

Yamazaki, A., Kumashiro, I. and Takenishi, T. (1968b). Synthesis of some N′-methyl-2-substituted inosines and their 5′-phosphates. *Chem. Pharm. Bull.*, **16**, 1561–1565.

10 Lipid-derived off-flavours in meat

L.H. SKIBSTED, A. MIKKELSEN and G. BERTELSEN

10.1 Introduction

With the continuing development of sensors and computer technology, quality control in food production will increasingly be based on concepts involving the entire production chain. For meat, improved hygiene standards and improved packaging systems allow products to retain their microbiological acceptability for longer periods and have made chemical changes during processing, storage and retail display more important for product quality. Chemical changes, and in particular oxidation, are responsible for changes in colour, flavour, odour and transformation of unsaturated fatty acids and cholesterol into compounds with negative effects on human health. Lipid oxidation in meat has accordingly been the subject of numerous studies over the past four decades (e.g. Gray and Crackel, 1992).

The new perspective for meat production is to evaluate the entire production chain and to identify the most critical steps and factors for initiation of oxidative changes and to develop procedures for protection against further oxidative damage. Identification of such critical control points for oxidative deterioration of meat and meat products is gaining increasing importance in the light of the increasing consumer demand for highly processed, precooked products in which convenience should ideally be combined with freshness. Also focus on more 'healthy' lipid profiles in the human diet is changing the consumer's preference towards meats with a higher proportion of unsaturated lipids, and efforts are being made to change the fatty acid profile of pork towards more monounsaturation through changes in the animal feed (Jensen *et al.*, 1997b; Morgan *et al.*, 1992). These trends all stress the importance of better control of lipid oxidation in meat. New quality systems need to be developed utilizing the emerging technologies where the quality of the final product can be related to data collected throughout the production chain. In order to contribute to this development, we have identified such critical control points for oxidation in meat and discuss their relative importance on the basis of the many available basic studies of lipid oxidation and antioxidant mechanisms in meat systems.

The contribution of lipids to flavour in raw meat and meat products is caused both by hydrolysis of triacylglycerols and phospholipids and by oxidation of fatty acids. Hydrolytic and oxidative rancidity are interrelated, as free fatty acids are oxidized more easily than triacylglycerols. Hydrolytic rancidity caused by free fatty acids is, however, significant only in products with microbial growth, since acid or base catalysis of ester bonds in triacylglycerols is minimal at pH values usually encountered in meat. In some fermented meat products, the microbiological conditions are controlled to enhance enzymatic hydrolysis. In this process lactic acid, in particular, is produced in larger amounts and pH decreases. The hydrolytic rancidity thus developed as the result of enzymatic processes and enhanced acid catalysis is an essential and characteristic part of the flavour in products such as sausages (Klettner and Baumgartner, 1980).

Oxidative decomposition of unsaturated lipids, resulting in oxidative rancidity, is the main factor in off-flavour generation and deterioration of muscle-based foods. However, it is important to consider the off-flavours developing in precooked meat on reheating. The term warmed-over flavour (WOF) has been coined for such off-flavours which develop rapidly when cooked meat is reheated after storage (Tims and Watts, 1958). Notably, the characteristic WOF aroma profile and the aroma profile related to lipid oxidation in raw meat are different, but it is believed that the flavour compounds involved are qualitatively the same but occur in different concentrations (St. Angelo *et al.*, 1987). Several aldehydes and ketones of chain length 6–12 carbon atoms or even smaller flavour compounds are formed in the oxidative reactions leading to lipid-derived off-flavours (Konopka and Grosch, 1991). The threshold values for these impact compounds are in the ppb range (Mottram, 1987).

The potential of lipid oxidation varies among animal species (Rhee *et al.*, 1996), reflecting the dependence of several controllable and noncontrollable factors such as structure of meat tissue, fat content, fatty acid composition, animal diet, content of prooxidants and antioxidants in the muscle, and conditions for slaughter, carcass chilling and storage. To obtain optimal products, the controllable factors must be identified and critical control points defined, as illustrated in Figure 10.1.

10.2. Lipid oxidation in muscle tissues

Lipid oxidation in meat products is recognized by the presence of secondary oxidation products such as hexanal and for precooked meat also by other more specific oxocompounds like the very potent *trans*-4,5-epoxy-(E)-2-decenal (Konopka and Grosch, 1991). However, prior to the formation of these volatile secondary oxidation products, lipid hydroperoxides are formed as primary lipid oxidation products. Lipid hydroperoxides have no impact on the meat flavour, but are precursors for the off-flavour compounds. For pork, the level of hydroperoxides has been found to decrease

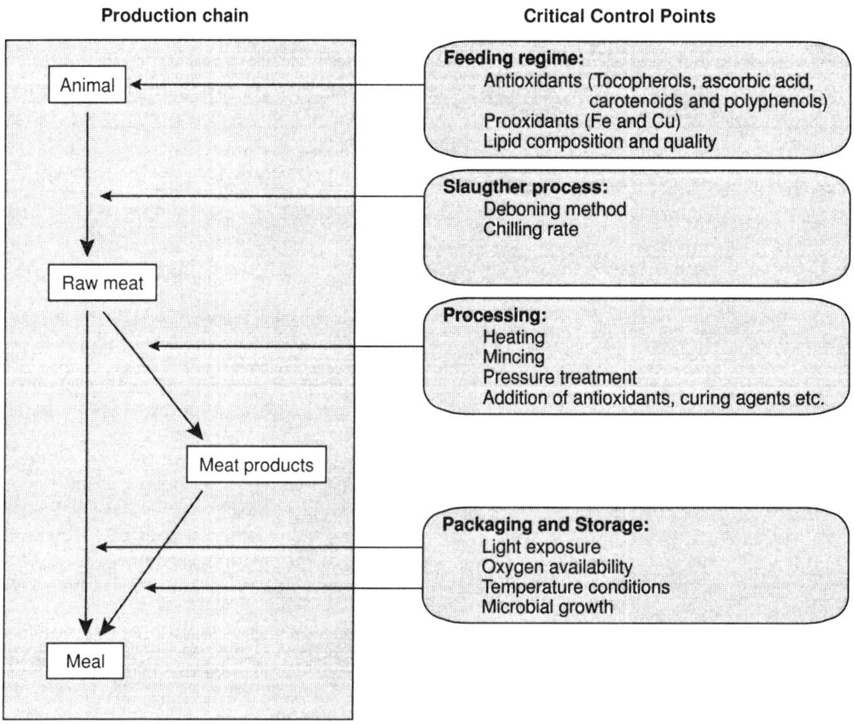

Figure 10.1 Critical control points along the production chain from animal to meal are identified in relation to oxidative deterioration of meat and meat products.

during cold storage with a concomitant increase in secondary lipid oxidation products measured as 2-thiobarbituric acid reactive substances (TBARS) (Nielsen *et al.*, 1997). This is in agreement with the generally accepted free radical mechanism for lipid oxidation, which involves an initiation phase, a propagation phase and a termination phase as outlined in Figure 10.2. The role of the primary lipid oxidation products, formed in the propagation phase, for development of secondary lipid oxidation products during processing, was further recognized, as storage prior to heating, depleting the lipid hydroperoxides, gave a precooked product with less development of secondary oxidation products (Nielsen *et al.*, 1997). The initial level of primary lipid oxidation products in the muscle is most likely related to the physiological status of the animal prior to slaughter and does define a critical control point, since high levels of hydroperoxides may warrant uses other than production of precooked meat.

An important difference between lipid oxidation in meat and in refined vegetable oils and lard is the presence of several catalytic systems for oxygen activation in meat. Activation of ground state oxygen into singlet

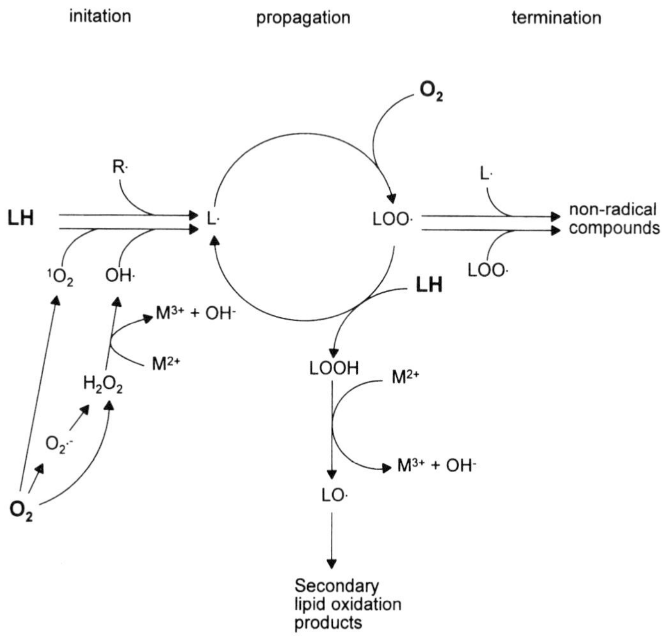

Figure 10.2 Initiation of lipid oxidation in meat systems depending on oxygen activation through enzyme activity, photochemical processes or metal catalysis, or on exposure to free radicals merging from other sourses such as desinfectants. LH is an unsaturated lipid and $R^{\bullet}$ is a free radical, such as a hypervalent iron protein radical like $^{\bullet}$MbFe(IV)=O. Chain-breaking antioxidants are active in the propagation step by reacting with peroxy radicals, $LOO^{\bullet}$, or alkoxy radicals, $LO^{\bullet}$.

oxygen, superoxide radical, the hydroxyl radical or peroxides, or transformation of unsaturated lipids into lipid radicals is required to overcome the spin restriction for the reaction between paramagnetic compounds such as ground state oxygen and diamagnetic compounds such as lipids. Singlet oxygen can form via photochemical reactions where a sensitizer transfers light energy to oxygen. Protoporphyrin and haematin, degradation products from meat pigments, but not myoglobin and apomyoglobin, are found to act as sensitizer in solutions (Whang and Peng, 1988). The superoxide anion is formed in meat by enzymes and during autoxidation of oxymyoglobin (Satoh and Shikama, 1981), and hydrogen peroxide is formed spontaneously or via enzyme-catalysed dismutation of superoxide anion and by microbial metabolism. Superoxide anion and hydrogen peroxide are not able to react directly with unsaturated fatty acids (Kaschnitz and Hatefi, 1975; Kanner *et al.*, 1987). However, in the presence of transition metal ions they act as precursors for hydroxy radical, which readily reacts with unsaturated fatty acids. Superoxide anion reacts with lipid hydroperoxides (Thomas *et al.*, 1978) or may indirectly promote lipid oxidation by inhibiting catalases and peroxidases (Kono and Fridovich, 1982).

Lipid hydroperoxides are formed in the propagation phase when lipid radicals react with oxygen. Pure hydroperoxides are relatively stable, but transition metal ions and haem compounds catalyse their decomposition both by oxidation and by reduction as seen in Figure 10.3. The decomposition of hydroperoxides involves further free radical reactions and formation of low-molecular-weight volatile compounds, including aldehydes, ketones, alcohols, hydrocarbons and esters.

The composition of compounds produced in lipid oxidation depends on the fatty acid profile of the meat. The more unsaturated the fatty acids in meat lipids, the more easily a hydrogen atom is abstracted and the oxidation process initiated.

Although other enzymes are involved in oxygen activation in meat systems, the term 'enzymatic lipid oxidation' is normally reserved to describe the action of lipoxygenases and cyclooxygenases which catalyse oxidation of unsaturated fatty acids and hence assists in the formation of stereospecific hydroperoxides and endoperoxides, respectively (Yamamoto, 1992). Cyclooxygenase incorporates two molecules of oxygen into the fatty acid and can be considered as a special case of lipoxygenase which incorporates one molecule of oxygen. The stereospecific elimination of hydrogen from the 1,4-pentadiene structure of an unsaturated fatty acid followed by oxygenation is the major action of lipoxygenases. However, several other activities have been identified, including oxygenation at other sites in the substrate, further oxygenation of the primary hydroperoxide, and transformation of the primary hydroperoxide into an epoxy acid (Yamamoto, 1992). The enzymatic lipid oxidation contributes to the lipid hydroperoxide pool, and when hydroperoxides decompose by homolytic cleavage or β-scission, secondary lipid oxidation products are formed. Lipoxygenases also contribute to generation of carbonyl compounds with specific flavours (Yamamoto, 1992; Hsieh and Kinsella, 1989).

Lipoxygenases have been recognized in several tissues of fish (German and Creveling, 1990; German and Kinsella, 1985; German *et al.*, 1986;

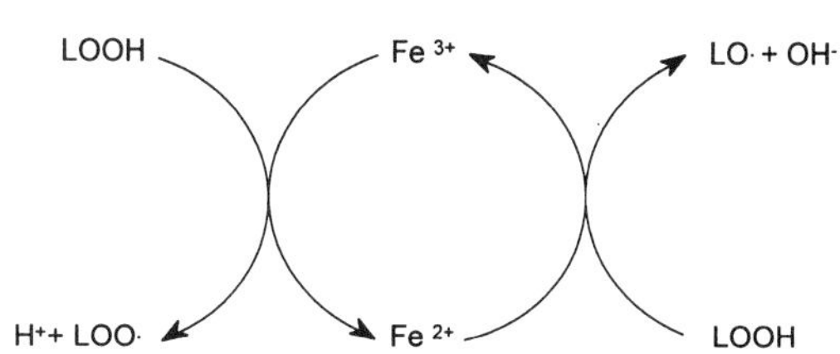

Figure 10.3 Transition metal catalysis involving both oxidation and reduction of lipid hydroperoxides to yield lipid peroxy and alkoxy radicals, which may initiate new chain reactions, such as those in Figure 10.2.

Hsieh *et al.*, 1988; McDonald and Hultin, 1987) and in skeletal muscles of poultry and mammals (Grossman *et al.*, 1988; Gata *et al.*, 1996). Most lipoxygenases use arachidonic acid as major substrate, but a lipoxygenase from pig skeletal muscle with preference for linoleic acid has been purified and characterized (Gata *et al.*, 1996). This higher activity toward linoleic acid probably reflects the much higher concentration of linoleic acid than arachidonic acid in the pig muscles. The enzyme activity of 15-lipoxygenase from chicken is preserved during frozen storage and it has been suggested that the enzyme is responsible for some of the lipid oxidation occurring in chicken meat during frozen storage (Grossman *et al.*, 1988). The role of enzymes for development of WOF is in preformation of lipid hydroperoxides prior to processing, since heat treatment will inactivate these and other enzymes of importance for lipid oxidation, including those protecting meat systems.

Initiation of oxidation in meat is balanced by the presence of natural antioxidants. The primary or chain-breaking antioxidants compete by reacting with the lipid peroxy and lipid alkoxy radicals (Figure 10.2). The latter reaction for phenolic antioxidants such as tocopherols (φ–OH) is gaining increased attention.

$$\varphi\text{–OH} + \text{LO}^{\bullet} \rightarrow \varphi\text{–O}^{\bullet} + \text{LOH}$$

This is because the alkoxy radical is present in larger steady state concentrations, and since it prevents the reaction which otherwise could initiate a new chain reaction (Frankel, 1995).

$$\text{LO}^{\bullet} + \text{LOOH} \rightarrow \text{LOH} + \text{LOO}^{\bullet}$$

The protective effect of phenolic antioxidants on primary lipid oxidation products resulting from this mechanism has been demonstrated for phospholipid hydroperoxides in turkey meat (Bruun-Jensen *et al.*, 1995).

10.3 Critical control points in prevention of lipid oxidation in meat and meat products

The oxidative processes and accumulation of lipid oxidation products start in the live animal and continue when the animal is slaughtered and the homeostatic systems of the body stop functioning. The rate of lipid oxidation at this and further stages in the meat production chain depends on the access to oxygen, on the content of endogenous prooxidants and antioxidants and on the handling, processing and storage conditions. With reference to the critical control points identified in Figure 10.1, the individual steps in the production chain will now be considered. The trend towards reduced use of synthetic food additives, including antioxidants, provides an example of how the different control points in the production chain should be considered together, since a higher level of endogenous antioxidant obtained through the use of a super-nutritional level in the animal feed may have even better effects than antioxidants added to the product at a later stage.

10.3.1 Animal feeding as a critical control point

During the last few decades an increasing number of investigations have been initiated in order to explore the beneficial effects of dietary vitamin E supplementation on lipid oxidation and pigment oxidation in post-mortem muscles and on lipid oxidation in adipose tissue. In general, very positive effects have been observed, as it appears from reviews of these studies, and the effects have primarily been traced to a direct incorporation of the tocopherol into subcellular membranes, where it maximizes the antioxidative capacity of the system (Monahan, 1995; Liu *et al.*, 1995; Buckley *et al.*, 1995). Studies with chicken (Webb *et al.*, 1972a; Brekke *et al.*, 1975; Lin *et al.*, 1989a,b; Jensen *et al.*, 1995), turkeys (Webb *et al.*, 1972b; Marusich *et al.*, 1975), pigs (Buckley and Connolly, 1980; Monahan *et al.*, 1992; Jensen *et al.*, 1997a), cattle (Faustman *et al.*, 1989), veal (Shorland *et al.*, 1981) and fish (Frigg *et al.*, 1990) have all demonstrated reduced lipid oxidation in muscles and adipose tissues from the supplemented animals compared to the same tissues from nonsupplemented animals. Accumulation of α-tocopherol in muscles has been found to take place in a dose and duration dependent manner. In general, the greater the amount of vitamin E fed and/or the longer supplementation period, the higher the tissue concentration of α-tocopherol (Arnold *et al.*, 1993), and it seems difficult to reach a saturation level within realistic feeding levels. Thus, deposition of α-tocopherol did not reach saturation in muscle tissues using dietary *all-rac*-α-tocopheryl acetate concentrations at up to 700 mg/kg feed for pigs (Jensen *et al.*, 1997a).

Tissue α-tocopherol concentrations in muscle tissues seem to be linearly related to the logarithm of the dietary intake (Jensen *et al.*, 1997a; Machlin and Gabriel, 1982), implying that costs are increasing rapidly for even moderate increases of tocopherol in the muscle beyond a certain level. For pigs, the level of endogenous α-tocopherol being deposited in *m. psoas major* (PM) and in *m. longissimus dorsi* (LD) could both be approximated by such logarithmic relationships, as illustrated in Figure 10.4. Moreover, it could be concluded that the slope is approximately the same for PM and for LD, showing that the total uptake of α-tocopherol in the two muscles can be considered as equal, except that PM has a consistently higher background level of α-tocopherol than LD.

Some 37 different muscles from porcine carcasses were further included to explore the differences in the response from different tissues on increased dietary vitamin E supplementation (O'Sullivan *et al.*, 1997). The mean α-tocopherol levels in the pork muscle tissues were for all muscles significantly higher in a high-supplemented group (160 mg α-tocopheryl acetate/kg feed) compared to a low-supplemented (20 mg/kg feed) and a basal group (unsupplemented). For pigs raised on the supplemented feed, the muscles of the thoracic limb (4.8–9.9 mg α-tocopherol/kg tissue) and neck and thorax (3.7–9.2 mg/kg) contained the highest levels of α-tocopherol compared to the muscles of the pelvic limb (4.0–5.6 mg/kg) and of the back (2.5–3.5 mg/kg). No single factor has, however, been identified to account for the large variation in skeletal muscle concentration of α-tocopherol in the same animal (Liu *et al.*, 1995). Variation in lipid content, fibre type distribution and metabolic characteristics, including uptake and excretion of tocopherols, of the individual muscles together seem to determine the steady state level (Buckley *et al.*, 1995; Jensen *et al.*, 1997a). From a comparison between different muscles, it can further be concluded that a higher content of α-tocopherol does not necessarily result in greater oxidative stability. In spite of a higher content of α-tocopherol in *m. psoas major* (PM) compared to both *m. gluteus medius* and *m. longissimus lumborum*, muscle lipids in PM were less protected against oxidative deterioration (Chan *et al.*, 1996), as is also evident from the significantly higher 2-thiobarbituric acid reactive substance (TBARS) values measured during chill storage for cooked *m. psoas major* compared to *m. longissimus dorsi* (Jensen *et al.*, 1997a). The higher content of haem

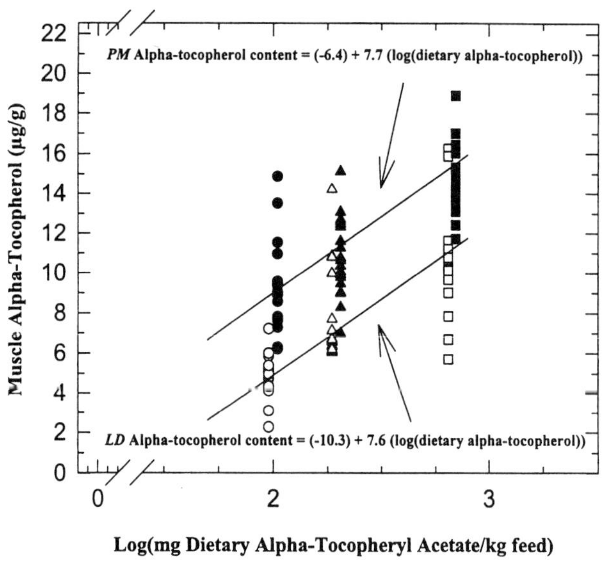

Figure 10.4 Linear dependence of α-tocopherol deposited in porcine *m. longissimus dorsi* (LD) and in *m. psoas major* (PM) on the logaritm of the amount of α-tocopheryl acetate (mg/kg feed) in the feed (adapted from Jensen *et al.*, 1997b).

and its role as an oxidation catalyst was suggested as an explanation for the lower oxidative stability of PM (Jensen *et al.*, 1997a). Such considerations are valuable, when the feeding regime is considered as a critical control point, especially in relation to optimal use of different muscles.

Recommendations for vitamin E supplementation established from a purely nutritional viewpoint are 15–40 mg *all-rac*-α-tocopheryl acetate/kg feed for finishing pigs and poultry. However, in order to optimize the quality and the storage life of meat and meat products a much higher level should be recommended. One Belgian supermarket chain has specified that cattle should be fed 1500 mg supplemented vitamin E for at least 100 days before slaughter (Anon., 1997). Such recommendations should, however, also be seen in relation to the lipid source in the feed. From a general quality consideration, the requirement for vitamin E supplementation depends on several parameters of which one of the most important is the quality and composition of the dietary fat. Today there is a considerable interest in modification of fatty acid composition of animal tissues in an attempt to produce new 'designer' or 'functional' food (Monahan, 1995). Use of elevated levels of unsaturated fatty acids increases the susceptibility to oxidation of meat and meat products (Monahan *et al.*, 1992; Rhee *et al.*, 1988). However, supplementation with vitamin E may be used to overcome the increased susceptibility as shown for pigs (Monahan *et al.*, 1992) and poultry (Sheehy *et al.*, 1992).

One of the consequences of using oxidized oil for animal feed is a significant reduction in the concentration of α-tocopherol content in pig and poultry muscle (Jensen *et al.*, 1998). If inferior quality lipids are fed to pigs or poultry, the dietary α-tocopherol should be increased in order to compensate for the increased oxidative stress and to maintain a favourable antioxidant/prooxidant balance in the muscle membranes and in the muscle food products (Liu *et al.*, 1995; Galvin *et al.*, 1993).

The effect of dietary α-tocopherol supplementation in increasing the oxidative stability of meats are particularly pronounced in products exposed to increased oxidative stress such as comminuted meats and meats held in frozen storage (Jensen *et al.*, 1997b; Monahan, 1995; Liu *et al.*, 1995). Precooked meat products are highly susceptible to lipid oxidation resulting in warmed-over flavour (WOF). Based on the results of Faustman *et al.* (1989), Liu *et al.* (1995) concluded that although a vitamin E effect was significant, it may have little practical significance in delaying WOF development. This 'low vitamin E effect' on WOF may, however, indicate that even higher levels of vitamin E supplementation are needed to suppress oxidation in the very vulnerable cooked meat products. Thus, in cooked pork meat significant improvement in oxidative stability, as measured by TBARS during chill storage, was achieved by supplementation of 700 mg α-tocopheryl acetate/kg feed (Jensen *et al.*, 1997a). The presence of 100 mg or even 200 mg α-tocopheryl acetate/kg feed had little, if any,

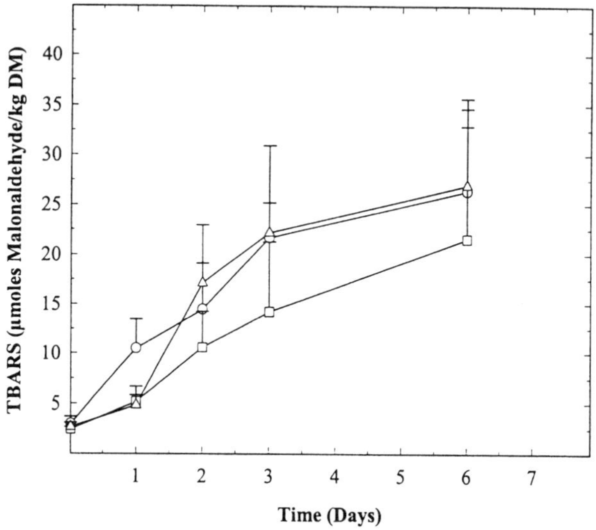

Figure 10.5 Effect of dietary α-tocopheryl acetate supplementation on lipid oxidation as reflected in 2-thiobarbituric acid reactive substances (TBARS) of chill stored cooked patties from *m. longissimus dorsi* of pigs fed different dietary levels of α-tocopheryl acetate: ○: 100 mg/kg feed; △: 200 mg/kg feed; and □: 700 mg/kg feed (adapted from Jensen *et al.*, 1997b).

effect (Figure 10.5). When using animal feeding regime as a critical control point, it should be realized that different uses of the meat may warrant different levels of feed additives.

10.3.2 Handling of fresh and frozen meat as a critical control point

For raw meat, quality changes may primarily be categorized as (a) bacterial growth resulting in possible pH changes, slime formation and off-odours, (b) oxidation of meat pigments causing discoloration, and (c) oxidation of lipids resulting in undesirable meat flavour together with formation of compounds with adverse biological effects. The latter two categories are 'purely' chemical in nature and will, keeping all other factors equal, dominate with decreasing temperature. However, at this critical control point a better understanding of the complex interaction between microbial growth, oxidative discoloration and lipid oxidation is required in order to limit quality deterioration caused by lipid oxidation during subsequent processing and storage.

The earlier studies focused more or less on the analytical aspects of the interaction between pigment and lipid oxidation in order to separate 'free iron' catalysis and 'haem pigment' catalysis using washed muscle fibre systems (Sato and Hegarty, 1971; Johns *et al.*, 1989). However, more recent investigations include a dynamic description using concepts from reaction kinetics in order to identify rate determining steps and active catalytic species in lipid oxidation (Mikkelsen *et al.*, 1992b; Chan *et al.*, 1997; Baron *et al.*, 1997; Nakayama *et al.*, 1997). It has been realized for some time that 'haem pigment' catalysis and 'free iron' catalysis are both important for lipid oxidation in distinct types of products and under different conditions (Love and Pearson, 1974). However, it also appears safe to conclude that for uncooked, raw meat, meat pigments are the more important catalysts (Johns *et al.*, 1989; Kanner *et al.*, 1991c; Hansen *et al.*, 1996) while 'free iron' liberated during heat treatment may exert a significant effect in the development of warmed-over flavour in precooked meat products. The high levels of carbon dioxide used in modified atmosphere packaging (MAP) of raw meat to control microbial growth may also play a role by increasing the mobility of iron between storage and transport proteins, thus increasing the iron catalysis (Pollack *et al.*, 1977). Storage experiments with both meat and model systems need to be designed to explore such effects together with possible effects from the ultraviolet component of fluorescent light on mobilization of iron from transferrin (Aubailly *et al.*, 1996).

For beef, the oxygen level is normally kept low during long-term storage and adjusted to atmospheric level or even higher during retail packaging. In relation to the penetration of oxygen into the meat and to the effect of decreasing oxygen pressure on lipid oxidation in the surface of meat, effects on (a) pigment oxidation and (b) microbial growth deserve further attention in future investigations. As for the spoilage flora, a surplus of oxygen will favour growth of *Pseudomonas* and related species on the product surface (Pierson *et al.*, 1970). In the deeper layers of minced meat, lactic acid bacteria may become dominant due to microaerophilic conditions resulting from oxygen depletion. Notably, lactic acid bacteria reduce oxygen to hydrogen peroxide with a simultaneous decrease in pH due to lactic acid production. Hydrogen peroxide production as well as the pH decrease in the deeper layers of a meat surface act synergistically with the decreasing oxygen pressure and hasten the oxidative deterioration of the meat. Further evidence for a direct coupling of pigment oxidation with lipid oxidation is being provided from both model system (Chan *et al.*, 1997; Nakayama *et al.*, 1997) and meat system studies (Hansen *et al.*, 1996).

The rate of autoxidation of $MbFe(II)O_2$ is at a maximum at the oxygen partial pressure at which the concentrations of $MbFe(II)O_2$ and MbFe(II) are equal (i.e. 6 mmHg at 0°C for beef; Ledward, 1970).

$$MbFe(II)O_2 \rightleftharpoons MbFe(II) + O_2$$
$$MbFe(II)O_2 + MbFe(II) + 2H^+ \rightarrow 2MbFe(III) + H_2O_2$$

Pigment oxidation appears first as a thin brown layer of MbFe(III) at the limit of oxygen penetration between the violet interior of the meat, where

MbFe(II) dominates, and the cherry-red surface of the meat due to MbFe(II)O_2. Notably, MbFe(II)O_2 autoxidation is subject to specific acid catalysis and to a primary positive kinetic salt effect (Andersen *et al.*, 1988b). A local decrease in pH of 0.3 by lactic acid formation will double the rate, and the brown layer is expected to extend towards the surface. However, the 'metmyoglobin reducing activity' also increases with decreasing pH, dependent on the reducing cofactor NADH and muscle type (Mikkelsen and Skibsted, 1992). In some muscles, brown discoloration is initiated rapidly because of lactic acid formation, while other muscles like *longissimus dorsi* are less affected (Renerre and Labadie, 1993). For all muscles, a local decrease in pH will, however, increase the turn-over number of the 'colour cycle of meat' with a faster NADH consumption and concomitant oxygen activation, as may be seen from the net stoichiometry of the colour cycle of Figure 10.6, when combined with superoxide dismutation.

$$\begin{array}{l} 2O_2 + NADH \rightarrow 2O_2^{\bullet -} + H^+ + NAD^+ \\ 2O_2^{\bullet -} + 2H^+ \rightarrow O_2 + H_2O_2 \\ \hline O_2 + NADH + H^+ \rightarrow H_2O_2\ NAD^+ \end{array}$$

Thus, a decrease in pH would increase the rate of formation of reactive oxygen species, such as superoxide and hydrogen peroxide, and with further consequences for membrane and lipid stability. Moreover, a permanent brown discoloration of meat is indicative of depletion of NADH.

Hydrogen peroxide, formed as the result of growth of lactic acid bacteria and in the colour cycle through dismutation of superoxide (Satoh and Shikama, 1981), is central to further oxidative degradation. In precooked

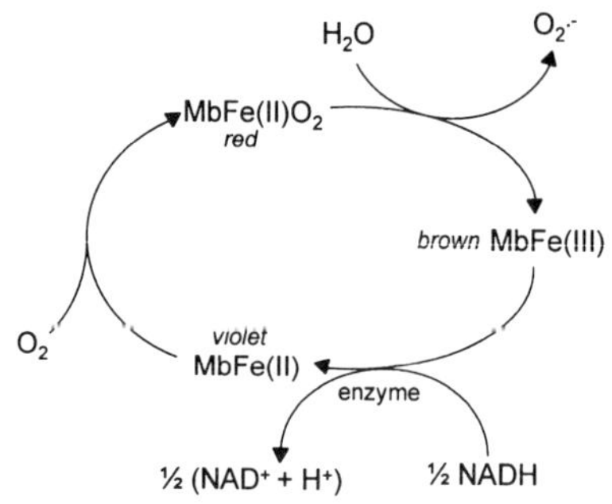

Figure 10.6. The colour cycle of raw meat. The balance between MbFe(II)O_2 autoxidation and enzymatic reduction of MbFe(III) determines the colour of the meat. MbFe(II)O_2 autoxidation depends on oxygen concentration, temperature, pH and light exposure, and enzymatic reduction of MbFe(III) depends on temperature, pH and level of reducing cofactors. The turn-over number of the colour cycle determines the rate of oxygen activation and the rate of depletion of reducing cofactors.

meat, in which iron has been liberated from proteins and in which catalase activity is reduced or absent due to enzyme denaturation, hydrogen peroxide may oxidize iron(II) in a one-electron process to yield the highly reactive hydroxyl radical or, more important in raw meat, react with MbFe(III). Hydrogen peroxide activation of myoglobin has received increasing attention since activated myoglobin was found to initiate membrane lipid oxidation (Harel and Kanner, 1985b). Several recent studies have confirmed that even minor changes in myoglobin, such as those resulting from heat treatment or from single amino acid substitution, transform this oxygen storage protein to a pseudo-peroxidase (Kristensen and Andersen, 1997; Ozaki *et al.*, 1996). The myoglobin activation involves formation of hypervalent myoglobin analogues of Compound I and Compound II, as are known for peroxidases, and is directly coupled to the colour of meat, as illustrated in Figure 10.7. For all conditions of relevance for biological systems, autoxidation of oxymyoglobin has been demonstrated to be slower than autoreduction of ferrylmyoglobin, $^{\bullet}$MbFe(IV)=O, and especially of the protein radical perferrylmyoglobin, $^{\bullet}$MbFe(VI)=O, initially formed (Mikkelsen and Skibsted, 1995):

$$\mathrm{MbFe(III)} + \mathrm{H_2O_2} \rightarrow \{{}^{\bullet}\mathrm{MbFe(IV)}{=}\mathrm{O} \rightleftarrows \mathrm{MbFe(V)}{=}\mathrm{O}\} + \mathrm{H_2O}$$

Thus for a given level of reductants in meat, the steady state concentration of MbFe(IV)=O depends on the production of H_2O_2 by sources other than autoxidation of MbFe(II)O_2, e.g. growth of lactic acid bacteria. The activated forms of myoglobin have been shown to initiate lipid oxidation

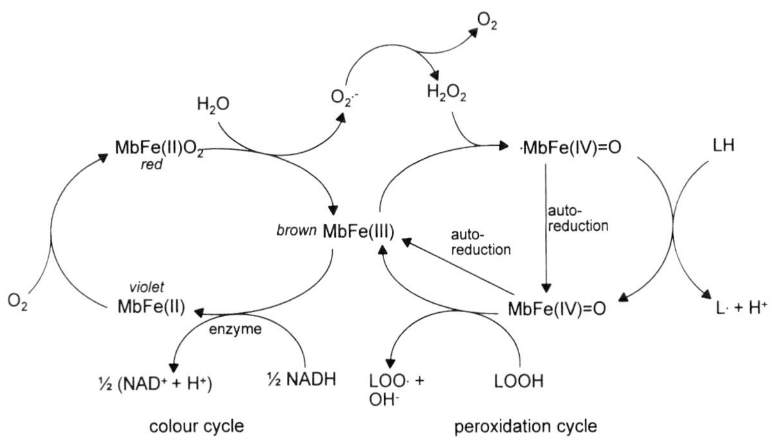

Figure 10.7 Pseudo-peroxidase cycle of myoglobin as linked to colour cycle of meat. The autoreduction of the protein radical, $^{\bullet}$MbFe(IV)=O, competing with initiation of lipid oxidation involving an unsaturated lipid, LH, is fast with a half-life of a few seconds; the autoreduction of MbFe(IV), competing with the cleavage of lipid hydroperoxides, LOOH, has a half-life of approximately 1 h.

(Kelman *et al.*, 1994; Rao *et al.*, 1994), and accordingly the presence of cytoplasmic reductants is important for the protection of membrane lipids against haem-catalysed oxidation. MbFe(IV)=O is reduced efficiently by NADH, and for a high level of NADH, MbFe(IV)=O will be kept at low concentrations, and membrane lipids will not be oxidized. However, when NADH is depleted, this kinetic protection is terminated, as the steady state concentration of MbFe(IV)=O increases. Lipid oxidation is initiated and off-flavours develop. NADH also serves as a cofactor for the metmyoglobin reducing enzyme complex, and depletion of NADH will thus be evident through local brown discoloration. Therefore, meat discoloration indicates NADH depletion concomitant with termination of the kinetic protection of pigment-catalysed oxidation of lipids. Accordingly, it should be possible to detect the very early stages of lipid oxidation in the brown layer developing under the meat surface. These conclusions, including the role of microbially produced hydrogen peroxide, appear to be important in the understanding of the coupling between meat discoloration and lipid oxidation (Figure 10.8).

The importance of hydrogen peroxide production for myoglobin catalysis of lipid oxidation was perhaps most clearly demonstrated in a phosphatidyl choline liposome system, where MbFe(II)O_2 was a better catalyst than MbFe(III), and where catalase (but not superoxide dismutase) was effective in inhibiting both lipid and MbFe(II)O_2 oxidation (Chan *et al.*, 1997). This study and that of linoleic acid oxidation by hydrogen peroxide in micelles (Nakayama *et al.*, 1997) represent important steps towards the kinetic modelling of myoglobin catalysis of lipid oxidation. Lipid peroxides were found not to be a prerequisite for catalysis in the liposome system, which may be explained in light of hydrogen peroxide generation in the autoxidation of MbFe(II)O_2. In contrast, myoglobin catalysis proved important in oxidation of linoleic acid in the micelle system and also in an oil-in-water emulsion of linoleic acid only with preformed hydroperoxides (Nakayama *et al.*, 1997; Baron *et al.*, 1997), in agreement with the classical Tappel mechanism, in which lipid oxidation is driven by homolytic cleavage of hydroperoxides (Tappel, 1955). Moreover, in the linoleic acid emulsion,

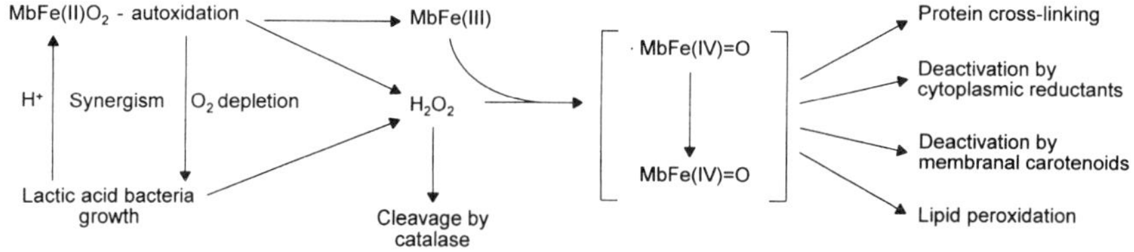

Figure 10.8 Hypervalent forms of myoglobin, formed due to synergism between lactic acid bacterial growth and oxymyoglobin autoxidation, are initiators of lipid peroxidation and cross-linking of proteins. Membrane and cytoplasmic antioxidants are important for protection of lipids and proteins against oxidative damage.

$MbFe(II)O_2$, MbFe(III) and MbFe(IV)=O were found to have similar efficiency as catalysts, whereas •MbFe(IV)=O did not show any catalytic activity at all (Baron *et al.*, 1997). Apparently, the protein radical underwent an intramolecular electron transfer reaction to yield the so-called green pigment, which is inefficient as a catalyst (Fox *et al.*, 1974). However, the apparently conflicting results from the two investigations (Chan *et al.*, 1997; Baron *et al.*, 1997) may be rationalized when considering that •MbFe(IV)=O, also formed during autoxidation of $MbFe(II)O_2$, may react directly with phospholipids present in the liposomes by electron transfer (Figure 10.7). The resulting carbon-centred phospholipid radical will then further react with oxygen to yield a peroxy radical, a precursor for a phospholipid hydroperoxide. Establishment of such differences in catalytic mechanism between •MbFe(IV)=O, active in initiation of lipid oxidation in the phospholipid, and MbFe(IV)=O, active in cleavage of preformed hydroperoxides, will, however, need further experimental verification. In the hydroperoxide free linoleic acid micelles, the induction period for oxygen consumption during which hydrogen peroxide and myoglobin generate linoleic hydroperoxides is increased by ascorbate, while the oxygen consumption rate following the induction period is not (Nakayama *et al.*, 1997). Thus, different catalytic mechanisms exists, and ascorbate is expected to be very effective in deactivating the protein radical (Kröger-Ohlsen and Skibsted, 1997).

Reductants other than NADH deactivate hypervalent myoglobins and protect the membrane lipids against oxidation, and they also protect proteins against oxidative cross-linking (Østdal *et al.*, 1996). For ascorbate, three reaction pathways are important in the reduction of MbFe(IV)=O in the pH region of relevance for meat (Kröger-Ohlsen and Skibsted, 1997). In moderate acidic solutions, deactivation of ferrylmyoglobin occurs mainly by intramolecular electron transfer in a ferrylmyoglobin/ascorbate complex from which an ascorbyl radical dissociates. In neutral solutions, reduction is the result of a bimolecular reaction between ascorbate and ferrylmyoglobin, and the rate and activation parameters, where available, are rather similar for inactivation by ascorbate and NADH, and for reaction with the whey protein β-lactoglobulin (Table 10.1). More surprising is the high rate at which the carotenoid crocin deactivates MbFe(IV)=O and especially •MbFe(IV)=O (Jørgensen *et al.*, 1997). The role of carotenoids in protection of lipids against pigment-mediated oxidation and proteins against cross-linking should be explored further. The binding of ascorbate to oxidized pigments is a highly exothermic process, and binding of ascorbate increases with decreasing temperature. This type of protein binding should also be studied in more detail as it could serve as a protective mechanism against long-lived protein radicals (Østdal *et al.*, 1997).

In conclusion, at this critical control point, colour preservation is important when considering the protection of lipids in both fresh and frozen

Table 10.1 Hypervalent myoglobin as catalyst. Kinetic data for reduction in neutral aqueous solution at 25°C

Reactants	k ($l\ mol^{-1}\ s^{-1}$)	ΔH^{a} ($kJ\ mol^{-1}$)	ΔS^{a} ($J\ mol^{-1}\ K^{-1}$)	Reference
•MbFe(IV)=O + crocin	~10^5	ND[a]	ND[a]	Jørgensen *et al.*, 1997
MbFe(IV)=O + crocin	183	55 ± 1	−17 ± 4	Jørgensen *et al.*, 1997
MbFe(IV) + NADH	~1	ND[a]	ND[a]	Mikkelsen and Skibsted, 1995
•MbFe(IV)=O + ascorbate	2.7	45 ± 13	−85 ± 46	Kröger-Ohlsen and Skibsted, 1997
•MbFe(IV)=O + β-lactoglobulin	1.2	45 ± 6	−93 ± 20	Østdal *et al.*, 1996

[a]ND, not determined.

meats due to the direct coupling between pigment oxidation and lipid oxidation. Bacterial growth is also critical in relation to oxygen activation and depletion of reducing cofactors in meat. The level of reducing cofactors should further be considered in relation to the physiological status of the animals.

10.3.3 Meat processing as a critical control point

Precooking. Manufacturing of precooked meats for food-service systems or for ready-to-eat dishes often involves heating as part of combined processes including mincing, slicing, marinating or curing. The heat treatment, as such, induces oxidative changes in meat, and upon reheating develops WOF. Exclusion of oxygen during heat treatment as in the *sous-vide* method (Bertelsen and Juncher, 1996) has been shown to be efficient in suppressing the formation of secondary lipid oxidation products in beef (Hansen *et al.*, 1995). Notably, upon introduction of oxygen to the *sous-vide* product, lipid oxidation starts to progress rapidly (Hansen *et al.*, 1995), reflecting the formation of catalytic species during the heating procedure and inactivation of protective enzymes such as catalase. For different types of meat, the sensory perception of WOF has been found to correlate differently with the level of secondary oxidation products as measured by determination of TBARS. For pork, a linear relationship existed (Shahidi *et al.*, 1987) while the sensory response for beef was found to be proportional to the logarithm of TBARS, in agreement with the Weber–Fechners law (Stapelfeldt *et al.*, 1992). The rate of oxidation, quantified as TBARS formation, has been shown to increase with increasing heating temperature and heating time (Smith *et al.*, 1987; Mielche and Bertelsen, 1993). Heating below a certain temperature causes a very slow development of TBARS in products (Figure 10.9). Thus, it seems possible to minimize WOF in meats by selecting a sufficiently low temperature and also by choosing muscles which are less susceptible to oxidation (Figure 10.9). However, the

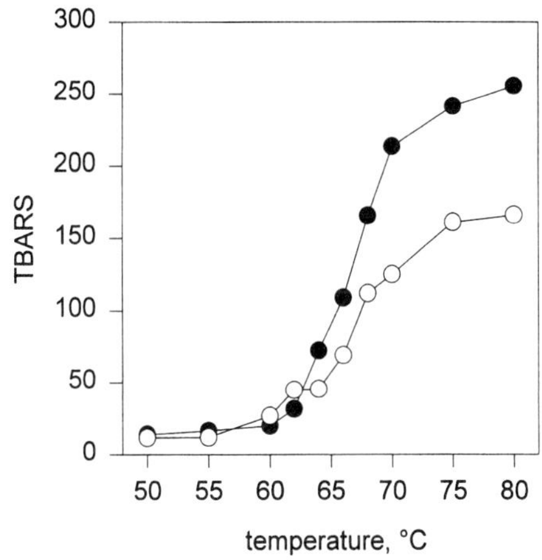

Figure 10.9 Changes in the contents of 2-thiobarbituric acid reactive substances (TBARS, μmole malonaldehyde equivalents per kg dry matter) in ground chicken thigh (●) and breast (○) heated at various temperatures and stored at 4°C for 2 days (Mielche *et al.*, unpublished results).

different threshold values have to be obtained experimentally under the actual process conditions. Thus, for chicken breast meat, a threshold value of 74°C has been reported (Smith *et al.*, 1987), whereas considerable oxidation of chicken meat when cooked below this temperature was noted (Yamauchi, 1972). Heating rate and holding time should also be considered as process parameters. However, use of the lowest temperature needed to minimize WOF development might pose problems when microbial safety of the final product must also be taken into consideration.

Above the threshold temperature a steep increase in subsequent development of WOF is observed, to which many factors contribute. These are (a) liberation of catalytically active iron from myoglobin and other iron-containing proteins (Sato and Hegarty, 1971; Igene *et al.*, 1979; Chen *et al.*, 1984; Rhee and Ziprin, 1987; Graf and Panter, 1991), (b) formation of prooxidative species of myoglobin (Harel and Kanner, 1985a), (c) denaturation of membranes that in the nonheated cell compartmentalize phospholipids and prooxidants (Yamauchi, 1972), and (d) denaturation of antioxidative enzyme systems. The relative importance of these factors has not been agreed upon and may also depend on the type of meat used.

Heat treatment at temperatures above 100°C reduces WOF development (Figure 10.10), as antioxidative Maillard reaction products (MRP) are formed at high cooking temperatures (Bailey, 1988). Maillard browning is an efficient way of preventing development of WOF. High cooking temperatures, however, result in other quality changes which might adversely affect the sensory quality, such as reducing water-holding

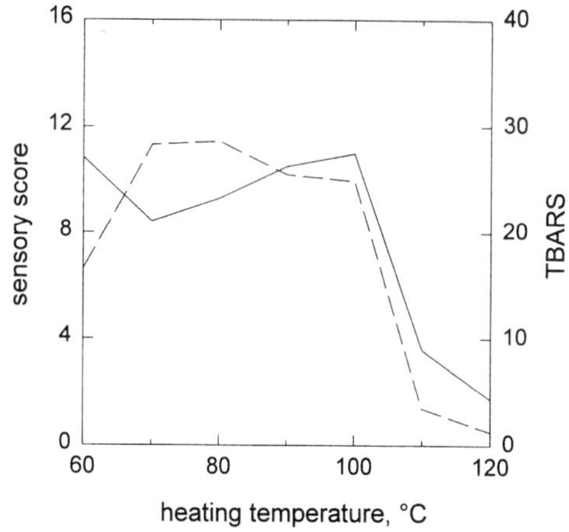

Figure 10.10 Least squares mean values of WOF taste (—) and mean values of TBARS (---) of beef heated at various temperatures for 30 min and stored at 4°C for 2 days. WOF taste ranges from 0 (no WOF) to 16 (extreme WOF), and TBARS are expressed as moles malonaldehyde equivalents/kg dry weight (Mielche and Bertelsen, 1995, reproduced with permission).

capacity and dark browning. The method of heat treatment also seems to affect the development of WOF. Thus, it has been shown that microwave cooking results in faster development of WOF compared to roasting or grilling (Gros *et al.*, 1986; Satynarayan and Honikel, 1992). These differences may, however, be explained by variation in cooking temperature and/or differences in the formation of Maillard reaction products (MRP). In conclusion, the heating procedure, including the temperature regime and heating rate, should be adjusted by consideration of the type of meat examined, as a critical control point.

Curing. Salt generally catalyses lipid oxidation in meat, and several mechanisms have been proposed to account for its prooxidative effect. An ion-exchange mechanism, entailing displacement of iron from proteins by sodium ions, depends on 'free ion' catalysis (Kanner *et al.*, 1991). Specific effects of the chloride anion focus on catalysis involving activation of peroxidases like myeloperoxidase, which in a hydrogen peroxide driven oxidation of chloride yields the potent hypochlorite oxidant (Kanner and Kinsella, 1983). Definitive proof for an actual contribution from any of these ion-specific mechanisms to lipid oxidation in meat systems has not yet been provided, and less specific effects from an increase in ionic strength entailing changes in protein solvation and membrane properties should also be considered (Shomer *et al.*, 1987). Addition of 1% NaCl to a meat product thus results in an increase in the ionic strength from the physiological value of 0.16 to approximately 0.35. Such changes in ionic strength have been shown to reduce the catalytic activity of the antioxidant enzymes, namely catalase, superoxide dismutase and especially glutathione peroxidase, and in effect markedly lower the oxidative stability of salted meat products (Lee *et al.*, 1997). The acid-catalysed autoxidation of oxymyoglobin is subject to a positive kinetic salt effect, and the increase in ionic strength corresponding to addition of 1% salt will increase the oxidation rate by 40% for traditionally deboned meat (Andersen *et al.*, 1988b). For high-pH meat resulting from a hot-deboning process, a similar rate increase in $MbFe(II)O_2$ autoxidation from salt addition is not expected, and was not found in a storage experiment with hot processed frozen minced beef (Andersen *et al.*, 1990b). According to the model proposed for the coupling between lipid oxidation and pigment oxidation in Figure 10.7, the kinetic salt effect on pigment oxidation could accordingly provide a rationale for the effect of salt on lipid oxidation and also for the observation, that hot-deboned beef is less exposed to salt-induced lipid oxidation (Andersen *et al.*, 1990b). For minced pork, light-induced and also salt-induced lipid oxidation were found to be preceded by discoloration during frozen storage, a finding which adds further support to such a direct coupling mechanism (Andersen and Skibsted, 1991).

Independent of the actual mechanism responsible for salt-induced lipid oxidation in meat products, correct addition of salt is critical for the control of lipid oxidation. The level of salt should always be considered in relation to the type of meat and muscle and in relation to the desired pH development, since the addition of salt prior to freezing has been shown to be important in maintaining the high pH and the high level of ATP characteristic of prerigor meat. Of equal importance is the quality of the salt used for food processing, and contamination with transition metals and other potential prooxidants should for obvious reasons be avoided. However, it is notable that the commonly used anticaking agent $K_4Fe(CN)_6$, even in 'normal' concentrations added to salt as a processing aid and transferred to the meat products, has been found to increase lipid oxidation (Hansen *et al.*, 1996). The $Fe(CN)_6^{3-}/Fe(CN)_6^{4-}$ couple is known as a redox mediator in the metmyoglobin reductase system (Faustman and Cassens, 1989), and the turn-over number of the colour cycle of Figure 10.7 will be increased, generating more activated oxygen species and in effect destabilizing lipids and membranal structure. For pork patties, progression of lipid oxidation further correlated with the rate of discoloration during frozen storage, confirming the enhanced electron transfer to metmyoglobin in the presence of the $Fe(CN)_6^{3-}/Fe(CN)_6^{4-}$ couple, and providing further evidence for a direct coupling between oxidative discoloration and lipid oxidation. It was recommended that the use of salt with $K_4Fe(CN)_6$ added should be reconsidered in meat processing for each individual product (Hansen *et al.*, 1996).

When salt is combined with the strong oxidant nitrite in the curing process, meat with enhanced oxidative stability is obtained, a somewhat surprising result for which the underlying mechanism is not yet fully understood. Nitrite is a unique additive that in addition to protection of meats against lipid oxidation also has important antimicrobial effects and contributes to colour and flavour formation (Perigo *et al.*, 1967; Cassens *et al.*, 1979). Meat curing with nitrite has from time to time been facing serious legislatory problems due to the toxicity of nitrite and to formation of carcinogenic *N*-nitrosamines in some products under certain conditions (Cassens, 1990). However, medical research is currently uncovering the fascinating role of nitric oxide and derived compounds in bioregulation, and the chemistry of meat curing continues to show surprising parallels to processes important in the regulation of free radicals in the human body (Feldman *et al.*, 1993). However, there is still a strong interest in developing nitrite-free meat curing systems (Shahidi and Pegg, 1992; Pegg *et al.*, 1996).

The question as to whether nitrite per se or a product from reaction of nitrite with reactants present in meat is the most important antioxidant has not yet been clarified (MacDonald *et al.*, 1980; Kanner *et al.*, 1980). For nitrite at least three mechanisms have been proposed:

(a) Nitrite stabilizes the lipid fraction by reaction with the unsaturated fatty acids (Goutefongea *et al.*, 1977; Zubillaga *et al.*, 1984).
(b) Nitrite chelates trace metals, thus preventing catalysis by free metals ions like iron(II) (MacDonald *et al.*, 1980).
(c) Nitrite forms stable complexes with haem pigments by coordination to the iron centre, in effect preventing catalytic breakdown of lipid peroxides (Zipser *et al.*, 1964).

As for (a), the binding of nitrite to the unsaturated lipids, a simple stoichiometric calculation shows that the amount of nitrite used in meat curing is insufficient to protects more than a small fraction of the unsaturated lipids (Erduran and Hotchkiss, 1995). Nitrite binding to simple iron(II) ions (b) has not been documented except in mixed-ligand complexes with strong π-accepting ligands such as cyanide, and calculations show that NO is a more likely ligand for iron(II) (Skibsted, 1992), leaving us with alternative (c), binding of nitrite to haem pigments, which remains for further consideration. By using differential scanning calorimetry, nitrite has been found to increase the temperature for thermal denaturation of MbFe(III) by several degrees, and also to protect the haem cavity against oxidative degradation by oxidizing lipids (Arendt *et al.*, 1997). Binding of nitrite as a 'radical guard' at the entrance to the haem cavity restricts the access of lipid peroxides and also of lipid radicals to the iron(III) centre (Arendt *et al.*, 1997). Polyphosphates also behave as antioxidants in meat systems, and polyphosphates have been found to bind in a way similar to nitrite in the haem cleft of MbFe(III) (Mikkelsen *et al.*, 1992a). The 'radical guard mechanism' proposed for nitrite may also apply to the polyphosphates, and it further finds an interesting parallel in the binding of ascorbate to hydrogen peroxide-activated myoglobin (Kröger-Ohlsen and Skibsted, 1997). However, much remains to be learned from the chemical model system of the role of binding of small molecules as 'radical guards' to otherwise prooxidative metalloproteins and long-lived protein radicals before meat processing can benefit from this finding.

Pediococcus pentosaceus, *Staphylococcus xylosus* and other bacteria with high lipolytic activity are important for aroma development in dry-cured meat products (Molina *et al.*, 1991). Free fatty acids are more vulnerable to oxidation than triacylglycerols, but dry-cured products still have a high stability even at room temperature. The free fatty acids may also act as 'radical guards', the hypothesis of binding of free fatty acids to MbFe(III) (Yackzan and Wingo, 1982) appears to gain support from experimental findings (Stewart, 1990). While free fatty acids enhance the autoxidation of MbFe(II)O_2, binding to the resulting MbFe(III) inhibits the peroxidase activity of MbFe(III). Lipolysis is accordingly expected to have the opposite effect on colour stability and on the oxidative stability of the meat lipids. In contrast to uncured meat, where a brown discoloration indicates loss of protection against lipid oxidation, a colour change induced by lipolysis in dry-cured meat may indicate a decrease in the haem pigment-catalysed lipid oxidation. The observation of a decrease in oxidation rate with increasing temperature for linoleic acid in the presence of MbFe(III) further supports the hypothesis of a hydrophobic interaction and entropy-controlled binding of free fatty acids to the catalytic centre of haem proteins (Mikkelsen *et al.*, 1992a). The complex interaction between microbial activity and hydrolytic and oxidative degradation of lipids deserves further attention in the production of dry-cured meat products, as a correct balance is critical in order to prevent the formation of lipid-derived off-flavours. However, again more basic research seems essential before the control methods can be optimized.

Thermodynamic studies predict that nitrite will react with a wide variety of reducing components present in meat, and the nitric oxide and some of the nitrosated products formed by reduction of nitrite are potential antioxidants (Skibsted, 1992). Phospholipids extracted from cured pork have thus been found to be active in *trans*-nitrosation reactions (Freybler *et al.*, 1989), and different pools of nitrite-modified compounds like nitrosated cytochromes and nitrosated tryptophyl moieties of the proteins seem to be rapidly interchanging the NO radical with each other and with residual nitrite and with N_2O_3 (Figure 10.11). As a free radical, NO is capable of terminating chain reactions in lipid oxidation, but it is also subject to reaction with ground state oxygen. NO is, however, protected against direct oxidation through coordination to the iron(II) centre of

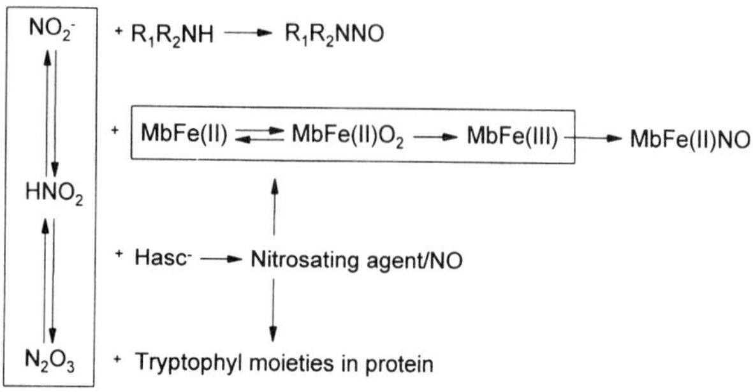

Figure 10.11 The nitric oxide radical in cured meat, transferred through rapid transnitrosation reactions between different pools of nitrite-modified compounds (Andersen *et al.*, 1990c). Reaction of nitrite with ascorbate ($Hasc^-$) inhibit the formation of nitrosamines (R_1R_2NNO).

myoglobin and other haem compounds. Meat curing is most easily recognized through the colour change, as the cured meat pigments are pink, and the chemistry behind the colour change seems to be central to our understanding of the improved oxidative stability. When establishing critical control points in meat processing, colour formation and colour preservation can easily be used to ensure antioxidative capacity of the product. The uncooked cured meat pigments nitrosylmyoglobin together with its heat-denaturated forms thus serve as an antioxidant buffer from which NO is readily mobilized (Kanner *et al.*, 1980; Andersen and Skibsted, 1992). The NO or nitrosated compounds like *S*-nitrosocysteine (Kanner, 1979) may be the actual chain-breaking antioxidant, but the colour of the products provides a direct indication of the oxidative status. The reaction of NO with lipid-derived free radicals and activated oxygen species yields mainly non-reactive products (Rubbo and Freeman, 1996), although the highly reactive peroxynitrite is a reaction intermediate which deserves further attention in food systems (Figure 10.12). The nitrosyl pigments may also have an antioxidative effect which does not depend on NO dissociation. This has been demonstrated for CCMP (cooked cured-meat pigment) in a β-carotene/linolate antioxidant assay (Pegg *et al.*, 1996). The CCMP is an ingredient of the nitrite-free meat curing system which, although synthesized outside the meat matrix, has been shown to be chemically identical to the cured meat pigment.

The MbFe(II)NO is oxidized in a rather slow and very temperature-dependent reaction yielding nitrate and MbFe(III) as shown in Figure 10.13. Brown discoloration of cured meat products, in which lipolysis is insignificant and where ascorbic acid has been added as reductant, indicates loss of resistance against development of oxidative rancidity. In contrast to $MbFe(II)O_2$, oxidation of MbFe(II)NO decreases with decreasing partial pressure of oxygen and is strongly accelerated by light (Andersen and

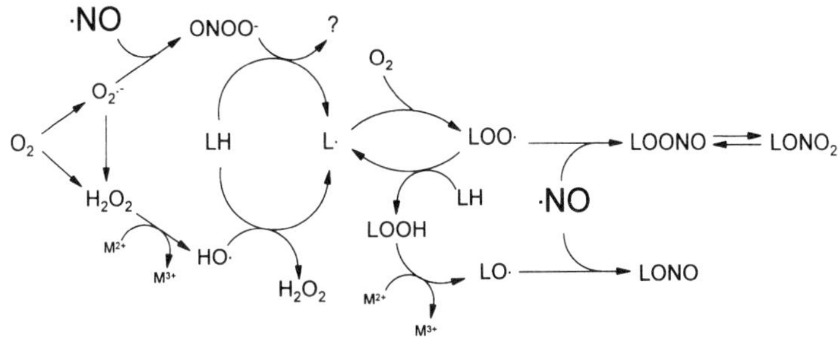

Figure 10.12 Nitric oxide in scavenging of lipid Peroxy and alkoxy radicals and reaction with the superoxide radical to form peroxynirite (adapted from Rubbo and Freeman, 1996).

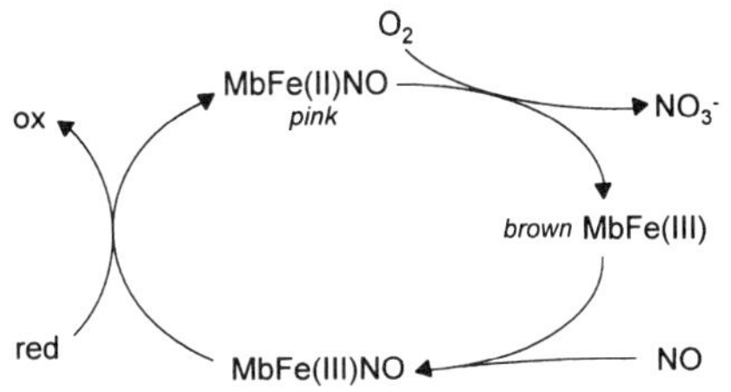

Figure 10.13 The colour cycle of cured meat. In vacuum-packed ham exposed to fluorescent light and residual oxygen, the pink nitrosylmyoglobin is oxidized to the brown metmyoglobin in a photochemical process. When oxygen is depleted, the pink colour is reestablished (Andersen *et al.*, 1988a).

Skibsted, 1992). During retail display, effective vacuum is mandatory in order to protect cured meat from the combined action of oxygen and light. Colour fading of sliced products is, however, often observed, although the pink colour is partially re-established, when oxygen is depleted in the product below a critical limit for photooxidation. An effective oxygen barrier in the packaging film together with excess ascorbate are essential for regeneration of MbFe(II)NO, and the protection of lipids in meats cured with nitrite and ascorbate is thus linked to the colour cycle (Figure 10.13).

The microbial changes in cured meat products are usually related to the growth of lactic acid bacteria, although *Brochotrix thermosphacta* can be an important spoilage bacteria. The interaction between microbial growth and the oxidative deterioration is less dependent on the lactic acid production and souring of the product, since both photooxidation and thermal oxidation of nitrosylmyoglobin show little, if any, dependence on pH. For cooked cured meat, the catalase activity has decreased significantly and the hydrogen peroxide produced by the lactic acid bacteria can oxidize nitrosylmyoglobin (Kanner *et al.*, 1991b):

$$\text{MbFe(II)NO} + H_2O_2 \rightarrow \text{MbFe(III)} + NO_2^- + H_2O$$

The finding that MbFe(III), rather than a hypervalent myoglobin, is the oxidation product, as in the raw meat, is of utmost importance for understanding of the stability of lipids in cured meats. For cooked products catalase is thermally inactivated, and microbial production of hydrogen peroxide could, if hydrogen peroxide was not otherwise deactivated, initiate lipid oxidation. The NO_2^-, the other oxidation product, is moreover available for regeneration of MbFe(II)NO as long as ascorbate or endogenous reductants are not depleted. As may be seen from Figure 10.14, MbFe(II)NO is involved in ascorbate deactivation of hydrogen peroxide, in effect acting as an ascorbate peroxidase. However, for the superoxide radical, similar information on the reaction with MbFe(II)NO is not available.

Meat curing has served mankind in producing meat products with keeping qualities for extended periods. For cooked cured meat products it is also critical to protect their pink colour against photooxidation since MbFe(II)NO is an antioxidant buffer protecting the lipids by deactivation of hydrogen peroxide. However, ultraviolet barriers in the packaging material yields little, if any, protection, contrary to findings for frozen uncured meat products (Andersen and Skibsted, 1991). For dry-cured meats, free fatty acids from microbial lipolysis further block catalytic iron(III) haem pigments and brown discoloration is less critical.

Use of antioxidants. In different cultures, different herbs have been added to food in order to improve the flavour. In the Nordic countries, thyme was thus an important ingredient for blood sausages and in the Mediterranean cultures rosemary and oregano have been widely used. Spices contribute a pleasant flavour and also exhibit an antioxidative effect through their active compounds. This serendipitous use of natural antioxidants has been important in food preservation throughout the history. The challenge for industrial food production today is to identify the herbs and spices appropriate for different meat products and to add combinations and amounts of spices or spice extract in order to secure a pleasant flavour and to obtain optimal oxidative stability.

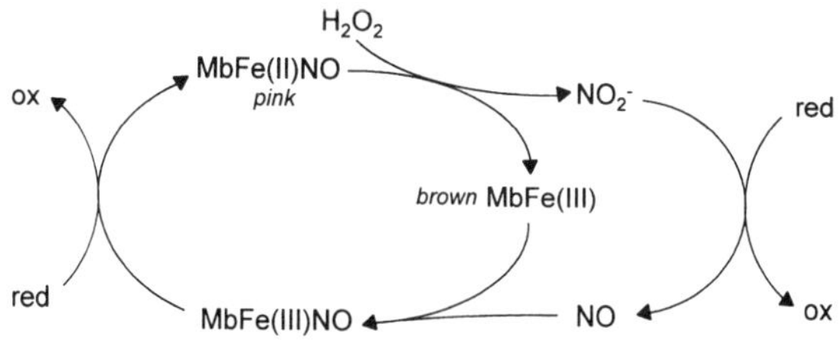

Figure 10.14 The colour cycle of cured meat as linked to the antioxidative activity of the nitric oxide free radical capacity. Lipid molecules are protected against peroxides and hypervalent iron.

Very few of the antioxidative compounds in spices have so far been isolated and identified. However, phenolic compounds have been found to constitute a larger proportion of compounds shown to have antioxidative properties. Thus, in rosemary (*Rosmarinus officinalis* L.), which has been the subject of numerous investigations, carnosol and carnosic acid together with rosmarinic acid have been identified as potent antioxidants (Brieskorn *et al.*, 1964; Brieskorn and Dömling, 1969; Wenkert *et al.*, 1965; Wu *et al.*, 1982). Carnosol and carnosic acid have also been isolated from sage (*Salvia officinalis* L.) (Brieskorn *et al.*, 1964; Brieskorn and Dömling, 1969) and from summer savory (*Satureja hortensis* L.) (Reschke, 1983; Bertelsen *et al.*, 1995), both of which as well as rosemary belong to the family Labiateae. This plant family is especially prosperous, as the 'pizza-spice' oregano (*Origanum vulgare* L.) also contains rosmarinic acid (Kikuzaki and Nakatani, 1989) as well as different flavonoids with antioxidative effect (Vekiari *et al.*, 1993). Thyme (*Thymus vulgaris* L.) is another spice belonging to the Labiateae family. Flavonoids may also serve as important antioxidative compounds in thyme as they are in black pepper (Nakatani *et al.*, 1986; Vösgen and Hermann, 1980). Attention is also being paid to the possible use of grape skin extracts, containing a variety of antocyanins (Mazza and Miniati, 1993) and tea extracts containing cathechins and epicathechins (Amarowicz and Shahidi, 1996) in foods.

Many of the antioxidants identified in spices and herbs such as carnosol and carnosic acid are both tasteless and odourless. However, some of the antioxidative compounds present in spices do have a strong flavour, which limits the use of their extracts as antioxidants in meat products. For instance, the volatiles carvacrol and thymol present in oregano and thyme (Deighton *et al.*, 1993) and eugenol present in clove (*Eugenia caryophyllata* Thumb.) (Kramer, 1985) have very strong impact on the flavour of products.

The early research recognized that the antioxidative activity of spices or spice extracts depended on the substrate (Chipault *et al.*, 1952; Chipault *et al.*, 1955). Rosemary and sage were thus found to be most effective in lard, while clove showed the highest activity in oil-in-water emulsions. Rosmarinic acid, carnosic acid and carnosol have further been evaluated as antioxidants. In bulk corn oil, the highest activity was due to the more polar compounds rosmarinic acid and carnosic acid, whereas the less polar compound carnosol showed the highest antioxidative activity in oil-in-water emulsions (Frankel *et al.*, 1996). In bulk oil, hydrophilic antioxidants are oriented in the oil–air interface providing optimal protection of the lipids, whereas hydrophobic antioxidants dissolve in oil and thus are less efficient, while in the oil-in-water emulsions hydrophobic antioxidants are most effective as they concentrate at the oil–water interface (Frankel *et al.*, 1996). This rationale for the 'polar paradox' emphasizes the necessity of evaluation of antioxidative activity in heterogenous model systems and in the actual food product prior to practical use. Recently a three-step procedure for evaluation of spices was suggested including (a) determination of radical scavenging using electron spin resonance spectroscopy, (b) test in model systems, and (c) final test in food storage experiments (Madsen *et al.*, 1997).

The 'novel' and increasing use of natural ingredients like spices by the food industry has somewhat decreased the interest in traditional antioxidants like tocopherol and ascorbates as additives to meat products. However, for certain products, these classical antioxidants are still important for use in combination in certain products in which they show important interactions (Lambelet *et al.*, 1985). In an effort to model the antioxidant interaction, the induction period for oxidation determined in an oxygen atmosphere at elevated temperatures was compared by measuring TBARS and depletion of different tocopherol homologues in cooked, minced turkey during chill storage (Bruun-Jensen *et al.*, 1996a,b). The effect of ascorbic acid, ascorbyl palmitate and tocopherols extracted from soybean oil on TBARS could be described by the maximum level of TBARS in products and by the first order rate constant for development of TBARS. Tocopherols and, to a lesser extent, ascorbyl palmitate reduced the maximal level, while ascorbate acted as a prooxidant. These effects, including the effect of concentration, could together be quantified by a protection factor obtained by multiplication of relative protection factors for the individual additives. The rate of development of TBARS was reduced more by ascorbyl palmitate than by tocopherol, while ascorbate had only small effect on the rate (Bruun-Jensen *et al.*, 1996a). Such mathematical modelling of antioxidant synergism should be further pursued and used as critical control point for prediction of shelf-life during subsequent storage.

Muscle tissue has an inherent protection against oxidative damage which, if understood adequately, may form the basis for adjustment of proteolytic ripening conditions for meat or development of new antioxidants for meat products. Several amino acids (Marcuse, 1960, 1962; Riisom *et al.*, 1980; Sims and Fioriti, 1977; Revankar, 1974; Karel *et al.*, 1966), peptides (Chen *et al.*, 1996; Shahidi and Amarowicz, 1996; Murase *et al.*, 1993; Yamashoji and Kajimoto, 1980; Kawashima *et al.*, 1979; Decker and Faraji, 1990) and protein hydrolysates (Shahidi and Amarowicz, 1996; Chen *et al.*, 1995; Yee *et al.*, 1980; Tsuge *et al.*, 1991; Amarowicz and Shahidi, 1997; Bishov and Henick, 1972, 1975) have thus been found to have antioxidative activities and may be used as additives in meat products. Among the amino acids tested in food systems, methionine showed antioxidant activity in vegetable oil (Sims and Fioriti, 1977), while methionine, threonine, lysine and histidine exhibited antioxidant activity in sunflower oil emulsions (Riisom *et al.*, 1980). Proline has been found to inhibit oxidation of sardine oil (Revankar, 1974). Again the antioxidative effect of amino acids depends on the food or model system used. For example, cysteine and histidine have been found to act as antioxidants at pH 9.5 and as prooxidants at pH 7.5 in a linoleic acid emulsion (Marcuse, 1960). Several amino acids have shown increasing antioxidative effects with increasing concentration up to a certain level above which the antioxidative effect decreases (Revankar, 1974). The active amino acids have been identified as primary antioxidants but may also act synergistically with α-tocopherol (Marcuse, 1960).

Among peptides, naturally occurring peptides, protein hydrolysates and designed peptides have been tested for their antioxidant activity. Carnosine (β-alanyl-L-histidine) is an endogenous skeletal muscle dipeptide which has gained special attention as a potential food antioxidant. Pure carnosine thus inhibits lipid oxidation in model systems (Kohen *et al.*, 1988; Decker and Faraji, 1990) and in uncooked (Decker and Crum, 1991) and cooked ground pork (Decker and Crum, 1993). Procedures for the production of an antioxidative carnosine-containing extract from beef muscle have been developed, and the antioxidative activity of the muscle extracts was tested in iron/ascorbate-catalysed oxidation of phosphatidyl choline liposomes (Chan *et al.*, 1993).

The antioxidative effect of carnosine appears to rely on a combination of chelation of metal ions and scavenging of reactive radicals, as carnosine binds copper but not iron, as shown by NMR spectroscopy (Decker *et al.*, 1992). However, carnosine inhibits both the copper- and iron-catalysed lipid oxidation (Decker *et al.*, 1992) as well as nonmetal-initiated oxidation processes including oxidation catalysed by azo compounds (Kohen *et al.*, 1988), H_2O_2-activated myoglobin, lipoxidase and photo-activated riboflavin (Decker and Faraji, 1990). The ESR spin-trapping studies have shown that carnosine, histidine-containing dipeptides and histidine are able to quench hydroxyl radicals, whereas β-alanine and glycine increased hydroxyl radical production in a Fenton reaction (Chan *et al.*, 1994). Carnosine has further been reported to scavenge peroxy radicals (Kohen *et al.*, 1988) and to quench singlet oxygen (Dahl *et al.*, 1988). However, carnosine seems to be unique since histidine and β-alanine do not possess antioxidative activity in copper- and iron-catalysed oxidation of phosphatidyl choline, whereas carnosine is effective (Decker *et al.*, 1992). Thus, the antioxidative effect of carnosine seems to be related to the peptide bond and the amino acid sequence of the peptide, as it has been further indicated for analogous peptides like anserine (β-alanyl-L-1-methylhistidine) and ophidine (β-alanyl-L-3-methylhistidine) (Chan and Decker, 1994).

Protein hydrolysates from different sources including vegetables (Chen *et al.*, 1995; Yee *et al.*, 1980), fish (Shahidi and Amarowicz, 1996; Amarowicz and Shahidi, 1997), egg (Tsuge *et al.*, 1991) and yeast proteins (Bishov and Henick, 1972, 1975) are also effective antioxidants in model systems. Antioxidant peptides composed of 5–16 amino acids have been isolated from a hydrolysed soybean protein and their amino acid sequences have been determined (Chen *et al.*, 1995). Further studies with several synthetic peptides, designed on the basis of the smallest peptide from the soybean hydrolysate, again indicated that histidine and proline are crucial to the antioxidative activity (Chen *et al.*, 1996). Hydrophobic interactions of the histidine imidazole group in the peptide region with lipids seems to be important in determining the antioxidative properties (Decker *et al.*, 1992). However, care must be exercised before application of protein hydrolysates in meat products, as peptide fractions promoting lipid oxidation have also been isolated (Amarowicz and Shahidi, 1997). As for other antioxidants, the overall effect depends on the interactions between components in the hydrolysates as well as the interactions with other components present in the meat system.

High-pressure treatment and other minimal processing techniques. New minimal processing techniques for meat products are currently being developed in order to accommodate consumer demands for products which seem fresh (Mertens and Knorr, 1992). Ultra-high-pressure technology and high-intensity light pulse technology, which in other food systems have been shown to retard spoilage processes, might in future find their way into meat processing. High-pressure treatment may also be used to change enzyme–substrate interaction for proteolytic enzymes and to modify functional properties with the further perspective of reducing the use of food additives. However, safety aspects of products and quality risk in relation to lipid oxidation must also be considered.

The use of intense light pulses for meat preservation will depend on quantitative studies of the effect of spectral distribution and relative light intensities on microbial inactivation, which should be maximized. At the same time, photooxidation of food pigments, proteins and lipids, if possible, should be minimized (Bertelsen and Skibsted, 1987); such information is not yet available.

High-pressure effects on oxidation processes in meat are complex and involve both meat pigments and lipids. High-pressure treatment of very fresh beef has thus been found to inhibit metmyoglobin formation during subsequent storage, while pressure treatment of aged meat had no effect on its colour stability (Cheah and Ledward, 1997). The kinetics of formation of metmyoglobin was found to depend on a pressure-sensitive catalytic system related to oxygen depletion at or near the surface of the meat, and on metmyoglobin reduction by enzymes, which are less pressure sensitive. The catalytic mechanism related to oxygen-utilizing enzymes in meat dominates in fresh meat, but the activity of these enzymes decreases rapidly, and in aged meat the activity of the metmyoglobin-reducing enzymes becomes the controlling factor (Ledward, 1991). Two opposing factors were found to affect the autoxidation rate of $MbFe(II)O_2$, when studied in a model system designed to explore pressure effects on meat systems. Pressure decreased the rate, corresponding to an expansion in the transition state of autoxidation, but at the same time pressure increased specific acid catalysis due to a pressure-induced decrease in solution pH (Bruun-Jensen *et al.*, 1997). For the cured meat pigment, MbFe(II)NO, autoxidation is independent of pH, and only the increased volume in the transition state is important (Bruun-Jensen and Skibsted, 1996). These findings provide explanation for the observation that the colour of minced beef was found to be affected by pressure (Carlez *et al.*, 1995) and that nitrite prevented such pressure-induced discoloration of minced meat (Goutefongea *et al.*, 1995). Moreover, for uncured meat, hot-deboned and other high-pH meats are expected to be less exposed to pressure-initiated oxidation since the pH decrease around neutral does not induce acid catalysis. The coupling of pigment oxidation to lipid oxidation in meat is evident (Figure 10.7), but the questions remain as to how pressure denaturation of protein and pressure disintegration of membrane structures affect oxidation of phospholipids.

High-pressure treatment of meat involves pressures of $\geq$100 MPa. An important goal prior to the design of processes in the meat industry is to define parameters that describe the effects of varying pressure and treatment time on the initiation of lipid oxidation. Pressure activation parameters, which further are expected to vary with temperature may, together with similar parameters for pressure-induced inactivation of microorganisms, be used to optimize product quality and to predict shelf-life. For vacuum-packed, pressure-treated turkey thigh muscle, formation of secondary lipid oxidation products was found to depend primarily on pressure (up to 500 MPa investigated) and to a lesser extent on pressurization time. Pressure treatment at $\leq$400 MPa for 30 min resulted in formation of lesser amounts of secondary oxidation products during 6 days of storage at 5°C compared to meat treated at 100°C for 10 min, while pressure treatment at 500 MPa for 30 min had a similar effect as did the heat treatment (Dissing *et al.*, 1997). The logarithm of the rate of development of rancidity during chill storage was found to depend linearly on the working pressure and relationships based on the thermodynamic equation:

$$\ln k_p = \ln k_1 - (\Delta V^{\#}/RT)\, p$$

was suggested for prediction of the rate of development of rancidity for treatment at varying working pressures. In this equation, k_1 is the rate (constant) for quality deterioration for nonpressure-treated product, and k_p is the rate (constant) for the same product treated under pressure p. $\Delta V^{\#}$, an apparent volume of activation for initiation of lipid oxidation (or other quality changes), is an empirical parameter which depends on the holding time. Although a final conclusion should be drawn based on a larger number of storage experiments for different types of meat, it was suggested that $\Delta V^{\#}$ will approach a limiting value for increasing holding time which is characteristic for each product. Thus, $\Delta V^{\#}$ may in future find uses to quantify the effect of pressure treatment similar to the use of energy of activation and Q-values for heat treatment.

10.3.4 Packaging and storage as a critical control point

The most important critical control points for the control of lipid-derived off-flavours in meat and meat products are the packaging and retail display conditions. The storage temperature certainly is important, especially in relation to the interaction between microbial growth and lipid oxidation (Figure 10.8). Packaging in a material with a low oxygen transmission rate combined with reduced partial pressure of oxygen obtained by vacuum packaging or by modified atmosphere packaging (MAP) in practical storage experiments have shown to be effective in retarding oxidative changes in meats (Stapelfeldt *et al.*, 1993; Bruun-Jensen *et al.*, 1994).

The exposure to light causes photooxidative changes in meat (Bekkölet, 1990). Storage in darkness or selection of less harmful lighting conditions for displayed products have proven effective in retarding oxidative changes in meats (Andersen *et al.*, 1988a, 1990a). Again optimal packaging and storage conditions depend on the actual meat product. Based on quantum yields determined for photooxidation of nitrosylmyoglobin and oxymyoglobin at the wavelength characteristic for fluorescent light and thermal activation parameters and rate constants, a comparison has been made between thermal pigment oxidation and light-induced pigment oxidation for realistic storage conditions of both fresh and cured meats (Skibsted, 1992; Andersen and Skibsted, 1992). Such mathematical modelling should be further encouraged and must include different oxygen levels. The predictions made so far have been confirmed qualitatively in practical storage experiments.

In fresh meat, the red surface of the meat is retained only if the microatmosphere surrounding the meat contains oxygen, as is obtained either by wrapping it in a packaging material with a relatively high oxygen transmission rate or by using high oxygen MAP. Use of packaging material with an oxygen transmission rate of 5000 $cm^3/m^2/24$ h/atm results in an initial depth of the oxymyoglobin layer responsible for the red colour of approximately 2 mm (Taylor, 1982). High oxygen MAP, as used to improve the colour stability, will result in a deeper oxymyoglobin layer of up to 12 mm (Taylor and McDougall, 1973; Clark and Lentz, 1973; Bartkowski *et al.*, 1982), but may result in formation of lipid-derived off-flavour (Taylor, 1985). Differences in the oxidative status of the meat prior to storage seems to be very important, however, as in other practical storage experiments no off-flavours were generated during high oxygen MAP storage (Ordonez and Ledward, 1977; Lopez-Lorenzo *et al.*, 1980;

Skibsted *et al.*, 1994), again stressing the importance of considering all critical control points together.

The development of WOF in precooked, chilled or frozen stored meat products can be retarded effectively by eliminating oxygen in the microatmosphere by vacuum packaging as shown for precooked chill-stored beef (Hsieh and Baldwin, 1984; McDaniel *et al.*, 1984; Stapelfeldt *et al.*, 1993) and for precooked chill or frozen stored pork (Nolan *et al.*, 1989). Anoxic MAP, such as that using a mixture of nitrogen and carbon dioxide, is an effective alternative, for ground turkey (Bruun-Jensen *et al.*, 1994), ground pork (Huisman *et al.*, 1994) and sliced beef (Stapelfeldt *et al.*, 1993) as it retards the development of WOF during chill storage. When MAP is employed, the major challenge is to obtain a sufficiently low residual oxygen level in the packages to reduce the development of WOF effectively (Stapelfeldt *et al.*, 1993; Bruun-Jensen *et al.*, 1994). The combination of MAP and packaging in a material containing oxygen absorbers (Rooney, 1995) may be effective for reducing WOF development. *Sous vide* technology, a further development for the conventional cook–chill system (Hansen *et al.*, 1995), has been shown to control the development of WOF during storage in the intact packages.

The first sign of oxidative change in cured meat products displayed in chill cabinets is colour fading which is caused by the combined action of light and oxygen (Andersen *et al.*, 1988a). Colour fading can be prevented by excluding either light or oxygen, and different packaging strategies for practical use have been described. High initial vacuum packaging in a material with a low oxygen transmission rate combined with cold storage in the dark for 4 days in order to reduce residual oxygen in the product below a critical limit has been shown to be effective (Andersen *et al.*, 1988a). Flushing with carbon dioxide, in effect diluting the residual oxygen below a critical value for photooxidation, has also been found to eliminate oxidative changes in displayed sliced ham, as have combined vacuum packaging and the use of oxygen scavengers (Andersen *et al.*, 1990a; Andersen and Rasmussen, 1992).

10.4 Conclusions and perspectives

The concept of critical control points for the control of lipid oxidation in meat and meat products is introduced. It is discussed how precautions at each critical control point can be used in order to improve the overall quality of the final product. A closer look at the different steps involved in the production chain indicate that it is important, when possible, to consider the entire chain as a whole.

It is also important to include microbiological aspects and interactions between oxidation and proteolytic ripening processes. Proteolytic, lipolytic and oxidative reactions all affect the overall quality of meat, and thus complex interactions between different types of reactions should be considered. Haem pigments, i.e. haemoglobin and myoglobin, play key roles in this interaction, since they promote lipid oxidation, but also inhibit muscle proteinases (Rosell *et al.*, 1996) and lipases (Rosell and Toldrá, 1996). The mechanism of this inhibition of proteinases and lipases is, however, largely unknown. Variation in myoglobin content between different muscles has led to the classification of 'red muscles' with high pigment content and 'white muscles' with lower pigment content. These differences in pigment content may be responsible for existing differences in rates of meat tenderization and lipid oxidation during storage. Thus, myoglobin inhibits the cathepsins, the calcium-dependent neutral proteinase system and the amino peptidases from pig skeletal muscles (Rosell *et al.*, 1996). This may further explain why 'red muscles' with a high concentration of proteolytic enzymes exhibit a low rate of tenderization compared to 'white muscles' with a low concentration of proteolytic enzymes.

References

Amarowicz, R. and Shahidi, F. (1996). A rapid chromatographic method for separation of individual catachins from green tea. *Food Res. Int.*, **29**, 71–76.

Amarowicz, R. and Shahidi, F. (1997). Antioxidant activity of peptide fractions of capelin protein hydrolysates. *Food Chem.*, **58**, 355–359.

Andersen, H. and Rasmussen, M.A. (1992). Interactive packaging as protection against photodegradation of the colour of pasteurized, sliced ham. *Int. J. Food Sci. Technol.*, **27**, 1–8.

Andersen, H.J. and Skibsted, L.H. (1991). Oxidative stability of frozen pork patties. Effect of light and added salt. *J. Food Sci.*, **56**, 1182–1184.

Andersen, H.J. and Skibsted, L.H. (1992). Kinetics and mechanism of thermal oxidation and photooxidation of nitrosylmyoglobin in aqueous solution. *J. Agric. Food Chem.*, **40**, 1741–1750.

Andersen, H.J., Bertelsen, G., Boegh-Soerensen, L., Shek, C.K. and Skibsted, L.H. (1988a). Effect of light and packaging conditions on the colour stability of sliced ham. *Meat Sci.*, **22**, 283–292.

Andersen, H.J., Bertelsen, G. and Skibsted, L.H. (1988b). Salt effect on acid-catalyzed autoxidation of oxymyoglobin. *Acta Chem. Scand.*, **A42**, 226–236.

Andersen, H.J., Bertelsen, G., Ohlen, A. and Skibsted, L.H. (1990a). Modified packaging as protection against photodegradation of the colour of pasteurized ham. *Meat Sci.*, **28**, 77–83.

Andersen, H.J., Bertelsen, G. and Skibsted, L.H. (1990b). Colour and colour stability of hot processed frozen minced beef. Results from chemical model experiments tested under storage conditions. *Meat Sci.*, **28**, 87–97.

Andersen, H.J., Johansen, H.S., Shek, C.K. and Skibsted, L.H. (1990c). Nitric oxide exchange in nitrosylmyoglobin. *Z. Lebensm. Unters. Forsch.*, **191**, 293–298.

Anon. (1997). Retail benefits from antioxidants. *Meat Int.*, **7**, 10–12.

Arendt, B., Skibsted, L.H. and Andersen, H.J. (1997). Antioxidative activity of nitrite in metmyoglobin induced lipid peroxidation. *Z. Lebensm. Unters. Forsch.*, **204**, 7–12.

Arnold, R.N., Arp, S.C., Schleller, K.K., Williams, S.N. and Schaefer, D.M. (1993). The equilibration and subcellular distribution of vitamin E relative to myoglobin and lipid oxidation in displayed beef. *J. Anim. Sci.*, **71**, 105–118.

Aubailly, M. Salmon, S., Morlière, D., Haigle, J. and Santus, R. (1996). Iron mobilization from ultraviolet-irradiated iron-saturated transferrin. *Redox Report*, **2**, 41–45.

Bailey, M.E. (1988). Inhibition of warmed-over flavour with emphasis on Maillard reaction products. *Food Technol.* **42**, 123–126.

Baron, C.P., Skibsted, L.H. and Andersen, H.J. (1997). Prooxidative activity of myoglobin species in linoleic acid emulsions. *J. Agric. Food Chem.*, **45**, 1704–1710.

Bartkowski, L., Dryden, F.D. and Marchello, J.A. (1982). Quality changes of beef steaks stored in controlled gas atmospheres containing high or low levels of oxygen. *J. Food Prot.*, **45**, 41–45.

Bertelsen, G. and Juncher, D. (1996). Oxidative stability and sensory quality of sous vide cooked products. In *Proceedings of the Second European Symposium on Sous Vide*, Leuven, Belgium, pp. 133–145.

Bertelsen, G. and Skibsted, L.H. (1987). Photooxidation of oxymyoglobin. Wavelength dependence of quantum yields in relation to light discoloration of meat. *Meat Sci.*, **19**, 243–251.

Bertelsen, G., Christophersen, C., Nielsen, P.H., Madsen, H.L. and Stadel, P. (1995). Chromatographic isolation of antioxidants guided by methyl linoleate assay. *J. Agric. Food Chem.*, **43**, 1272–1275.

Bishov, S.J. and Henick, A.S. (1972). Antioxidant effect of protein hydrolyzates in a freeze-dried model system. *J. Food Sci.*, **37**, 873–875.

Bishov, S.J. and Henick, A.S. (1975). Antioxidant effect of protein hydrolyzates in freeze-dried model systems. Synergistic action with a series of phenolic antioxidants. *J. Food Sci.*, **40**, 345–348.

Brekke, C.J., Al-Hakim, S.H., Highlands, M.E. and Hogan, J.M. (1975). The relation of *in vivo* and *in vitro* tocopherol supplementation to stability of fowl fat. *Poult. Sci.*, **54**, 2019–2024.

Bekkölet, M. (1990). Light effects on food. *J. Food Prot.*, **53**, 430–440.

Brieskorn, C.H. and Dömling, H.-J.Z. (1969). Carnosolsäure, der wichtige antioxydativ wirksame inhaltsstoff des rosmarin- und salbeiblattes. *Z. Lebensm. Unters. Forsch.*, **141**, 10–16.

Brieskorn, C.H., Fuchs, A., Bredenberg, J.B., McChesney, J.D. and Wenkert, E.J. (1964). The structure of carnosol. *J. Org. Chem.*, **29**, 2293–2298.

Bruun-Jensen, L. and Skibsted, L.H. (1996). High-pressure effects on oxidation of nitrosylmyoglobin. *Meat Sci.*, **44**, 145–149.

Bruun-Jensen, L., Skovgaard, I.M., Skibsted, L.H. and Bertelsen, G. (1994). Antioxidant synergism between tocopherols and ascorbyl palmitate in cooked, minced turkey. *Z. Lebensm. Unters. Forsch.*, **199**, 210–213.

Bruun-Jensen, L., Colarow, L. and Skibsted, L.H. (1995). Detection and quantification of phospholipid hydroperoxides in turkey meat extracts by planar chromatography. *J. Planar Chromatogr.* **8**, 475–479.

Bruun-Jensen, L., Skovgaard I.M., Madsen, E.A., Skibsted, L.H. and Bertelsen, G. (1996a). The combined effect of tocopherols, L-ascorbyl palmitate and L-ascorbic acid on the development of warmed-over flavour in cooked, minced turkey. *Food Chem.*, **55**, 41–47.

Bruun-Jensen, L., Skovgaard, I.M. and Skibsted, L.H. (1996b). The antioxidative activity of RRR-α-tocopherol vs RRR-δ-tocopherol in combination with ascorbyl palmitate in cooked, minced turkey. *Food Chem.*, **56**, 347–354.

Bruun-Jensen, L., Sosniecki, L. and Skibsted, L.H. (1997). Pressure effects on acid-catalysed autoxidation of oxymyoglobin. *Z. Lebensm. Unters. Forsch.*, **205**, 407–410.

Buckley, D.J. and Connolly, J.F. (1980). Influence of alpha-tocopherol (vitamin E) on the storage stability of pork and bacon. *J. Food Protect.*, **43**, 265–267.

Buckley, D.J., Morrissey, P.A. and Gray, J.I. (1995). Influence of dietary vitamin E on the oxidative stability and quality of pig meat. *J. Anim. Sci.*, **73**, 3122–3130.

Carlez, A., Veciana-Nogues, T. and Cheftel, J.-C. (1995). Change in colour and myoglobin of minced beef meat due to high pressure processing. *Lebensm. Wiss. u. Technol.*, **28**, 528–538.

Cassens, R.G. (1990). *Nitrite-cured Meat: A Food Safety Issue in Perspective.* Food and Nutrition Press, Trumbull, CT.

Cassens, R.G., Greaser, T., Itoh, T. and Lee, M. (1979). Reactions of nitrite in meat. *Food Technol.*, **7**, 46–57.

Chan, K.M. and Decker, E.A. (1994). Endogenous skeletal muscle antioxidants. *CRC Crit. Rev. Food Sci. Nutr.*, **34**, 403–426.

Chan, K.M., Decker, E.A. and Means, W.J. (1993). Extraction and activity of carnosine, a naturally occurring antioxidant in beef muscle. *J. Food Sci.*, **58**, 1–4.
Chan, W.K.M., Decker, E.A., Lee, J.B. and Butterfield, D.A. (1994). EPR spin-trapping studies of the hydroxyl radical scavenging activity of carnosine and related dipeptides. *J. Agric. Food Chem.*, **42**, 1407–1410.
Chan, W.K.M., Hakkarainen, K., Faustman, C., Schaefer, D.M., Scheller, K.K. and Liu, Q. (1996). Dietary vitamin E effect on colour stability and sensory assessment of spoilage in three beef muscles. *Meat Sci.*, **42**, 387–399.
Chan, W.K.M., Faustman, C., Yin, M. and Decker, E.A. (1997). Lipid oxidation induced by oxymyoglobin and metmyoglobin with involvement of H_2O_2 and superoxide anion. *Meat Sci.*, **46**, 181–190.
Cheah, P.B. and Ledward, D.A. (1997). Inhibition of metmyoglobin formation in fresh beef by pressure treatment. *Meat Sci.*, **45**, 411–418.
Chen, C.C., Pearson, A.M., Gray, J.I., Fooladi, M.H. and Ku, P.K. (1984). Some factors influencing the nonheme iron content of meat and its implications in oxidation. *J. Food Sci.*, **49**, 581–584.
Chen, H.M., Muramoto, K. and Yamauchi, F. (1995). Structural analysis of antioxidative peptides from soybean β-conglycinin. *J. Agric. Food Chem.*, **43**, 574–578.
Chen, H.M., Muramoto, K., Yamauchi, F. and Nokihara, K. (1996). Antioxidant activity of designed peptides based on the antioxidative peptide isolated from digest of a soybean protein. *J. Agric. Food Chem.*, **44**, 2619–2623.
Chipault, J.R., Mizuno, G.R., Hawkins, J.M. and Lundberg, W.O. (1952). The antioxidant properties of natural spices. *Food Res.*, **17**, 46–55.
Chipault, J.R., Mizuno, G.R. and Lundberg, W.O. (1955). Antioxidant properties of spices in oil-in-water emulsions. *Food Res.*, **20**, 443–448.
Clark, D.S. and Lentz, C.P. (1973). Use of mixtures of carbon dioxide and oxygen for extending shelf-life of prepackaged fresh beef. *Can. Inst. Food Sci. Technol. J.*, **6**, 194–196.
Dahl, T.A., Midden, W.R. and Hartman, P.E. (1988). Some prevalent biomolecules as defenses against singlet oxygen damage. *Photochem. Photobiol.*, **47**, 357–362.
Decker, E.A. and Crum, A.D. (1991). Inhibition of oxidative rancidity in salted ground pork by carnosine. *J. Food Sci.*, **56**, 1179–1181.
Decker, E.A. and Crum, A.D. (1993). Antioxidant activity of carnosine in cooked ground pork. *Meat Sci.*, **34**, 245–253.
Decker, E.A. and Faraji, H. (1990). Inhibition of lipid oxidation by carnosine. *J. Am. Oil Chem. Soc.*, **67**, 650–652.
Decker, E.A., Crum, A.D. and Calvert, T. (1992). Differences in the antioxidant mechanism of carnosine in the presence of copper and iron. *J. Agric. Food Chem.*, **40**, 756–759.
Deighton, N., Glidewell, S.M., Deans, S.G. and Goodman, B.A. (1993). Identification by EPR spectroscopy of carvacrol and thymol as the major sources of free radicals in the oxidation of plant essential oils. *J. Sci. Food Agric.*, **63**, 221–225.
Dissing, J. Bruun-Jensen, L. and Skibsted, L.H. (1997). Effect of high pressure treatment on lipid oxidation in turkey thigh muscle during chill storage. *Z. Lebensm. Unters. Forsch.*, **205**, 11–13.
Erduran, S. and Hotchkiss, J.H. (1995). Nitrite effect on formation of volatile oxidation products from triolein. *J. Food Sci.*, **60**, 946–948.
Faustman, C. and Cassens, R.G. (1989). Strategies for improving fresh meat colour. In *Proceedings of the 35th International Congress of Meat Science and Technology.* Danish Meat Research Institute, Copenhagen, pp. 446–453.
Faustman, C., Cassens, R.G., Schaefer, D.M., Buege, D.R., Williams, S.N. and Scheller, K.K. (1989). Improvement of pigment and lipid stability in Holstein steer beef by dietary supplementation with vitamin E. *J. Food Sci.*, **54**, 858–862.
Feldman, P.L., Griffith, O.W. and Stuehr, D.J. (1993). The surprising life of nitric oxide. *Chemical and Engineering News*, **December 20**, 26–38.
Fox, J.B., Nicholas, R.A., Ackerman, S.A. and Swift, C.E. (1974). A multiple wavelength analysis of the reaction between hydrogen peroxide and metmyoglobin. *Biochemistry*, **13**, 5178–5186.
Frankel, E.N. (1995). Natural and biological antioxidants in foods and biological systems. Their mechanism of action, applications and implications. *Lipid Technol.*, **7**, 77–80.
Frankel, E.N., Huang, S.-W., Aeschbach, R. and Prior, E. (1996). Antioxidant activity of a

rosemary extract and its constituents, carnosic acid, carnosol, and rosmarinic acid, in bulk oil and oil-in-water emulsion. *J. Agric. Food Chem.*, **44**, 131–135.

Freybler, L.A., Gray, J.I., Asghar, A., Booren, A.M., Pearson, A.M. and Buckley, D.J. (1989). Mechanism of nitrite stabilization of meat lipids and heme pigments. In *Proceedings of the 35th International Congress of Meat Science and Technology*. Danish Meat Research Institute, Copenhagen, pp. 903–908.

Frigg, M., Prabucki, A.L. and Ruhdel, E.U. (1990). Effect of dietary vitamin E levels on oxidative stability of trout fillets. *Aquaculture*, **84**, 145–158.

Galvin, K., Morrissey, P.A., Buckley, D.J. and Frigg, M. (1993). Influence of oil quality and α-tocopheryl acetate supplementation of α-tocopherol and lipid oxidation in chicken tissues. In *Proceedings of the 11th European Symposium on Quality of Poultry Meat*, France, eds. P. Colin, J. Culidi and F.H. Ricard. World Poultry Science Association, French Branch, Tours, France, pp. 423–429.

Gata, J.L., Pinto, M.C. and Macías, P. (1996). Lipoxygenase activity in pig muscle: Purification and partial characterization. *J. Agric. Food Chem.*, **44**, 2573–2577.

German, J.B. and Creveling, R.K. (1990). Identification and characterization of a 15-lipoxygenase from fish gills. *J. Agric. Food Chem.*, **38**, 2144–2147.

German, J.B. and Kinsella, J.E. (1985). Lipid oxidation in fish tissue. Enzymatic initiation via lipoxygenase. *J. Agric. Food Chem.*, **33**, 680–683.

German, J.B., Bruckner, G.G. and Kinsella, J.E. (1986). Lipoxygenase in trout gill tissue acting on arachidonic, eicosapentaenoic and docosahexaenoic acids. *Biochim. Biophys. Acta*, **875**, 12–20.

Goutefongea, R., Cassens, R.G. and Woolford, G. (1977). Distribution of sodium nitrite in adipose tissue during curing. *J. Food Sci.*, **42**, 1637–1641.

Goutefongea, R. Rampon, V. Nicolas, N. and Dumont, J.P. (1995). Meat colour changes under high pressure treatment. In *Proceedings of the 41st International Congress of Meat Science and Technology*. American Meat Science Association, pp. 384–385.

Graf, E. and Panter, S. (1991). Inhibition of warmed-over flavour development by polyvalent cations. *J. Food Sci.*, **56**, 1055–1058, 1067.

Gray, J.I. and Crackel, R.L. (1992). Oxidative flavour changes in meats: Their origin and prevention. In *The Chemistry of Muscle-based Foods*, eds. D.E. Johnston, M.K. Knight and D.A. Ledward. Royal Society of Chemistry, London. pp. 145–168.

Gros, J.N., Howat, P.M., Younathan, M.T., Saxton, A.M. and McMillin, K.W. (1986). Warmed-over flavour development in beef patties prepared by three dry heat methods. *J. Food Sci.*, **51**, 1152–1155.

Grossman, S., Bergman, M. and Sklan, D. (1988). Lipoxygense in chicken muscle. *J. Agric. Food Chem.*, **36**, 1268–1270.

Hansen, T.B., Knøchel, S., Juncher, D. and Bertelsen, G. (1995). Storage characteristics of sous vide cooked roast beef. *Int. J. Food Sci. Technol.*, **30**, 365–378.

Hansen, T.B., Skibsted, L.H. and Andersen, H.J. (1996). The influence of the anticaking agent potassium ferrocyanide and salt on the oxidative stability of frozen minced pork meat. *Meat Sci.*, **43**, 135–144.

Harel, S. and Kanner, J. (1985a). Hydrogen peroxide generation in ground muscle tissues. *J. Agric. Food Chem.*, **33**, 1188–1192.

Harel, S. and Kanner, J. (1985b). Muscle membranal lipid peroxidation initiated by H_2O_2-activated metmyoglobin. *J. Agric. Food Chem.*, **33**, 1193–1198.

Hsieh, J.-H. and Baldwin, R.E. (1984). Storage and microwave reheating effects on lipid oxidation of roast beef. *J. Microwave Power*, **19**, 187–194.

Hsieh, R.J. and Kinsella, J.E. (1989). Lipoxygenase generation of specific volatile flavour carbonyl compounds in fish tissues. *J. Agric. Food Chem.*, **37**, 279–286.

Hsieh, R.J., German, J.B. and Kinsella, J.E. (1988). Lipoxygenase in fish tissues: Some properties of the 12-lipoxygenase from trout gill. *J. Agric. Food Chem.*, **36**, 680–685.

Huisman, M., Madsen, H.L., Skibsted, L.H. and Bertelsen, G. (1994). The combined effect of rosemary (*Rosmarinus officinalis* L.) and modified atmosphere packaging as protection against warmed over flavour in cooked minced pork meat. *Z. Lebensm. Unters. Forsch.*, **198**, 57–59.

Igene, J.O., King, J.A., Pearson, A.M. and Gray J.I. (1979). Influence of heme pigments, nitrite, and non-heme iron on development of warmed-over flavor (WOF) in cooked meat. *J. Agric. Food Chem.*, **27**, 838–842.

Jensen, C., Skibsted, L.H., Jakobsen, K. and Bertelsen, G. (1995). Supplementation of broiler diets with *all-rac*-α- or a mixture of natural source *RRR*-α,γ-,δ-tocopheryl acetate. 2. Effect of oxidative stability of raw and precooked broiler meat products. *Poult. Sci.*, **74**, 2048–2056.

Jensen, C., Guidera, J., Skovgaard, I.M., Staun, H., Skibsted, L.H., Jensen, S.K., Møller, A.J., Buckley, J. and Bertelsen, G. (1997a). Effects of dietary α-tocopheryl acetate supplementation on α-tocopherol deposition in porcine *m. psoas major* and *m. longissimus dorsi* and on drip loss, colour stability and oxidative stability of pork meat. *Meat Sci.*, **45**, 491–500.

Jensen, C., Skibsted, L.H. and Bertelsen, G. (1997b). Influence of dietary fat source and vitamin E on quality of frozen pork meat products. In *Proceedings of 43rd International Congress of Meat Science and Technology*, Auckland, New Zealand, pp. 402–403.

Jensen, C., Engberg, R., Jakobsen, K., Skibsted, L.H. and Bertelsen, G. (1997c). Influence on the oxidative quality of dietary oil on broiler meat storage stability. *Meat Sci.*, **27**, 211–222.

Johns, A.M., Birkinshaw, L.H. and Ledward, D.A. (1989). Catalysis of lipid oxidation in meat products. *Meat Sci.*, **25**, 209–220.

Jørgensen, L.V., Andersen, H.J. and Skibsted, L.H. (1997). Kinetics of reduction of hypervalent iron in myoglobin by crocin in aqueous solution. *Free Rad. Res.*, **27**, 73–87.

Kanner, J. (1979). S-nitrosocysteine (RSNO), an effective antioxidant in cured meat. *J. Am. Oil Chem. Soc.*, **56**, 74–76.

Kanner, J. and Kinsella, J.F. (1983). Lipid deterioration initiated by phagocytic cells in muscle foods: β-carotene destruction by a myeloperoxidase–hydrogen peroxide–halide system. *J. Agric. Food Chem.*, **31**, 370–376.

Kanner, J., Ben-Gera, I. and Berman, S. (1980). Nitric-oxide myoglobin as an inhibitor of lipid oxidation. *Lipids*, **15**, 944–948.

Kanner, J., German, J.B. and Kinsella, J.E. (1987). Initiation of lipid peroxidation in biological systems. *CRC Crit. Rev. Food Sci. Nutr.*, **25**, 317–364.

Kanner, J., Harel, S. and Jaffe, R. (1991a). Lipid peroxidation of muscle foods as affected by NaCl. *J. Agric. Food Chem.*, **39**, 1017–1021.

Kanner, J. Harel, S. and Granit, R. (1991b). Nitric oxide as an antioxidant. *Arch. Biochem. Biophys.*, **289**, 130–136.

Kanner, J., Salan, M.A., Harel, S. and Shegalovich, I. (1991c). Lipid peroxidation of muscles food. The role of the cytosolic fraction. *J. Agric. Food Chem.*, **30**, 242–246.

Karel, M., Tannembaum, S.R., Wallace, D.H. and Maloney, H. (1966). Autoxidation of methyl linoleate in freeze-dried model systems. III. Effects of added amino acids. *J. Food Sci.*, **31**, 892–896.

Kaschnitz, R.M. and Hatefi, Y. (1975). Lipid oxidation in biological membranes. Electron transfer proteins as initiators of lipid autoxidation. *Arch. Biochem. Biophys.*, **171**, 292–304.

Kawashima, K., Itoh, H., Miyoshi, M. and Chibata, I. (1979). Antioxidant properties of branched-chain amino acid derivatives. *Chem. Pharm. Bull.*, **27**, 1912–1916.

Kelman, D.J., DeGray, J.A. and Mason, R.P. (1994). Reaction of myoglobin with hydrogen peroxide forms a peroxyl radical which oxidizes substrates. *J. Biol. Chem.*, **269**, 7458–7463.

Kikuzaki, H. and Nakatani, N. (1989). Structure of a new antioxidative phenolic acid from oregano (*Origanum vulgare* L.). *Agric. Biol. Chem.*, **53**, 519–524.

Klettner, P.G. and Baumgartner, P.A. (1980). The technology of raw dry sausage manufacture. *Food Technol. Aust.*, **32**, 380–384.

Kohen, R., Yamamoto, Y., Cundy, K.C. and Ames, B.N. (1988). Antioxidant activity of carnosine, homocarnosine, and anserine present in muscle and brain. *Proc. Natl Acad. Sci.*, **85**, 3175–3179.

Kono, Y. and Fridovich, I. (1982). Superoxide radical inhibits catalase. *J. Biol. Chem.*, **257**, 5751–5754.

Konopka, U.C. and Grosch, W. (1991). Potent odorants causing the warmed-over flavour in boiled beef. *Z. Lebensm. Unters. Forsch.*, **193**, 123–125.

Kramer, R.E. (1985). Antioxidants in clove. *J. Am. Oil Chem. Soc.*, **62**, 111–113.

Kristensen, L. and Andersen, H.J. (1997). Effect of heat denaturation on the pro-oxidative activity of metmyoglobin in linoleic acid emulsions. *J. Agric. Food Chem.*, **45**, 7–13.

Kröger-Ohlsen, M., and Skibsted, L.H. (1997). Kinetics and mechanism of reduction of ferrylmyoglobin by ascorbate and D-isoascorbate. *J. Agric. Food Chem.*, **45**, 668–676.

Lambelet, P., Saucy F. and Löliger J. (1985). Chemical evidence for interactions between vitamins E and C. *Experientia*, **41**, 1284–1388.

Ledward, D.A. (1970). Metmyoglobin formation in beef stored in carbon dioxide enriched and oxygen depleted atmosphere. *J. Food Sci.*, **35**, 33–37.
Ledward, D.A. (1991). Meat colour stability. *J. Food Sci.*, **56**, viii.
Lee, S.K., Mei, L. and Decker, E.A. (1997). Influence of sodium chloride on antioxidant enzyme activity and lipid oxidation in frozen ground pork. *Meat Sci.*, **46**, 349–356.
Lin, C.F., Asghar, A., Gray, J.I., Buckley, D.J., Booren, A.M., Crackel, R.L. and Flegal, C.H. (1989a). Effects of oxidized dietary oils and antioxidant supplementation on broiler growth and meat stability. *Br. Poult. Sci.*, **30**, 855–864.
Lin, C.F., Gray, J.I., Asghar, A., Buckley, D.J., Booren, A.M. and Flegal, C.J. (1989b). Effects of dietary oils and α-tocopherol supplementation on lipid composition and stability of broiler meat. *J. Food Sci.*, **54**, 1457–1460, 1484.
Liu, Q., Lanari, M.C. and Schaefer, D.M. (1995). A review of dietary vitamin E supplementation for improvement of beef quality. *J. Anim. Sci.*, **73**, 3131–3140.
Lopez-Lorenzo, P., Hernandez, P., Sanz-Perez, B. and Ordonez, J.A. (1980). Effect of O_2- and CO_2-enriched atmospheres on shelf-life extension of refrigerated ground pork. *Meat Sci.*, **4**, 89–94.
Love, J.D. and Pearson, A.M. (1974). Metmyoglobin and non-heme iron as prooxidants in cooked meats. *J. Agric. Food Chem.*, **22**, 1032–1034.
Machlin, L.H. and Gabriel, E. (1982). Kinetics of tissue α-tocopherol uptake and depletion following administration of high levels of vitamin E. *Ann. N.Y. Acad. Sci.*, **393**, 48–60.
Madsen, H.L., Bertelsen, G. and Skibsted, L.H. (1997). Antioxidative activity of spices and spice extracts. In *Spices. Flavor Chemistry and Antioxidant Properties*, eds. S.J. Risch and C.-T. Ho., American Chemical Society, ACS Symposium Series 660, Washington, DC, pp. 176–187.
Marcuse, R. (1960). Antioxidative effect of amino acids. *Nature*, **186**, 886–887.
Marcuse, R. (1962). The effect of some amino acids on the oxidation of linoleic acid and its methyl ester. *J. Am. Oil Chem. Soc.*, **39**, 97–103.
Marusich, W.L., De Ritter, E., Ogrinz, E.F., Keating, J., Mitrovic, M. and Bunnell, R.H. (1975). Effect of supplemental vitamin E in control of rancidity in poultry meat. *Poult. Sci.*, **54**, 831–844.
Mazza, G. and Miniati, E. (1993). *Anthocyanins in Fruits, Vegetables, and Grains*. CRC Press, Boca Raton, FL, pp. 85–130.
MacDonald, B., Gray, J.I. and Gibbins, L.N. (1980). Role of nitrite in cured meat flavor: Antioxidant role of nitrite. *J. Food Sci.*, **45**, 893–897.
McDaniel, M.C., Marchello, J.A. and Tinsley, A.M. (1984). Effect of different packaging treatments on microbiological and sensory evaluation of precooked beef roasts. *J. Food Prot.*, **47**, 23–26.
McDonald, R.E. and Hultin, H.O. (1987). Some characteristics of the enzymic lipid peroxidation system in the microsomal fraction of flounder skeletal muscle. *J. Food Sci.*, **52**, 15–21, 27.
Mertnes, B. and Knorr, D. (1992). Developments of nonthermal processes for food preservation. *Food Technol.*, **46**, 124, 126–133.
Mielche, M. and Bertelsen, G. (1993). Effects of heat treatment on warmed-over flavour in ground beef during aerobic chill storage. *Z. Lebensm. Unters. Forsch.*, **197**, 8–13.
Mielche, M. and Bertelsen, G. (1995). Evaluation of a mathematical model for prediction of lipid oxidation in heat-treated beef during chill storage. *Z. Lebensm. Unters. Forsch.*, **200**, 415–419.
Mikkelsen, A. and Skibsted, L.H. (1992). Kinetics of enzymatic reduction of metmyoglobin in relation to oxygen activation in meat products. *Z. Lebensm. Unters. Forsch.*, **194**, 9–16.
Mikkelsen, A. and Skibsted, L.H. (1995). Acid-catalysed reduction of ferrylmyoglobin: product distribution and kinetics of autoreduction and reduction by NADH. *Z. Lebensm. Unters. Forsch.*, **200**, 171–177.
Mikkelsen, A., Jun, H. and Skibsted, L.H. (1992a). Effects of polyphosphates on reactions of metmyoglobin related to oxidative changes in meat products. *Z. Lebensm. Unters. Forsch.*, **194**, 317–321.
Mikkelsen, A., Sosniecki, L. and Skibsted, L.H. (1992b). Myoglobin catalysis in lipid oxidation. *Z. Lebensm. Unters. Forsch.*, **195**, 228–234.
Molina, I., Nieto, P., Flores, J., Silla, H. and Bermell, S. (1991). Study on the microbial flora of dry-cured ham. V. Lipolitic activity. *Fleischwirtsch.*, **71**, 906–908.

Monahan, F.J. (1995). Effects of feed on the oxidative stability of raw and processed meat. In *Meat Quality and Safety as Affected by Primary Production*, eds. A.J. Møller, M.M. Mielche and P. Barton-Gade. European Consortium for Continuing Education in Advanced Meat Science and Technology (ECCEAMST), Copenhagen, June 1995.

Monahan, F.J., Buckley, D.J., Morrissey, P.A., Lynch, P.B. and Gray, J.I. (1992). Influence of dietary fat and α-tocopherol supplementation on lipid oxidation in pork. *Meat Sci.*, **31**, 229–241.

Morgan, C.A., Noble, R.C., Cocehi, M. and McCartney, R. (1992). Manipulation of the fatty acid composition of pig meat lipids by dietary means. *J. Sci. Food Agric.*, **58**, 357–368.

Mottram, D. (1987). Lipid oxidation and flavour in meat and meat products. *Food Sci. Technol. Today*, **1**, 159–162.

Murase, H., Nagao, A. and Tera, J. (1993). Antioxidant and emulsifying activity of N-(long-chain-acyl)histidine and N-(long-chain-acyl)carnosine. *J. Agric. Food Chem.*, **41**, 1601–1604.

Nakatani, N., Inatani, R., Ohta, H. and Nishioka, A. (1986). Chemical constituents of peppers (*Piper* spp.) and application to food preservation: Naturally occurring antioxidative compounds. *Environ. Health Perspect.*, **67**, 135–142.

Nakayama, T., Chiba, Y. and Hashimoto, K. (1997). Lipid peroxidation in linoleic acid micelles caused by H_2O_2 in the presence of myoglobin. *Biosci. Biotechnol. Biochem.*, **61**, 817–820.

Nielsen, J.H., Sørensen, B., Skibsted, L.H. and Bertelsen, G. (1997). Oxidation in pre-cooked minced pork as influenced by chill storage of raw muscle. *Meat Sci.*, **46**, 191–197.

Nolan, N.L., Bowers, J.A. and Kropf, D.H. (1989). Lipid oxidation and sensory analysis of cooked pork and turkey stored under modified atmosphere. *J. Food Sci.*, **54**, 846–849.

Ordonez, J.A. and Ledward, D.A. (1977). Lipid and myoglobin oxidation in pork stored in oxygen- and carbon dioxide-enriched atmospheres. *Meat Sci.*, **1**, 41–48.

O'Sullivan, M.G., Kerry, J.P., Buckley, D.J., Lynch, P.B. and Morrissey, P.A. (1997). The distribution of dietary vitamin E in the muscles of porcine carcass. *Meat Sci.*, **45**, 297–305.

Ozaki, S., Matus, T. and Watanabe, Y. (1996). Conversion of myoglobin into a highly stereo-specific peroxygenase by the L29H/H64L mutation. *J. Am. Chem. Soc.*, **118**, 9784–9785.

Østdal, H., Daneshvar, B. and Skibsted, L.H. (1996). Reduction of ferrylmyoglobin by β-lactoglobulin. *Free Rad. Res.*, **24**, 429–438.

Østdal, H., Skibsted, L.H. and Andersen, H.J. (1997). Formation of long-lived protein radicals in the reaction between H_2O_2-activated metmyoglobin and other proteins. *Free Rad. Biol. Med.*, **23**, 754–761.

Pegg, R.B., Shahidi, F., Gogan, N.J. and DeSilva, S.I. (1996). Elucidation of the chemical structure of preformed cooked cured-meat pigment by electron paramagnetic resonance spectroscopy. J. *Agric. Food Chem.*, **44**, 416–421.

Perigo, J.A., Whiting, E. and Bashford, T.E. (1967). Observations on the inhibition of vegetative cells of *Clostridium sporogenes* by nitrite which has been autoclaved in a laboratory medium, discussed in the context of sub-lethally processed cured meats. *J. Food Technol.*, **2**, 377–397.

Pierson, M.D., Collins-Thompsen, D.L. and Ordal, Z.J. (1970). Microbiological, sensory and pigment changes of aerobically and anaerobically packed beef. *Food Technol.*, **24**, 1171–1175.

Pollack, S., Vanderhoff, G. and Lasky, F. (1977). Iron removal from transferrin. An experimental study. *Biochem. Biophys. Acta*, **497**, 481–487.

Rao, S.I., Wilks, A., Hamberg, M. and Ortiz de Montellano, P.R. (1994). The lipoxygenase activity of myoglobin. *J. Biol. Chem.*, **269**, 7210–7216.

Renerre, M. and Labadie, J. (1993). Fresh red meat packaging and meat quality. In *Proceedings of the 39th International Congress of Meat Science and Technology*, Calgary, Canada, p. 361.

Reschke, A.Z. (1983). Capillargaschromatographische bestimmung der rosmarinsäure in blattgewürzen. *Z. Lebensm. Unters. Forsch.*, **176**, 116–119.

Revankar, G.D. (1974). Proline as an antioxidant in fish oil. *J. Food Sci. Technol.*, **11**, 10–11.

Rhee, K.S. and Ziprin, Y.A. (1987). Lipid oxidation in retail beef, pork and chicken muscles as affected by concentrations of heme pigments and nonheme iron and microsomal enzymic lipid peroxidation activity. *J. Food Biochem.*, **11**, 1–15.

Rhee, K.S., Ziprin, Y.A., Ordonez, G. and Bohac, C.E. (1988). Fatty acid profiles of the total lipids and lipid oxidation in pork muscles as affected by canola oil in the animal diet and muscle location. *Meat Sci.*, **23**, 201–210.

Rhee, K.S., Anderson, L.M. and Sams, A.R. (1996). Lipid oxidation potential of beef, chicken, and pork. *J. Food Sci.*, **61**, 8–12.
Riisom, T., Sims, R.J. and Fioriti, J.A. (1980). Effect of amino acids on the autoxidation of safflower oil in emulsions. *J. Am. Oil Chem. Soc.*, **57**, 354–359.
Rooney, M.L. (1995). Overview of active food packaging. In: *Active Food Packaging*, ed. M.L. Rooney. Blackie Academic & Professional, Glasgow, pp. 1–37.
Rosell, C.M. and Toldrá, F. (1996). Effect of myoglobin on the muscle lipase system. *J. Food Biochem.*, **20**, 87–92.
Rosell, C.M., Flores, M. and Toldrá, F. (1996). Myoglobin as an endogenous inhibitor of proteolytic muscle enzymes. *J. Agric. Food Chem.*, **44**, 3453–3456.
Rubbo, H. and Freeman, B.A. (1996). Nitric oxide regulation of lipid oxidation reactions: Formation and analysis of nitrogen-containing oxidized lipid derivatives. In *Methods in Enzymology. Vol. 269. Nitric Oxide. Part B. Physiological Processes*, ed. L. Packer. Academic Press, London, pp. 385–394.
Sato, K. and Hegarty, G.R. (1971). Warmed-over flavor in cooked meats. *J. Food Sci.*, **36**, 1098–1102.
Satoh, Y. and Shikama, K. (1981). Autoxidation of oxymyoglobin. *J. Biol. Chem.*, **256**, 10 272–10 275.
Satynarayan, V.T. and Honikel, K.-O. (1992). Effect of different cooking methods on warmed-over flavour development in pork. *Z. Lebensm. Unters. Forsch.*, **194**, 422–425.
Shahidi, F. and Pegg, R.B. (1992). Nitrite-free meat curing systems: Update and review. *Food Chem.*, **43**, 185–191.
Shahidi, F. and Amarowicz, R. (1996). Antioxidant activity of protein hydrolyzates from aquatic species. *J. Am. Oil Chem. Soc.*, **73**, 1197–1199.
Shahidi, F., Yun, J. and Rubin, L.J. (1987). The hexanal content as an indicator of oxidative stability and flavour acceptability in cooked ground pork. *Can. Inst. Food Sci. Technol. J.*, **20**, 104–106.
Sheehy, P.J.A., Morrissey, P.A. and Flynn, A. (1992). Influence of heated vegetable oils and α-tocopheryl acetate on susceptibility of chicken tissues to lipid peroxidation. *Proc. Nutr. Soc.*, **51**, 13A.
Shomer, I., Weinberg, Z.G. and Vasiliver, R. (1987). Structural binding properties of silver-carp (*Hypophthalmichthys molitrix*) muscle affected by NaCl and $CaCl_2$ treatments. *Food Microstr.*, **6**, 199–207.
Shorland, F.B., Igene, J.O., Pearson, A.M., Thomas, J.W., McGuffey, R.K. and Aldridge, A.E. (1981). Effects of dietary fat and vitamin E on the lipid composition and stability of veal during frozen storage. *J. Agric. Food Chem.*, **29**, 863–871.
Sims, J.R. and Fioriti, J.A. (1977). Methional as an antioxidant for vegetable oils. *J. Am. Oil Chem. Soc.*, **54**, 4–7.
Skibsted, L.H. (1992). Cured meat products and their oxidative stability. In *The Chemistry of Muscle-based Foods*, eds. D.E. Johnston, M.K. Knight and D.A. Ledward. The Royal Society of Chemistry, London, pp. 266–286.
Skibsted, L.H., Bertelsen, G. and Qvist, S. (1994). Quality changes during storage of meat and slightly preserved meat products. In *Proceedings of the 40th International Congress of Meat Science and Technology*, The Hague, Netherlands.
Smith, D.M., Salih, A.M. and Morgan, R.G. (1987). Heat treatment effects on warmed-over flavor in chicken breast meat. *J. Food Sci.*, **52**, 842–845.
St. Angelo, A.J., Vercellotti, J.R., Legendre, M.G., Vinnet, C.H., Kuan, J.W., James, C. and Dupuy, H.P. (1987). Chemical and instrumental analyses of warmed-over flavour in beef. *J. Food Sci.*, **52**, 1163–1168.
Stapelfeldt, H., Bjørn, H., Skovgaard, I.M., Skibsted, L.H. and Bertelsen, G. (1992). Warmed-over flavour in cooked sliced beef. Chemical analysis in relation to sensory evaluation. *Z. Lebensm. Unters. Forsch.*, **195**, 203–208.
Stapelfeldt, H., Bjørn, H., Skibsted, L.H. and Bertelsen, G. (1993). Effect of packaging and storage conditions on development of warmed-over flavour in sliced, cooked meat. *Z. Lebensm. Unters. Forsch.*, **196**, 131–136.
Stewart, J.M. (1990). Free fatty acid enhance the oxidation of oxymyoglobin and inhibit the peroxidase activity of metmyoglobin. *Biochem. Cell. Biol.*, **68**, 1096–1102.
Tappel, A.L. (1955). Unsaturated lipid oxidation catalysed by hematin compounds. *J. Biol. Chem.*, **217**, 721–733.

Taylor, A.A. (1982). Retail packaging systems for fresh meat. In *Proceedings of the International Symposium Meat Science and Technology*. Lincoln, Nebraska, pp. 353–366.

Taylor, A.A. (1985). Packaging fresh meat. In *Developments in Meat Science*, ed. R.A. Lawrie. Elsevier Applied Science, Oxford.

Taylor, A.A., and McDougall, D.B. (1973). Fresh beef packed in mixtures of oxygen and carbon dioxide. *J. Food Technol.*, **8**, 453–461.

Thomas, M.J., Mehl, K.A. and Pryor, W.A. (1978). The role of the superoxide anion in the xanthine oxidase-induced autoxidation of linoleic acid. *Biochem. Biophys. Res. Comm.*, **83**, 927–932.

Tims, M.J. and Watts, B.M. (1958). Protection of cooked meats with phosphates. *Food Technol.*, **12**, 240–243.

Tsuge, N., Eikawa, Y., Nomura, Y., Yamamoto, M. and Sugisawa, K. (1991). Autoxidative activity of peptides prepared by enzymatic hydrolysis of egg-white albumin. *Nippon Nogei Kagaku Kaishi*, **65**, 1635–1641.

Vekiari, S.A.; Oreopoulou, V., Tzia, C. and Thomopoulus, C.D. (1993). Oreogano flavonoids as lipid antioxidants. *J. Amer. Oil Chem. Soc.*, **70**, 484–487.

Vösgen, B. and Hermann, K.Z. (1980). Flavonoglykoside von pfeffer (Piper nigrum. L.), gewürznelken (*Syzygium aromaticum* (L) Merr. Et Perry) und piment (*Pimenta dioica* (L.) Merr.). *Z. Lebensm. Unters. Forsch.*, **170**, 204–207.

Webb, J.E., Brunson, C.C. and Yates J.D. (1972a). Effects of feeding antioxidants on rancidity development in pre-cooked, frozen broiler parts. *Poult. Sci.*, **51**, 1601–1605.

Webb, R.W., Marion, W.W. and Hayse, P.L. (1972b). A research note: Tocopherol supplementation and lipid stability in the turkey. *J. Food Sci.*, **37**, 496.

Wenkert, E., Fuchs, A. and McChesney, J.D. (1965). Chemical artifacts for the family Labiatae. *J. Org. Chem.*, **30**, 2931–2934.

Whang, K. and Peng, I.C. (1988). Electron paramagnetic resonance studies of the effectiveness of myoglobin and its derivatives as photosensitizers in singlet oxygen generation. *J. Food Sci.*, **53**, 1863–1865.

Wu, J.W., Lee, M.-H., Ho, C.-T. and Chang, S. (1982). Elucidation of the chemical structures of natural antioxidants isolated from rosemary. *J. Amer. Oil Chem. Soc.*, **59**, 339–345.

Yackzan, K.S. and Wingo, W.J. (1982). Transport of fatty acids by myoglobin – a hypothesis. *Med. Hypotheses*, **8**, 613–618.

Yamamoto, S. (1992). Mammalian lipoxygenases: molecular structures and functions. *Biochim. Biophys. Acta*, **1128**, 117–131.

Yamashoji, S. and Kajimoto, G. (1980). Antioxidative effect of Gly-Gly-His on Cu(II)-catalyzed autoxidation and photosensitized oxidation of lipids. *Agric. Biol. Chem.*, **44**, 2735–2736.

Yamauchi, K. (1972). Effect of heat treatment on the development of oxidative rancidity in meat and its isolated tissue fraction. *Bull. Faculty Agric. Miyazaki Univ.*, **19**, 147–154.

Yee, J.J., Shipe, F. and Kinsella, J.E. (1980). Antioxidant effects of soy protein hydrolysates on copper-catalyzed methyl linoleate oxidation. *Food Sci.*, **45**, 1082–1083.

Zipser, M.W., Kwon, T.W. and Watt, B.N. (1964). Oxidative changes in cured and uncured frozen cooked pork. *J. Agric. Food Chem.*, **12**, 105–109.

Zubillaga, M.P., Maerker, G. and Foglia, T.A. (1984). Antioxidant activity of sodium nitrite. *J. Amer. Oil Chem. Soc.*, **61**, 772–776.

11 Lipid-derived off-flavours in meat by-products: effect of antioxidants and Maillard reactants

S.M. VAN RUTH, T. CHERAGHI and J.P. ROOZEN

11.1 Introduction

The rendering industry salvages various by-products from meat processing as a source of fat. These raw materials include viscera, fat, trimmings, bone scraps, and fallen and unsound animals from the livestock dressing process. After the fat has been removed by a combination of cooking and applied pressure, the remaining residue is ground and sold to the animal feed industry (Nash and Mathews, 1971). Meat meal is interesting for the pet food industry as it is an inexpensive protein source, which contains about 70% protein and 15% crude fat (Greenberg, 1981a).

The pet food industry works on a large scale and represents an important part of the prepared food industry in many countries. Pet food units generally may be regarded as processors of many materials which are not utilized within the human food sector, among which are meat by-products. There are several facets which can be identified for a successful pet food product, namely palatability, nutrition, convenience in use, economy and acceptability to the pet owner. Palatability is of prime importance because if the animal will not eat the food, the other aspects are irrelevant (Booth, 1979). Utilization of meat by-products in pet foods is, however, limited because of a high content of unsaturated fatty acids (50%), which renders them susceptible to rancidity and off-flavour development (Hsieh and Kinsella, 1989). Many processed foods undergo flavour changes during storage, resulting in the formation of characteristic off-flavours associated with lipid oxidation. Beside lipid oxidation, Maillard reaction is an important pathway for the formation of flavour in cooked foods; both of these have been the subject of extensive research (Danehy, 1986; Chan, 1987). However, such reactions do not occur in isolation; all foods comprise a complex mixture of components, including lipids, sugars and amino acids. Thus cooking (processing) of foods would be expected to facilitate interactions between the Maillard and lipid oxidation pathways.

For utilization of meat by-products, formation of off-flavours has to be prevented in order to produce a palatable pet food. The oxidation can be controlled in many ways; these include the removal of oxygen and possible addition of antioxidants and/or reducing agents. Furthermore, interactions between lipid oxidation and Maillard reaction products are of interest for flavour improvement.

11.2 Fatty acid and amino acid composition of meat by-products

The composition of meat and bone meal has been studied by Nash and Mathews (1971), who reported that meat by-products contain 54–59% protein, 10–12% crude fat, 28–29% ash and 3–6% moisture. In the authors' laboratory the fatty acid composition of a commercially available meat meal was analysed (Table 11.1), results of which resembled the composition of pork fat. The high content of unsaturated fatty acids (58%) made this meat by-product very susceptible to lipid oxidation. Table 11.1 shows also the major amino acids, namely alanine (9.8%), glycine (23.1%), proline (13.7%), hydroxyproline (9.7%), lysine (6.1%) and glutamic acid (10.3%). These results clearly demonstrate that the origin of this meat meal is from connective and adipose tissues (Ward and Courts, 1977). This is in agreement with studies of Nash and Mathews (1971), who reported glutamic acid and glycine as the most abundant amino acids in meat and bone meal.

11.3 Volatile compounds

The flavour composition of meat by-products, which are utilized in pet foods, has hardly been subjected to extensive studies. This is in contrast to

Table 11.1 Fatty acid and amino acid composition of meat meal[a]

Fatty acid		Amino acid	
Compound	Weight %	Compound	Weight %
C12:0	0.2	Alanine	9.8
C14:0	1.6	Glycine	23.1
C16:0	23	Valine	3.9
C16:1	2.3	Threonine	2.7
C17:0	1.7	Serine	3.4
C18:0	15	Leucine	5.3
C18:1	40	Proline	13.7
C18:2	10	Hydroxyproline	9.7
C18:3	1.5	Aspartic acid	3.6
C20:0	1.2	Phenylalanine	4.0
C20:1	2.0	Glutamic acid	10.3
C20:2	2.1	Lysine	6.1
		Tyrosine	1.8
		Arginine	2.7

[a]Fat content 11.5%, protein content 75%, moisture content 2.7%

the flavour of prime meats which are used in human consumption and to a limited extent in pet foods. For example, studies of the flavour of prime meats have been reviewed by Shibamoto (1980) and by Shahidi *et al.* (1986) Greenberg (1981a,b,c) reported the volatile composition of meat and bone meal and poultry by-product meal. The volatiles were isolated by high vacuum degassing using cryogenic traps and were analysed by gas chromatography–mass spectrometry (GC–MS). The poultry by-product meal and the meat and bone meal showed some similarities in their volatile profiles. Both products had large relative concentrations of hexanal, 3-octen-2-one and heptanal. Both foods had unsaturated ketones, methyl ketones, aliphatic alcohols, as well as 2-(n-alkyl) furans. Meat and bone meal showed larger relative concentrations of pentanal, heptanal and octanal. Furthermore, meat and bone meal contained alkadienals, *trans*-2-nonenal, decanal, undecenal and several alkyl pyrazines which were not detected in the poultry by-product meal. The majority of the compounds detected by Greenberg have been found in lipid systems and can arise through autoxidative degradations. The two by-product meals mentioned did not contain volatile compounds usually associated with pleasant cooked meat odours. For example, they lack the presence of pyrazines, thiazoles, thiazolines, trithiolanes, oxazoles, thiophenes, furanones and other compounds that contribute to the savoury meat-like aromas in prime meats, such as beef or pork and their livers.

11.4 Changes in volatile composition of meat by-products during storage

The meal type of meat by-products can be stored without mould development due to their low moisture content (Table 11.1). However, the flavour of meat meals can vary considerably due to processing and storage conditions (Greenberg, 1981a).

In the authors' laboratory changes in volatile composition of meat meal during storage were studied. When the product was stored in bulk containers or paper bags, the shelf-life was found to be rather short. Rancid odours were developed within a few weeks, which indicated lipid oxidation and quality deterioration (peroxide value was approximately 20 meq/kg of extracted fat). Exclusion of oxygen by vacuum packaging in plastic aluminium laminates prevented oxidation to a large extent (Table 11.2). Peroxide and *p*-anisidine values (IUPAC, 1979) gave more distinct differences than the 2-thiobarbituric acid method (Roozen, 1987). This might be caused by the monounsaturated nature of the majority of the fatty acids (Table 11.1) or by a delay in the formation of secondary lipid oxidation products of the polyunsaturated fatty acids to form malondialdehyde and other TBA reactive substances (Kim and LaBella, 1987).

Table 11.2 Lipid oxidation in meat meal stored in vacuum pack or paper bags for 110 days at room temperature

Methods	Vacuum pack	Paper bag
p-Anisidine value	10.5	117
Peroxide value	3	240
TBARS (direct)	7	7
TBARS (distillation)	0	1

TBARS = thiobarbituric acid reactive substances.

Volatile compounds of meat meal and of a processed Maillard meat meal–water mixture (PMMWM) were analysed by combined GC–MS. Volatiles were extracted with a continuous steam distillation extraction technique (Likens–Nickerson) for 3 h. PMMWM was prepared by adding 80 g meat meal to 165 ml boiling water and was then cooked at 90°C for 2.5 min. Afterwards, 150 ml water were added as well as cysteine (2.92 g) and/or xylose (3.62 g) and each sample was filled into 450 ml tin cans and processed at 127°C for 35 min. A control containing no cysteine or xylose was also used.

More than 60 different compounds were identified in these samples, namely aldehydes, ketones, pyrazines, S-compounds and furans as presented in Table 11.3. The compounds 2- and 3-methylbutanal, pentanal, hexanal, heptanal, octanal, nonanal and decanal were detected in relatively high concentrations. They are known to be the secondary products of lipid oxidation at moderate temperature, i.e. the scission of products of monohydroperoxides of unsaturated fatty acids (Chan, 1987).

Table 11.3 Volatile constituents of stored meat meal and processed Maillard meat meal–water mixtures identified by gas chromatography–mass spectrometry

Aldehydes	**Ketones**	**Alcohols**
2-Methylpropanal	2-Butanone	2-Methylpropanol
Butanal	2-Pentanone	1-Butanol
2-Methylbutanal	2-Heptanone	1-Penten-3-ol
3-Methylbutanal	2-Octanone	1-Pentanol
Pentanal	2-Nonanone	1-Hexanol
Hexanal	3-Octen-2-one	1-Octanol
Heptanal	2-Decanone	1-Nonanol
Octanal	3,5-Octadien-2-one	1-Penten-3-ol
2-Octenal	2,3-Pentanedione	1-Octen-3-ol
Nonanal		1-Heptanol
2-Nonenal		2-Ethyl-1-hexanol
Decanal		2-Octen-1-ol
2-Decenal		3-Methyl-1-butanol
Benzaldehyde		1-Nonen-3-ol
Pyrazines	**Sulphur compounds**	**Furans**
2,5-Dimethylpyrazine	Dimethyl trisulphide	2-Ethylfuran
2-Methyl-5-ethylpyrazine	Methyl thiophene	2-Butylfuran
Trimethylpyrazine	2-Pentyl thiophene	2-Pentylfuran
Dimethyl-ethylpyrazine	Dimethyl disulphide	2-Hexylfuran

In Figure 11.1 two reconstructed ion chromatograms are presented. They differ mainly at the higher retention times. This difference is caused by storage of the meal before processing of the PMMWM. Volatile heterocyclic sulphur compounds (probably 3-thiazolines; Table 11.4) were ten times higher in the 'fresh meat meal' PMMWM than in the 'old meat meal' PMMWM. A rather simple explanation is the oxidation of cysteine into cystine in the 'old meat meal' PMMWM. In that case, cysteine is not available in PMMWM for producing fragments like hydrogen sulphide, acetaldehyde and ammonia, which react with acetoin from sugar–amino browning reactions to give 3-thiazolines (Lindsay, 1996).

The stage of oxidation of meat meal is extremely important for its application: PMMWM with 'fresh' meat meal had a stronger meaty aroma than with 'old' meat meal, probably associated to the presence of heterocyclic sulphur compounds (Cramer, 1983; Gasser and Grosch, 1988). As stated above, the 3-thiazoline peaks in the 'fresh' chromatogram and their near

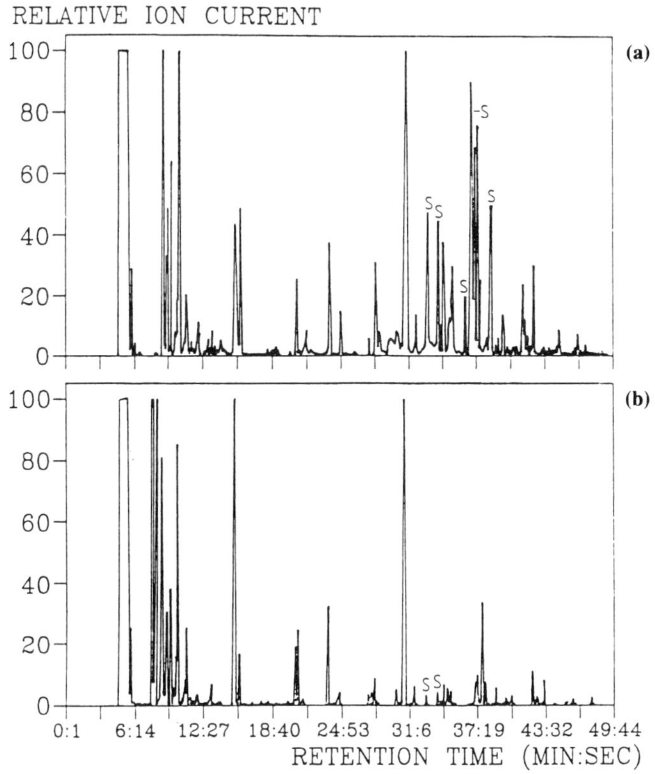

Figure 11.1 Reconstructed ion chromatograms of volatiles extracted from processed Maillard water mixtures with meat meal stored in (a) vacuum-packs (fresh) or (b) paper bags (old) for 110 days. S = heterocyclic sulphur–nitrogen compound (3-thiazoline?).

Table 11.4 Listing of mass spectral data of two 3-thiazolines found and 2,4,5-trimethylthiazoline[a]

M/Z (relative intensity) of main fragments of mass spectrum
41 (35), 42 (55), 60 (30), 68 (25), 69 (100), 102 (83), 114 (27), 128 (23), 143 (57)
42 (59), 55 (38), 68 (30), 69 (56), 88 (100), 114 (26), 129 (52)
[a]42 (82), 55 (46), 68 (33), 69 (58), 88 (100), 114 (28), 129 (45)

[a]Mussinan *et al.* (1976).

absence in the 'old' one are of practical interest. The aliphatic aldehydes, alcohols, and ketones are considered to contribute primarily to the fatty odour and flavour, whereas the pyrazines and some of the sulphur-containing compounds are more closely associated with the basic meaty flavour of products (Table 11.3). Many carbonyl compounds formed by lipid oxidation reactions are not important contributors, whereas certain aldehydes and ketones of Maillard reactions contribute to desirable roasted flavour quality of a product; e.g. 3-methylbutanal, which is the Strecker degradation product of leucine formed during roasting (MacLeod and Coppock, 1977).

11.5 Effect of antioxidants on volatile composition

We have shown that the oxidative state of meat meal is important for its application as pet food. Therefore, the influence of antioxidants on the volatile compounds of meat meal was studied. The antioxidant/chelators, namely critic acid (0.01%) or ascorbic acid (0.05%) were added to 150 g meat meal and 450 ml water. After mixing and freeze drying, the dry matter was pulverized and stored in paper bags. The major aldehydes were extracted by a Likens–Nickerson procedure and analysed by gas chromatography (Figure 11.2).

Both the heavy metal ion chelating agent citric acid and the oxygen scavenger ascorbic acid diminished the total amount of aldehydes by 70%. The aldehydes pentanal, hexanal, heptanal and octanal showed similar differences for the samples. Although citric acid and ascorbic acid were rather effective in prevention of development of aldehydes, exclusion of oxygen by vacuum packaging immediately after the production of meat meal was even more effective ('fresh' in Figure 11.2). In that case the total amount of aldehydes decreased by 95%. In conclusion, vacuum packaging and addition of antioxidants reduced lipid oxidation markedly, which is of importance in the production of a high-quality processed meat meal.

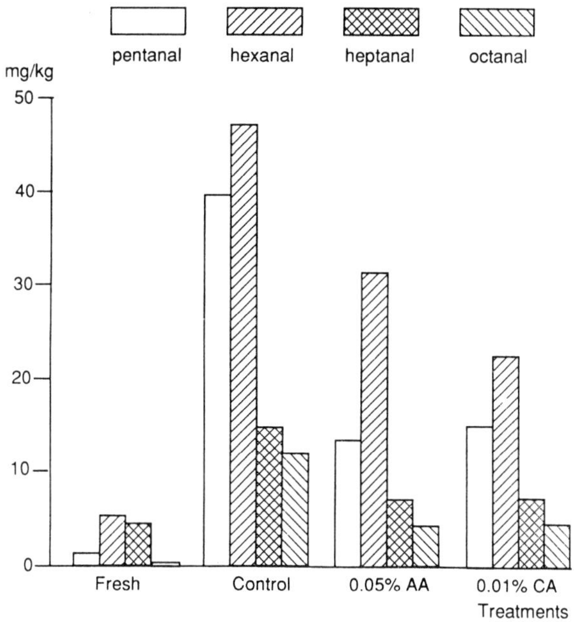

Figure 11.2 Effect of packaging and antioxidants on the content of four aldehydes in meat meal stored for 100 days at room temperature. Fresh = stored vacuum-packed (PV = 3 meq/kg extracted fat); Control = stored in paper bag (PV = 240 meq/kg extracted fat); AA = ascorbic acid; CA = citric acid.

11.6 Effect of Maillard reactants on volatile composition

The meaty aroma of a processed meat meal is an attractive property for its application as pet food. Therefore, we studied the effect of added Maillard reactants (cysteine and/or xylose) on the volatile composition of processed meat meal–water mixtures (PMMWM). Preparation was described in section 11.4. The PMMWM samples were evaluated for their content of volatiles, in particular the four major aldehydes, namely pentanal, hexanal, heptanal and octanal (Figure 11.3). Volatiles were extracted by a Likens–Nickerson steam distillation and analysed by gas chromatography. The concentrations of aldehydes in the control sample (no reactants added) hardly differed from that of the 'fresh' sample, which indicates that the volatile composition was not changed much by the preparation procedure (one part meat meal in three parts PMMWM).

When Maillard reactants, a combination of cysteine and xylose, were present in PMMWM, the concentration of aldehydes associated with oxidation decreased by 31%. Xylose was as effective as cysteine/xylose but cysteine alone had no significant effect on the reduction in total aldehydes. Meat meal seems to contain sufficient amino groups for starting

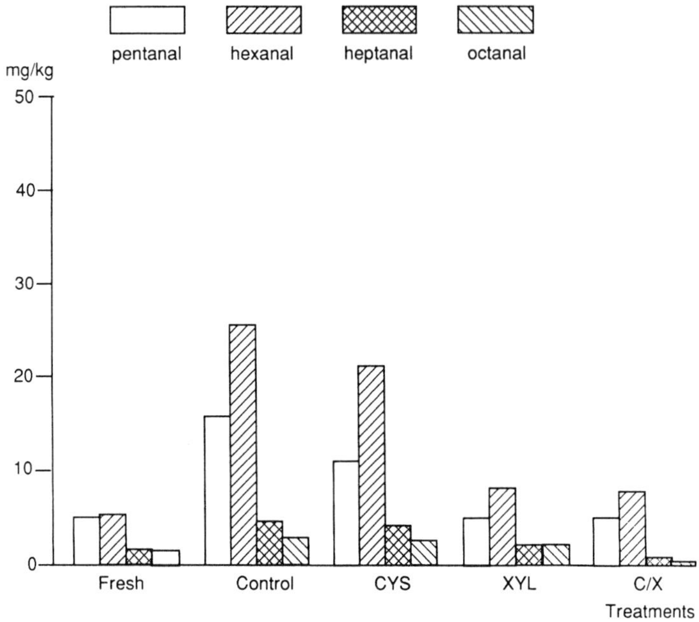

Figure 11.3 The content of four aldehydes determined in processed Maillard meat meal–water mixtures. Fresh = stored vacuum-packed (PV = 3 meq/kg extracted fat); Control = stored in paper bag (PV = 240 meq/kg extracted fat); CYS = cysteine (2.92 g) added; XYL = xylose (3.63 g) added; C/X = cysteine (2.92 g) + xylose (3.62 g) added.

up the Maillard reaction with xylose. The intermediates formed can react then with the aldehydes from the lipid oxidation reaction. Addition of cysteine does not change much in this respect, especially when cysteine is oxidized into cysteine immediately after mixing.

The antioxidative effect of the Maillard reaction products has been well documented (Lingnert and Erikson, 1980; Bailey *et al.*, 1987). It was suggested that the cause of reductions in aldehydes by addition of Maillard reactants is the suppression of individual pathways rather than of free radical oxidation as a whole. Among the volatile products of model reactions containing both Maillard reactants and lipids, a number of compounds have been reported which are not formed from either the Maillard reaction or lipid oxidation alone (Whitfield, 1992; Farmer and Whitfield, 1993). Thus, lipid–Maillard interactions produce a number of compounds that include long-chain heterocyclic compounds and alkanethiols. The routes of formation proposed for lipid–Maillard products illustrate some of the mechanisms by which the lipid oxidation and Maillard pathways may interact. Long-chain alkylthiophenes, alkylthiapyrans and alkylpyridines can all be formed by the reaction of H_2S or NH_3 with alkadienals (Farmer *et al.*, 1989). Volatile compounds likely to originate from lipid–Maillard interactions have also been identified in prime foods such as French-fried potatoes and cooked meats (Whitfield, 1992).

11.7 Concluding remarks

Meat meal has a good balance of protein and fat content for nutritional purposes and, together with its water holding capacity, may serve as a potential ingredient for pet food. The state of oxidation of meat by-products is important for their use in such applications. Antioxidants (reducing and chelating agents) are often added to products with a high content of unsaturated fatty acids to prevent the formation of undesirable volatiles. It was shown that the level of aldehydes associated with oxidation can be reduced by addition of ascorbic or citric acid, but vacuum packaging was shown to be more effective. Addition of Maillard reactants before processing the meat meal resulted in a reduction of aldehydes. In addition, volatile compounds were formed in Maillard reactions, such as Strecker degradation products, which can contribute to a desirable roasted flavour of the product.

References

Bailey, M.E., Shin-Lee, S.Y., Dupuy, H.P., St. Angelo, A.J. and Vercellotti, J.R. (1987). Inhibition of warmed-over flavour by Maillard reaction products. In *Warmed-over Flavour of Meat*, eds. A.J. St. Angelo and M.E. Bailey. Academic Press, London, pp. 237–266.

Booth, P. (1979). An approach to meat flavour research with evaluation by the dog and cat. In *Progress in Flavour Research*, eds. P.G. Land and H.E. Nursten. Applied Science, London, pp. 75–77.

Chan, H.W.-S. (1987). *Autoxidation of Unsaturated Lipids*. Academic Press, London.

Cramer, D.A. (1983). Chemical compounds implicated in lamb flavour. *Food Technol.*, **37**, 249–257.

Danehy, J.P. (1986). Maillard reactions: Non-enzymatic browning in food systems with special reference to the development of flavour. *Adv. Food Res.*, **30**, 77–138.

Farmer, L.J. and Whitfield, F.B. (1993). Some aroma compounds formed from the interaction of lipid in the Maillard interaction. In *Proceedings of Progress in Flavour Precursor Studies*, ed. P. Schreier. Allured, Carol Stream, IL, pp. 387–390.

Farmer, L.J., Mottram, D.S. and Whitfield, F.B. (1989). Volatile compounds produced in Maillard reactions involving cysteine, ribose and phospholipid. *J. Sci. Food Agric.*, **49**, 347–368.

Gasser, U. and Grosch, W. (1988). Identification of volatile flavour compounds with high aroma values from cooked beef. *Z. Lebensm. Unters. Forsch.*, **186**, 489–494.

Greenberg, M.J. (1981a). Characterisation and comparison of flavour volatiles in meat by-products. In *Flavour '81*, ed. P. Schreier. Walter de Gruyter, Berlin, pp. 599–608.

Greenberg, M.J. (1981b). Characterization of meat and bone meal flavour volatiles. *J. Agric. Food Chem.*, **29**, 1276–1280.

Greenberg, M.J. (1981c). Characterization of poultry byproduct meal flavour volatiles. *J. Agric. Food Chem.*, **29**, 831–834.

Hsieh, R.J. and Kinsella, J.E. (1989). Oxidation of polyunsaturated fatty acids: mechanisms, products, and inhibition with emphasis on fish. In *Advances in Food and Nutrition Research 33*, ed. J.E. Kinsella. Academic Press, San Diego, pp. 233–341.

IUPAC (1979). *Standard Methods for the Analysis of Oils, Fats and Derivates*. 6th edn, Part 1. International Union of Pure and Applied Chemistry, Pergamon Press, Oxford.
Kim, R.S. and LaBella, F.S. (1987). Comparison of analytical methods for monitoring autoxidation profiles of authentic lipids. *J. Lipid Res.*, **28**, 1110–1117.
Lindsay, R.C., 1996. Flavours. In *Food Chemistry*, 3rd edn, ed. O.R. Fennema. Marcel Dekker, New York, p. 758.
Lingnert, H. and Eriksson, C.E. (1980). Antioxidative Maillard reaction products. 1. Products from sugars and amino acids. *J. Food Process. Preserv.*, **4**, 161–172.
MacLeod, G. and Coppock, B.M. (1977). A comparison of the chemical composition of boiled and roasted aromas of heated beef. *J. Agric. Food Chem.*, **25**, 113–117.
Mussinan, C.J., Wilson, R.A., Katz, I., Hruza, A. and Vock, M.H. (1976). Identification and flavour properties of some 3-oxazolines and 3-thiazolines isolated from cooked beef. In *Phenolic, Sulphur, and Nitrogen Compounds in Food Flavours*, eds. G. Charalambous and I. Katz. ACS Symposium Series 26, American Chemical Society, Washington, DC, pp. 133–145.
Nash, H.A. and Mathews, R.J. (1971). Food protein from meat and bone meal. *J. Food Sci.*, **36**, 930–934.
Roozen, J.P. (1987). Effects of types I, II and III antioxidants on phospholipid oxidation in a meat model for warmed-over flavour. *Food Chem.*, **24**, 167–185.
Shahidi, F., Rubin, L.J. and D'Souza, L.A. (1986). Meat flavour volatiles: A review of the composition, techniques of analysis, and sensory evaluation. *CRC Crit. Rev. Food Sci. Nutr.*, **24**, 141–243.
Shibamoto, T. (1980). Heterocyclic compounds found in cooked meats. *J. Agric. Food Chem.*, **28**, 237–243.
Ward, A.G. and Courts, A. (1977). *The Science and Technology of Gelatin*. Academic Press, London.
Whitfield, F.B. (1992). Volatiles from interactions of Maillard reactions and lipids. *CRC Crit. Rev. Food Sci. Nutr.*, **31**, 1–58.

12 Maillard reactions and meat flavour development

M.E. BAILEY

12.1 Introduction

The flavour of raw fresh meat is bland, metallic and slightly salty, whereas desirable meat flavour is apparent only after heating. As with the flavour of most foods, both nonvolatile and volatile components are essential. The nonvolatile ingredients consist of taste compounds, flavour enhancers, or precursors for reaction products which may be responsible for desirable volatile compounds.

Obviously, thermal degradation is responsible for the formation of volatile components which are formed from water-soluble precursors, such as thiamine, glycogen, glycoproteins, nucleotides, nucleosides, free sugars, amino acids, peptides, sugar phosphates, amines and organic acids. These natural precursors react in meat products during heating in primary reactions to form intermediate products which can further react with other degradation products to form a complex mixture of volatiles responsible for meat flavour.

Almost 1000 volatile compounds have been identified from meat or from model systems consisting of meat ingredients and one might predict that many more thousands of volatile compounds will be identified from these systems in the future. Many reaction mechanisms have been proposed for the formation of these numerous compounds but the primary ones involve: (a) the degradation of vitamins, particularly thiamine; (b) the thermal degradation of carbohydrates and amines; and (c) the Maillard reaction, including Strecker degradation.

The importance of the chemistry of meat flavour to academia and the food industry is reflected in the number of reviews that have appeared in the literature since 1980 (Ho, 1980; Shibamoto, 1980; Katz, 1981; MacLeod and Seyyedain-Ardebili, 1981; Lawrie, 1982; Bailey, 1983; Cramer, 1983; IFT Symposium, 1983; Moody, 1983; Baines and Mlotkiewicz, 1984; MacLeod, 1984; MacLeod, 1986; Shahidi *et al.*, 1986; Horstein and Wasserman, 1987; Parliment *et al.*, 1989; Rhee, 1989; Shahidi, 1989; Mottram, 1991; MacLeod, 1994; Mottram, 1994), along with many other associated articles, particularly those concerned with the Maillard reaction and thermal reaction flavours.

Although many meat flavour compounds can be made by heating thiamine or a mixture of carbohydrates, ammonia and hydrogen sulphide, the Maillard reaction remains a focal point for the formation of most recognized meat flour compounds. The water-soluble extracts of desirable meat flavour are well suited for the Maillard reaction. Wood and Bender (1957) and Bender *et al.* (1958) were the first to thoroughly examine the water-soluble extracts of boiled beef, which they believe contained precursors of meat flavour. Bender *et al.* (1958) concluded, 'it is beyond reasonable doubt that development of the brown colour and meaty flavour characteristics of these extracts is a result of the Maillard reaction.' Hornstein and Crowe (1960) first demonstrated that lyophilized diffusate from cold water extracts the raw beef and pork produced meaty odours upon heating.

Macy *et al.* (1964a,b) extended these studies of the diffusates of beef, pork and lamb and demonstrated that amino acids, sugars, sugar phosphates, nucleotides and nucleosides are all decreased in concentration during heating at 100°C for 1 h. Wasserman and Spinellli (1970) confirmed these results after heating beef diffusates in water at 125°C for 1 h. Wasserman (1979) also concluded that Maillard browning (nonenzymatic browning) was important for the formation of desirable meat-flavoured compounds. This is a marked contrast to enzymatic browning which has no significance in the flavour of meat, but may be important for the flavour of certain sea foods, plants and vegetables (Lee and Whitaker, 1995).

12.2 The Maillard reaction

The Maillard reaction (nonenzymatic browning) involves the reaction of aldehydes with amines and through numerous reactions, food flavour compounds and dark pigments (melanoidins) are formed. Louis-Camille Maillard, at the University of Nancy, France, published a series of papers on the reaction (Maillard, 1916), and scientists continue to study the many ramifications of the reaction.

The importance and complexity of the Maillard reaction is revealed by the large number of different review articles published. The most important early reviews related to food were those of Hodge (1953, 1967), Anet and Reynolds (1957) and Reynolds (1963, 1965). Many other reviews have been written on the chemistry of the Maillard reaction since 1970 (Hodge *et al.*, 1972; Williams, 1976; Hurrell and Carpenter, 1977; Mabrouk, 1979; Paulsen and Pflughaupt, 1980; Feather, 1981; Mauron, 1981; Nursten, 1981; Feeney and Whitaker, 1982; Hurrell, 1982; Vernin and Parkayi, 1982; Namiki and Hayashi, 1983; Feather, 1985; Danehy, 1986; Heath and Reineccius, 1986; Nursten, 1986; Yaylayan and Sporns, 1987; Namiki, 1988; Baltes *et al.*, 1989; Ledl *et al.*, 1989; O'Brien and Morrissey, 1989; Eskin,

1990; Ledl, 1990; Lee and Whitaker, 1995; Ikan, 1996), and there have been five symposia (Eriksson, 1981; Waller and Feather, 1983; Fujimaki *et al.*, 1986; Finot *et al.*, 1990; Labuza *et al.*, 1994).

Factors affecting the progress of the Maillard reaction include temperature, time, moisture content, pH, and the concentration and nature of the reactants. The initial reactions between the amines and aldehydes have been studied by numerous investigators and are reasonably well understood. The sugar reacts reversibly with an amine group to form a glycosylamine. The glucosylamines from aldose sugar undergo Amadori rearrangement to yield Amadori compounds such as 1-amino-1-deoxy-2-ketoses; the reaction between ketoses (fructose) and amines usually involves the formation of ketosylamines, followed by the Heynes rearrangement to form 2-amino-2-deoxyaldolases (Reynolds, 1965).

The most widely accepted mechanisms for degradation pathways of Amadori compounds by dehydration and fission were suggested by the brilliant work of Hodge (1953) and the actual products formed depends on the basicity of the amine, the pH of the reaction mixture and the temperature.

Hodge (1967) emphasized that, although browning flavours in most foods are most likely to arise from the Maillard reaction, many other food components, particularly sugars, could interact to form carboxylic and other flavour constituents. Much recent evidence supports his ideas that oxidized component of lipids such as aldehydes, α,β-unsaturated ketones and acrolein participate in browning reactions to produce meat flavour compounds. Most importantly, he discussed mechanisms whereby sugar caramelization at high temperature produced many of the important intermediates in meat flavour that can likewise form by Maillard reaction at lower temperatures.

In an excellent paper, Vernin and Parkanyi (1982) summarized the Maillard reaction involving sugars and α-amino acids and outlined how reductones and dehydroreductones could be degraded into a number of flavour compounds by dehydration, retro-aldolization and Strecker degradation. The end products were *N,S,O*-heterocyclics. They also outlined how Amadori and Heynes intermediates were rearranged to form reductones. The primary reductones were 3-deoxyosones formed by 1,2-enolization of the carbonyl-amine compounds, 1-deoxyosones and their equilibrium 1-deoxyreductones formed by 2,3-enolization of the carbonyl-amine compounds (Figure 12.1). Apparently the 3-deoxyosones are sufficiently stable to be separated by HPLC and characterized (Weenan, 1991). These compounds can be isolated and reacted to form many flavour intermediates, including 2-furfural derivatives which can react with ammonia and hydrogen sulphide to form *N,S,O*-heterocyclics (Barone and Chanon, 1982).

A simplified version of the formation of rearrangement products from Amadori compounds and their possible degradation by dehydration and

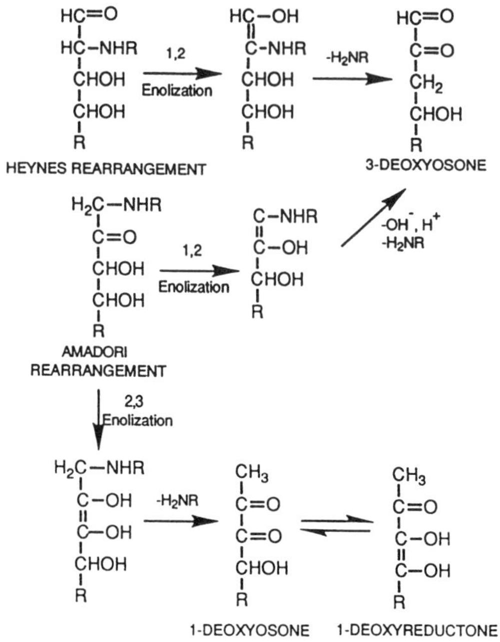

Figure 12.1 Rearrangement of Amadori and Heynes intermediates from Maillard reaction of carbonyl-amine compounds to form reductones (adapted from Vernin and Parkanyi, 1982).

retro-aldolization (fission) is shown in Figure 12.2. The structures labelled flavour compounds are important intermediates that can react with other Maillard reaction degradation products to form meat flavour compounds. A modified version of the suggested reaction routes (Baltes *et al.*, 1989) for the degradation via 1-deoxyosones to important meat flavour intermediates is presented in Figure 12.3. The compounds formed are key precursors involved in reactions responsible for meat flavour.

This pathway shows the formation of a number of cyclic oxygen intermediates important to meat flavour production. These compounds are 4-hydroxy-5-methyl-3(2*H*)-furanone from pentoses, 4-hydroxy-2,5-dimethyl-3(2*H*)-furanone (furaneol) from hexoses, along with isomaltol, maltol, 4-hydroxymaltol and 2-hydroxy-3-methylcyclopentene-2-one (cyclotene). The most important reaction product of the 1-deoxyosone pathway is 5-hydroxy-5,6-dehydromaltol. This compound represents about 30% of volatiles formed by heating glucose at 120°C and it disappears at higher temperatures with the appearance of cyclotene and furaneol (Baltes *et al.*, 1989). Cyclotene is also formed by condensation of hydroxyacetone (Vernin and Parkanyi, 1982).

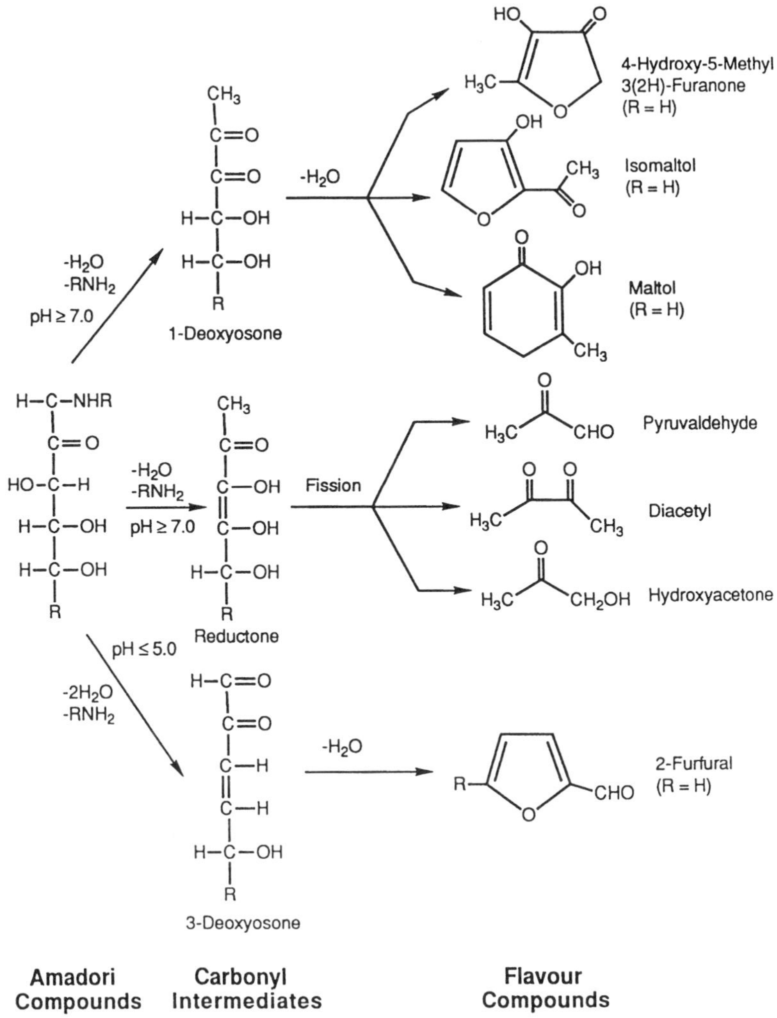

Figure 12.2 Deamination and dehydration of Amadori compounds to form impotant meat flavour intermediates.

Feather (1981) evaluated the evidence for the existence of the 1-deoxyosone and its enolic form and concluded that it might be involved in the formation of isomaltol and maltol by cyclization–dehydration under basic conditions as hypothesized by Hodge *et al.* (1953, 1967). A similar pathway for the formation of maltol and isomaltol was recently confirmed by Ledl (1990).

Under the appropriate conditions for formation, it seems likely that 4-hydroxy-5-methyl-3(*2H*)-furanone could be formed by condensation of

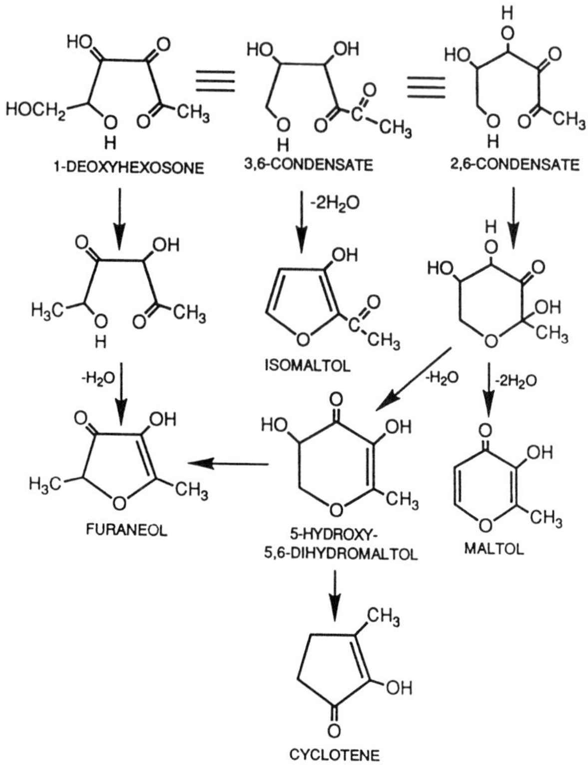

Figure 12.3 Formation of specific meat flavour intermediates by cyclization and dehydration of 1-deoxyhemosones (modified from Baltes, 1989).

pentose with amine, formation of Amadori compound and dehydration via 2,3-enolization to form the 1-deoxyosone. Hicks and Feather (1975) conclusively showed that pentose-derived Amadori compound 1-benzylamino-1-deoxy-D-*thrio*-pentalose dehydrates to 4-hydroxy-5-methyl-3(2*H*)-furanone. Hicks *et al.* (1974) also found that the furanone was formed from a 1-amino-1-deoxy-D-fructuronic acid by decarboxylation. It has also been formed by heating D-xylose (Severin and Seilmeier, 1968), D-ribose and D-ribose phosphate (Peer and van den Ouweland, 1968). Furaneol (4-hydroxy-2,5-dimethyl-3(2*H*)-furanone) was concluded to be involved in meat flavour by Tonsbeek *et al.* (1968), who isolated it along with 4-hydroxy-5-methyl-3(2*H*)-furanone from beef. Hexoses form 5-methylfurfurals and furaneol, while pentoses form furfural and 4-hydroxy-5-methyl-3(2*H*)-furanone, although there is some evidence that the furaneol can lose formaldehyde to yield the furanone during the Maillard reaction (Ledl, 1990). Furaneol is also formed by heating precursors in yeast and cereal extracts (Schieberle, 1991). Ching (1979) identified three 3(*2H*) furanones by heating low-molecular-weight dialysable diffusate at 155°C in a closed system for 30 min. Thus, the compounds can be formed with low levels of sugars from extracts of meat.

1,2-Enolization of the Amadori compound is favoured by acid conditions, and following deamination and rearrangement, yields 3-deoxyosones as intermediates. These degrade to yield 5-hydroxymethyl-2-furfural for hexoses and 2-furfural from pentoses. These compounds are also produced by acid decomposition of sugars or caramelization (Feather and Harris, 1973). The furfural derivatives are very reactive with ammonia and hydrogen sulphide to form heterocyclic compounds. Barone and Chanon (1982), using a computer program simulating organic synthesis in the Maillard reaction, proposed more than 1000 structures from this reaction. Heterocyclics predicted in the reaction include furans, pyrroles, thiophenes, oxazoles, thiazoles, imidazoles, pyrans, pyridines and pyrazines, many of which have been identified by Shibamoto (1977).

The 1-deoxyreductone (equilibrium product of 1-deoxyosone) in a basic medium ($pH > 5.0$) can degrade by retroaldo reactions to yield a number of very reactive carbonyl compounds such as pyruvaldehyde, diacetyl, dihyroxyacetone, glyoxal and hydroxyacetol and acetic acid (Hodge, 1967; Feather, 1985). These compounds are particularly reactive with amino acids in Strecker degradation.

A fourth arm of the Maillard reaction is Strecker degradation by oxidation of the amino acids. This is undoubtedly one of the most important steps in the formation of meat flavour. This reaction involves the interaction of dicarbonyls, such as diacetyl, pyruvaldehyde, hydroxyacetone and hydroxydiacetyl, to degrade amino acids to aldehydes with one less carbon atom than the original amino acid, carbon dioxide, α-amino ketones, ammonia and dihydrogen sulphide. Some degradation products include acetaldehyde from α-alanine, isobutyraldehyde from valine, isovaleraldehyde from leucine, 2-methylbutanal from isoleucine, benzaldehyde from phenylglycine and acetaldehyde from cysteine.

12.3 The Maillard reaction and meat flavour compounds

The above evidence reveals that the Maillard reaction is responsible for the formation of many meat flavour compounds, both in the cooking of meat and in model systems during formation of synthetic meat flavours. These compounds consist predominantly of *N,S,O*-heterocyclic compounds and other sulphur-containing constituents. The list of compounds includes furans, pyrazines, pyrroles, thiophenes, thiazoles (thiazolines), imidazoles, pyridines, oxazoles and cyclic ethylene sulphides.

Some of the many reviews on meat flavour chemistry were enumerated above and several of the authors have continued their discussions of meat flavour formation by reporting mechanisms involving the Maillard reaction. Only a few aspects of this broad topic can be considered here.

12.3.1 Low-molecular-weight precursors of meat flavour

The concentrated diffusate from water extracted meat (Wood, 1961; Hornstein and Crowe, 1960; Macy *et al.*, 1964a,b; Wasserman and Gray, 1965; Wasserman and Spinelli, 1972; Ching, 1979; Einig, 1983; Shin-Lee, 1988) is a natural starting place for the study of meat flavour chemistry. The heating of this nonproteinaceous meat flavour concentrate in water gives an aroma of boiled beef upon boiling and a stronger odour of broiled steak when heated to higher temperatures (150°C). This concentrate contains many precursors essential for meat flavour production as sugars, amino acids, sugar phosphates and nucleic acid components.

Macy *et al.* (1964a,b) demonstrated that many of these precursors from meat diminish in concentration during heating. Ching (1979), Einig (1983) and Shin-Lee (1988) separated and identified many volatile compounds formed by heating diffusate under various conditions. These compounds have been enumerated previously (Bailey and Einig, 1989), and the important compounds are summarized in Table 12.1.

Although 167 compounds were identified, the most important volatiles related to meat flavour were furanones, ketones, sulphur compounds (sulphides, thiophenes and thiazoles), pyrones, pyrroles and pyrazines. The latter compounds were a prominent group of volatiles found by heating beef diffusates.

Table 12.1 Important meat flavour compounds identified from heated beef diffusate[a]

Class of compound	Examples of compounds within classes
Furanones	2-Methyl-4,5-dihydro-3(2*H*)-furanone 2,5-Dimethyl-4-hydroxy-3(2*H*)-furanone 4-Hydroxy-5-methyl-3(2*H*)-furanone
Ketones	2-Hydroxy-3-methyl-2-cyclopentenone (cyclotene) Other cyclic ketones Acyclic ketones
Sulphur compounds	Sulphides, disulphides, trisulphides Thiophenes Thiazoles
Pyrones	3,5-Dihydroxy-2-methyl-4*H*-pyran-4-one
Pyrazines	37 Pyrazines, including 5 cyclopenta-pyrazines
Pyrroles	8 Pyrroles, including 2 acetyl pyrroles

[a]Compounds identified by Bailey and Einig (1989).

12.3.2 Pyrazines

From the above discussion it is obvious that heterocyclic volatile compounds are important meat flavorants, and many of these can be formed by the interaction of lipids and Maillard reaction products as discussed by Mottram (1994). Many of these compounds are pyrazines and sulphur-containing heterocyclics.

Pyrazines constitute a major class of volatiles formed from the Maillard reaction, particularly if Strecker degradation is considered as part of these reactions. Reactant conditions, such as moisture content, temperature, pH and time, are important in pyrazine formation (Vernin and Parkanyi, 1982). van den Ouweland *et al.* (1989) surmised that a low percentage of volatile identified from natural beef aroma have their origin from the Maillard reaction and that 50% of these are pyrazine derivatives.

A number of mechanisms have been proposed for the formation of pyrazines and an important pathway would be the condensation of dicarbonyl compounds formed by Strecker degradation to form alkylpyrazines. Cyclopentapyrazines can be formed by the condensation of cyclic ketones such as cyclotene (2-hydroxy-3-methyl-cyclopentanone). Pathways proposed by Vernin and Parkanyi (1982) and Flament *et al.* (1976) are shown in Figure 12.4.

There is recent evidence that 3-deoxyglucosone can be a primary precursor for pyrazines. Data have been presented that it is degraded by

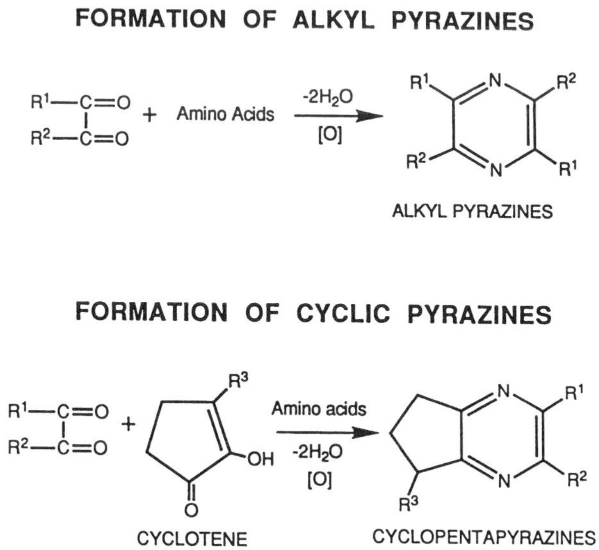

Figure 12.4 Reaction pathways proposed for the formation of pyrazines (modified from Vernin and Parkanyi, 1982).

retro-aldolization and a 2,4-scission to form pyruvaldehyde, which can be involved in the formation of dimethylpyrazine by Strecker degradation (Weeman, 1991).

Pyrazines have been found in all meat species following cooking. Shahidi *et al.* (1986), who have done an excellent job in cataloguing volatile constituents from cooked meat, listed 48 pyrazines from beef, 36 from pork and 16 from lamb. Bailey and Einig (1989) listed 37 pyrazines identified in heated systems of low-molecular-weight diffusates from beef. Twenty-five were alkylpyrazines and five were cyclopentapyrazines. The latter group of compounds are interesting since they have been reported to have roasted, grilled and other animal notes from roast meat (Ohloff and Filament, 1978). Bodrero *et al.* (1981) obtained a strong relationship between sensory scores for meaty odour and a reaction mixture of 2-acetyl-3-methylpyrazine and H_2S. Mottram *et al.* (1984) found 27 pyrazines in well-done grilled pork. Pyrazines accounted for 77% of the total volatiles found in pork and these compounds appear to be extremely important constituents of meat cooked at high temperature.

12.3.3 Sulphur-containing heterocyclics

Meat would indeed have an entirely different flavour in the absence of sulphur compounds. Large quantities of H_2S are produced during the heating of meat, which becomes more evident when the volatiles from meat cookery are filtered through acid to remove the ammonia. Under these conditions, the H_2S concentration is very prominent. Most researchers agree that sulphur compounds are the most important volatiles formed during meat cookery and sulphur precursors are used in essentially all synthetic meat flavour mixtures.

Sulphur-containing amino acids, cysteine and cystine, and the peptide glutathione, are indispensable compounds for generating meat-like aromas during thermal processing. They participate in the Maillard reaction and Strecker degradation to form volatile sulphur-containing compounds. Several meat-like compounds can be formed by heating cysteine, glutathione, alanine and sodium sulphide with dimethyl-4-hydroxy-3(2*H*)-furanone (furaneol) (Zang *et al.*, 1997).

MacLeod (1986) has done a superlative job of identifying and describing the many volatile compounds in the literature described as meaty. Obviously most of these compounds contain sulphur. She listed 78 compounds as having meaty-like aromas; 65 are heterocyclic sulphur compounds, seven are aliphatic sulphur compounds, and six are nonsulphur heterocyclics. MacLeod (1994) discussed the formation of 25 cyclic-sulphur compounds identified by various investigations of cooked beef aromas. She specified in detail how most of them could be formed by Maillard reactions or by the thermal degradation of thiamine. The most important meat flavour compounds from the Maillard reaction appear to be furans or thiophenes having methyl or sulphur substituents in the 1, 2 or 5 positions. This is confirmed by the generalized molecular structure of Dimoglo *et al.* (1988) for compounds having meaty odours. The C_2 and C_5 methyl groups and C_3 sulphur groups are important for furan and thiophene derivatives. Additionally, two furan rings and two or more sulphur atoms increase meaty odour.

Acetaldehydes formed by Strecker degradation of alanine and other aldehydes are important since these compounds can react with H_2S. Ammonia and methanethiol are also formed by Strecker degradation to yield dithiozines, thiols, sulphides and trithianes. A scheme for the formation of these sulphur compounds is given in Figure 12.5. These reactions can be extrapolated to the formation of many similar compounds as has been diagrammed in previous reviews of meat flavour.

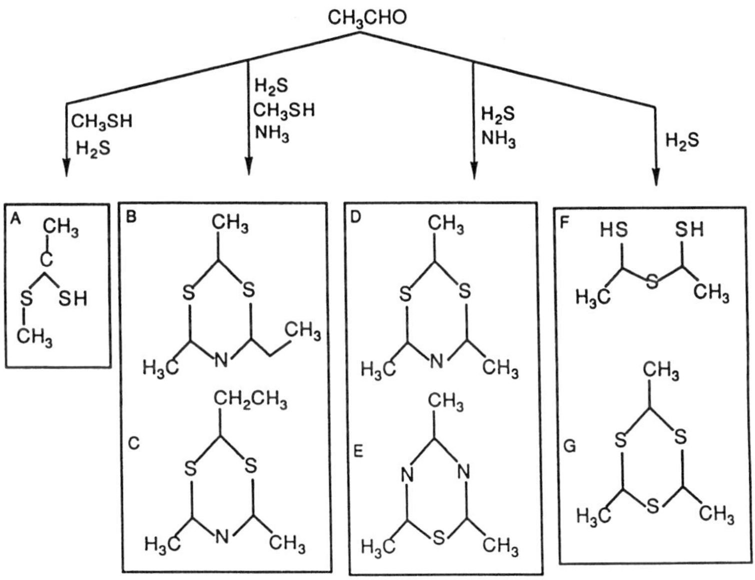

A. 1-methylthio-1-ethanethiol
B. 4-ethyl-2,6-dimethyldihydro-1,3,5-dithiazine
C. 2-ethyl-4,6-dimethyldihydro-1,3,5-dithiazine
D. 5,6-dihydro-2,4,6-trimethyl-1,3,5-dithiazine
E. 5,6-dihydro-2,4,6-trimethyl-1,3,5-thiadiazine
F. *Bis*-(1-mercaptoethyl) sulphide
G. 2,4,6-trimethyl-1,3,5-trithiane

Figure 12.5 Sulphur compounds formed from reaction of acetaldehyde, methane, thiol, ammonia and hydrogen sulphide.

12.3.4 Sulphur compounds from furan-like components

As outlined above, furan-like derivatives are very prominent products of Maillard reactions carried out in model systems, and the compounds emphasized form secondary reaction products with H_2S and ammonia to form meat flavours. Sugar degradation products like maltol, isomaltol, 4-hydroxy-5-methyl-3(2*H*)-furanone, 2,5-dimethyl-4-hydroxy-3(2*H*)-furanone and 2 hydroxy-3-methyl-2-cyclopentene-1-one (cyclotene) can exchange oxygen in the ring with nitrogen or sulphur.

van den Ouweland and Peer (1975) contributed significantly to knowledge of sulphur heterocyclic formation when they reacted H_2S with 4-hydroxy-5-methyl-3(2*H)*-furanone to form a number of 'meat-like' mercapto-substituted furan and thiophenone derivatives (Figures 12.6 and 12.7). They concluded that the reaction proceeds via a 2,4-diketone intermediate which then reacts with H_2S to form thiophenones (Figure 12.6). These workers strongly believe, however, that these compounds are not formed via a Maillard-type reaction in meat, but are derived from nucleic acid derivatives and ribose-5-phosphate. One of these compounds, 4-mercapto-5-methyl-tetrahydro-3-furanone, has a very strong meaty odour and was identified by Ching (1979) after heating freeze-dried defatted beef with triacylglycerols. It has a meaty or maggi odour and could be considered to be a meat flavour impact compound.

The importance of the reaction between furanones and sulphur-containing compounds in meat flavour was demonstrated by Bodrero *et al.* (1981), who used surface response methodology to demonstrate that this type of reaction gave the highest predictive score perceived for cooked beef aroma. 4-Hydroxy-2,5-dimethyl-3(2*H)*-furanone, a Maillard reaction product also formed by cyclization and by reacting L-rhamnose with piperidine acetate (Hodge and Osman, 1976), was reacted with cystine by

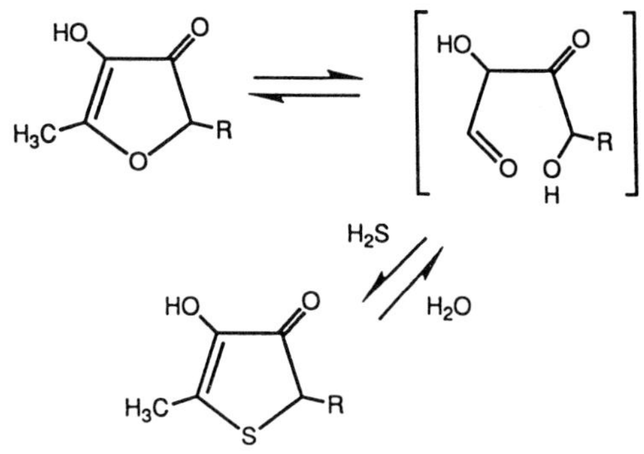

4-HYDROXY-3(2H)-FURANONES (R = H, CH_3)

Figure 12.6 Initial reaction between furanones and hydrogen sulphide for form *S*-heterocyclics (adapted from van den Ouweland and Peer, 1975).

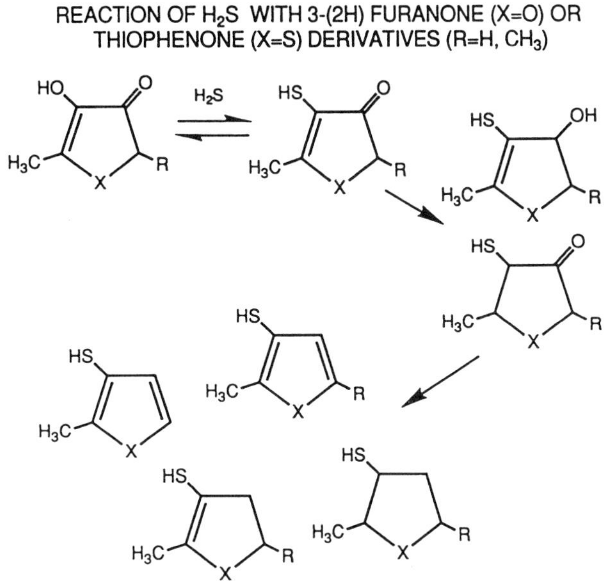

Figure 12.7 Meat flavour compounds formed by reacting H_2S with furanones or thiophenones (adapted from van den Ouweland and Peer, 1975).

Shu and Ho (1989) under various conditions and found to produce numerous sulphur-containing compounds with meaty aroma. The major ones were 3,5-dimethyl-1,2,4-trithiolanes, thiophenones and thiazoles. The amount of moisture in a reaction mixture of glycerol and water, the oxygen content and the pH of the reaction were found to be important parameters in the formation of these volatiles.

Shu *et al.* (1985) proposed a mechanism similar to that of van den Ouweland and Peer (1975) for the formation of sulphur derivatives, including 2,5-dimethyl-2,4-dihydroxy-3(2*H*)-thiophenone, which they describe as having a post roast aroma and flavour. Seventy-five percent moisture resulted in a maximum yield of 3,5-dimethyl-1,2,4-trithiolane and thiophenones at pH 4.7. Greater quantities of 3,5-dimethyl-1,2,4-trithiolane was formed anaerobically compared to oxygen environments.

Similar compounds that might possibly have their derivation from reaction of 4-hydroxy-5-methyl-3(2*H*)-furanone with H_2S have recently been identified by MacLeod and Ames (1986) as meat flavour character impact compounds. These compounds are 2-methyl-3-(methylthio)-furan and 2-methyl-3-(methyldithio)-furan, which have odour thresholds in water of 0.05 and 0.01 ppm, respectively, and both have meaty aromas below 1 ppb. As described later, disulphides having similar structures are also considered to be meat flavour impact compounds.

Cyclotene (2-hydroxy-3-methylcyclopent-2-enone) formed from 5-hydroxy-5,6-dihydromaltol (Figure 12.3) and from hydroxyacetone condensation is also a precursor of volatile compounds having meaty aromas. Nishimura *et al.* (1980) described a reaction between ammonia, H_2S and cyclotene as having meaty odours and containing 1,2,4-trithiolane, 5-trithiane and 1,2,4,6-tetrathiepane, as well as 2-methylcyclopentanone and 3-methyl-cyclopentanone, which were described as having a roasted beef odour (Nishimura *et al.*, 1980). Tricyclic pyrazines and dihydro-cyclopentapyrazine are also formed by heating cyclotene and alanine (Rizzi, 1976). Other prominent intermediates from Amadori rearrangements products that can react further to form compounds with meat-like odours are furfural and isomaltol. Furfural, when heated with H_2S and ammonia, also formed sulphur heterocyclics having meaty odour, probably being first degraded to furan aldehyde (Shibamoto, 1977). Isomaltol is also a prominent precursor of heterocyclic compounds related to cooked meat flavour (Tonsbeek *et al.*, 1968).

Mention must be made of the excellent studies by Werkhoff *et al.* (1989; 1990) of the formation of numerous sulphur-containing compounds with meaty odour obtained by heating cystine, thiamine, ascorbic acid and MSG in a model system. Although many of the constituents identified undoubtedly are derived from thiamine degradation (MacLeod, 1986; Werkhoff *et al.*, 1990), Maillard-type reactions could be involved in the formation of some compounds, particularly compounds like 1-(2-methyl-2-thienylthio)-ethanethiol and 1-(2-methyl-3-furylthio)-ethanethiol which can be formed respectively from 2-methyl-3-furathiol and 2-methyl-3-thiophenethiol (Figure 12.8). Some of these compounds originate from furans and thiophenes that could be derived from Maillard reaction, which are then substituted with sulphur in the 2 and 3 positions. Several of these sulphur-substituted compounds have characteristic meat flavour notes and are likely to have importance in meat flavour.

Farmer and Patterson (1991) identified five important disulphides having meaty odour from the headspace of freshly cooked (80°C) beef. Two were the same as those identified by Werkhoff *et al.* (1990). Two of these compounds, *bis*-(2-methyl-3-furyl) disulphide and 2-furfuryl-2-methyl-3-furyl disulphide (Figure 12.9), both have strong meaty/roasted/burnt odours. *Bis*-(2-methyl-3-furyl) disulphide, 2-methyl-3-furanthiol and 2-furfurylthiol have been identified as important beef odour constituents (Gasser and Grosch, 1988) and have a very low threshold of 2 parts in 10^{14} parts of water (Buttery *et al.*, 1984). This is one of the lower thresholds for flavour compounds so far identified. Grosch (1991) reported that the meat flavour impact compound, 2-methyl-3-furanthiol, previously identified in beef and pork could be synthesized by heating meat precursors, and that oxidation products from this compound (*bis*-(2-methyl-3-furyl) disulphide and 2-furfurylthiol) should also be considered meat flavour

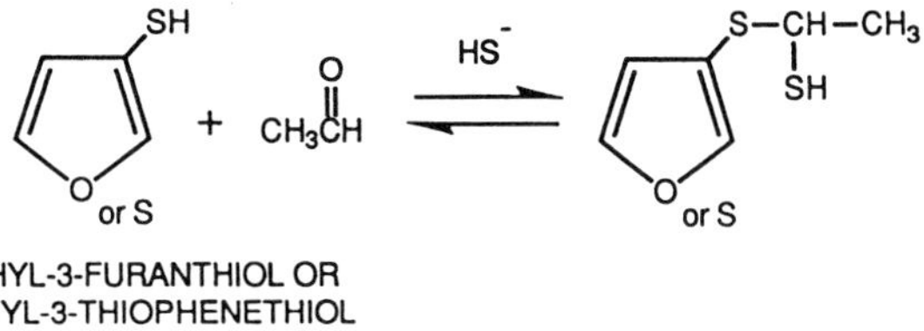

Figure 12.8 Synthesis of 1-[(2-methyl-3-thienyl)thio] ethanethiol and 1-[(2-methyl-2-furyl)thio] ethanethiol from acetaldehyde hydrogen sulphide and methyl furan derivatives.

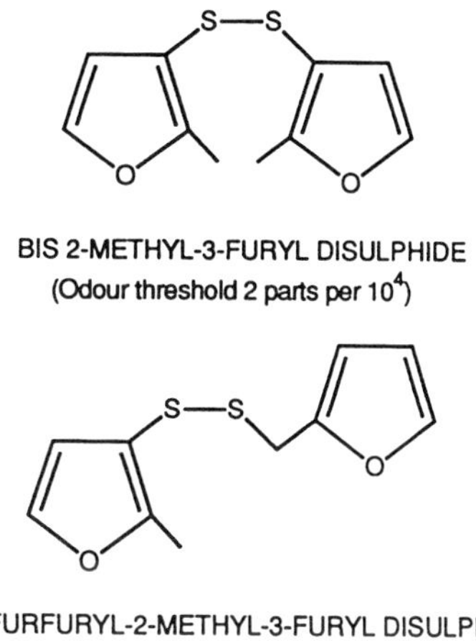

Figure 12.9 Meat flavour impact compounds identified by Werkhoff *et al.* (1989) and found in meat by Farmer and Patterson (1991).

impact compounds. Preliminary results by Farmer and Patterson (1991) also indicate that the threshold of 2-furfuryl-2-methylfuryl disulphide is very low. It is likely that other 3-methyl-3-furylthio compounds similar to those synthesized by Werkhoff *et al.* (1990) will also be found in meat volatiles at low levels. MacLeod (1994) listed several other Maillard type compounds with high sensory significance associated with beef aroma.

12.3.5 Synthetic flavours and antioxidants from Maillard reactions

The patent literature abounds with formulations for synthetic meat flavour. Most contain precursors that participate in Maillard reactions. Buckholz (1989) stated that several hundred patents have appeared in the literature based on nonenzymatic browning technology for meat flavour production, and at least 50 patents which include amino acids and sugars in their reaction mixtures. Most of the patents for the formulation of meaty flavours utilize cysteine, cystine or methionine as the source for sulphur, but more recently thiamine has also been incorporated into reaction flavours (Buckholz, 1989).

Bailey and Um (1991) recently prepared a synthetic mixture of natural products that served both as an antioxidant and as a meat flavour enhancer during storage of cooked beef and pork. The antioxidant was prepared by heating 0.2 M glucose and 0.2 M histidine at 150°C for 2 h in 40% glycerol. This mixture was then used to disperse 0.5% cysteine, 0.5% thiamine, 1.0% glycine, 3.0% autolysed yeast and 2% beef fat, which was then heated at 125°C for 1 h. Excess fat was removed after chilling, but the lipids are important ingredients for reasons discussed by Mottram (1994) and because of the unique flavours supplied by the beef fat. The resulting synthetic meat flavour, when added to beef roast prior to cooking, was found superior to all commercial samples tested for protecting meaty flavour of cooked roast beef during storage. Roast beef prepared with this flavour mixture was accepted at a high level by consumers during storage at 4°C.

12.4 Maillard reaction products as preservatives of fresh meat flavour

An important aspect of food preservation of lipid-containing foods such as meat is the oxidation which results in a characteristic undesirable flavour. A specific flavour developed in cooked fresh meat during refrigerated or frozen storage is warmed-over flavour (WOF), which is undesirable to most consumers. This flavour quality reduces the sale of cooked fresh meat as an easily prepared, convenient dish, and most precooked meat is processed by curing using nitrite.

A number of different antioxidants have been used to prevent WOF in fresh precooked meat, but the most useful ones are synthetic meat flavour compounds formed by the Maillard reaction. The antioxidant properties of Maillard reaction products have recently been discussed by Bailey *et al.* (1997).

The desirable flavours of cooked fresh beef or pork roasts and ground meat from beef and pork have been stabilized during storage at 4°C for a week or more by adding a synthetic flavour mixture formed by heating precursors for the Maillard reaction (Bailey *et al.*, 1991). A Maillard reaction mixture was described to prepare a synthetic meat flavour (SMF) that was used to prolong the refrigerated storage life of cooked beef roast.

The SMF was prepared similar to that described above, by heating a 0.3 M solution of glucose with 0.3 M histidine and 0.2 mM cysteine at 150°C for 2 h in 40% glycerol. This was used as a solvent to disperse 0.5% cysteine, 0.5% thiamine, 1.0% glycine, 3.0% autolysed yeast and 2% melted beef fat which was heated at 125°C, pH 6.6 for 1 h in an oil bath.

This mixture when added at the level of 2% was found to improve the flavour and prevent WOF of cooked fresh beef and pork for up to

28 days during storage at 4°C. Similar flavour mixtures also improved the flavour of refrigerated beef and pork following roasting or broiling as judged by consumer panels. The cooked beef products were found to be better than similar products presently available on the market.

12.5 Summary

Reducing sugars can react with amino acids to produce *N*-glucosylamines or *N*-fructosylamines which rearrange to Amadori or Heynes intermediates that deaminate to form carbonyls such as 1-deoxyreductones, 1-deoxysones and 3-deoxyosones that form many precursors for meaty flavour. Precursors for the 1-deoxyreductones can react with Strecker degradation products such as aldehydes, ammonia and hydrogen sulphide to form pyrazines, nitrogen–sulphur heterocyclics such as thiazoles and thiazolines, cyclic sulphur compounds such as dithianes, trithianes and trithiolanes, and mono, di-, tri- and tetrasulphides. Meat flavour precursors from the 1-deoxysones, such as HO-Fone (4-hydroxy-5-methyl-3(2*H*)-furanone), HM-Fone (4-hydroxy-2,5-dimethyl-3(2*H*)-furanone), isomaltol, maltol or cyclotene, can react with Strecker degradation products to form many meat flavour impact compounds. These include 4-mercapto-5-methyl-tetrahydro-3-furanone (maggi odour); 2,5-dimethyl-2,4-dihyroxy-3-(2*H*)-thiophenone (pot roast); 2-methyl-3-(methylthio)-furan (meaty); *bis*-(2-methyl-3-furyl) disulphide (2 parts/10^4 parts water; meaty); 1-(2-methyl-2-thienylthio)-ethanethiol (meaty); 1-(2-methyl-3-furylthiol-ethanethiol (meaty); and some non-aromatic cyclic sulphur compounds formed by heating cyclotene and H_2S.

Important chemical structures for meaty–roasted flavours formed from Maillard reactions appear to be aromatic nitrogen derivatives such as pyrazines; *N,S*-heterocyclics such as thiazoles, 2-methyl-3-thio (or sulphide)-furans; 2-methyl-5-thio (or sulphide)-thiophenes; sulphur-substituted tetrahydrofuranones; and nonaromatic ring sulphur derivatives containing two or more sulphur moieties.

The discovery of the elixir of meat flavour is just on the horizon, but this mixture may be very unstable and we will continue to be dependent upon Maillard reaction intermediates as essential ingredients for the production and stabilization of desirable meaty flavours. A resume of these activities in the formation of meat flavour impact compounds is given in Figure 12.10.

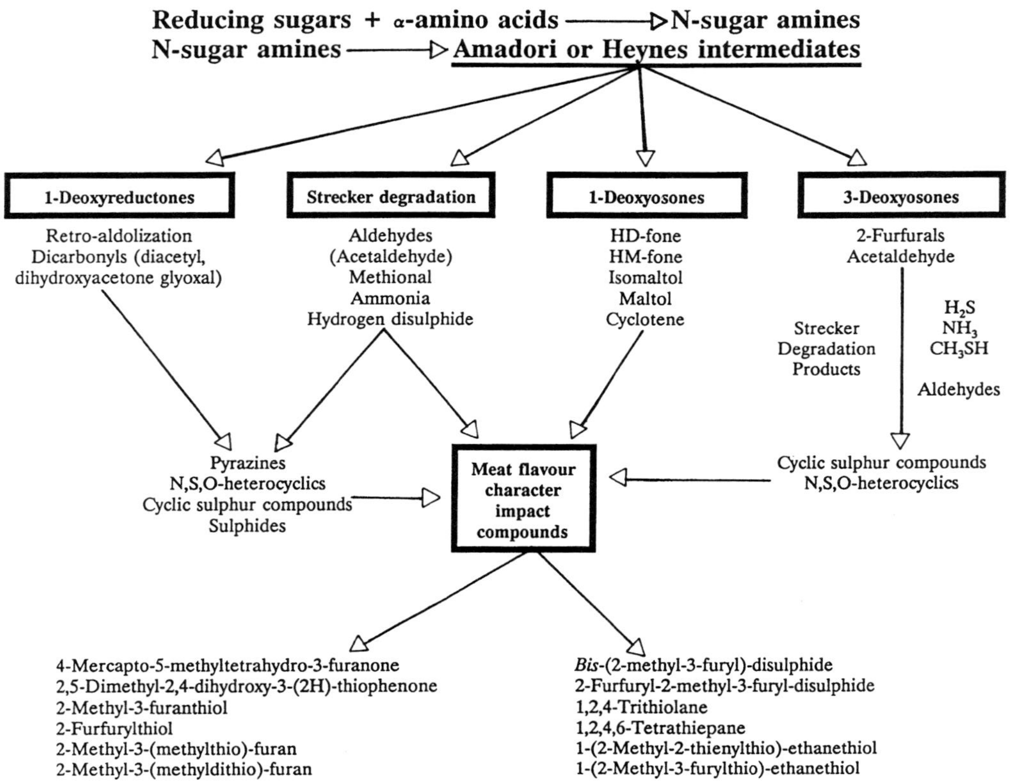

Figure 12.10 Formation of meat flavour impact compounds by Maillard reaction of sugars and amines.

References

Anet, E.F.L.J. and Reynolds, T.M. (1957). Chemistry of non-enzymic browning. I. Reactions between amino acids, organic acids and sugars in freeze-dried apricots and peaches. *Aust. J. Chem.*, **10**, 182–192.

Bailey, M.E. (1983). The Maillard reaction and meat flavor. In *The Maillard Reaction in Foods and Nutrition*, ed. G.R. Waller and M.S. Feather. ACS Symposium Series 215, Washington, DC, pp. 169–184.

Bailey, M.E. and Einig, R.G. (1989). Reaction flavors of meat. In *Thermal Generation of Aromas*, ed. T.H. Parliment, R.J. McGorrin and C.-T. Ho. ACS Symposium Series 409, American Chemical Society, Washington, DC, pp. 421–432.

Bailey, M.E. and Um, K.-W. (1991). Maillard reaction products and lipid oxidation. Paper 119, Agric. Food Chem. Div., *Fourth Chemical Congress of North America*, New York, 25–30 August.

Bailey, M.E., Clarke, A.D., Hedrick, H.B., Shin-Lee, S.Y., and Choate, G.K. (1991). *Proceedings of the 37th International Congress of Meat Science and Technology, II,* Kulmbach, Germany, pp. 684–688.

Bailey, M.E., Clarke, A.D., Kim, Y.S., and Fernado, L. (1997). Antioxidant properties of Maillard reaction products as meat flavor compounds. In *Natural Antioxidants*, ed. F. Shahidi. AOCS Press, Champaign, IL., pp. 296–310.

Baines, D.A. and Mlotkiewicz, J.A. (1984). The chemistry of meat flavor. In *Recent Advances in the Chemistry of Meat*, ed. A.J. Bailey. Royal Society Chemistry, London, pp. 119–164.

Baltes, W., Kunert-Kirchhoff, J. and Reese, G. (1989). Model reactions on generation of thermal aroma compounds. In *Thermal Generation of Aromas*, ed. T.H. Parliment, R.J. McGorrin and C.-T. Ho. ACS Symposium Series 409, American Chemical Society, Washington, DC, pp. 143–155.

Barone, R. and Chanon, M. (1982). Computer application of non-interactive program of simulation organic synthesis in Maillard's reaction: A proposition for new heterocyclic compounds for flavours. In *Chemistry of Heterocyclic Compounds in Flavours and Aromas*, ed. G. Vernin. Ellis Horwood, Chichester, pp. 249–261.

Bender, A.E., Wood, T. and Palgrove, J.A. (1958). Analyses of tissue constituents. Extract of fresh ox-muscle. *J. Sci. Food Agric.*, **9**, 812–817.

Bodrero, K.O., Pearson, A.M. and Magee, W.T. (1981). Evaluation of the contribution of flavor volatiles to the aroma of beef by surface response methodology. *J. Food Sci.*, **46**, 26–31.

Buckholz, L.L. Jr. (1989). Maillard technology as applied to meat and savory flavors. In *Thermal Generation of Aromas*, ed. T.H. Parliment, R.J. McGorrin and C.-T. Ho. ACS Symposium Series 409, American Chemical Society, Washington, DC, pp. 406–420.

Buttery, R.G., Haddon, W.F., Seifert, R.M. and Turnbaugh, J.G. (1984). Thiamine odor and *bis*-(2-methyl-3-furyl) disulfide. *J. Agric. Food Chem.*, **32**, 674–676.

Ching, J.C.Y. (1979). Volatile flavor compounds from beef and beef constituents. PhD thesis, University of Missouri, Columbia, MO.

Cramer, D. (1983). Chemical compounds implicated in lamb flavor. Overview. *Food Technol.*, **37**, 249–257.

Danehy, J.P. (1986). Maillard reactions: Nonenzymatic browning in food systems with special reference to the development of flavor. *Adv. Food Res.*, **30**, 77–138.

Dimoglo, A.S., Gorbachov, M.Y., Bersuker, I.B., Greni, A.I., Vysotskaya, L.E., Stepanova, O.V. and Lukash, E.Y. (1988). Structural and electronic origin of meat odour of organic heteroatomic compounds. *Nahrung,* **32**, 461–473.

Einig, R. (1983). Interaction of meat aroma volatiles with soy proteins. PhD thesis, University of Missouri, Columbia, MO.

Eriksson, C. (1981). *Maillard Reactions in Food.* Pergamon Press, Oxford.

Eskin, N.A.M. (1990). *Biochemistry of Foods.* Academic Press, San Diego.

Farmer, L.J. and Patterson, R.L.S. (1991). Compounds contributing to meat flavour. *Food Chem.*, **40**, 201–205.

Feather, M.S. (1981). Amine-assisted sugar dehydration reactions. *Prog. Food Nutr. Sci.*, **5**, 37–45.

Feather, M.S. (1985). Some aspects of the chemistry of nonenzymatic browning (the Maillard reaction). In *Chemical Changes in Food During Processing*, eds. T. Richardson and J.W. Finley. AVI Publishing Co., Westport, CT, pp. 289–303.

Feather, M.S. and Harris, J.F. (1973). Dehydration reactions of carbohydrates. *Adv. Carbohydr. Chem.*, **28**, 161–224.

Feeney, R.E. and Whitaker, J.R. (1982). The Maillard reaction and its prevention. In *Food Protein Deterioration. Mechanisms and Functionality*, ed. J.P. Cherry. ACS Symposium Series 206, American Chemical Society, Washington, DC, pp. 201–229.

Finot, P.A., Aeschbacher, H.U., Hurrell, R.F. and Liardon, R. (1990). *The Maillard Reaction in Food Processing, Human Nutrition and Physiology*. Birkhäuser Verlag, Basel.

Flament, I., Kohler, M. and Aschiero, R. (1976). Sur l'arome de viande de boeuff grillée. II. Dehydro-6,7–5H-cyclopenta [b] pyrazines, identification et mode de formation. *Helv. Chim. Acta*, **59**, 2308–2313.

Fujimaki, M., Namiki, M. and Kato, N. (1986). *Amino Carbonyl Reactions in Food and Biological Systems*. Elsevier, Amsterdam.

Gasser, U. and Grosch, W. (1988). Identification of volatile flavour compounds with high aroma values from cooked beef. *Z. Lebensm. Unters. Forsch.*, **186**, 489–494.

Grosch, W. (1991). Studies on the formation of meat-like flavour compounds. Paper 70, Agric. Food Chem. Div., *Fourth Chemical Congress of North America*, New York, 25–30 August.

Heath, H.B. and Reineccius, G.A. (1986). *Flavor Chemistry and Technology*. AVI Publishing Co., Westport, CT.

Hicks, K.B. and Feather, M.S. (1975). Studies on the mechanism of formation of 4-hydroxy-5-methyl-3(2*H*)-furanone, a component of beef flavor, from Amadori products. *J. Agric. Food Chem.*, **23**, 957–960.

Hicks, K.B., Harris, D.W., Feather, M.S. and Loeppky, R.N. (1974). Production of 4-hydroxy-5-methyl-3(2*H*)-furanone, a component of beef flavour, from 1-aino-1-deoxy-D-fructuronic acid. *J. Agric. Food Chem.*, **22**, 724–725.

Ho, C.-T. (1980). Meat flavor: Past, present and future. In *Proceedings of the Meat Industry Research Conference*. American Meat Institute Foundation, Arlington, VA, pp. 41–49.

Hodge, J.E. (1953). Chemistry of browning reactions in model systems. *J. Agric. Food Chem.*, **1**, 928–943.

Hodge, J.E. (1967). Origin of flavors in food. Nonenzymatic browning reactions. In *Symposium on Foods: Chemistry and Physiology of Flavors*, ed. H.W. Schultz, E.A. Day and L.M. Libbey. AVI Publishing Co., Westport, CT, pp. 465–491.

Hodge, J.E. and Osman, E.M. (1976). Carbohydrates. In *Principles of Food Science. Part I. Food Chemistry*, ed. O.R. Fennema. Marcel Dekker, New York, pp. 41–138.

Hodge, J.E., Mills, F.D. and Fisher, B.E. (1972). Compounds of browned flavor derived from sugar–amine reactions. *Cereal Sci. Today*, **17**, 34–40.

Hornstein, I. and Crowe, P.F. (1960). Flavor studies on beef and pork. *J. Agric. Food Chem.*, **8**, 494–498.

Hornstein, I. and Wasserman, A. (1987). Chemistry of meat flavour. In *The Science of Meat and Meat Products*, 3rd edn, eds. J.F. Price and B.S. Schwigert. Food and Nutrition Press, Westport, CT, pp. 329–347.

Hurrell, R.F. (1982). Maillard reaction in flavour. In *Food Flavours. Part A. Introduction*, eds. I.D. Morton and A.J. MacLeod. Elsevier, Amsterdam, pp. 399–437.

Hurrell, R.F. and Carpenter, K.J. (1977). Maillard reactions in foods. In *Physical, Chemical and Biological Changes in Food Caused by Thermal Processing*, eds. T. Hoyem and O. Kvàle. Applied Science, London, pp. 168–184.

IFT Symposium (1983). 42nd Annual Meeting. *Food Technol.*, **37**, 225–268.

Ikan, R. (1996). The Maillard Reaction, *Consequences for the Chemical and Life Sciences*. Wiley, New York.

Katz, I. (1981). Recent progress in some aspects of meat flavor chemistry. In *Flavour Research: Recent Advances*, eds. R. Teranishi, R.A. Flath and H. Sugisawa. Marcel Dekker, New York, pp. 217–229.

Labuza, T.P., Reineccius, G.A., Monnier, V., O'Brien, J. and Baynes, J.W. (1994). *Maillard Reactions in Chemistry, Food and Health*. The Royal Society of Chemistry, London.

Lawrie, R.A. (1982). The flavour of meat and meat analogues. *Food Flavor Int.*, **4**(2), 10–13, 15.

Ledl, F. (1990). Chemical pathways of the Mallard reaction. In *The Maillard Reaction in Food Processing, Human Nutrition and Physiology*, eds. P.A. Finot, H.U. Aeschbacher, R.R. Hurrell and R. Liardon. Birkhauser Verlag, Basel, pp. 19–42.
Ledl, F., Beck, J., Sengl, M., Osiander, H., Estendorfer, S., Severin, T. and Huber, B. (1989). Chemical pathways of the Maillard reaction. In *The Maillard Reaction in Aging, Diabetes and Nutrition*, eds. J.W. Baynes and V.M. Monnier. Alan R. Liss, New York, pp. 23–42.
Lee, C.Y. and Whitaker, J.R. (1995). *Enzymatic Browning and its Prevention.* ACS Symposium Series 600. American Chemical Society, Washington, DC.
Mabrouk, A.F. (1979). Flavor of browning reaction products. In *Food Taste Chemistry*, ed. J.C. Boudreau. ACS Symposium Series 115, American Chemical Society, Washington, DC, pp. 205–245.
MacLeod, G. (1984). The flavour of meat. *Proc. Inst. Food Sci. Technol.*, **17**(4), pp. 184–197.
MacLeod, G. (1986). The scientific and technological basis of meat flavours. In *Food Flavors*, eds. G.G. Birch and M.G. Lindley. Elsevier, London, pp. 191–223.
MacLeod, G. (1994). In *Flavor of Meat and Meat Products*, ed. F. Shahidi. Blackie Academic and Professional, London, pp. 4–37.
MacLeod, G. and Ames, J.M. (1986). The effect of heat on beef aroma: Comparisons of chemical composition and sensory properties. *Flavor Fragrance*, **1**, 91–104.
MacLeod, G. and Seyyedain-Ardebili, M. (1981). Natural and simulated meat flavors (with particular reference to beef). *CRC Crit. Rev. Food Sci. Nutr.*, **14**, 309–437.
Macy, R.L. Jr., Naumann, H.D. and Bailey, M.E. (1964a). Water-soluble flavor and odor precursors of meat. I. Qualitative study of certain amino acids, carbohydrates, non-amino compounds and phosphoric acid esters of beef, pork and lamb. *J. Food Sci.*, **29**, 131–141.
Macy, R.L. Jr., Naumann, H.D. and Bailey, M.E. (1964b). Water-soluble flavor and odor precursors of meat. II. Effects of heating on amino acids, nitrogen constituents and carbohydrates in lyophilized diffusates from aqueous extracts of beef, pork and lamb. *J. Food Sci.*, **29**, 142–148.
Maillard, L.-C. (1916). Synthesis des matières humiques par action des acides amines sur les sucres reducteurs. *Ann. Chim. (Paris)*, **5**(9), 258–317.
Mauron, J. (1981). The Maillard reaction in food: A critical review from the nutritional standpoint. In *Maillard Reactions in Food*, ed. C. Eriksson. Pergamon Press, Oxford, pp. 5–35.
Moody, W.G. (1983). Beef flavor – a review. *Food Technol.*, **37**(5), 227–232.
Mottram, D.S. (1991). Meat. In *Volatile Compounds in Food and Beverages*, ed. H. Maarse. Marcel Dekker, New York, pp. 107–177.
Mottram, D.S. (1994). In *Flavor of Meat and Meat Products*, ed. F. Shahidi. Blackie Academic and Professional, London, pp. 210–230.
Mottram, D.S., Croft, S.E. and Patterson, R.L.S. (1984). Volatile components of cured and uncured pork: The role of nitrite and the formation of nitrogen compounds. *J. Sci. Food Agric.*, **35**, 233–239.
Namiki, M. (1988). Chemistry of Maillard reactions: Recent studies on the browning reaction mechanism and the development of antioxidants and mugaens, *Adv. Food Res.*, **32**, 115–184.
Namiki, M. and Hayashi, T. (1983). In *The Maillard Reaction in Foods and Nutrition*, ed. G.R. Waller and M.S. Feather. ACS Symposium Series 215, American Chemical Society, Washington, DC, pp. 21–46.
Nishimura, G., Mihara, S. and Shibamoto, T. (1980). Compounds produced by the reaction of 2-hydroxy-2-cyclopentene-1-one with ammonia and hydrogen sulfide. *J. Agric. Food Chem.*, **28**, 39–43.
Nursten, H.E. (1981). Recent developments in studies of the Maillard reaction. *Food Chem.*, **6**, 263–277.
Nursten, H.E. (1986). Maillard browning reactions in dried foods. In *Concentration and Drying of Foods*, ed. D. MacCarthy. Elsevier, London, pp. 53–68.
O'Brien, J. and Morrissey, P.A. (1989). Nutritional and toxicological aspects of the Maillard reaction in foods. *CRC Crit. Rev. Food Sci. Nutr.*, **28**, 210–248.
Ohloff, G. and Filament, I. (1978). Heterocyclic constituents of meat aroma. *Heterocycles*, **11**, 663–695.
Parliment, T.H., McGorrin, R.J. and Ho, C.-T. (1989). *Thermal Generation of Aromas.* ACS Symposium Series 409, American Chemical Society, Washington, DC.

Paulsen, H. and Pflughaupt, K.W. (1980). Glycosylamines. In *The Carbohydrates, Chemistry and Biochemistry,* 2nd edn, Vol. 1B, eds. W. Pigman, D. Horton and J.D. Wander. Academic Press, New York, pp. 881–927.

Peer, H.A. and van den Ouweland, G.A.M. (1968). Synthesis of 4-hydroxy-5-methyl-3(2*H*)-furanone from D-ribose 5-phosphate. *Rec. Trav. Chim.*, **87**, 1017–1020.

Reynolds, T.M. (1963). Chemistry of nonenzymic browning. I. The reaction between aldoses and amines. *Adv. Food Res.*, **12**, 1–52.

Reynolds, T.M. (1965). Chemistry of nonenzymic browning II. *Adv. Food Res.*, **14**, 167–283.

Rhee, K.S. (1989). Chemistry of meat flavor. In *Flavor Chemistry of Lipid Foods*, eds. D.B. Min and T.H. Smouse. American Oil Chemists' Society, Champaign, IL, pp. 166–189.

Rizzi, G.P. (1976). Non-enzymic transmination of unsaturated carbonyls: A general source of nitrogenous flavor compounds in foods. In *Phenolic, Sulfur and Nitrogen Compounds in Food Flavors*, ed. G. Charalambous and I. Katz. ACS Symposium Series 26, American Chemical Society, Washington DC, pp. 122–132.

Schieberle, P. (1991). Studies on the formation of furanol. Paper 52, Agric. and Food Chem. Div., *Fourth Chemical Congress of North America*, New York, 25–30 August.

Severin, T.H. and Seilmeier, W. (1968). Studien zur Maillard reaktion. III. Unwandlung von geniose under einfluss von methylammoniumacetat. *Z. Lebensm. Unters. Forsch.*, **137**, 4–6.

Shahidi, F. (1989). Flavor of cooked meats. In *Flavor Chemistry: Trends and Developments*, ed. R. Teranishi, R.G. Buttery and F. Shahidi. ACS Symposium Series 388, American Chemical Society, Washington, DC, pp. 188–201.

Shahidi, R., Rubin, L.J. and D'Souza, L.A. (1986). A review of the composition, techniques, analyses and sensory evaluation. *CRC Crit. Rev. Food Sci. Nutr.*, **24**(2), 141–243.

Shibamoto, T. (1977). Formation of sulfur- and nitrogen-containing compounds from the reaction of furfural with hydrogen sulfide and ammonia. *J. Agric. Food Chem.*, **25**, 206–208.

Shibamoto, T. (1980). Heterocyclic compounds found in cooked meats. *J. Agric. Food Chem.*, **28**, 237–243.

Shin-Lee, S.Y. (1988). Warmed-over flavor and its prevention by Maillard reaction products. PhD thesis, University of Missouri, Columbia, MO.

Shu, C.-K. and Ho, C.-T. (1989). Parameter effects on the thermal reaction of cystine and 2,5-dimethyl-4-hydroxy-3(2*H*)furanone. In *Thermal Generation of Aromas*, eds. T.H. Parliment, R.J. McGorrin and C.-T. Ho. ACS Symposium Series 409, American Chemical Society, Washington, DC, pp. 229–241.

Shu, C.-K., Hagedorn, M.L., Mookherjee, B.D. and Ho, C.-T. (1985). Two novel 2-hydroxy-3(2*H*)-thiophenones from the reaction between cysteine and 2,5-dimethyl-4-hydroxy-3(2*H*)-furanone. *J. Agric. Food Chem.*, **33**, 638–641.

Teranishi, R., Flath, R.A. and Sugisawa, H. (1971). *Flavour Research: Recent Advances.* Marcel Dekker, New York.

Tonsbeek, C.H.T., Plancken, A.J. and v.d. Weerdhof, T. (1968). Components contributing to beef flavour. Isolation of 4-hydroxy-5-methyl-3(2*H*)-furanone and its 2,5-dimethyl homolog from beef broth. *J. Agric. Food Chem.*, **16**, 1016–1021.

van den Ouweland, G.A.M. and Peer, H.G. (1975). Components contributing to beef flavor. Volatile compounds produced by the reaction of 4-hydroxy-5-methyl-3(2*H*)-furanone and its thio analog with hydrogen sulfide. *J. Agric. Food Chem.*, **23**, 501–505.

van den Ouweland, G.A.M., Demole, E.D. and Enggist, P. (1989). Processed meat flavor development and the Maillard reaction. In *Thermal Generation of Aromas*, eds. T.H. Parliment, R.J. McGorrin and C.T. Ho. ACS Symposium Series 409, American Chemical Society, Washington, DC, pp. 433–441.

Vernin, G. and Parkanyi, C. (1982). Mechanisms of formation of heterocyclic compounds in Maillard and pyrolysis reactions. In *Chemistry of Heterocyclic Compounds in Flavors and Aromas*, ed. G. Vernin. Halsted Press, New York, pp. 151–207.

Waller, G.R. and Feather, M.S. (1983). *The Maillard Reaction in Food and Nutrition.* ACS Symposium Series 215, American Chemical Society, Washington, DC.

Wasserman, A.E. (1979). Chemical basis for meat flavor: A review. *J. Food Sci.*, **44**, 6–11.

Wasserman, A.E. and Gray, N. (1965). Meat flavor. I. Fractionation of water-soluble flavor precursors of beef. *J. Food Sci.*, **30**, 801–807.

Wasserman, A.E. and Spinelli, A.M. (1970). Sugar–amino acid interaction in the diffusate of water extract of beef and model systems. *J. Food Sci.*, **35**, 328–332.

Wasserman, A.E. and Spinelli, A.M. (1972). Effect of some water-soluble compounds on aroma of heated adipose tissue. *J. Agric. Food Chem.*, **20**, 171–174.

Weenan, H. (1991). Analysis and reactivity of glucosones. Paper 51, Agric. and Food Chem. Div., *Proceedings of the Fourth Chemical Congress of North America*, New York, 25–30 August.

Werkhoff, P., Emberger, R., Güntert, M. and Köpsel, M. (1989). Isolation and characterization of volatile sulfur-containing meat flavor components in model systems. In *Thermal Generation of Aromas*, eds. T.H. Parliment, R.J. McGorrin and C.T. Ho. ACS Symposium Series 409, American Chemical Society, Washington, DC, pp. 460–478.

Werkhoff, P., Brüning, J., Emberger, R., Güntert, M., Köpsel, M., Kuhn, W. and Surburg, H. (1990). Isolation and characterization of volatile sulfur-containing meat flavor components in model systems. *J. Agric. Food Chem.*, **38**, 777–791.

Williams, J. C. (1976). Chemical and non-enzymatic changes in intermediate moisture foods. In *Intermediate Moisture Foods*, eds. R. Davies, G.G. Birch and K.J. Parker. Applied Science, London, pp. 100–119.

Wood, T. (1961). The browning of ox-muscle extracts. *J. Sci. Food Agric.*, **12**, 61–69.

Wood T. and Bender, A.E. (1957). Analysis of tissue constituents. Commercial ox-muscle extract. *Biochem. J.*, **67**, 366–373.

Yaylayan, V. and Sporns, P. (1987). Novel mechanisms for the decomposition of 1-(amino acid)-1-deoxy-D-fructoses (Amadori compounds): A mass spectrometric approach. *Food Chem.*, **26**, 283–305.

Zang, Y., Brown, S., Ledig, W.O., Mussinan, C. and Ho, C.-T. (1997). Formation of sulfur-containing flavour compounds from reactions of furanese, cysteine, glutathione, hydrogen sulfide and alanine/hydrogen sulfide. *J. Agric. Food Chem.*, **45**, 894–897.

13 Flavour of cured meat

N. RAMARATHNAM*

13.1 Introduction

The origin of the use of nitrate/nitrite in the curing of meat is lost in history, but it is strongly believed that the preservation of meat with salt preceded the intentional use of nitrate by many centuries. It appears that meat preservation was first practised in the saline deserts of Hither Asia and on coastal areas (Binkerd and Kolari, 1975). Salt was in common use in ancient Palestine as early as 1600 BC because of its availability from the salt-rich Dead Sea. The technology of sea-salt production was also known by at least 1200 BC by the Chinese, who early made salt from drilled wells (Jensen, 1953, 1954). Salt from the sea, desert, and as found as an efflorescence on the walls of caves and stables, was used by ancient peoples in the curing of meat. These salts contained nitrates as impurities, and though saltpetre or 'nitre' was gathered in ancient China and India long before the Christian era, the reddening effect of nitrates on meat was not mentioned until late Roman times.

The curing of meat as practised today is based upon the ancient art known through aeons of time, and perhaps to a far greater extent upon the scientific principles developed since about the turn of the twentieth century. In the USA, permission to conduct the first and subsequent series of experiments with the direct use of nitrite in meat curing, under federal inspection, was given in 1923 by the Bureau of Animal Industry of the United States Department of Agriculture (USDA) (Kerr *et al.*, 1926). On the basis of the results obtained in those experiments, the use of sodium nitrite in meat-curing was formally regulated by the USDA in 1925, and this authorization by the Bureau of Animal Industry restricted the amount of sodium nitrate used in the cured-meat product to 200 parts per million (mg/kg). Present day meat-curing practice involves the addition of sodium nitrite and salt along with other additives, such as sugar, certain reducing agents, phosphates and, where appropriate, seasonings, to impart characteristic properties to the end product. It is estimated that over 70% of pork is cured with nitrite. In the curing pickle, the function of salt, which

*This chapter is dedicated to the memory of Dr L.J. Rubin from the Department of Chemical Engineering and Applied Chemistry of the University of Toronto, Canada, where the work presented here was carried out.

constitutes the bulk of the mixture, is to contribute to taste, act as a preservative, and enhance the functional properties of meat protein; sugar enhances the flavour of the cured product; the reducing agents, such as ascorbates and erythorbates, accelerate the development of the characteristic pink colour of cured meat; phosphates help in retaining the moisture of the cured product, and therefore play an important role in contributing to the mouthfeel and juiciness of the processed product; and seasonings and spices impart additional aroma and taste.

13.2 Nitrite curing of meat

Nitrite is a unique and multifunctional ingredient in the meat-curing system. It imparts the characteristic pink colour to the cured meat (Eakes *et al.*, 1975, 1977) and provides oxidative stability to meat by preventing lipid oxidation (Pearson *et al.*, 1977; Fooladi *et al.*, 1979; MacDonald *et al.*, 1980; Shahidi *et al.*, 1987a; Yun *et al.*, 1987). This effect is complex, but it is believed to be associated with bringing forth the cured-meat flavour and the prevention of the warmed-over flavour (WOF) in meat (Mottram and Rhodes, 1974; Skjelvale and Tjaberg, 1974; Rubin and Shahidi, 1988). Nitrite has an antimicrobial effect which is important in preventing the outgrowth of *Clostridium botulinum* and the formation of deadly toxin (Hauschild *et al.*, 1982; Pierson and Smoot, 1982; Wood *et al.*, 1986), particularly under conditions of product mishandling.

Although a variety of factors may influence the flavour of meats, no single group of factors can be assigned the principal role. In their review on meat-flavour volatiles, Shahidi *et al.* (1986) described the qualitative differences in the nature of carbonyl compounds among the different species. The distribution of carbonyls varies with the lipid composition of the original meat – pork, beef, lamb or poultry. Since the lipids which constitute the fat of different animals are composed of different fatty acids, species differences probably arise by the formation of carbonyl compounds that differ in their qualitative and quantitative composition (Gray *et al.*, 1981).

13.3 Antioxidant role of nitrite in meat curing

The overall acceptance of meat products depends, to a large extent, on their flavour quality. When meat is cooked under very mild conditions, such as by heating in water, it develops a characteristic desirable flavour attributable to the animal species. The final flavour spectrum of each species depends mainly on the nature and amount of the nonvolatile flavour precursors, such as the amino acids, amines, sugars and fatty acids, present in the raw meat.

Lipids in meat make an important contribution toward the overall flavour and mouthfeel of the cooked meat. They serve as a reservoir of fat-soluble compounds that volatilize upon heating to form aroma compounds, and can themselves undergo degradation and autoxidation to produce a wide range of carbonyl compounds. This class of compounds has been implicated as a significant contributor to the flavour of uncured meat, but not in cured meat.

In their study of cured and uncured ham, Cross and Ziegler (1965) reported that the volatiles of cooked cured and uncured ham were qualitatively similar but quantitatively very different. Striking differences were observed, especially in pentanal and hexanal concentrations. They were present in appreciable quantities in the uncured product but were barely detectable in the volatiles of the cured meat.

Shahidi *et al.* (1987b) demonstrated that the flavour acceptability of cooked pork decreased as the TBA number of hexanal content increased. A linear relationship also existed between TBA and hexanal contents (Shahidi *et al.*, 1987; Wettasinghe and Shahidi, 1997). On storage, uncured cooked meats develop an unpleasant warmed-over flavour (WOF) which is not observed in cured meats due to the potent antioxidant effect of nitrite (Pearson *et al.*, 1977; Fooladi *et al.*, 1979; MacDonald *et al.*, 1980).

When meat is nitrite-cured, the flavour of the resultant product, though desirable, is not the same as the flavour of its uncured counterpart. Thus, volatiles derived from cured and uncured ham, beef or chicken, after passage through a solution of 2,4-dinitrophenylhydrazine, possessed a characteristic cured-ham aroma in the effluent stream in all of these systems (Cross and Ziegler, 1965; Minor *et al.*, 1965). This observation by Cross and Ziegler indicated that 'cured-meat' flavour, irrespective of the meat source, was comprised of essentially the same constituents as were generated by a combination of several reactions in addition to the suppression of lipid oxidation. In these studies, beef, pork and chicken meat were included.

Although the nature of cured-meat flavour seems to be much simpler than that of uncured meat, and is postulated to be the basic flavour of meat from different species (Rubin and Shahidi, 1988), with the exception of sheep meat as discussed in Chapter 6, the elucidation of the compounds which are responsible for the cured-meat flavour is not easy. Minute amounts of compounds can be aroma-effective, creating enormous analytical difficulties for their isolation and identification (MacLeod and Ames, 1986).

While the flavour of freshly cooked meat is generally described as 'meaty', this flavour deteriorates upon storage for an extended period. Storage of cooked meat, or prolonged storage of raw meat prior to cooking, gives rise to 'old, stale, oxidized, rancid, or warmed-over' flavour, caused by the oxidation of its unsaturated fatty acids. Nitrite has been shown to retard lipid oxidation and inhibit the development of warmed-over flavour in cooked meat and meat products. Younathan and Watts (1959) studied lipid oxidation in cured and uncured pork and found the highest TBA (2-thiobarbituric acid) values in uncured samples regardless of storage period. Sato and Hegarty (1971) were able to eliminate WOF in cooked ground beef using 50 mg/kg of nitrite as indicated by TBA values. The antioxidant effect of nitrite in the meat-curing process, using TBA values and sensory scores, was also demonstrated by other workers (Hadden *et al.*, 1975; Love and Pearson, 1976; MacDonald *et al.*, 1980). Yun *et al.* (1987) demonstrated that the flavour acceptability of cooked pork decreased as the TBA number or hexanal content of samples increased. While the TBA values of the control (uncured cooked meat) stored at 4°C increased from 4.2 (0 week) to 10.8 (4 weeks), the TBA values for meat cured with 150 mg/kg nitrite remained constant at 0.1 throughout the 4 weeks of storage. Sensory evaluation of the nitrite-cured and uncured meat samples after storage for 24 h at 4°C indicated differences in their acceptability scores.

Initial studies on the oxidation of unsaturated lipids in meat products implicated the haem proteins to be the major prooxidants (Tappel, 1952; Watts, 1954; Maier and Tappel, 1959; Younathan and Watts, 1959). In recent years, however, data have been presented to demonstrate that the nonhaem iron released during cooking, from the bound haem pigments, accelerated lipid oxidation in cooked meat (Sato and Hegarty, 1971; Love and Pearson, 1976). The exact mechanism of action of nitrite as an antioxidant in the elimination of WOF in cooked cured meats is not yet thoroughly understood. It has, however, been suggested by Pearson *et al.* (1977) that nitrite may either inhibit the action of natural pro-oxidants in the muscle or stabilize the lipid components of the membranes.

14.4 Chemistry of cured-meat flavour

Although nitrite is closely associated with cured-meat flavour, the chemistry behind the formation and composition of this unique flavour is not clearly understood (Gray and Pearson, 1984). There have been only a few reports focusing attention mainly on the composition of cured-meat flavour (Bailey and Swain, 1973; MacDougall *et al.*, 1975). Some of the recent reviews on meat flavour chemistry have given special emphasis to the discussion of the nature of cured-meat flavour (Gray *et al.*, 1981; Rhee, 1989; Shahidi, 1989).

Nearly 1000 compounds have been so far identified in the volatiles of pork, beef, chicken and lamb. The general chemical composition of volatiles in uncured and cured pork, as surveyed by Shahidi *et al.* (1986), is listed in Table 13.1. While the total number of carbonyl compounds

Table 13.1 Composition of volatile components found in uncured and cured pork

Class	Uncured pork	Cured pork
Aldehydes	35	29
Alcohols	24	9
Carboxylic acids	5	20
Esters	20	9
Ethers	6	–
Furans	29	5
Hydrocarbons	45	4
Ketones	38	12
Lactones	2	–
Oxazoles/oxazolines	4	–
Phenols	9	1
Pyrazines	36	1
Pyridines	5	1
Pyrroles	9	–
Thiazoles/thiazolines	5	–
Thiophenes	11	–
Other nitrogen compounds	6	3
Other sulphur compounds	20	31
Halogenated compounds	4	1
Miscellaneous	1	11
Total	314	137

and hydrocarbons identified in uncured pork (118) is higher compared with that in cured pork (45), the number of sulphur compounds identified in each of them is the same, i.e. 31. Hornstein and Crowe (1960) were among the first to report that the fat, and more specifically the carbonyl compounds derived from fat oxidation, contributed to differences in flavour among species. In their study, initiated to develop satisfactory methods for the isolation, separation and identification of volatile compounds in dry-cured country-style hams, Ockerman *et al.* (1964) tentatively identified numerous aldehydes, ketones, acids, bases and sulphur compounds. Cross and Ziegler (1965) concluded from gas chromatographic examination that the composition of volatiles of cured and uncured ham was qualitatively quite similar, but there were distinct quantitative differences. Pentanal and hexanal were present in appreciable quantities in the uncured product, but were barely detectable in the volatiles of the cured meat. It was suggested that the absence of these aldehydes in the cured meat was responsible for the flavour differences between cured and uncured ham, and that it was brought about by the inhibition of the autoxidation of unsaturated lipids in the presence of nitrite. It was also demonstrated by Cross and Ziegler (1965) that when volatiles from uncured and cured ham, beef and chicken were passed through a solution of 2,4-dinitrophenylhydrazine, the effluents from all samples had aromas similar to cured ham, as mentioned previously. This observation led them to postulate that carbonyl components were not essential for cured meat flavour, and that the cured-ham flavour represented the basic meat flavour derived from precursors other than the triglycerides. They further suggested that the differences in the cooked-meat flavour depended on the spectrum of carbonyls generated during oxidation of lipids, which in themselves differed in composition depending on species. This preliminary observation on volatiles from uncured and cured meat emphasizing the importance of carbonyls in causing species-specific differences did not receive the attention it deserved.

Attempts to identify the flavour compounds in country-cured hams, isolated by steam-distillation under reduced pressure, showed the presence of various carbonyls, alcohols and esters (Lillard and Ayres, 1969). Piotrowski *et al.* (1970) studied various fractions of ham to isolate and identify the cured-ham aroma and to follow changes occurring in pork during curing, cooking and smoking. It was found that components of precursors of cured and smoky aroma could be extracted from hams with a mixture of chloroform–methanol (2:1 v/v). Further examination of this extract did not show the presence of any single compound that had 'meaty' or 'cured-ham' aroma. A list of individual constituents identified so far in the volatiles of cured ham is given in Table 13.2. In contrast with the large number of papers published on the chemistry of uncured-meat flavour, the information available on the chemistry of cured-meat flavour has been contributed by a small group of researchers listed in Table 13.2. Most of the compounds given in this table have, however, been identified and reported previously in other cooked meats.

A large number of components in the volatiles of meat have been isolated and identified in the past two decades, and exhaustive review papers on meat flavour have been published (Herz and Chang, 1970; Bailey and Swain, 1973; Dwivedi, 1975; Chang and Peterson, 1977; Wasserman, 1979; Gray *et al.*, 1981; MacLeod and Seyyedain-Ardebili, 1981; Ramaswamy and Richards, 1982; Moody, 1983; Baines and Mlotkiewicz, 1984; Shahidi *et al.*, 1986; Rhee, 1989). Although over 700 components in beef and half that number in pork, chicken and lamb have been characterized, the search for individual character-impact components possessing notes specific for beef, pork, chicken or lamb has remained largely unsuccessful. Numerous reports have indicated that carbonyl components make a significant contribution to the flavour of uncured meat (Hornstein and Crowe, 1960; Jacobsen and Koehler, 1963; Sanderson *et al.*, 1966; Langer *et al.*, 1970). However, no attempt has been made to prevent the formation of such carbonyl compounds by using antioxidants such as nitrite or other specific reagents, and to study the organoleptic properties of the resulting aroma mixture in depth. Thus, the nature of cured-meat flavour, which is made apparent by suppression of lipid oxidation by nitrite and which is the basic flavour of cooked meat (Rubin and Shahidi, 1988), remains a mystery so far.

Table 13.2 Volatile constituents identified in nitrite-treated ham

Aldehydes
Methanal (formaldehyde)[b,c]
Ethanal (acetaldehyde)[a,b,c,d]
Propanal[a,b,c]
β-Methylthiopropanal (methional)[d]
2-Methylpropanal[b,c]
Propenal (acrolein)[d]
Butanol[a,b,c]
2-Methylbutanal[a]
3-Methylbutanal[a,b,d]
Pentanal[a,b,c,d]
Hexanal[a,b,c]
2-Hexenal[c,d]
Heptanal[c,d]
2-Heptenal[c]
2,4-Heptadienal[c]
Octanal[c,d]
2-Octenal[c]
Nonanal[c,d]
2-Nonenal[c,d]
2,4-Nonadienal[c,d]
Decanal[c,d]
2-Decenal[c]
2,4-Decadienal[c,d]
Undecanal[c,d]
2-Undecenal[c]
2,4-Undecadienal[c]
Dodecanal[c,d]
2-Dodecenal[c,d]
2,4-Dodecadienal[c]

Carboxylic acids
Methanoic (formic)[b]
Ethanoic (acetic)[b,d]
Propanoic[b]
Butanoic[b,d]
4-Methylpentanoic[b]
Hexanoic[d]
Octanoic[c,d]
Decanoic[c]
Dodecanoic[c]
Tetradecanoic[c]
Hexadecanoic[c]
Hexadecenoic[c]
Octadecanoic[c]
Octadecenoic[c]
Octadecadienoic[c]
Octadecatrienoic[c]
Eicosanoic[c]
Eicosenoic[c]
Eicosadienoic[c]
Eicosatrienoic[c]

Alcohols
Methanol[c]
Ethanol[b]
Propanol[c]
3-Methylbutanol[d]
Hexanol[c,d]
4-Hexanol[d]
Heptanol[c]
Octanol[c,d]
2-Octanol[c]

Esters
Ethylmethanoate[c]
Methylethanoate[c]
Methylpropanoate[c]
Ethylbutanoate[c]
Methylhexanoate[c]
Ethylhexanoate[c]
Methyloctanoate[c]
Methyldecanoate[c]
Methyldodecanoate[c]

Furans
Methylfuran[d]
Pentylfuran[d]
Heptylfuran[d]
2-Acetylfuran[d]
Furyl alcohol[d]

Hydrocarbons
Heptane
Pentadecane (and isomers)[d]
1-Pentadecene[d]
Hexadecane (and isomers)[d]

Ketones
2-Propanone (acetone)[a,b,d]
2-Butanone[b,c]
2,3-Butanedione[b,c,d]
2-Pentanone[c]
2,3-Pentanedione[d]
2,6-Hexanedione[c]
2-Octanone[d]
2-Nonanone[d]
2-Undecanone[d]
2-Dodecanone[d]
2-Tridecanone[d]
Pentadecanone[d]

Phenols
Dimethylphenol[d]

Pyridines
2-Methylpyridine[d]

Pyrazines
2-Methylpyrazine[d]

Other nitrogen compounds
Ammonia[b]
Methylamine[b]
N-ethylpyrrolidone[d]

Sulphur compounds
Hydrogen sulphide[b]
Methylmercaptan[d,e]
Ethylmercaptan[e]
n-Propylmercaptan[e]
n-Butylmercaptan[e]
3-Methylbutylmercaptan[e]
4-Methyl-4-mercaptan-pentane-2-one[e]
Methyl ethylsulphide[e]
Ethylenesulphide[e]
Propylenesulphide[e]
Ethyl isobutylsulphide[e]
2,2′-*bis*(Ethylthio)-propane
Thiacyclohexane[d]
Ethylthioacetate[e]
n-Butylthioacetate[e]
n-Butylthiopropionate[e]
Isobutylthiobutanoate[e]
Methylthioacetate[e]
Dimethyldisulphide[d,e]
Ethyl methyldisulphide[d,e]
Diethyldisulphide[e]
1,3-Thioxalane[e]
Dimethyltrisulphide[d,e]
Methyl ethyltrisulphide[e]
Diethyltrisulphide[e]
3,5-Dimethyl-1,2,4-trithiolane[d]
3,6-Dimethyl-1,2,4-trithiolane[d]
Thiophene[e]
2-Methylthiophene[e]
2-Formylthiophene[e]
2,4,6-Triethyl perhydro-1,3,5-dithiazine (thialdine)[d]

Halogenated compounds
Dichlorobenzene[d]

Miscellaneous compounds
Benzonitrile[f]
Phenylacetonitrile[f]
Nonanenitrile[f]
Decanenitrile[f]
Undecanenitrile[f]
Dodecanenitrile[f]
Tridecanenitrile[f]
Pentylnitrate[f]
Hexylnitrate[f]
Heptylnitrate[f]
Octylnitrate[f]

[a]Cross and Ziegler (1965); [b]Ockerman *et al.* (1964); [c]Lillard and Ayres (1969); [d]Bailey and Swain (1973); [e]Golovnya *et al.* (1982); [f]Mottram (1984).

In our attempts to unravel the complex nature of cured-meat flavour, we have taken a stepwise approach. We have kept the cooking conditions mild, so as to limit the formation of heterocyclic nitrogen and sulphur compounds, and also have used a rather modest-sized sample of 250–400 g of ground meat. As the first step, volatile components from uncured and cured pork isolated by two classical procedures, steam distillation followed by solvent extraction and continuous steam distillation–extraction (SDE), were compared (Ramarathnam *et al.*, 1991a). In this study we deliberately focused our attention on characterization of carbonyls and hydrocarbons that usually make up the bulk of the meat-flavour spectrum. The aroma concentrates isolated by the two methods were analysed by gas chromatography, and the individual components were further identified and quantified using GC–MS with hexanal and decanal as internal standards.

Typical gas chromatograms of the uncured and cured pork aroma concentrates extracted by the conventional steam distillation method and SDE method using the modified Likens–Nickerson flavour-extraction apparatus are given in Figures 13.1 and 13.2. It was observed that the aroma concentrates from cured meat isolated by either technique had fewer constituents than the uncured-meat sample. It was also found that many of the volatile constituents present in the aroma concentrate of uncured pork in the retention time region of 15–40 min were either absent or present in much lower concentrations in the cured-meat aroma concentrate. This result is attributable to the suppression of lipid oxidation due to the presence of nitrite in cured meat. It was observed that the relative concentrations of the individual components in the aroma concentrates isolated by the SDE method were higher than those present in the aroma concentrates prepared by the conventional steam-distillation method. This could be due to the partial loss of volatiles during the extraction and concentration steps of the latter method, which involves twice the volume of extraction solvent as the SDE method. Also, the SDE method, being a closed system, was more effective in preventing the loss of the volatiles.

The separated constituents in uncured and cured pork are reported in Table 13.3. In all 50 hydrocarbons, 37 carbonyls, 6 acids and 2 alcohols were identified. Table 13.3 lists these components and also shows in which of the samples prepared by the two isolation methods the components were identified. It was observed that the aroma concentrates of uncured and cured pork isolated by the SDE method were resolved into 77 and 72 components, while those extracted by the conventional steam distillation contained only 59 and 51 components, respectively. This shows that the Likens–Nickerson flavour extraction apparatus, being a closed system, was more effective in extracting aroma components than the conventional steam-distillation method. Of the components identified, hexanal was found to be present in uncured meat at a concentration of 12.66 $\pm$ 0.08 mg/kg, while in the cured meat it was found to be present as a

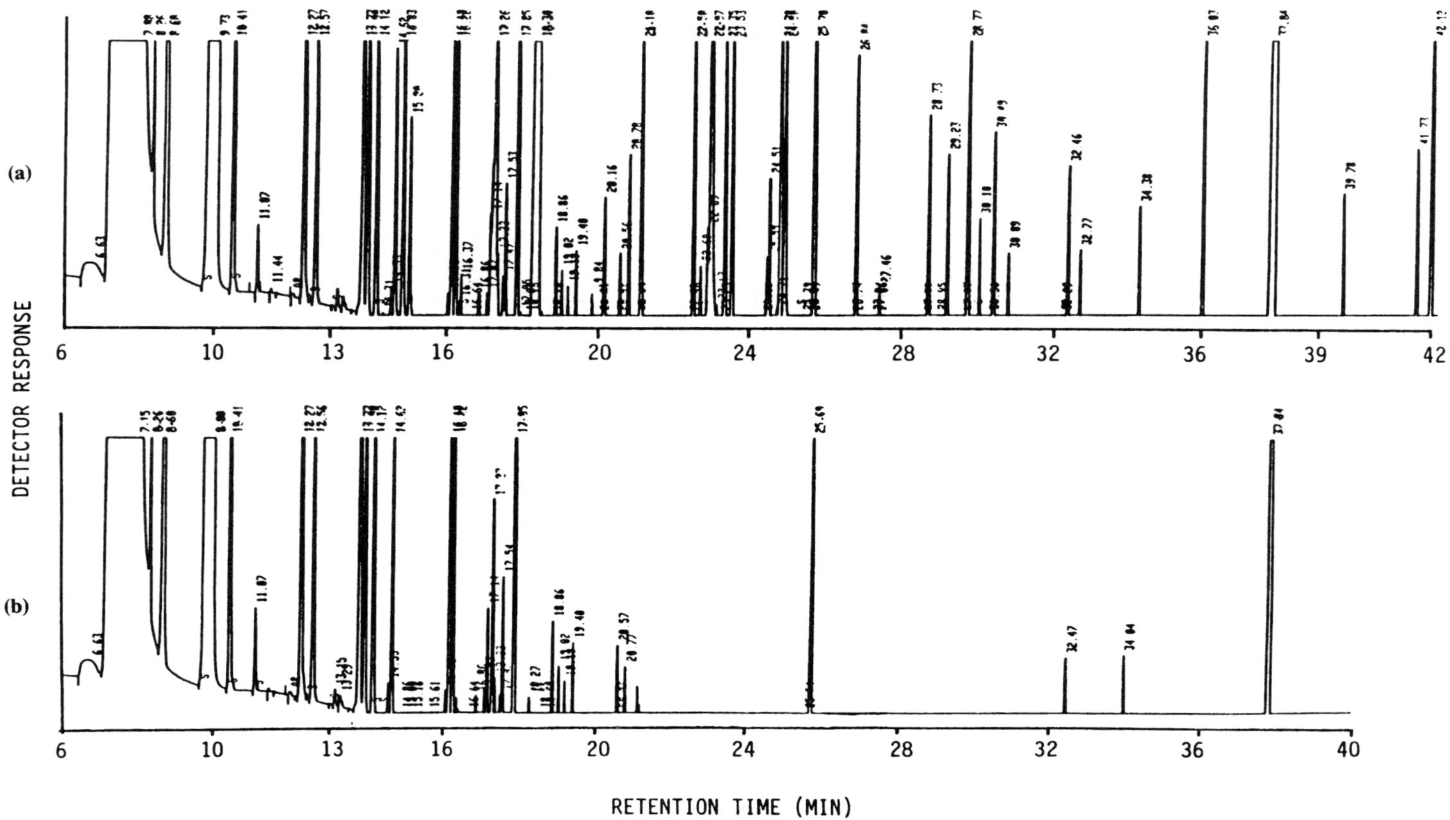

Figure 13.1 Typical gas chromatograms of (a) uncured-pork-flavour, and (b) cured-pork-flavour concentrates isolated by steam distillation.

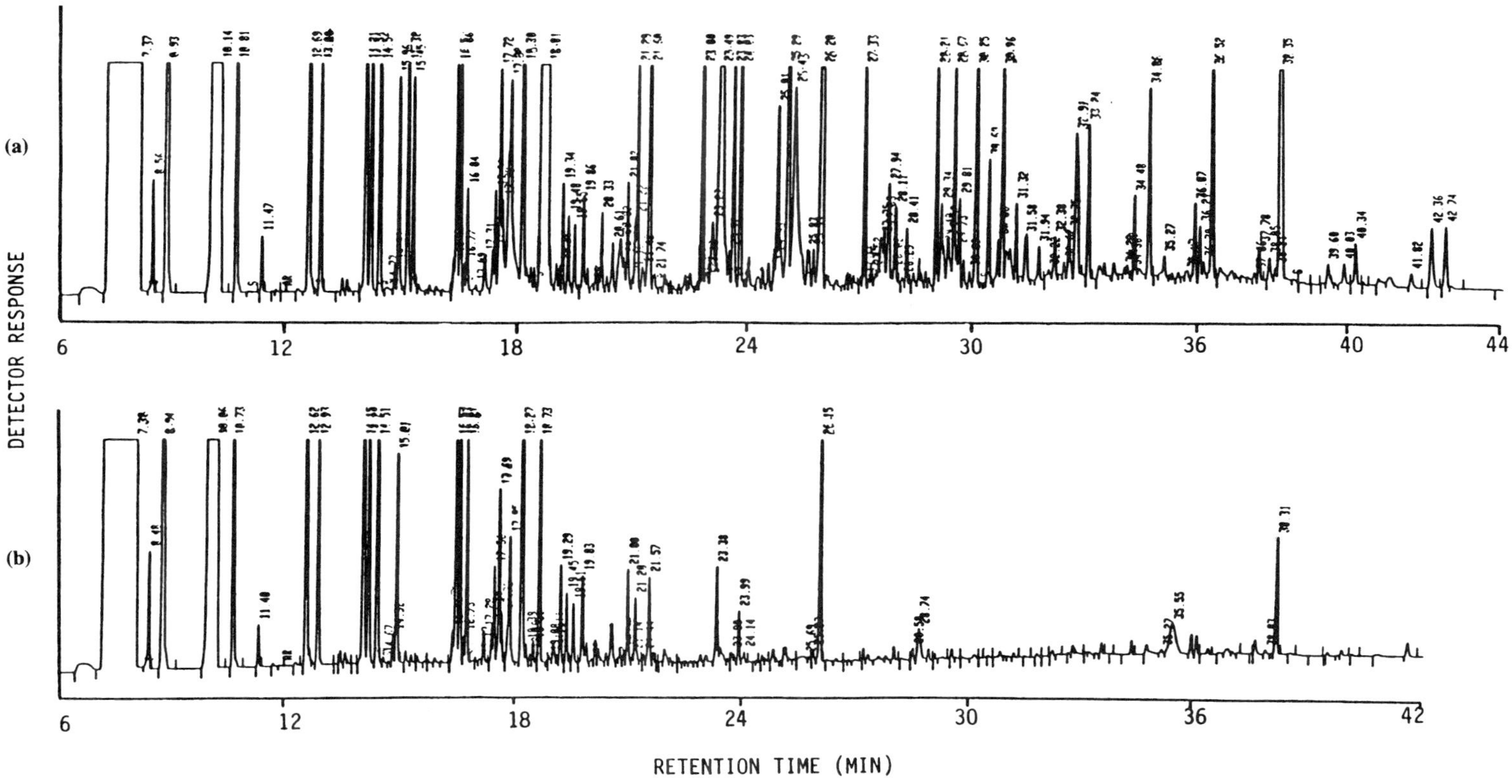

Figure 13.2 Typical gas chromatograms of (a) uncured-pork-flavour and (b) cured-pork-flavour concentrates isolated by the SDE methods.

minor component to the extent of 0.030 ± 0.004 mg/kg, which amounts to only 0.24% of that present in uncured meat. Shahidi *et al.* (1987b) found the hexanal content of cured pork to be 2% of the value observed for uncured pork, which is quantitatively similar to the values obtained by us. The difference in magnitude is probably due to differences in the extraction technique and sample size.

As the next step, using the SDE method as the choice of extraction technique, we have prepared aroma concentrates from uncured and cured beef and chicken, and by use of GC–MS compared the qualitative and quantitative differences in them. We have compared the data for carbonyls in beef and chicken with those of pork we reported earlier (Ramarathnam *et al.*, 1991a), and have made an attempt to identify the components responsible for species differences (Ramarathnam *et al.*, 1991b). Analysis of the aroma concentrates isolated from uncured and cured beef and chicken using GC–MS showed that the aroma concentrates isolated from uncured and cured beef had 59 and 40 components, respectively (Figures 13.3(a) and (b)), while those of chicken were resolved into 48 and 36 components (Figures 13.4(a) and (b)). Of the separated constituents, 31 hydrocarbons, 26 carbonyls, three alcohols, and two acids were identified in uncured and cured beef (Table 13.4). The corresponding figures for chicken (Table 13.5) were 29 hydrocarbons, 26 carbonyls and two alcohols. Our earlier work on pork volatiles showed that aroma concentrates from cooked uncured and cured pork resolved into 77 and 72 components, respectively (Ramarathnam *et al.*, 1991a). The differences in the total number of components and individual carbonyls identified among the three species could be attributed to the differences in their fat content, and also to a great extent to the differences in their fatty acid compositions. Pork used in our investigation had a fat content of 10.4 ± 0.1% (Ramarathnam *et al.*, 1991a), while beef and chicken had fat contents of 6.5 ± 0.2% and 2.4 ± 0.3%, respectively. It is also well known that the composition of polyunsaturated fatty acids (PUFA) differs widely among the three species (Fogerty *et al.*, 1990).

The separated constituents in beef and chicken volatiles are reported in Tables 13.4 and 13.5 respectively. Among the carbonyl components identified in beef and chicken, a distinct difference was observed in the hexanal content of uncured and cured meat. Hexanal content in uncured beef was 8.15 ± 0.17 mg/kg (peak 15, Table 13.4), whereas the content of this lipid oxidation product in uncured chicken was 9.84 ± 0.17 mg/kg (peak 14, Table 13.5). The corresponding values for cured beef and chicken were 0.05 ± 0.02 mg/kg and 0.11 ± 0.04 mg/kg, respectively. A comparison of differences in the content of carbonyl components in the three species – beef, chicken and pork – is summarized in Table 13.6. The concentration of 3-hexanone in uncured and cured beef was 1.08 ± 0.09 mg/kg and 0.57 ± 0.02 mg/kg, respectively, while the corresponding levels in cooked uncured and cured chicken were 5.78 ± 0.13 mg/kg and 1.45 ± 0.07 mg/kg, respectively. This component was found in uncured pork at a level of 0.42 ± 0.06 mg/kg, while in cured pork it was present only in small traces. The absence of 4-methyl-2-pentanone in the uncured meat of all the three species and its presence in small amounts, 0.03 ± 0.01 mg/kg in beef and 0.06 ± 0.02 mg/kg in chicken, and traces in pork, indicate that this component may be one of the typical constituents of the 'cured-meat' volatiles. Whether it is a constituent of the spectrum of volatiles which forms the 'cured-meat' flavour remains to be established experimentally. The fact that 4-methyl-2-pentanone is absent in uncured meat and present in small amounts in the cured product indicates that this compound is not derived from lipid oxidation. It may be formed as a result of a Maillard reaction.

3,3-Dimethylhexanal was present in uncured and cured beef at a concentration of 0.01 ± 0.04 mg/kg and 0.03 ± 0.02 mg/kg, respectively, and was not detected in chicken or pork. 3-Methyl-4-heptanone was present in chicken at a fairly high level of 0.44 ± 0.05 mg/kg, while it was absent in both pork and beef. It may be an important component of uncured cooked-chicken flavour. Octanal was not detected in cured beef, cured chicken, and in cured or uncured pork, while in uncured chicken it was present as a major component (5.08 ± 0.14 mg/kg), and in uncured beef it was detected to the extent of 0.69 ± 0.07 mg/kg.

Striking differences were also observed between the contents of nonanal and 16-octadecenal. Nonanal was absent in uncured pork, but was present in uncured chicken at the very high concentration of 11.59 ± 0.12 mg/kg, and in uncured beef the level of this compound was only 1.44 ± 0.07 mg/kg. 16-Octadecenal was present in uncured chicken at a concentration of 5.86 ± 0.18 mg/kg. This component was present in uncured beef at a concentration of only 1.81 ± 0.14 mg/kg, while in uncured pork the concentration of 16-octadecenal was found to be 8.34 ± 0.35 mg/kg. Nonanal was absent in the cured meat of all the three species, while 16-octadecenal was detected in cured pork at a concentration of 2.20 ± 1.26 mg/kg.

Thus, the presence and absence of certain carbonyls, or the differences in their concentration in the volatiles among the three species, can be a major contributory factor to the differences in the aroma nuances observed in them. Carbonyl compounds, which are formed due to the oxidation of unsaturated lipids and during the nonenzymatic amino-carbonyl reactions, have been implicated as significant contributors to the flavour of uncured meat, but not of cured meat. Since the aroma concentrates of cured pork, beef and chicken are similar and the concentration of the individual carbonyls, with the exception of 3-hexanone and 16-octadecenal, is less than 1 mg/kg, it is therefore evident that the 'cured-meat flavour' or the 'basic flavour of cooked meat', which is devoid of any lipid-oxidation product, should originate from nontriglyceride precursors. Removal of carbonyls by the use of carbonyl-specific reagents

Table 13.3 Components of the aroma concentrates of uncured and cured pork

Peak no.	RT (min)	Component	Detected in[a]				Content[b] (mg/kg)	
			A	**B**	**A+**	**B+**	**A**	**B**
1	2.47	2-Methylhexane	+	+	+	+	1.20 ± 0.18	1.01 ± 0.06
2	2.56	3-Methylhexane	+	+	+	+	0.69 ± 0.04	0.59 ± 0.01
3	2.70	2,2-Dimethylhexane	+	+	+	+	0.33 ± 0.04	0.28 ± 0.03
4	2.88	3-Hexanone	+	+	–	–	0.42 ± 0.06	tr
5	3.23	Unidentified	–	+	–	+	–	tr
6	3.41	2,4-Dimethylhexane	+	+	+	+	0.93 ± 0.08	0.68 ± 0.15
7	3.55	4-Methyl-2-pentanone	–	+	–	–	–	tr
8	3.68	2,3-Dimethylhexane	–	+	–	+	–	tr
9	3.76	3,3-Dimethylhexane	–	+	–	+	–	0.03 ± 0.01
10	3.90	4-Methylheptane	–	+	+	+	–	0.09 ± 0.01
11	4.02	2,5-Dimethylhexane	+	+	+	+	0.23 ± 0.09	0.17 ± 0.03
12	4.16	3-Methylheptane	+	+	+	+	0.21 ± 0.06	0.09 ± 0.01
13	4.35	2,2,5-Trimethylhexane	+	+	+	+	1.27 ± 0.09	1.08 ± 0.06
14	4.44	2,2,4-Trimethylhexane	–	+	–	+	–	0.09 ± 0.06
15	4.69	Hexanal	+	+	+	+	12.66 ± 0.08	0.03
16	4.76	Unidentified	–	–	–	+	–	tr
17	4.96	2,3,5-Trimethylhexane	+	+	+	+	0.12 ± 0.02	0.10 ± 0.02
18	5.10	2,4-Dimethylheptane	–	+	+	+	–	0.07 ± 0.01
19	5.26	2,6-Dimethylheptane	+	+	+	+	tr	0.07 ± 0.02
20	5.35	2,5-Dimethylheptane	+	+	+	+	0.17 ± 0.02	0.15 ± 0.03
21	5.63	1,2,4-Trimethylcyclohexane	–	+	–	+	–	0.03 ± 0.01
22	5.76	2-Hexanal	+	+	+	+	tr	tr
23	5.92	3-Methyl-4-heptanone	–	+	–	+	–	tr
24	6.02	1,3-Dimethylbenzene	–	+	–	+	–	tr
25	6.09	2,5-Dimethyloctane	–	+	–	+	–	0.04 ± 0.02
26	6.17	4-Ethyl-2,2-dimethylhexane	–	+	–	+	–	0.12 ± 0.02
27	6.20	2,2,3-Trimethylhexane	+	–	+	–	0.23 ± 0.11	–
28	6.30	2,2,4-Trimethylheptane	+	+	+	+	0.12 ± 0.01	0.10 ± 0.03
29	6.46	1,2-Dimethylbenzene	–	+	–	–	–	0.04 ± 0.03
30	6.50	2-Heptanone	+	–	+	–	0.20 ± 0.06	–
31	6.55	3,3,5-Trimethylheptane	–	+	+	+	–	0.05 ± 0.01
32	6.66	3-Methyl-2-nonene	–	+	–	+	–	0.04 ± 0.01
33	6.74	3-Methylhexanal	+	–	+	–	0.65 ± 0.14	–
34	6.89	3,5-Dimethyloctane	–	+	–	+	–	tr
35	7.74	2,4,6-Trimethyloctane	–	+	–	+	–	tr
36	7.81	2-Heptenal	+	–	+	–	0.34 ± 0.04	–
37	7.83	Benzaldehyde	+	+	+	+	0.11 ± 0.01	0.04 ± 0.05
38	8.12	3-Methyloctane	+	–	–	–	0.11 ± 0.01	–
39	8.18	Unidentified	+	–	–	–	1.77 ± 0.015	–
40	8.33	2,3-Octanedione	+	–	–	–	0.88 ± 0.09	–
41	8.36	1,3,5-Trimethylbenzene	–	+	–	–	–	tr
42	8.40	1-Nonen-3-ol	+	–	+	–	0.75 ± 0.05	–
43	8.46	3,6-Dimethyloctane	–	+	–	–	–	tr
44	8.56	Unidentified	–	+	–	+	–	0.04 ± 0.01
45	8.66	3-Ethoxy-2-methyl-l-propene	+	–	+	–	0.66 ± 0.01	–
46	8.85	2,3,4-Trimethyloctane	–	+	–	–	–	tr
47	9.00	D-Limonene	–	+	–	–	–	0.02
48	9.13	3-Ethyl-2-methyl-1,3-hexadiene	+	–	+	–	0.14 ± 0.01	–
49	9.40	4,4,5-Trimethyl-2-hexene	+	–	–	–	0.13 ± 0.02	–
50	9.47	5-Methylundecane	–	+	–	–	–	tr
51	9.57	5,5-Dimethyl-2-hexene	–	+	–	–	–	tr
52	9.65	(*E*)-2-octenal	+	–	+	–	0.99 ± 0.01	–
53	9.86	2-Octen-l-ol	+	–	+	–	0.69 ± 0.15	--
54	10.20	3,7-Dimethylnonane	+	+	–	–	tr	tr
55	10.40	Methylcyclohexane	+	+	+	+	2.10 ± 0.33	0.33 ± 0.08
56	11.30	2-Nonenal	+	–	+	–	0.39 ± 0.05	–
57	11.34	4-Ethylbenzaldehyde	+	–	+		tr	–
58	11.45	5-Undeca-3(*Z*),5-diyne	+	–	–	–	tr	–
59	11.60	5-Undeca-3(*E*),5-diyne	+	–	–	–	tr	–
60	11.68	Naphthalene	+	+	–	–	0.12 ± 0.03	0.04 ± 0.01

Table 13.3 *continued*

Peak no.	RT (min)	Component	Detected in[a]				Content[b] (mg/kg)	
			A	**B**	**A⁺**	**B⁺**	**A**	**B**
61	11.81	Dodecane	+	+	+	–	0.28 ± 0.06	tr
62	11.96	Decanal	+	+	+	–	tr	tr
63	12.13	2,4-Nonadienal	+	–	+	–	tr	–
64	12.40	Unidentified	+	–	–	–	tr	–
65	12.87	2-Undecenal	+	–	+	–	0.39 ± 0.07	–
66	13.13	Unidentified	+	–	–	–	tr	–
67	13.27	2-Undecanone	–	+	–	–	–	0.02
68	13.32	4,6-Dimethylundecane	–	+	–	–	–	tr
69	13.40	Tridecane	+	+	+	–	0.49 ± 0.04	tr
70	13.49	Undecanal	+	–	–	–	tr	–
71	13.70	(*E,E*)-2,4-decadienal	+	–	+	–	0.69 ± 0.16	–
72	13.75	(*E,Z*)-2,4-decadienal	+	–	+	–	0.41 ± 0.15	–
73	14.11	Unidentified	+	–	+	–	tr	–
74	14.22	5-Tridecanone	–	+	–	–	–	tr
75	14.35	2-Dodecenal	+	–	+	–	0.43 ± 0.08	–
76	14.76	Tetradecane	+	–	–	–	0.14 ± 0.04	–
77	14.92	Dodecanal	+	–	–	–	tr	–
78	15.10	2,4-Undecadienal	+	–	–	–	tr	–
79	15.70	4-Pentylbenzaldehyde	+	–	+	–	tr	–
80	15.87	1-Pentadecene	+	–	–	–	tr	–
81	16.07	Pentadecane	+	+	+	+	0.19 ± 0.05	0.08 ± 0.07
82	16.29	Tridecanal	+	+	+	+	0.25 ± 0.05	0.09 ± 0.01
83	16.97	Dodecanoic acid	–	+	–	–	–	tr
84	17.31	Unidentified	–	+	–	–	–	0.14 ± 0.01
85	17.36	Hexadecane	+	+	+	+	tr	0.04 ± 0.01
86	17.49	3-Tridecen-l-yne	+	–	–	–	tr	–
87	17.55	Tetradecanal	+	+	+	+	0.40 ± 0.14	0.03 ± 0.01
88	18.51	Heptadecane	+	+	–	–	0.13 ± 0.02	0.11 ± 0.03
89	18.55	2-Pentadecanone	+	+	–	+	tr	0.06 ± 0.02
90	18.75	Hexadecanal	+	+	+	+	0.65 ± 0.05	0.06 ± 0.02
91	19.37	Tridecanoic acid	–	–	+	+	–	tr
92	19.48	Unidentified	+	–	–	–	0.12 ± 0.02	–
93	19.64	1,14-Tetradecanediol	+	+	+	+	tr	0.05 ± 0.01
94	19.88	17-Octadecenal	+	–	+	+	tr	–
95	20.01	16-Octadecenal	+	+	+	+	8.34 ± 0.35	2.20 ± 1.26
96	20.56	Unidentified	+	–	–	–	0.17 ± 0.03	–
97	20.78	Pentadecanitrile	+	+	–	+	0.21 ± 0.05	0.12 ± 0.04
98	20.99	15-Octadecenal	+	+	+	+	0.70 ± 0.04	0.14 ± 0.04
99	21.52	Hexedecanoic acid	+	+	+	+	0.97 ± 0.07	0.14 ± 0.02
100	21.81	9-Octadecenal	+	+	+	+	0.81 ± 0.06	0.14 ± 0.02
101	21.86	5-Octadecenal	–	+	+	+	–	0.05 ± 0.01
102	22.05	Octacanal	+	+	+	+	1.19 ± 0.11	0.19 ± 0.09
103	23.17	9,12-Octadecadienoic acid	+	+	–	–	tr	0.13 ± 0.03
104	23.22	9-Octadecenoic acid	+	+	+	+	0.16 ± 0.05	tr
105	23.37	Octadecanoic acid	–	+	–	+	–	tr

[a]Qualitative information only.
+, detected; –, not detected.
A and **A⁺** are uncured meat flavour constituents isolated by SDE and steam distillation methods.
B and **B⁺** are cured meat flavour constituents isolated by SDE and steam distillation methods.
[b]Concentration of constituents in uncured (**A**) and cured meat (**B**), isolated by the SDE method.
Reported values are mean ± S.D., *n* = 3.
tr, trace amount (< 0.01mg/kg).

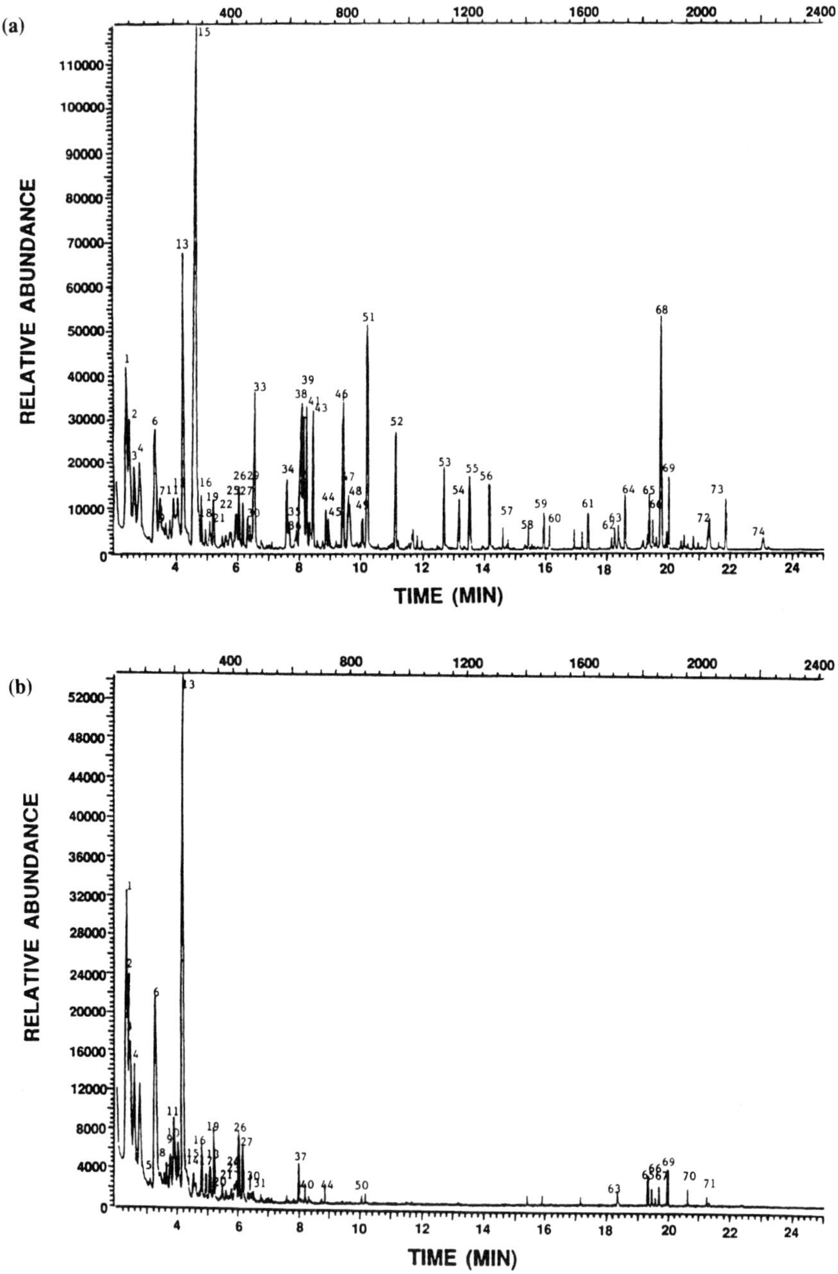

Figure 13.3 Total ion chromatograms (TIC) of (a) uncured-beef-flavour and (b) cured-beef-flavour concentrates isolated by the SDE method.

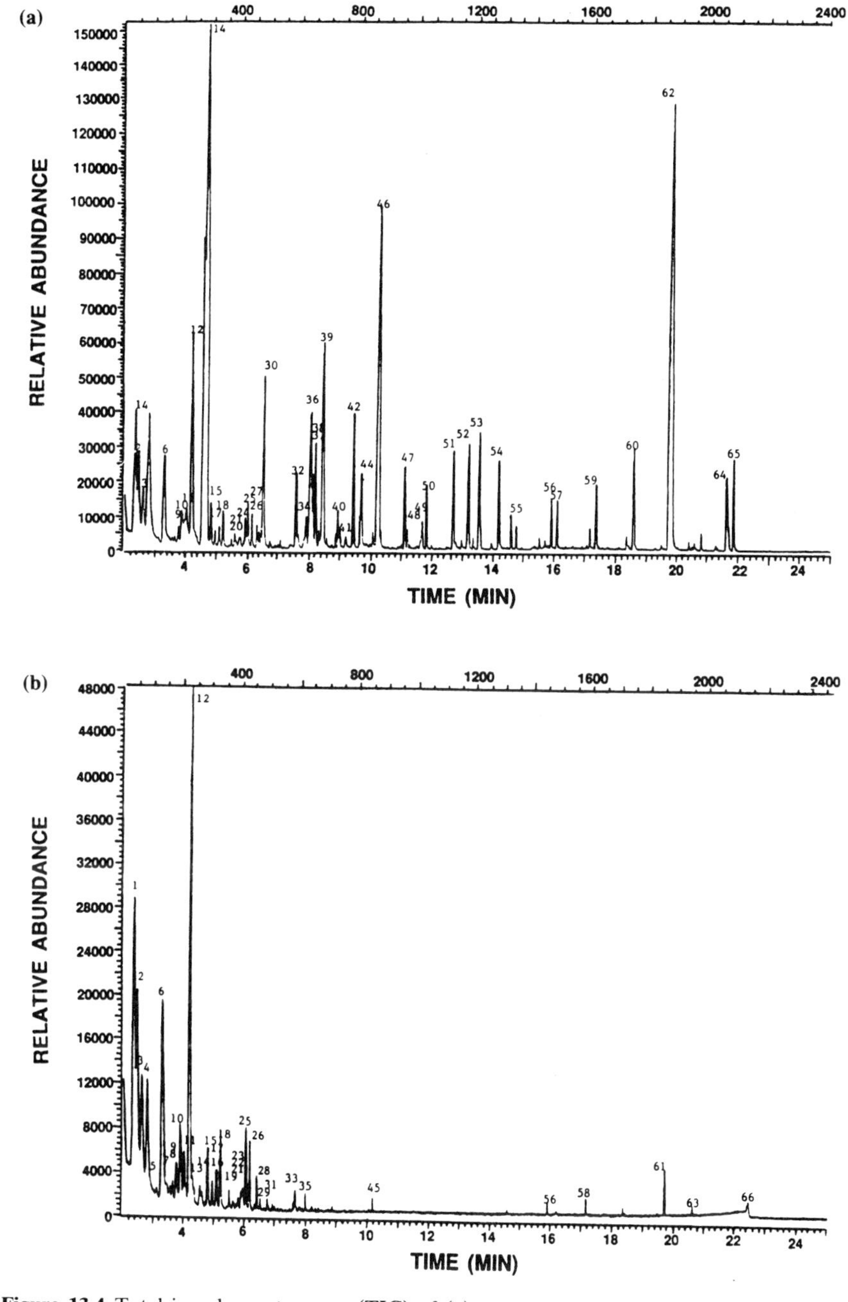

Figure 13.4 Total ion chromatograms (TIC) of (a) uncured-chicken-flavour and (b) cured-chicken-flavour concentrates isolated by the SDE method.

Table 13.4 Components of the aroma concentrates of uncured and cured beef, isolated by the SDE method

Peak no.	RT (min)	Component	Content (mg/kg)	
			Uncured	Cured
1	2.35	2-Methylhexane	1.82 ± 0.12	1.58 ± 0.09
2	2.45	3-Methylhexane	1.11 ± 0.08	0.94 ± 0.04
3	2.60	2,2-Dimethylhexane	0.71 ± 0.07	0.60 ± 0.03
4	2.79	3-Hexanone	1.08 ± 0.09	0.57 ± 0.02
5	3.12	Unidentified	–	0.04 ± 0.02
6	3.30	2,4-Dimethylhexane	1.46 ± 0.12	1.22 ± 0.11
7	3.47	2-Hexanone	0.38 ± 0.06	–
8	3.57	4-Methyl-2-pentanone	–	0.03 ± 0.01
9	3.65	3,3-Dimethylhexane	0.12 ± 0.04	0.10 ± 0.02
10	3.77	4-Methylheptane	0.13 ± 0.05	0.12 ± 0.02
11	3.89	2,5-Dimethylhexane	0.42 ± 0.09	0.04 ± 0.02
12	4.03	3-Methylheptane	0.36 ± 0.08	0.14 ± 0.04
13	4.21	2,2,5-Trimethylhexane	2.27 ± 0.12	1.91 ± 0.09
14	4.55	2,2,4-Trimethylhexane	–	0.12 ± 0.04
15	4.65	Hexanal	8.15 ± 0.17	0.05 ± 0.02
16	4.81	2,3,5-Trimethylhexane	0.27 ± 0.08	0.19 ± 0.08
17	4.96	2,3,4-Trimethylhexane	0.11 ± 0.05	0.07 ± 0.02
18	5.09	2,6-Dimethylheptane	0.20 ± 0.11	0.12 ± 0.05
19	5.21	2,5-Dimethylheptane	0.37 ± 0.15	0.24 ± 0.09
20	5.50	1,2,4-Trimethylcyclohexane	–	0.05 ± 0.02
21	5.61	3,3-Dimethylhexanal	0.11 ± 0.04	0.03 ± 0.02
22	5.75	Unidentified	0.10 ± 0.01	–
23	5.80	3-Methyl-4-heptanone	–	0.04 ± 0.01
24	5.87	1,3-Dimethylbenzene	–	0.04 ± 0.02
25	5.95	2,5-Dimethyloctane	0.30 ± 0.05	0.10 ± 0.04
26	6.04	4-Ethyl-2,2-dimethylhexane	0.33 ± 0.11	0.19 ± 0.07
27	6.17	2,2,4-Trimethylhexane	0.28 ± 0.12	0.15 ± 0.07
28	6.33	1,2-Dimethylbenzene	–	0.03 ± 0.01
29	6.34	2-Heptanone	0.21 ± 0.07	–
30	6.41	3,3,5-Trimethylheptane	0.10 ± 0.02	0.07 ± 0.02
31	6.47	Unidentified	–	0.03 ± 0.01
32	6.53	Unidentified	–	0.04 ± 0.01
33	6.56	3-Methylhexanal	1.09 ± 0.07	–
34	7.60	(*E*)-2-heptenal	0.45 ± 0.04	–
35	7.67	Benzaldehyde	0.20 ± 0.02	–
36	7.90	3-Methyloctane	0.14 ± 0.05	–
37	8.00	1,3,5-Trimethylbenzene	–	0.11 ± 0.02
38	8.10	1-Hepten-3-ol	1.73 ± 0.15	–
39	8.16	2,3-Octanedione	0.65 ± 0.11	–
40	8.21	3,6-Dimethyloctane	–	0.04 ± 0.02
41	8.24	Unidentified	0.75 ± 0.06	–
42	8.35	3-Ethoxy-2-methyl-1-propene	0.11 ± 0.04	–
43	8.46	Octanal	0.69 ± 0.07	–
44	8.87	D-Limonene	0.18 ± 0.05	0.04 ± 0.02
45	8.95	3-Ethyl-2-methyl-1,3-hexadiene	0.15 ± 0.11	–
46	9.45	(*E*)-2-octenal	1.07 ± 0.15	–
47	9.62	2-Octen-1-ol	0.38 ± 0.09	–
48	9.66	Unidentified	0.23 ± 0.05	–
49	10.07	3,7-Dimethylnonane	0.17 ± 0.04	–
50	10.18	Unidentified	–	0.03 ± 0.01
51	10.23	Nonanal	1.44 ± 0.07	–
52	11.13	2-Nonenal	0.68 ± 0.11	–

Table 13.4 *continued*

Peak no.	RT (min)	Component	Content (mg/kg)	
			Uncured	Cured
53	12.71	2-Undecenal	0.44 ± 0.05	–
54	13.19	Tridecane	0.31 ± 0.03	–
55	13.53	(*E*,*E*)-2,4-decadienal	0.42 ± 0.08	–
56	14.18	2-Dodecenal	0.35 ± 0.06	–
57	14.60	Tetradecane	0.10 ± 0.02	–
58	15.43	Unidentified	0.10 ± 0.05	–
59	15.92	Pentadecane	0.17 ± 0.08	–
60	16.12	Tridecanal	0.12 ± 0.04	–
61	17.38	Tetradecanal	0.17 ± 0.03	–
62	18.25	Unidentified	0.12 ± 0.02	–
63	18.37	2-Pentadecanone	0.19 ± 0.08	0.05 ± 0.02
64	18.60	Hexadecanal	0.28 ± 0.07	–
65	19.38	1,14-Tetradecanediol	0.27 ± 0.09	0.07 ± 0.01
66	19.50	Octadecane	0.15 ± 0.09	0.05 ± 0.02
67	19.73	17-Octadecenal	–	0.04 ± 0.01
68	19.79	16-Octadecenal	1.81 ± 0.14	–
69	20.02	Unidentified	0.37 ± 0.15	0.09 ± 0.02
70	20.65	Pentadecanitrile	–	0.04 ± 0.01
71	21.27	15-Octadecenal	–	0.03 ± 0.02
72	21.35	Hexadecanoic acid	0.32 ± 0.04	–
73	21.87	Octadecanal	0.26 ± 0.11	–
74	23.06	9-Octadecenoic acid	0.15 ± 0.03	–

–, not detected.
Reported values are mean ± S.D., $n = 3$.

should result essentially in a simplified basic meat flavour mixture (Cross and Ziegler, 1965; Minor *et al.*, 1965). Although the nature of such a mixture seems to be much simpler than that of uncured meat, the elucidation of the compounds which are responsible for the cured-meat flavour is not easy. This simplified mixture still has a second major group of volatiles, the hydrocarbons, that make practically no contribution to the 'meaty note' detectable in cured meat. Minute traces of aroma-effective heterocyclic components having very low flavour threshold values can present enormous difficulties in their isolation and identification (MacLeod and Ames, 1986). The TIC profiles have clearly shown that the flavour spectrum of cured meat is indeed simple (Figures 13.3(b) and 13.4(b)). The components identified, however, do not show the presence of sulphur and nitrogenous substances. Suitable modifications to the existing isolation and analytical techniques should be helpful in overcoming this problem. Preliminary experiments on the isolation of volatiles from the three meat species using the purge-and-trap technique, which is milder than the SDE method, have been successful in identifying certain heterocyclic compounds (data not shown). It is also believed that the heterocyclic compounds could be preferentially extracted from cooked meat by using supercritical carbon dioxide at relatively low temperatures

Table 13.5 Components of the aroma concentrates of uncured and cured chicken, isolated by the SDE method

Peak no.	RT (min)	Component	Content (mg/kg)	
			Uncured	Cured
1	2.36	2-Methylhexane	4.27 ± 0.28	3.03 ± 0.04
2	2.46	3-Methylhexane	2.55 ± 0.06	1.80 ± 0.11
3	2.61	2,2-Dimethylhexane	1.57 ± 0.08	1.17 ± 0.14
4	2.80	3-Hexanone	5.78 ± 0.13	1.45 ± 0.07
5	3.13	Unidentified	–	0.09 ± 0.02
6	3.30	2,4-Dimethylhexane	3.76 ± 0.12	2.78 ± 0.09
7	3.58	4-Methyl-2-pentanone	–	0.06 ± 0.02
8	3.66	3,3-Dimethylhexane	–	0.11 ± 0.05
9	3.78	4-Methylheptane	0.38 ± 0.11	0.29 ± 0.04
10	3.89	2,5-Dimethylhexane	0.94 ± 0.08	0.70 ± 0.06
11	4.04	3-Methylheptane	0.92 ± 0.12	0.48 ± 0.08
12	4.22	2,2,5-Trimethylhexane	5.81 ± 0.16	4.37 ± 0.09
13	4.55	2,2,4-Trimethylhexane	–	0.20 ± 0.02
14	4.60	Hexanal	9.84 ± 0.17	0.11 ± 0.04
15	4.82	2,3,5-Trimethylhexane	0.67 ± 0.05	0.42 ± 0.04
16	4.95	2,3,4-Trimethylhexane	–	0.17 ± 0.05
17	5.10	2,6-Dimethylheptane	0.46 ± 0.07	0.29 ± 0.03
18	5.22	2,5-Dimethylheptane	0.78 ± 0.08	0.63 ± 0.06
19	5.50	1,2,4-Trimethylcyclohexane	–	0.11 ± 0.02
20	5.62	2-Hexenal	0.40 ± 0.04	–
21	5.79	3-Methyl-4-heptanone	0.44 ± 0.05	0.09 ± 0.04
22	5.88	1,3-Dimethylbenzene	–	0.11 ± 0.03
23	5.92	Unidentified	–	0.11 ± 0.02
24	5.96	2,5-Dimethyloctane	0.86 ± 0.07	0.13 ± 0.05
25	6.04	2,2,3-Trimethylhexane	0.83 ± 0.06	0.51 ± 0.07
26	6.17	2,2,4-Trimethylheptane	0.65 ± 0.05	0.40 ± 0.03
27	6.34	2-Heptanone	0.54 ± 0.05	–
28	6.41	3,3,5-Trimethylheptane	–	0.20 ± 0.02
29	6.53	Unidentified	–	0.09 ± 0.02
30	6.58	3-Methylhexanal	4.22 ± 0.04	–
31	6.76	3,5-Dimethyloctane	–	0.11 ± 0.05
32	7.60	(*E*)-2-heptenal	1.78 ± 0.05	–
33	7.66	Benzaldehyde	–	0.12 ± 0.04
34	7.93	3-Methyloctane	0.96 ± 0.09	–
35	8.00	Unidentified	–	0.11 ± 0.04
36	8.10	1-Hepten-3-ol	4.91 ± 0.05	–
37	8.15	2,3-Octanedione	1.23 ± 0.01	–
38	8.23	Unidentified	1.98 ± 0.12	–
39	8.49	Octanal	5.08 ± 0.14	–
40	8.96	3-Ethyl-2-methyl-1,3-hexadiene	0.64 ± 0.06	–
41	9.20	4,4,5-Trimethyl-2-hexene	0.37 ± 0.03	–
42	9.46	(E)-2-octenal	3.07 ± 0.11	–
43	9.65	2-Octen-1-ol	0.69 ± 0.07	–
44	9.72	Unidentified	1.60 ± 0.09	–
45	10.18	Unidentified	–	0.08 ± 0.02
46	10.28	Nonanal	11.59 ± 0.12	–
47	11.13	2-Nonenal	1.46 ± 0.09	–
48	11.21	4-Ethylbenzaldehyde	0.36 ± 0.04	–
49	11.69	Dodecane	0.45 ± 0.05	–
50	11.83	Decanal	1.05 ± 0.06	–
51	12.72	2-Undecenal	1.94 ± 0.09	–
52	13.20	Tridecane	2.21 ± 0.11	–

Table 13.5 *continued*

Peak no.	RT (min)	Component	Content (mg/kg)	
			Uncured	Cured
53	13.56	(*E*,*E*)-2,4-decadienal	2.48 ± 0.15	–
54	14.20	2-Dodecenal	1.90 ± 0.09	–
55	14.60	Tetradecane	0.52 ± 0.07	–
56	15.92	Pentadecane	0.82 ± 0.09	0.07 ± 0.01
57	16.12	Tridecanal	0.85 ± 0.08	–
58	17.17	Hexadecane	–	0.09 ± 0.03
59	17.40	Tetradecanal	1.14 ± 0.08	–
60	18.61	Hexadecanal	2.13 ± 0.05	–
61	19.73	17-Octadecenal	–	0.23 ± 0.03
62	19.89	16-Octadecenal	5.86 ± 0.18	–
63	20.65	Pentadecanitrile	–	0.06 ± 0.04
64	21.67	9-Octadecenal	1.85 ± 0.09	–
65	21.90	Octadecanal	1.88 ± 0.04	–
66	22.46	Unidentified	–	0.18 ± 0.04

–, not detected.

Reported values are mean ± S.D., $n = 3$.

and in a completely inert atmosphere. Work is currently being planned in this direction.

Among the hydrocarbons identified, cured and uncured chicken had the highest concentration of low-boiling homologues of branched hexane, heptane and octane (Table 13.5), while the levels of such components in cured and uncured beef (Table 13.4) were only slightly higher than those of pork (Table 13.3). Hydrocarbons of specific interest, which were absent in the uncured meat of all the three species but present in the cured meat, are 2,2,4-trimethylhexane, which was present in cured beef to the extent of 0.12 ± 0.04 mg/kg (peak 14, Table 13.4), 0.20 ± 0.02 mg/kg in cured chicken (peak 13, Table 13.5) and 0.09 ± 0.06 mg/kg in cured pork (peak 14, Table 13.3); 1,2,4-trimethylcyclohexane, detected in cured beef to the extent of 0.05 ± 0.02 mg/kg (peak 20, Table 13.4), 0.11 ± 0.02 mg/kg in cured chicken (peak 19, Table 13.5) and 0.03 ± 0.01 mg/kg in cured pork (peak 21, Table 13.3); 1,3-dimethylbenzene, present in cured beef to the extent of 0.04 ± 0.02 mg/kg (peak 24, Table 13.4), 0.11 ± 0.03 mg/kg in cured chicken (peak 22, Table 13.5) and in small traces in cured pork (peak 24, Table 13.3). D-Limonene (peak 44, Table 13.4) was detected both in uncured beef (0.18 ± 0.05 mg/kg) and cured beef (0.04 ± 0.02 mg/kg), and was absent in chicken. This compound was also absent in uncured pork, while in cured pork it was present to the extent of 0.02 mg/kg (peak 47, Table 13.3). Hydrocarbons are formed due to the breakdown of unsaturated fatty acids during autoxidation of lipids. The differences in the concentration of most of the hydrocarbons detected in pork, chicken and beef can be attributed to the differences in the contents of total fat and unsaturated fatty acids.

Table 13.6 Carbonyls in the aroma concentrates of uncured and cured beef, chicken and pork, isolated by the SDE method

Component	Content (mg/kg)					
	Beef		Chicken		Pork	
	Uncured	Cured	Uncured	Cured	Uncured	Cured
3-Hexanone	1.08 ± 0.09	0.57 ± 0.02	5.78 ± 0.13	1.45 ± 0.07	0.42 ± 0.06	tr
2-Hexanone	0.38 ± 0.06	–	–	–	–	–
4-Methyl-2-pentanone	–	0.03 ± 0.01	–	0.06 ± 0.02	–	tr
Hexanal	8.15 ± 0.17	0.05 ± 0.02	9.84 ± 0.17	0.11 ± 0.04	12.66 ± 0.08	0.03 ± 0.05
2-Hexenal	–	–	0.40 ± 0.04	–	tr	tr
3,3-Dimethylhexanal	0.11 ± 0.04	0.03 ± 0.02	–	–	–	–
3-Methyl-4-heptanone	–	0.04 ± 0.01	0.04 ± 0.05	0.09 ± 0.04	–	tr
2-Heptanone	0.21 ± 0.07	–	0.54 ± 0.05	–	0.20 ± 0.06	–
3-Methylhexanal	1.09 ± 0.07	–	4.22 ± 0.04	–	0.65 ± 0.14	–
(*E*)-2-heptenal	0.45 ± 0.04	–	1.78 ± 0.05	–	0.34 ± 0.04	–
Benzaldehyde	–	–	–	–	0.11 ± 0.01	0.04 ± 0.05
2,3-Octanedione	–	–	–	–	0.88 ± 0.09	–
Octanal	0.69 ± 0.07	–	5.08 ± 0.14	–	–	–
(*E*)-2-octenal	1.07 ± 0.15	–	3.07 ± 0.11	–	0.99 ± 0.10	–
Nonanal	1.44 ± 0.07	–	11.59 ± 0.12	–	–	–
2-Nonenal	0.68 ± 0.11	–	1.46 ± 0.09	–	0.39 ± 0.05	–
4-Ethylbenzaldehyde	–	–	0.36 ± 0.04	–	tr	–
Decanal	–	–	1.05 ± 0.06	–	tr	tr

Table 13.6 *continued*

2-Undecenal	0.44 ± 0.05	–	1.94 ± 0.09	–	0.39 ± 0.07	–
(*E*,*E*)-2,4,decadienal	0.42 ± 0.08	–	2.48 ± 0.15	–	0.69 ± 0.16	–
(*E*,*Z*)-2,4,decadienal	–	–	–	–	0.41 ± 0.15	–
2-Dodecenal	0.35 ± 0.06	–	1.90 ± 0.09	–	0.43 ± 0.08	–
Tridecanal	0.12 ± 0.04	–	0.85 ± 0.08	–	0.25 ± 0.05	0.09 ± 0.01
Tetradecanal	0.17 ± 0.03	–	1.14 ± 0.08	–	0.40 ± 0.14	0.03 ± 0.01
2-Pentadecanone	0.19 ± 0.08	0.05 ± 0.02	–	–	tr	0.06 ± 0.02
Hexadecanal	0.28 ± 0.07	–	2.13 ± 0.05	–	0.65 ± 0.05	0.06 ± 0.02
17-Octadecenal	–	0.04 ± 0.01	–	0.23 ± 0.03	tr	–
16-Octadecenal	1.81 ± 0.14	–	5.86 ± 0.18	–	8.34 ± 0.35	2.20 ± 1.26
15-Octadecenal	–	0.03 ± 0.02	–	–	0.70 ± 0.04	0.14 ± 0.04
9-Octadecenal	–	–	1.85 ± 0.09	–	0.81 ± 0.06	0.14 ± 0.02
Octadecanal	0.26 ± 0.11	–	1.88 ± 0.04	–	1.19 ± 0.11	0.19 ± 0.09

–, not detected.
tr, trace amount (< 0.01 mg/kg).
Reported values are mean ± S.D., $n = 3$.

13.5 Flavour of defatted meat

Dehydrated and partially defatted (hexane-extracted) as well as fully defatted (first hexane- and then chloroform–methanol-extracted) samples of ground pork loin were rehydrated to the original moisture content of the meat. Samples were then cooked, as such (uncured) or cooked with 150 mg/kg of sodium nitrite (nitrite-cured). Flavour concentrates were prepared, as described below. The cooked meat was homogenized with distilled water and the slurry in the extraction jar was subjected to heating at 65 ± 5°C. A slow stream of nitrogen was used to purge the volatiles from the headspace and the effluent stream was passed through a solution of 2,4-dinitrophenylhydrazine to remove the carbonyl compounds. The effluent from this treatment was taken into n-pentane at –60°C. This cold trap was connected to an aspirator. The volatiles collected over a 10 h period were dried over anhydrous sodium sulphate and concentrated to 250 μl using a stream of nitrogen gas.

The gas chromatograms of the separated constituents from the nitrogen purge-and-trap (NPT) technique, from the above preparation, of cured pork aroma concentrates of pork, partially defatted pork and fully defatted pork are shown in Figure 13.5. Removal of the depot fats (partially defatted cured pork) did not alter the volatiles of the cooked meat markedly (Figure 13.5(b)). However, fully defatted cured pork, which had a pale colour due to the removal of pigments during the extraction, showed that carbonyls with retention times in the region of 7–15 min were completely removed. In addition, there is an increase in the relative concentration of less volatile substances with retention times of 15–30 min which belong to heterocyclic and phenolic compounds.

Similar results were obtained when partially defatted and fully defatted pork samples were cooked in the absence of nitrite. Organoleptic properties of uncured cooked pork and partially defatted cooked pork and the gas chromatographic profiles of the aroma concentrates from these two meat products were found to be nearly identical (results not shown). Striking similarities were also noted in the odour descriptions of cured pork and fully defatted pork aroma concentrates with 'cured-ham' characteristics. Similar results were obtained when completely defatted beef and chicken were used. Thus, we concluded that 'cured-ham' aroma or the basic 'meaty' aroma of the cooked pork, beef and poultry should essentially comprise of a collection of selected heterocyclic components.

Quantitative differences in the composition of aroma concentrates prepared by the NPT method were determined by using pentanal and hexanal as internal standards. The components identified in the aroma concentrates of defatted uncured and cured pork, beef and chicken, prepared by the NPT method, are listed in Table 13.7. A total of 57 compounds were detected in the three meat species, of which there were

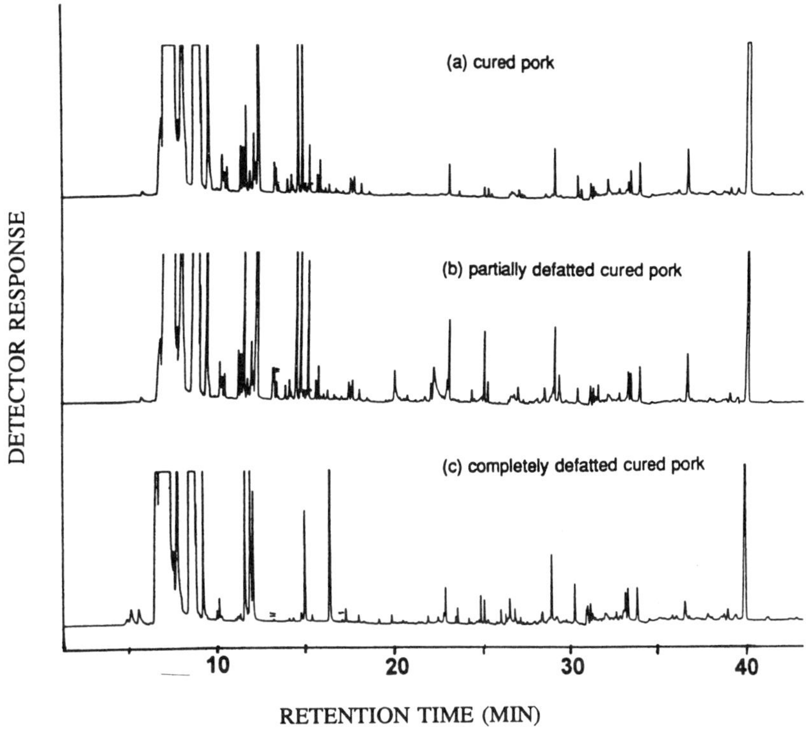

Figure 13.5 Gas chromatograms of volatile components in the aroma concentrates of cured pork, prepared by the NPT method: (a) nitrite cured pork; (b) partially defatted (hexane extraction) and cured pork; (c) completely defatted ($CHCl_3$–MeOH extraction) and nitrite-cured pork.

30 hydrocarbons, three carbonyls, five alcohols, six phenols, five esters and eight heterocyclic compounds. The results shown in Table 13.7 are clearly indicative of the mild nature of the NPT method that limits the breakdown of volatile components and protects them from undergoing further oxidation due to the use of the inert nitrogen gas. Furthermore, the use of defatted meat and the carbonyl-specific reagent also drastically reduced the formation and presence of carbonyls.

Using the same NPT technique, we found, for the first time, heterocyclic compounds tetrahydro-*cis*-2,4-dimethylfuran, 3-propyl-1*H*-1,2,4-triazole and 2,4,6-trimethylpyridine in the aroma concentrates of pork. Hydrocarbons such as 1,1-dimethylcyclopentane, 2-methylundecane and 4-methyl-1-decene were also present uniquely in the pork aroma concentrates. In addition, 4-ethylbenzaldehyde, diethylphthalate, 1,12-dodecanediol and 2,6-bis(1,1-dimethylethyl)-4-methylethyl-4-methylphenol were uniquely identified in chicken. Meanwhile, turpene hydrocarbons such as α-pinene and caphene were present only in beef aroma concentrates

Table 13.7 Concentration of individual compounds identified in the aroma concentrates of defatted pork, beef and chicken isolated by the nitrogen purge-and-trap method

RT (min)	Compound	Kovats index[a]	Pork (mg/kg)[b]		Beef (mg/kg)[b]		Chicken (mg/kg)[b]	
			Uncured	Cured	Uncured	Cured	Uncured	Cured
3.59	2-methyl-3-hexanone	734	0.05	0.08	0.10	0.22	– –	
3.64	2,4-dimethylhexane	736	0.22	0.25	0.56	0.41	0.21	0.25
4.16	methylbenzene	763	5.25	5.01	6.22	6.06	5.14	3.76
4.48	2,2,4-trimethylhexane	777	2.66	2.47	4.08	2.76	0.72	0.96
5.08	2,3,5-trimethylhexane	810	1.02	1.16	0.96	0.86	0.41	0.52
5.47	4-ethylheptane	831	–	–	0.11	0.14	–	–
5.50	4-ethyl-2-methylhexane	833	0.18	0.21	–	–	0.06	0.12
6.00	1,2-dimethylbenzene	860	–	–	0.16	–	–	–
6.11	1,3-dimethylbenzene	866	0.29	–	0.36	0.56	–	–
6.18	2,2,5,5-tetramethylhexane	870	0.07	0.21	0.25	0.29	–	–
6.28	2,2,4-trimethylheptane	875	0.18	0.25	1.10	0.86	0.38	0.29
6.58	1,4-dimethylbenzene	891	–	–	0.25	–	–	–
7.04	3,3-diethylpentane	917	–	–	0.29	0.21	–	–
7.40	α-pinene	937	–	–	0.17	–	–	–
7.88	2,2,6-trimethyloctane	963	–	–	0.42	0.29	–	–
8.23	7-octen-4-ol	983	2.64	–	4.01	–	2.66	–
8.50	1,3,5-trimethylbenzene	998	–	–	0.38	–	–	–
8.62	tetrahydro-*cis*-2,4-dimethyfuran	1004	0.66	–	–	–	–	–
9.02	2,2-diethyldecane	1028	–	–	0.16	0.04	–	–
9.12	D-limonene	1034	–	0.25	0.35	0.59	–	0.11
9.73	2,2,4,6,6-pentamethylheptane	1070	–	–	0.25	0.31	0.15	–
9.84	octanol	1078	3.56	0.07	4.22	0.11	3.81	0.16
10.13	4-ethyl-1,2-dimethylbenzene	1094	–	–	0.15	–	–	–
10.38	(*trans*)-1,2-dimethylcyclopentane	1110	0.95	0.53	1.78	0.40	2.09	–
11.32	3-propyl-1*H*-1,2,4-triazole	1172	0.86	–	–	–	–	–
11.46	4-ethylbenzaldehyde	1181	–	–	–	–	0.49	–
11.74	1,1-dimethylcyclopentane	1199	1.05	–	–	–	–	–
11.85	camphene	1206	–	–	0.25	0.19	–	–
11.91	3,6-dimethylundecane	1210	–	–	–	0.08	–	–

Table 13.7 *continued*

11.94	2-methylundecane	1213	0.07	0.03	–	–	–	–
12.02	2-methylcyclopentanol	1217	–	0.45	–	0.37	–	0.16
13.00	4-methyl-1-decene	1278	0.05	–	–	–	–	–
13.11	nonylcyclopropane	1285	2.11	–	2.76	–	2.46	–
13.23	1-nonen-3-ol	1293	0.61	–	–	–	–	–
13.50	5-propyldecane	1311	0.96	–	0.31	–	0.35	–
14.58	2-butyl-2-octenal	1389	–	1.62	1.47	–	–	2.95
14.88	2,3,5-trimethyldecane	1411	0.44	0.14	0.82	0.25	0.26	0.14
15.93	(1,1-dimethylethyl)-4-methoxyphenol	1486	0.45	–	0.56	–	0.34	0.49
16.21	pentadecane	1500	0.11	–	–	–	0.21	–
16.49	2,6-bis(1,1-dimethylethyl)-4-methylphenol	1527	–	–	–	–	–	1.25
17.47	diethylphthalate	1603	–	–	–	–	0.29	0.16
18.86	4-(2,2,3,3-tetramethylbutyl)phenol	1725	1.46	0.76	0.96	1.22	–	–
18.89	1,12-dodecanediol	1728	–	–	–	–	1.25	0.74
18.95	4-nonylphenol	1733	2.26	1.67	2.81	2.45	3.11	2.62
19.03	1,3-dihydro-2*H*-imidazo(4,5-*b*)pyridin-2-one	1740	2.54	1.82	2.21	1.76	1.44	1.16
19.08	2,4,6-trimethylpyridine	1744	0.86	0.24	0.96	0.42	–	–
19.14	4-(1-methylpropyl)phenol	1750	1.16	0.96	1.21	1.05	0.78	0.45
19.24	3-amino-5,6-dimethyltriazolo-(4,3-*a*)pyrazine	1758	0.46	0.22	1.45	1.01	0.96	0.45
19.30	3-methyl-1,2-benzisothiazole	1763	0.32	0.21	0.27	0.15	0.31	0.22
19.40	4-ethyl-2,6-dimethylpyridine	1772	0.82	0.45	0.61	0.74	0.95	0.74
19.54	2-butylphenol	1784	0.21	0.11	0.45	0.27	0.19	0.12
19.61	2,4-diphenyl-1*H*-pyrrole	1790	0.64	0.48	0.62	0.39	0.58	0.47
20.69	(*E*)-5-octadecene	1894	0.11	0.25	–	0.15	–	–
21.62	bis(2-methoxyethyl)phthalate	1985	3.21	2.95	3.68	3.14	3.45	2.72
22.87	methyl-11,14-eicosadienoate	2118	1.05	0.72	0.94	0.86	0.95	0.72
22.98	methyl-14-hydroxy-5-tetradecenoate	2130	1.25	0.84	2.42	1.92	1.86	0.92
26.90	bis(2-ethylhexyl)phthalate	2563	0.75	0.22	0.34	0.39	1.05	0.82

[a]Kovats indices calculated for the DB-5 capillary column of the GC–MS system.

[b]Average of two determinations; –, not detected.

(Ramarathnam *et al.*, 1993a,b). Other compounds that were present only in the volatiles of beef were 4-ethylheptane, 1,2-dimethylbenzene, 1,4-dimethylbenzene, 3,3-diethylpentane, 2,2,6-trimethyloctane, 1,3,5-trimethylbenzene, 2,2-diethyldecane, 4-ethyl-1,2-dimethylbenzene and 3,6-dimethylundecane. The compounds 7-octen-4-ol, nonylcyclopropane and 5-propyldecane were found in the aroma concentrates of all uncured meat samples, while 2-methylcyclopentanol was unique to the cured-meat aroma. The list of compounds that have been uniquely identified in one species, but absent in the other two, is highlighted in Table 13.8.

13.6 Conclusions

In the study of meat flavour volatiles, much attention has been focused on the characterization of the key components responsible for the flavour of meat from different species. Though higher in fat content, the number of volatiles detected in pork is far fewer than in beef, in which more than 700 components have been detected in the past two decades (Shahidi *et al.*, 1986). This could be due to the extensive investigations carried out on beef mainly because of its commercial importance and consumer preference (Baines and Mlotkiewicz, 1984). Nevertheless, the current literature available on meat flavour does not provide a clear path to the formulation of essences that could impart ‘meaty’ or ‘cured-meat’ type flavour notes.

Of the various components identified in the present investigation in the three meat species 4-methyl-2-pentanone, 2,2,4-trimethylhexane and 1,3-dimethylbenzene could be contributing either directly as individual constituents or indirectly as synergists in the formation of the ‘cured-meat’ aroma. Though these components were detected in small amounts in the cured-meat flavour concentrates of all three species, they were, however, absent in the cooked uncured meat. 16-Octadecenal, benzaldehyde, 2,3-octanedione and (*E*,*Z*)-2,4-decadienal may be responsible for the species-specific flavour notes in pork, while 2-hexanone and 3,3-dimethylhexanal have been uniquely identified in beef. The characteristic ‘chicken-like’ flavour perhaps includes a complex mixture of 3-hexanone, 3-hexenal, 3-methyl-4-heptanone, 3-methylhexanal, (*E*)-2-heptenal, octanal, (*E*)-2-octenal, nonanal, 16-octadecenal, 4-ethylbenzaldehyde and decanal. The preparation of a ‘nature-identical’ chicken flavour would involve a sophisticated methodology, both in the sensory evaluation and in the formulation.

Furthermore, by employing the nitrogen purge-and-trap technique and using defatted meat samples, we have quantified the levels of heterocyclic and phenolic components in the three main species. Irrespective of the species, the aroma extracts of fully defatted and cooked meat, both cured

Table 13.8 Volatile components uniquely identified in the aroma concentrates of defatted pork, beef and chicken isolated by the nitrogen purge-and-trap method

RT (min)	Compound	Kovats Index[a]	Content (mg/kg)[b]	
			Uncured	Cured
Pork				
8.62	tetrahydro-*cis*-2,4-dimethyfuran	1004	0.66	–
11.32	3-propyl-*H*-1,2,4-triazole	1172	0.86	–
11.74	1,1-dimethylcyclopentane	1199	1.05	–
11.94	2-methylundecane	1213	0.07	0.03
13.00	4-methyl-1-decene	1278	0.05	–
13.23	1-nonen-3-ol	1293	0.61	–
Beef				
5.47	4-ethylheptane	831	0.11	0.14
6.00	1,2-dimethylbenzene	860	0.16	–
6.58	1,4-dimethylbenzene	891	0.25	–
7.04	3,3-diethylpentane	917	0.29	0.21
7.40	α-pinene	937	0.17	–
7.88	2,2,6-trimethyloctane	963	0.42	0.29
8.50	1,3,5-trimethylbenzene	998	0.38	–
9.02	2,2-diethyldecane	1028	0.16	0.04
10.13	4-ethyl-1,2-dimethylbenzene	1094	0.15	–
11.85	camphene	1206	0.25	0.19
11.91	3,6-dimethylundecane	1210	–	0.08
Chicken				
11.46	4-ethylbenzaldehyde	1181	0.49	–
16.49	2,6-bis(1,1-dimethylethyl)-4-methylphenol	1527	–	1.25
17.47	diethylphthalate	1603	0.29	0.16
18.89	1,12-dodecanediol	1728	1.25	0.74

[a]Kovats indices calculated for the DB-5 capillary column of the GC–MS system.
[b]Average of two determinations; –, not detected.

and uncured, had identical organoleptic properties. The extracts were also found to contain higher levels of heterocyclic and phenolic compounds than those published earlier (Ramarathnam, 1993a). Thus, results of our work demonstrates that formulation of 'cured-meat' flavour is indeed impossible.

References

Bailey, M.E. and Swain, J.W. (1973). Influence of nitrite on meat flavour. In *Proceedings of the Meat Industry Research Conference*, American Meat Science Association, Chicago, pp. 29–45.

Baines, D.A. and Mlotkiewicz, J.A. (1984). The chemistry of meat flavour. In *Recent Advances in the Chemistry of Meat*, ed. A.J. Bailey. Royal Society of Chemistry, London, pp. 119–164.

Binkerd, E.F. and Kolari, O.E. (1975). The history and use of nitrate and nitrite in the curing of meat. *Food Cosmet. Toxicol.*, **13**, 655–661.

Chang, S.S. and Peterson, R.J. (1977). Symposium: the basis of quality in muscle foods. Recent developments in the flavour of meat. *J. Food Sci.*, **42**, 298–305.

Crocker, E.C. (1948). The flavour of meat. *Food Res.*, **13**, 179–183.

Cross, C.K. and Ziegler, P. (1965). A comparison of the volatile fractions from cured and uncured meats. *J. Food Sci.*, **30**, 610–614.
Dwivedi, B.K. (1975). Meat flavour. *CRC Crit. Rev. Food Technol.*, **5**, 487–535.
Eakes, B.D., Blumer, T.N. and Monroe, R.J. (1975). Effect of nitrate and nitrite on colour and flavour of country-style hams. *J. Food Sci.*, **40**, 973–976.
Fogerty, A.C., Whitfield, F.B., Svoronos, D. and Ford, G.L. (1990). Changes in the composition of the fatty acids and aldehydes of meat lipids after heating. *Int. J. Food Sci. Technol.*, **25**, 304–312.
Fooladi, M.H., Pearson, A.M., Coleman, T.H. and Merkel, R.A. (1979). The role of nitrite in preventing development of warmed-over flavour. *Food Chem.*, **4**, 283–292.
Giddings, C.G. (1977). The basis of colour in muscle foods. *CRC Crit. Rev. Food Sci. Nutr.*, **9**, 81–114.
Golovnya, R.V., Garbuzov, V.G., Grigor'eva, I.Ya., Zharich, S.L. and Bol'shakov, A.S. (1982). Gas chromatographic analysis of volatile sulphur-containing compounds of salted and salted-boiled pork. *Nahrung*, **26**, 89–96.
Gray, J.I., MacDonald, B., Pearson, A.M. and Morton, I.D. (1981). Role of nitrite in cured meat flavour. A review. *J. Food Prot.*, **44**, 302–312.
Hadden, J.P., Ockerman, H.W., Cahill, V.R., Parrett, N.A. and Borton, R.J. (1975). Influence of sodium nitrite on the chemical and organoleptic properties of comminuted pork. *J. Food Sci.*, **40**, 626–630.
Hauschild, A.H.W., Hilsheimer, R., Jarvis, G. and Raymond, D.P. (1982). Contribution of nitrite to the control of *Clostridium botulinum* in liver sausage. *J. Food Prot.*, **45**, 500–506.
Herz, K.O. and Chang, S.S. (1970). Meat flavor. *Adv. Food Res.*, **18**, 1–83.
Hornstein, I. and Crowe, P.F. (1960). Flavour studies on beef and pork. *J. Agric. Food Chem.*, **8**, 494–498.
Jacobsen, M. and Koehler, H.E. (1963). Components of the flavour of lamb. *J. Agric. Food Chem.*, **11**, 336–339.
Jensen, L.B. (1953). Early preparations of foods. In *Man's Foods*. The Garrard Press, Champaign, IL, pp. 159–170.
Jensen, L.B. (1954) Introduction and history. In *Microbiology of Meats*. The Garrard Press, Champaign, IL, pp. 1–11.
Kerr, R.H., Marsh, C.T.N., Shroeder, W.F. and Boyer, E.A. (1926). The use of sodium nitrite in curing of meat. *J. Agric. Res.*, **33**, 541–551.
Langer, H.J., Heckel, V. and Malek, E. (1970). Aroma substances in ripening dry sausage. *Fleischwirtschaft*, **50**, 1193–1199.
Lillard, D.A. and Ayres, J.C. (1960). Flavour compounds in country-cured hams. *Food Technol.*, **23**, 251–254.
Love, J.D. and Pearson, A.M. (1976). Metmyoglobin and nonheme iron as prooxidants in egg-yolk phospholipid dispersions and cooked meat. *J. Agric. Food Chem.*, **24**, 494–498.
MacDonald, B., Gray, J.I. and Gibbins, L.N. (1980). Role of nitrite in cured meat flavour: antioxidant role of nitrite. *J. Food Sci.*, **45**, 893–897.
MacDougall, D.B., Mottram, D.S. and Rhodes, D.N. (1975). Contribution of nitrite and nitrate to the colour and flavour of cured meats. *J. Sci. Food Agric.*, **26**, 1743–1754.
MacLeod, G. and Ames, J.M. (1986). Capillary gas chromatography-mass spectrometric analysis of cooked ground beef aroma. *J. Food Sci.*, **51**, 1427–1434.
MacLeod, G. and Seyyedain-Ardebili, M. (1981). Natural and simulated meat flavours (with particular reference to beef). *CRC Crit. Rev. Food Sci. Nutr.*, **14**, 309–437.
Maier, V.P. and Tappel, A.L. (1959). Studies of unsaturated fatty acid oxidation catalyzed by hematin compounds. *J. Am. Oil Chem. Soc.*, **36**, 12–15.
Minor, L.J., Pearson, A.M., Dawson, L.E. and Schweigert, B.S. (1965). Chicken flavour: the identification of some chemical components and the importance of sulphur compounds in the cooked volatile fraction. *J. Food Sci.*, **30**, 686–696.
Moody, W.G. (1983). Beef flavour – a review. *Food Technol.*, **37**, 227–232, 238.
Mottram, D.S. (1984). Organic nitrates and nitriles in the volatiles of cooked cured pork. *J. Agric. Food Chem.*, **32**, 343–345.
Mottram, D.S. and Rhodes, D.N. (1974). Nitrite and the flavour of cured meat. In *Proceedings of the International Symposium of Nitrite Meat Products*, Zeist, Pudoc, Wageningen, Holland, p. 161.

Ockerman, H.W., Blumer, T.W. and Craig, H.B. (1964). Volatile chemical compounds in dry-cured hams. *J. Food Sci.*, **29**, 123–129.
Pearson, A.M., Love, J.D. and Shorland, F.B. (1977). Warmed-over flavour in meat, poultry and fish. *Adv. Food Res.*, **23**, 1–74.
Pierson, M.D. and Smoot, L.A. (1982). Nitrite, nitrite alternatives, and the control of *Clostridium botulinum* in cured meats. *CRC Crit. Rev. Food Sci. Nutr.*, **17**, 141–187.
Piotrowski, E.G., Zaika, L.L. and Wasserman, A.E. (1970). Studies on aroma of cured ham. *J. Food Sci.*, **35**, 321–325.
Ramarathnam, N., Rubin, L.J. and Diosady, L.L. (1991a). Studies on meat flavour. 1. Qualitative and quantitative differences in uncured and cured pork. *J. Agric. Food Chem.*, **39**, 344–350.
Ramarathnam, N., Rubin, L.J. and Diosady, L.L. (1991b). Studies on meat flavour. 2. A quantitative investigation of the volatile carbonyls and hydrocarbons in uncured and cured beef and chicken. *J. Agric. Food Chem.* **39**, 1839–1847.
Ramarathnam, N., Rubin, L.J. and Diosady, L.L. (1993a). Studies on meat flavor. 3. A novel method for trapping volatile components from uncured and cured pork. *J. Agric. Food Chem.*, **41**, 933–938.
Ramarathnam, N., Rubin, L.J. and Diosady, L.L. (1993b). Studies on meat flavor. 4. Fractionation, characterization, and quantitation of volatiles from uncured and cured beef and chicken. *J. Agric. Food Chem.*, **41**, 939–945.
Ramaswamy, H.S. and Richards, J.F. (1982). Flavour of poultry meat – a review. *Can. Inst. Food Sci. Technol. J.*, **15**, 7–18.
Rhee, K.S. (1989). Chemistry of meat flavour. In *Flavour Chemistry of Lipid Foods*, eds. D.B. Min and T.H. Smouse. American Chemical Society, Champaign, IL, pp. 166–189.
Rubin, L.J. and Shahidi, F. (1988). Lipid oxidation and the flavour of meat products. In *Proceedings of the 34th International Congress of Meat Science Technology*, Brisbane, Australia, pp. 295–301.
Sanderson, A., Pearson, A.M. and Schweigert, B.S. (1966). Effect of cooking procedure on flavour components of beef – carbonyl compounds. *J. Agric. Food Chem.*, **14**, 245–247.
Sato, K. and Hegarty, G.R. (1971). Warmed-over flavour in cooked meats. *J. Food Sci.*, **36**, 1098–1102.
Shahidi, F. (1989). Flavour of cooked meats. In *Flavour Chemistry – Trends and Developments*, eds. R. Teranishi, R.G. Buttery and F. Shahidi. American Chemical Society, Washington, DC, pp. 188–201.
Shahidi, F., Rubin, L.H. and D'Souza, L.A. (1986). Meat flavour volatiles: a review of the composition, techniques of analysis, and sensory evaluation. *CRC Crit. Rev. Food Sci. Nutr.*, **24**, 141–243.
Shahidi, F., Rubin, L.J. and Wood, D.F. (1987a). Control of lipid oxidation in cooked ground pork with antioxidants and dinitrosyl ferrohemochrome. *J. Food Sci.*, **52**, 564–567.
Shahidi, F., Yun, J., Rubin, L.J. and Wood, D.F. (1987b). The hexanal content as an indicator of oxidative stability and flavour acceptability in cooked ground pork. *Can. Inst. Food Sci. Technol. J.*, **20**, 104–106.
Skjelvale, R. and Tjaberg, T.B. (1974). Comparison of salami sausage produced with and without addition of sodium nitrite and sodium nitrate. *J. Food Sci.*, **39**, 520–524.
Tappel, A.L. (1952). Linoleate oxidation catalyzed by hog muscle and adipose tissue extract. *Food Res.*, **17**, 550–559.
Wasserman, A.E. (1979). Symposium on meat flavour. Chemical basis for meat flavour: a review. *J. Food Sci.*, **44**, 6–11.
Watts, B.M. (1954). Oxidative rancidity and discolouration in meat. *Adv. Food Res.*, **5**, 1–52.
Wettasinghe, M. and Shahidi, F. (1997). Oxidative stability of cooked comminuted lean pork as affected by alkali and alkali-earth halides. *J. Food Sci.*, **61**, 1160–1164.
Wood, D.S., Collins-Thompson, D.L., Usborne, W.R. and Picard, B. (1986). An evaluation of antibotulinal activity in nitrite-free curing systems containing dinitrosyl ferrohemochrome. *J. Food Prot.*, **49**, 691–695.
Younathan, M.T. and Watts, B.M. (1959). Relationship of meat pigments to lipid oxidation. *Food Res.*, **24**, 728–734.
Yun, J., Shahidi, F., Rubin, L.J. and Diosady, L.L. (1987). Oxidative stability and flavour acceptability of nitrite-free meat-curing systems. *Can. Inst. Food Sci. Technol. J.*, **20**, 246–251.

14 Flavour analysis of dry-cured ham

M. FLORES, A.M. SPANIER and F. TOLDRÁ

14.1 Introduction

Cured products represent a large portion of processed meats usually classified as wet- or dry-cured (Flores and Toldrá, 1993). In wet-cured products the curing ingredients are dissolved in water to form a pickle or brine which is introduced or injected into the meat. On the other hand, dry curing is a traditional process where the curing ingredients are rubbed onto the surface of the meat. In general, the dry-curing process consists of several stages: salting (about 9–11 days at 2–4°C), washing, post-salting (20–40 days at 2–4°C) for salt equalization, and ripening–drying (7–12 months at 12–22°C). These dry-cured products are subjected to extensive ripening or aging periods where the generation of dry-cured flavour takes place through the action of biochemical reactions of a proteolytic and lipolytic nature. The main curing ingredients are salt, nitrate and/or nitrite. The primary role of salt in curing is to act as a bacteriostatic agent but it also affects the flavour of the product and controls the muscle enzyme system (Toldrá *et al.*, 1997a).

Dry-cured hams are products mostly typical from the Mediterranean area such as the Spanish Serrano, Italian Parma and French Bayonne hams. The processing of drying or ripening has many variations depending on the traditions of the countries. Other products from northern European countries and American Country-style ham are subjected to a final smoking stage.

The high quality of dry-cured ham depends on its unique flavour. However, increased production cost of long-term dry curing makes the product less competitive in the market. Several studies have attempted to reduce the processing time (Marriot *et al.*, 1987, 1992), but the length of the ripening–drying stage is necessary in order to have a complete cured colour formation and dry-cured flavour development (Toldrá *et al.*, 1997b).

The components responsible for the characteristic aroma and taste for dry-cured ham by sensory and instrumental methods were studied. Data from sensory analysis were related to flavour components in order to fully understand the nature of dry-cured flavour. Spanish 'Serrano' dry-cured hams were processed in a local factory in Spain and utilized in this study. Twenty hams were processed under traditional practices of salting, post-salting and ripening–drying. Ten of the hams were dried for 7 months (short process), and the other ten hams were dried for a total of 12 months (long process).

14.2 Sensory characteristics of dry-cured ham

Flavour, defined as the impressions perceived via the chemical senses from a product in the mouth (Caul, 1957), includes the aromatics (olfactory perceptions), tastes (gustatory perceptions) and chemical feeling factors (Meilgaard *et al.*, 1991).

The studies that have attempted to described the dry-cured ham flavour on Italian (Careri *et al.*, 1993; Virgili *et al.*, 1995; Hinrichsen and Pedersen, 1995), French (Berdague *et al.*, 1993; Buscailhon *et al.*, 1994a) and Spanish type dry-cured ham (Motilva *et al.*, 1994) have used terms that have not been well defined; these terms include dry-cured flavour, aged taste, and aroma typical of dry-ham, all of which are very subjective depending on the origin of the product.

The need for a lexicon of precise descriptors representing the flavour of dry-cured ham encouraged us to apply descriptive sensory methods to dry-cured ham. Descriptive analysis is a sensory method by which the attributes of a food or product are identified and quantified, using a panel trained specifically for this task. In order to have a complete, detailed and accurate descriptive characterization of a product's sensory attributes, the spectrum descriptive analysis method (Muñoz and Civille, 1992) was applied to two different processes of Spanish 'Serrano' dry-cured ham (Flores *et al.*, 1997a).

During training, the panel described the flavours detected in dry-cured products, such as Italian-type dry-cured ham, Country-style ham and Spanish 'Serrano' ham. A lexicon for dry-cured ham flavour was developed (Flores *et al.*, 1997a) and those attributes considered as most representative of desirable and off-flavours of dry-cured ham were selected for analysis (Table 14.1).

Sensory analysis of the processed ham samples from the two processes resulted in the observation of a significant differences in four of the eight attributes studied. Boar taint, barnyard, sour and salty were higher in the long-processed hams (12 months) than in the short-processed hams (7 months). These changes may be explained by the longer processing time

Table 14.1 Spanish 'Serrano' dry-cured flavour descriptors used in this analysis

Descriptor	Description
Fat complex (FCX)	The aromatics associated with lipid products such as animal fat
Boar taint (BOR)	The aromatic associated with boar meat; hormone-like (skatole)
Barnyard (BYD)	The aromatics associated with free fatty acids
Pork (PRK)	The aromatics associated with cooked pork muscle meat
Sour (SOR)	The taste on the tongue associated with citric acid
Salty (SLT)	The taste on the tongue associated with sodium ions
Bitter (BTR)	The taste on the tongue associated with caffeine
Mouthfilling (MTH)	Mouthfeel associated with monosodium glutamate (MSG)

resulting in an increase in volatile fatty acids (Motilva *et al.*, 1993) and thus in a higher barnyard attribute. The higher sour taste observed in the long process could be due to an increase in free amino acids, such as aspartic and glutamic acids (Toldrá *et al.*, 1995; Flores *et al.*, 1997c). The higher salty taste detected in the long-processed hams was due to a lower water activity. Interestingly, a smoky descriptor was defined by the panellists although the processing of 'Serrano' dry-cured ham does not include a smoking stage.

Boar taint is a flavour attribute defined by the aromatics associated with hormone-like odours. Pearson *et al.* (1995) have shown that the boar taint descriptor is not only associated with hormone-like compounds such as androstenones, C_{19}-Δ^{16} steroids, but also with a compound named skatole (3-methylindole) that is produced in the digestive tract of pigs by microbial breakdown of tryptophan. Androstenones possess a penetrating, perspiration- or urine-like odour; the odour of skatole has been defined as fecal. Due to the lipophilicity and hydrophobicity of androstenones these substances are good predictors of boar taint in backfat (Lundström *et al.*, 1988) while skatole, which is both fat and water soluble, is a good determinant for boar taint in fat and also for the taste of lean muscle. The large increment of the amino acid tryptophan during the dry-curing process may be responsible for the increase of boar taint observed in the long process.

14.3 Aroma contributors in dry-cured ham

Meat flavour quality is affected by several factors including antemortem and postmortem factors. However, the most remarkable factors are those involved in further processing of the meat (Toldrá *et al.*, 1997a) which, in the case of dry-curing processing, are due to traditional practices specific to each country.

The study of the volatile compounds characteristic of dry-cured ham flavour aroma is affected by many factors. Not only does processing affect volatile composition, but also different analytical methodologies such as solvent extraction, vacuum distillation, dynamic headspace, etc., influence the results. Previous studies on Country-style hams (Lillard and Ayres, 1969; Ockerman *et al.*, 1964) have indicated the presence of several carbonyls, alcohols and esters. Recent investigations were performed to identify and quantify the volatile compounds in French (Berdague *et al.*, 1991, 1993; Buscailhon *et al.*, 1993), Italian (Barbieri *et al.*, 1992; Careri *et al.*, 1993, Hinrichsen and Pedersen, 1995; Bolzoni *et al.*, 1996), and Spanish (Garcia *et al.*, 1991, Lopez *et al.*, 1992) dry-cured hams. Around 261 compounds were already detected (Table 14.2). The studies of volatile compounds by Hinrichsen and Pedersen (1995), Barbieri *et al.* (1992),

Buscailhon *et al.* (1993) and Bolzoni *et al.* (1996) were obtained by dynamic headspace analysis, while those of Garcia *et al.* (1991) and Berdague *et al.* (1991) were obtained by vacuum distillation. The difference in the extraction technique resulted in more carboxylic acids, lactones and aliphatic hydrocarbons found in the case of the vacuum distillation technique (Table 14.2) than by dynamic headspace. The number of identified aldehydes and alcohols was important in the three different dry-cured ham products (Italian, French and Spanish) while the total number of identified ester compounds was higher in the Italian product (Table 14.2). The volatile compounds consisted of a wide range of organic classes, but only a few correlations have been established between these volatile compounds and the sensory characteristics such as the Italian (Careri *et al.*, 1993; Hinrichsen and Pedersen, 1995) and French (Buscailhon *et al.*, 1994a) and none in the Spanish dry-cured ham. Moreover, there has been no indication that any specific compound or components of dry-cured ham possess the characteristic 'cured' aroma.

Dynamic headspace analysis of Spanish 'Serrano' dry-cured ham prepared by the two curing processes (short and long) was performed by purge-and-trap concentration of the volatile compounds on Tenax. Over 100 volatile compounds were detected (Figure 14.1) and 84 were identified (Table 14.3; Flores *et al.*, 1997b). Molecular weights of the identified compounds ranged from 46 for ethanol to 184 for branched-chain hydrocarbons. The headspace volatiles of the short and long curing processes contained essentially the same components. The total quantity of extracted volatiles were similar. Changes affecting chemical classes cannot be considered globally because of individual variations (Table 14.3). Ranking of the chemical classes following their quantitative importance was: 17 branched-chain hydrocarbons (47–48% of the total volatile area); 17 alcohols (16–17%); 11 aldehydes (13–15%); 9 ketones (6–7%); 7 aliphatic hydrocarbons (6.0–6.6%); 10 esters (3–3.5%); 4 aromatic hydrocarbons (1.3–1.4%); 2 furans (0.75%); 2 halides (0.6–0.9%); 3 nitrogen compounds (0.5%); and others (1.7–2.2%).

The impact of an odour component on the total aroma bouquet depends on a number of factors, such as odour threshold, concentration in the measured material, solubility in water or fat, and temperature. During the olfactory test 44 odour descriptors were repeatedly reported by the four trained panellists (Table 14.3). Many of the compounds responsible for these odours were identified, except odours corresponding to peaks 29, 54, 57, 74 and 141 due to their low concentrations or because they eluted in the area between 33 and 47 min where the branched-chain hydrocarbons also eluted thereby interfering in the identification of these compounds. These branched-chain hydrocarbons represented the group with the highest concentration (approximately 48% of the total area in both processes). The high flavour thresholds of hydrocarbons make minimal contribution to desirable and undesirable flavours (Min *et al.*,

Table 14.2 Volatile compounds previously reported in dry-cured ham

Aldehydes
(E)-2-Heptenal[4]
(E)-2-Nonenal[2]
(E)-2-Octenal[4]
(E)-2-Pentenal[4]
(E,E)-2,4-Decadienal[2]
(E,E)2,4-Pentadienal[4]
(E,Z)-2,4-Decadienal[2]
(Z)-2-Decenal[4]
(Z)-2-Nonenal[4]
2,4-Octadienal[3]
2,4-Nonadienal[3]
2-Decenal[1]
2-Dodecenal[1]
2-Furaldehyde[1]
2-Heptenal[3,7]
2-Hexenal[1,3,4,6]
2-Methyl-2-butenal[3]
2-Methyl-2-pentenal[1]
2-Methyl-2-propenal[3]
2-Methylbutanal[2,3,4,7]
2-Methylpropanal[3,6]
2-Nonenal[5]
2-Octenal[3,7]
3-Methyl-2-butenal[1]
3-Methylbutanal[2–7]
3-Methylhexanal[6]
4-Methyl-2-pentenal[1]
9-Octadecanal[5]
Acetaldehyde[3]
Benzaldehyde[1,3,4]
Benzenacetaldehyde[3,5]
Butanal[2,4]
Decanal[1–6]
Dodecanal[1,5]
Heptanal[1–4,7]
Hexadecanal[2,5]
Hexanal[1–7]
Hexenal[3]
Nonanal[1–3,5,6]
Octanal[1,3–7]
Octadecanal[5]
Octadecenal[5]
Pentanal[1–4,6,7]
Pentadecanal[5]
Phenylacetaldehyde[1,2]
Propanal[3]
Tetradecanal[5]
Undecanal[5]

Alcohols
(E)-2-Octen-1-ol[4]
(Z)-2-Octen-1-ol[4]
1,4-Butanediol[1]
1,2-Propanediol[4]
1-Butanol[4,6,7]
1-Butoxyethoxyethanol[5]
1-Butoxy-2-propanol[4,7]
1-Decanol[1]
1-Dodecanol[1]
1-Ethylcyclopropanol[3]
1-Heptanol[2,4,7]
1-Hexanol[1–5,7]
1-Octanol[1,4,5]
1-Octen-3-ol[2–7]
1-Pentanol[1–7]
1-Penten-3-ol[3,4,6,7]
1-Propanol[3,4,7]
1-Tetradecanol[2]
2-Butanol[3,4,7]
2-Butoxyethanol[2,4,6,7]
2-Ethyl-1-hexanol[4]
2-Ethoxyethoxyethanol[2]
2-Hepten-1-ol[4]
2-Hexanol[4]
2-Methyl-1-propanol[4,6,7]
2-Methyl-2-buten-1-ol[4,7]
2-Methyl-3-buten-2-ol[2,4,7]
2-Methylbutan-1-ol[1–3,5,6]
2-Pentanol[3,4,7]
2-Propanol[3]
3,7-Dimethyl-1-octen-3-ol[4]
3-Methyl-3-buten-1-ol[4,6,7]
3-Methyl-butan-1-ol[1–7]
3-Penten-1-ol[3]
5-Methylheptan-2-ol[1]
cis-3-Hexen-1-ol[1]
Epoxydihydrolinalool[4]
Ethanol[3,6]
Furfuryl alcohol[1]
Methanol[3]
Phenylethanol[2,5]

Aliphatic hydrocarbons
Alkyl cyclopentane[6]
1-Heptene[4,7]
1-Octene[4,6,7]
2,2,3-Trimethylpentane[4,5]
2,2,4,6,6-Pentamethylheptane[6]
3-Methylnonane[6]
4-Methylheptane[5]
4,6,8-Trimethylnonene[4]
Branched alkane[1]
Decane[2,5,6]
Decene[3]
Dimethylundecane[3]
Docosane[2]
Dodecane[2,5]
Heneicosane[2,5]
Heptadecane[2,5]
Heptane[3,4,6,7]
Hexane[6]
Hexadecane[2]
Myrcene[1]
Methyldecane[5]
Nonadecane[2]
Nonane[1,3,5,6]
Nonene[3]
Octadecane[2,5]
Octane[3,4,6,7]
Pentane[4,6]
Pentadecane[2,5]
Tetradecane[2]
Tridecane[2,5]
Undecane[2,5]

Aromatic hydrocarbons
1,2,3-Trimethylbenzene[1,4,6]
1,2-Dimethylbenzene[1,3–6]
1,2-Dimethyl-3-ethylbenzene[1]
1,2,4-Trimethylbenzene[5]
1,3-Dimethylbenzene[4–7]
1,4-Dimethylbenzene[1,4,6,7]
1-Ethyl-2-methylbenzene[4,6]
1-Methoxyhexane[4]
1-Methyl-3-methylethylbenzene[1]
Benzene[3,4,7]
Benzonitrile[4]
Ethenylbenzene[1,4,6]
Ethylbenzene[1,4,5,7]
Isopropylbenzene[4]
Methylvinylbenzene[4]
tert-Butylbenzene[4]
Toluene[3–7]

Ketones
1-Hydroxy-2-propanone[4]
2-Hexanone[6]
2-Hydroxy-3-pentanone[6]
2-Methyloctan-3-one[2]
2-Propanone[3,4,6,7]
2-Undecanone[4]
3-Hexanone[6]
3-Hydroxybutan-2-one[2,4–7]
3-Methyl-2-butanone[2,4–7]
3-Methyl-2-pentanone[6]
3-Octen-2-one[1]
4-Octen-3-one[3,4,7]
4-Methyl-2-pentanone[7]
6-Methyl-5-hepten-2-one[4,6,7]
Branched ketone[1]
Butan-2-one[2–7]
Butan-2,3-dione[2,3,6,7]
Cyclohexanone[4]
Heptan-2-one[2,3,5,6]
Nonan-2-one[3]
Octan-2-one[5,6]
Pentan-2,3-dione[2–4,6]
Pentan-2-one[1,3,4,6,7]
trans-Geranylacetone[5]

Table 14.2 *continued*

Carboxylic acids	Methyl decanoate[4]	**Chloride compounds**
2-Methylpropanoic acid[3]	Methyl hexanoate[4,7]	2-Chloronaphthalene[2]
3-Methylbutanoic acid[3]	Methyl hexadecanoate[5]	2,2-Dichloroethanol[6]
9-Hexadecenoic acid[5]	Methyl heptanoate[4,7]	Dichloromethane[6]
9,12-Octadecadienoic acid[5]	Methyl nonanoate[4]	Dichlorobenzene[6]
9-Octadecenoic acid[5]	Methyl octanoate[4,7]	Tetrachloroethene[6]
Acetic acid[1,3,4]	Methyl pentanoate[4,7]	Tricloromethane[2,5,6]
Butanoic acid[1,4]	Methyl propanoate[4,7]	**Furans**
Hexanoic acid[1,4,5]	Methyl 2-methylbutanoate[4,6,7]	2-Ethylfuran[4,6]
Hexadecanoic acid[5]	Methyl 2-methylpropanoate[4,7]	2-Pentylfuran[4,7]
Heptanoic acid[5]	Methyl 3-hexenoate[4]	2-Methyl-4,5-dihydrofurane[1]
Isohexanoic acid[1]	Methyl 3-methylbutanoate[4,7]	2,2,4-Trimethyl-2,5-dihydrofurane[1]
Isooctanoic acid[1]	Methyl 3-octenoate[4]	2,5-Dimethyltetrahydrofurane[1]
Octanoic acid[1,5]	Methyl 4-methylhexanoate[4,7]	**Pyrazines**
Pentadecanoic acid[2,5]	Methyl 4-methylpentanoate[4]	Pyrazine[4]
Pentanoic acid[1,3,4]	Methyl 4-decenoate[4]	2-Methylpyrazine[4]
Propanoic acid[3,4]	Methyl 5-hexenoate[4,7]	2,6-Dimethylpyrazine[4–7]
Nonanoic acid[5]	Methyl 6-methylheptanoate[4]	Trimethylpyrazine[4]
Undecanoic acid[5]	Propyl acetate[3]	**Miscellaneous**
Esters	**Lactones**	1-Methyl-2-pyrrolidinone[5]
Alkyl phathalate[2]	δ-Butyrolactone[1,3]	1,2-Benzenedicarboxylic acid[5]
Butyl acetate[6]	γ-Butyrolactone[2]	2,4,6-Trimethyl-1,3,5-trioxane[5]
Ethyl acetate[3,4,6,7]	γ-Hexalactone[2,5,6]	3-Vinylpyridine[4]
Ethyl butanoate[3,4,7]	γ-Nonalactone[2,5]	BHT[2,5]
Ethyl decanoate[2,4]	γ-Octalactone[2]	Dimethylphenol[6]
Ethyl 2-methylbutanoate[2–4,6,7]	γ-Pentalactone[6]	Farnesol[5]
Ethyl 3-methylbutanoate[3,4,7]	γ-Valerolactone[2]	Isovaleramide[1]
Ethyl pentanoate[3]	**Sulphur compounds**	Limonene[4,6]
Ethyl propanoate[3,4]	3-Methylthiopropanol[2]	Pentyloxyrane[1]
Ethyl hexanoate[1–4,7]	Dimethyl disulphide[3,4,6,7]	Phenylethylamine[1]
Ethyl hexadecanoate[2]	Dimethyl trisulphide[3,4,7]	Propanodiamide[1]
Ethyl heptanoate[3]	Dimethyl tetrasulphide[4]	Pyrrole lactone[2,4]
Ethyl octanoate[2–4,7]	Methyl 3-(methylthio)propanoate[4]	
Hexyl butyrate[1]	Methyl ethyl disulphide[3]	
Linalyl acetate[1]	Methyl n-pentyl disulphide[3]	
Methyl acetate[4]		
Methyl benzoate[4]		
Methyl butanoate[4,7]		
Methyl carbamate[1]		

[1]Lopez *et al.* (1992); [2]Garcia *et al.* (1991); [3]Hinrichsen and Pedersen (1995); [4]Barbieri *et al.* (1992); [5]Berdague *et al.* (1991); [6]Buscailhon *et al.* (1993); [7]Bolzoni *et al.* (1996).

1979; Chang and Petersen, 1977), although some unsaturated hydrocarbons are moderately potent odorants (Forss, 1972). On the other hand, the aromatic hydrocarbons have completely different characteristics such as *o*-xylene and *p*-xylene that give sweet and fruity odours, respectively (Shahidi *et al.*, 1986). In 'Serrano' dry-cured ham, (*m*- or *p*-)xylene gave an aroma described as smoked-phenolic while *o*-xylene gave a sweet-fruit candy flavour (Table 14.3).

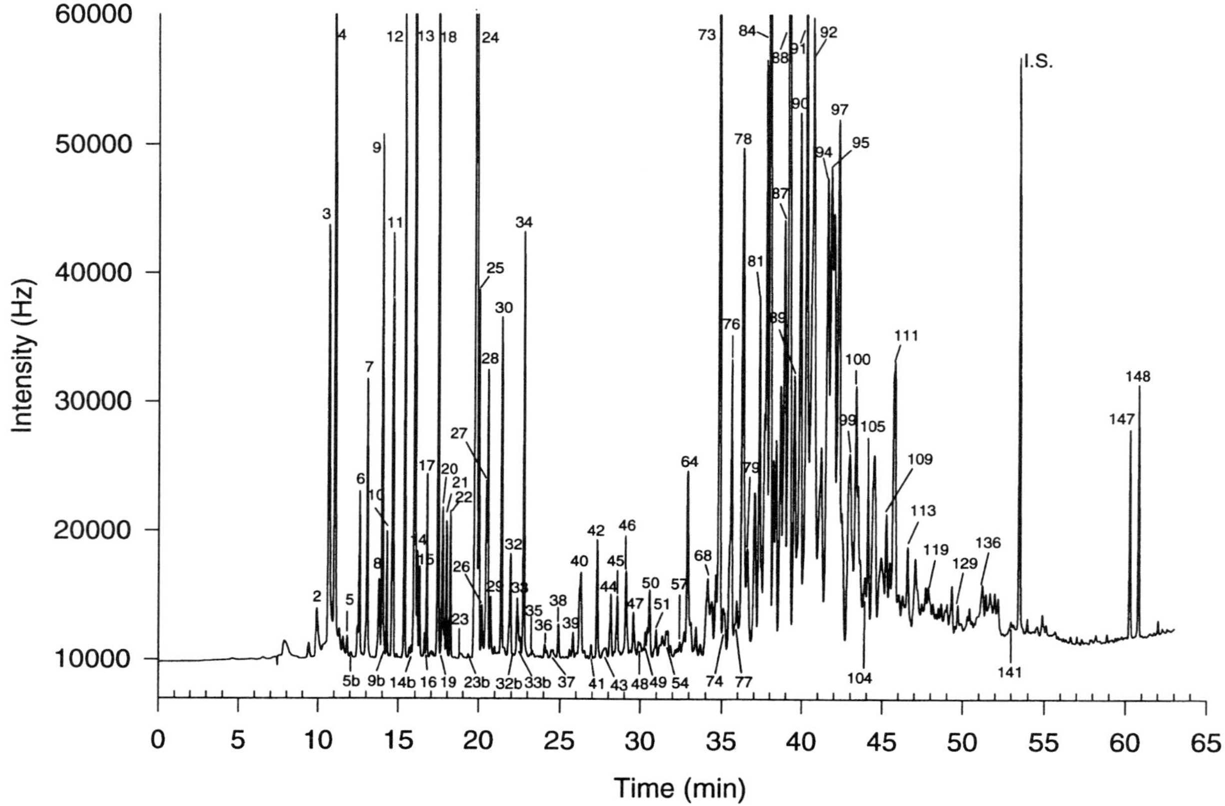

Figure 14.1 Typical GC–FID chromatogram of Spanish 'Serrano' dry-cured ham. Numbers refer to the compounds identified in Table 14.3.

The long processing of the dry-cured hams (12 months) provides ample time for the occurrence of lipolytic and oxidative degradation of unsaturated fatty acids which are found in abundance in both intramuscular and adipose tissues of swine. Flores *et al.* (1985) have reported that the generation of the characteristic aroma of dry-cured ham was in agreement with the beginning of lipid oxidation. Aliphatic hydrocarbons are derived from oxidative decomposition of lipids (Shahidi *et al.*, 1986). Most of the 16 alcohols identified are oxidative decomposition products of lipids. For example, 1-propanol and 1-butanol may be derived from miristoleic acid, 1-pentanol from linoleic acid, 1-hexanol may be formed from palmitoleic and oleic acids, and 1-octanol from oleic oxidation (Forss, 1972). Linolenate oxidation is the origin of 1-penten-3-ol which has a penetrating, grassy etheral odour (Shahidi *et al.*, 1986). The methyl-branched alcohols are probably derived from the Strecker degradation of amino acids. Isopropanol (2-propanol) can be distinguished from the other alcohols because of its high concentration, although its level was increased considerably in the long process (Table 14.3) in contrast with 2-methylpropanol, 1-butanol, 1-penten-3-ol, 2-pentanol, and 1-hexanol that were reduced in the long process. The flavour of alcohols in meat products was considered unimportant due to their relatively higher threshold value as compared to other carbonyl compounds (Drumm and Spanier, 1991). It has been shown that the straight-chain primary alcohols (C_2 and C_3 1-alkanols and 2-alkanols) are relatively flavourless, but as the carbon chain increases, the flavour becomes stronger (Forss 1972; Shahidi *et al.*, 1986) giving greenish, woody and fatty floral notes. In Serrano dry-cured ham, only three alcohols were related to specific aromas, e.g. 1-penten-3-ol gave an onion–toasted aroma, 3-methyl-1-butanol gave a penetrating green aroma odour and 3-methyl-2-hexanol was defined as having potato–wheat aroma (Table 14.3).

Many of the aldehydes are formed by oxidation of unsaturated fatty acids such as hexanal that comes from linoleic acid oxidative decomposition. Aldehydes have been shown to be contributors to the overall loss of desirable flavour in meats because of their low threshold for eliciting a flavour response and their high rate of formation during lipid oxidation (Frankel, 1984). This is the case of the high concentration of octanal and 3-methylbutanal, although the content of both compounds decreased during long processing. Drumm and Spanier (1991) reported that unsaturated aldehydes undergo further oxidation to shorter-chain aldehydes as happens with 2,4-decadienal and 2-undecenal, which were higher in the short than in the long process (Table 14.3). St. Angelo *et al.* (1980) suggested that 2,4-decadienal is an oxidation product of linoleic acid. On the other hand, branched aldehydes such as 2-methylpropanal, 2-methylbutanal and 3-methylbutanal are not usually derived from lipid oxidation but often arise from Strecker degradation of the amino acids

Table 14.3 Volatile compounds and aromas of Spanish 'Serrano' dry-cured ham

N[a]	Compound	RI[b]	KI[c]	Short[d]	Long	P[e]	Aroma
2	Ethanol	a	522	0.83 ± 0.04	0.83 ± 0.04	ns	
3	2-Propanone	a	542	1.03 ± 0.07	1.34 ± 0.09	*	
4	2-Propanol	a	555	1.55 ± 0.13	3.66 ± 0.36	***	
5	Methyl acetate	a	569	0.20 ± 0.01	0.19 ± 0.01	ns	
5b	2-Methylpropanol	a	594	nq	nq		
6	Hexane	a	600	0.83 ± 0.09	0.71 ± 0.03	ns	Green–mould
7	1-Propanol	a	613	0.59 ± 0.05	0.52 ± 0.05	ns	
8	2,3-Butanedione	a	633	0.44 ± 0.02	0.37 ± 0.02	*	Buttery
9	2-Butanone	a	638	1.04 ± 0.06	1.06 ± 0.06	ns	
9b	Ethyl acetate	a	641	nq	nq		
10	2-Butanol	a	647	0.54 ± 0.03	0.53 ± 0.05	ns	
11	Chloroform	a	657	0.57 ± 0.04	0.37 ± 0.02	***	Green
12	2-Methylpropanol	a	677	1.03 ± 0.08	0.48 ± 0.03	***	
13	2-Methyl-3-buten-2-ol	c	689	0.09 ± 0.01	0.09 ± 0.01	ns	
14	3-Methylbutanal	a	694	3.38 ± 0.36	2.32 ± 0.26	*	Cheesy–green
14b	Acetic acid	a	681	nq	nq		Vinegar
15	Heptane	a	700	0.59 ± 0.13	1.53 ± 0.22	***	
16	2-Methylbutanal	a	702	0.44 ± 0.03	0.66 ± 0.12	ns	
17	Methyl-2-methyl-propanoate	b	709	0.14 ± 0.01	0.12 ± 0.01	ns	
18	1-Butanol	a	712	0.39 ± 0.02	0.28 ± 0.01	***	
19	2,5-Dimethylfuran	a	721	0.06 ± 0.01	0.06 ± 0.01	ns	Sulphury–fishy
19b	2-Pentanone	a	726	nq	nq		
20	1-Penten-3-ol	a	728	1.54 ± 0.15	0.62 ± 0.04	***	Onion–toasted
21	Pentanal	a	734	0.53 ± 0.03	0.53 ± 0.05	ns	
22	2-Pentanol	a	740	0.84 ± 0.07	0.46 ± 0.02	***	
23	Methyl butanoate	a	745	0.38 ± 0.02	0.28 ± 0.02	***	Sweet-caramel
23b	3-Hydroxy-2-butanone	a	780	nq	nq		Fruit red-jello
24	Dimethyl disulphide	a	782	2.92 ± 0.25	1.49 ± 0.12	***	Dirty socks
25	3-Methyl-1-butanol	a	786	1.03 ± 0.11	1.07 ± 0.10	ns	Penetrating green
26	2-Methyl-1-butanol	a	790	0.42 ± 0.02	0.42 ± 0.04	ns	
27	Toluene	a	797	0.41 ± 0.02	0.36 ± 0.02	ns	
28	Octane	a	800	0.49 ± 0.03	0.48 ± 0.02	ns	
29	Unknown	–	802	0.44 ± 0.03	0.43 ± 0.03	ns	Sweet–tuti fruity
30	1-Octene	b	814	1.21 ± 0.11	0.53 ± 0.02	***	
32	Ethyl butanoate	a	823	0.52 ± 0.02	0.44 ± 0.01	**	Fruity
32b	3-Methyl-2-butenol	a	824	nq[f]	nq		
33	2-Hexanone	a	831	0.64 ± 0.02	0.54 ± 0.02	**	Floral–apple
33b	1-*H*-pyrrole	a	835	nq	nq		Meaty
34	Hexanal	a	838	0.91 ± 0.05	0.66 ± 0.03	***	Green–grassy
35	Methylpyrazine	a	861	0.09 ± 0.01	0.08 ± 0.01	*	Nutty
36	2,2-Dichloroethanol	b	861	0.10 ± 0.01	0.08 ± 0.01	**	Citrus-lemon
37	Ethyl 1-methylbutanoate	c	868	0.12 ± 0.01	0.11 ± 0.01	ns	
38	Ethyl 2-methylbutanoate	a	875	0.28 ± 0.01	0.32 ± 0.01	*	Fruity–strawberry
39	Ethyl 3-methylbutanoate	c	892	0.15 ± 0.01	0.12 ± 0.01	**	
40	Nonane	a	900	0.75 ± 0.10	0.61 ± 0.07	ns	Crackers
41	(*p*- or *m*-)Xylene	a	910	0.20 ± 0.01	0.17 ± 0.01	**	Smoked–phenolic
42	1-Hexanol	a	915	0.77 ± 0.05	0.45 ± 0.02	***	
43	Styrene	a	923	0.10 ± 0.01	0.10 ± 0.01	ns	Medicinal
44	*o*-Xylene	a	928	0.39 ± 0.02	0.35 ± 0.02	ns	Sweet–fruit candy
45	2-Heptanone	a	934	0.66 ± 0.03	0.53 ± 0.03	**	
46	Heptanal	a	942	0.78 ± 0.04	0.66 ± 0.03	*	
47	3-methyl-2-hexanol	b	947	0.23 ± 0.01	0.25 ± 0.03	ns	Potato–wheat
48	Methyl hexanoate	a	954	0.58 ± 0.02	0.55 ± 0.03	ns	Boiled meat
49	2,6-Dimethylpyrazine	a	955	0.28 ± 0.01	0.28 ± 0.02	ns	Toasted nuts

Table 14.3 *continued*

N[a]	Compound	RI[b]	KI[c]	Short[d]	Long	P[e]	Aroma
50	2-Butoxyethanol	a	960	0.54 ± 0.03	0.60 ± 0.04	ns	Dark toast–meaty
51	Hexylpropanoate	b	964	0.34 ± 0.02	0.30 ± 0.02	ns	
54	Unknown	–	973	0.30 ± 0.02	0.25 ± 0.01	*	Nutty–pecans
57	Unknown	–	981	0.42 ± 0.01	0.38 ± 0.02	ns	Boiled potatoes
64	Decane	a	1000	0.76 ± 0.13	0.48 ± 0.04	*	
68	2-Pentylfuran	a	1014	0.51 ± 0.02	0.50 ± 0.02	ns	Ham-like
73	Branch hydrocarbon (MW 156)	c	1028	3.86 ± 0.24	3.24 ± 0.21	ns	Green
74	Unknown	–	1035	1.29 ± 0.22	1.86 ± 0.26	ns	Musty–earthy
76	Branch hydrocarbon (MW 156)	c	1040	0.39 ± 0.06	0.53 ± 0.07	ns	Putrid–sulphur
77	6-Methyl-5-heptan-2-one	a	1044	0.64 ± 0.02	0.65 ± 0.04	ns	Citrus–candy
78	Octanal	a	1049	3.23 ± 0.17	2.65 ± 0.15	*	Green–fresh
79	Branch hydrocarbon (MW 170)	c	1055	1.56 ± 0.16	1.94 ± 0.18	ns	Mouldy nuts
84	Branch hydrocarbon (MW 170)	c	1068	2.51 ± 0.15	1.99 ± 0.13	*	Toasted–smoky
87	2,2-Dimethyloctanol	c	1077	1.17 ± 0.05	1.20 ± 0.05	ns	
90	Branch hydrocarbon (MW 170)	c	1088	4.88 ± 0.25	4.56 ± 0.22	ns	Boiled meat
97	1-Octanol	a	1122	2.15 ± 0.09	1.97 ± 0.08	ns	
99	Branch hydrocarbon (MW 170)	c	1128	2.73 ± 0.12	2.67 ± 0.11	ns	Nutty
100	Branch hydrocarbon (MW 170)	c	1135	1.20 ± 0.10	1.51 ± 0.12	*	Floral–talc
104	5-Ethyl-3-methyl-5-hepten-2-one	b	1148	0.58 ± 0.03	0.66 ± 0.04	ns	
105	Nonanal	a	1151	0.77 ± 0.03	0.73 ± 0.03	ns	Green
111	Branch hydrocarbon (MW 170)	c	1177	1.93 ± 0.07	1.78 ± 0.07	ns	Oxone–fresh
119	Dodecane	a	1200	0.54 ± 0.02	0.49 ± 0.02	*	
136	Decanal	a	–	0.44 ± 0.02	0.41 ± 0.02	ns	
141	Unknown	–	–	0.67 ± 0.02	0.64 ± 0.03	ns	Brothy–meaty
147	2,4-Decadienal (*E*,*Z*)	c	–	0.51 ± 0.02	0.44 ± 0.02	**	
148	2-Undecenal	c	–	0.60 ± 0.02	0.50 ± 0.02	**	

[a]*N*: number of peak as seen in Figure 14.1. [b]RI: Reliability of identification. [c]KI: Kovats index. [d]Results expressed as means of ten samples ± sem of the area of GC–FID peak normalized to the area of the internal standard. [e]*P*: ns, not significant, *significant at < 0.05, **significant at $P < 0.01$, ***significant at $P < 0.001$. [f]nq: not quantified. (Adapted from Flores *et al.* 1997b.)

valine, isoleucine and leucine, respectively (Forss, 1972). The intense proteolytic activity produced during the dry-curing process results in an increased concentration of free amino acids (Toldrá *et al.*, 1995) that serves as a pool for the Strecker reaction. C_3 and C_4 aldehydes have sharp and irritating flavours; intermediate (C_5–C_9) aldehydes have green, oily, fatty, tallowy flavours and the higher (C_{10}–C_{12}) aldehydes have citrus, orange peel flavours (Forss, 1972). In the headspace of the 'Serrano' dry-cured ham four aldehydes were identified as responsible for four aroma descriptors, such as 3-methylbutanal (cheesy–green, aromatics associated with the smell of cheese and green cut grass), hexanal (green–grassy, aromatics associated with the smell of green cut grass), octanal (green–fresh) and nonanal (green).

Aliphatic esters are very important constituents of foods, particularly in fruits. In meats, the intramuscular fat plays an important role in flavour retention. It acts as a solvent for the aroma volatiles formed in the muscle lean tissue and serves as a site for further reactions. The esters are formed from the interaction of free fatty acids and alcohols generated by lipid oxidation in the intramuscular tissue (Baines and Mlokiewicz, 1984; Shahidi *et al.*, 1986). Baines and Mlokiewicz (1984) reported that esters from C_1–C_{10} acids tend to impart a fruity sweet note to pork meat whereas esters derived from long-chain fatty acids give a more fatty flavour character as found in beef. In the headspace of 'Serrano' dry-cured ham, esters from C_1–C_{10} acids were mainly found that gave fruity (methyl 2-methylpropanoate, ethyl butanoate, ethyl 2-methylbutanoate) and sweet-caramel (methyl butanoate) aromas. Only methyl hexanoate was responsible for the boiled meat aroma (Table 14.3).The esters have been found in greater amounts and are characteristic in the aroma of Italian dry-cured ham (Barbieri *et al.*, 1992) but have not been previously found in Spanish-type hams (Garcia *et al.*, 1991; Lopez *et al.*, 1992) or at a very low percentage in French hams (Berdague *et al.*, 1991; Buscailhon *et al.*, 1993). It should be noted that nitrate is not used in the processing of Italian (Parma) ham (Parolari, 1996), while it is usually added to Spanish and French dry-cured hams. The inhibitory effect of nitrate/nitrite on lipid oxidation is the most likely reason for the lower concentration of esters found in Spanish and French hams. Microbial influence on lipid oxidation is considered minimal, based on the low microbial counts associated with Spanish and Italian products (Molina and Toldrá, 1992; Baldini *et al.*, 1992).

Seven saturated ketones were found in the Serrano dry-cured ham (Table 14.3). Acetone was present at the highest concentration and represented 1.81%, compared to 36–78% of the total volatile area that it constituted in French dry-cured ham (Buscailhon *et al.*, 1993; Berdague *et al.*, 1993). This important difference in the content of acetone could be due to the starting material and processing technique used. Moreover, the French dry-cured hams have been characterized for their high content of ketones, in contrast with the 7% of the total volatile area found in the Italian type (Barbieri *et al.*, 1992) that is very close to the 6–7% found in the Spanish 'Serrano' ham. Ketones are also produced via lipid oxidation. One of them, 2-heptanone, has been found to be an oxidation product of linoleic acid (St. Angelo *et al.*, 1980), although its mechanism of formation is not clear (Forss, 1972; Shahidi *et al.*, 1986). Unsaturated ketones are responsible for characteristic flavour notes in animal and vegetable fats (Forss, 1972). 3-Hydroxy-2-butanone gives a buttery note to cooked meat (Shahidi *et al.*, 1986), but in 'Serrano' dry-cured ham it was identified as being responsible for a fruit red-jello aroma (aromatics associated with the smell of commercial red-strawberry jello) and 2,3-butanedione was responsible for a buttery note. Two other ketones, namely 2-hexanone and 6-methyl-5-heptan-2-one, were responsible for a floral–apple and a citrus–candy aroma, respectively (aromatics associated with the smell of candies with citrus aromas; Table 14.3).

Two furans were identified in the volatiles of dry-cured ham (Table 14.3) although as a chemical class they are considered to contribute little to the basic meaty taste. However, they may contribute to the overall odour of broiled and roasted meats. One of them, 2-pentylfuran, has been suggested to be an oxidation product of linoleic acid (Forss, 1972; St. Angelo *et al.*, 1980; Baines and Mlotkiewicz, 1984).

Of the total volatile compounds in Serrano dry-cured ham only acetic acid, a carboxylic acid, was found; however, it was not possible to quantify it accurately (Table 14.3). This is remarkable in comparison to the large number of carboxylic acids identified in other hams, such as French (Berdague *et al.*, 1991) and Italian (Barbieri *et al.*, 1992; Hinrichsen and Petersen, 1995) ham. The free fatty acids originate from the action of enzymes on triacylglycerol and phospholipids (Motilva *et al.*, 1992, 1993), as these enzymes are active during the curing process.

Two pyrazines were found in the headspace volatiles of 'Serrano' dry-cured ham (Table 14.3). Pyrazines are products of Maillard reactions that occur extensively during meat cooking (Mottram and Edwards, 1983; Mottram, 1985). Moreover, the proportion of pyrazines increased in cooked meat as a function of curing time, with longer curing resulting in higher values, probably due to increases in the content of sugars and free amino acids associated with aging (Maga, 1982). The temperature used during the dry-curing process is not as high as in cooking, thereby fewer pyrazines are found in the cured products. These pyrazines have been recognized as the volatiles contributing to roasted aromas of cooked foods, such as nutty, green, earthy, potato-like, etc. (Maga, 1982). The processing of Serrano dry-cured ham resulted in the generation of two pyrazines that gave nutty (methylpyrazine) and toasted nuttty (2,6-dimethylpyrazines) aromas (Table 14.3).

Sulphides are formed principally from sulphur-containing amino acids, such as methionine, cysteine and cystine via Strecker degradations to thiols (Shahidi *et al.*, 1986). Dimethyl disulphide is an oxidation product of methanethiol, and can react to form dimethyl trisulphide and dimethyl sulphide (Drumm and Spanier, 1991). These sulphur compounds are important contributors to meat flavour because of their low flavour threshold (Chang and Petersen, 1977; Drumm and Spanier, 1991). Only dimethyl disulphide was detected giving an unpleasant aroma defined as dirty socks, but it was reduced to half the concentration during the long-curing process (Table 14.3).

Two halides (chloride compounds) were detected both at lower concentrations in the long than in the short process (Table 14.3). Their origin probably arises from pesticide residues ingested by the pigs (Buscailhon *et al.*, 1993).

In summary, the drying-ripening processes may be differentiated by 3-methylbutanal, octanal and dimethyl disulphide which were present at higher levels in the short (7 months) than in the long (12 months) processed hams. The short process was also characterized by a high content of 2-methylpropanol, 1-penten-3-ol and 1-octene, while the long process was characterized by high contents of 2-propanol, 2-propanone and heptane. On the other hand, none of the volatile compounds identified had the characteristic cured aroma, although some of the meaty aromas detected were due to 2-pentylfuran (ham-like), 1*H*-pyrrole (meaty) and 2-butoxyethanol (dark toast–meaty) (Table 14.3).

14.4 Taste contributors in dry-cured ham

The contributions of proteinaceous components, peptides and amino acids to the improvement of meat taste have been shown to occur both during postmortem aging (Nishimura *et al.*, 1988; Kato *et al.*, 1989; Spanier *et al.*, 1990) and at different cooking temperatures (Spanier *et al.*, 1988; Spanier and Miller, 1993). The increase in the content of free amino acids has been attributed to the action of muscle aminopeptidases which are active at neutral pH (Nishimura *et al.*, 1988, 1990). On the other hand, Spanier *et al.* (1990) have found that thiol proteinases, cathepsins B and L, are the best candidates for involvement in the production of peptide flavour precursors during beef postmortem aging and both retained significant activity even after cooking. These peptides can be further degraded by aminopeptidase activity, thereby adding to the pool of free amino acids available for flavour production.

Many of the changes produced during the processing of dry-cured ham are the result of endogenous hydrolytic activity, since low amounts of microorganisms were found inside the hams (Toldrá and Etherington, 1988; Molina and Toldrá, 1992). Many of these proteolytic changes were attributed to enzymes such as cathepsins, calpains and aminopeptidases (Toldrá *et al.*, 1997a). Cathepsins were found to be active through the entire process (Toldrá and Etherington, 1988; Toldrá *et al.*, 1993) while the activity of calpains was restricted to the initial curing stage (Rosell and Toldrá, 1996). The last step in the proteolytic process was the generation of free amino acids by the action of aminopeptidases on peptides (Flores *et al.*, 1993, 1996b). These aminopeptidases have also been found to be active during the processing of dry-cured ham (Toldrá *et al.*, 1992, 1995).

The effect of the two different curing processes (short and long) on the concentration of free amino acid content of Spanish 'Serrano' dry-cured ham is shown in Table 14.4. All amino acids exhibited increased concentrations during the long curing process when compared with the short curing process. The additional curing time (5 months) allows the proteolytic system, which is active during almost all the dry-cured process, to continue its action (Toldrá *et al.*, 1993). The generation of free amino acids is attributed to the action of exopeptidases especially alanyl aminopeptidase and aminopeptidase B which have been characterized from porcine skeletal muscle (Flores *et al.*, 1993, 1996a). Alanyl aminopeptidase may be mainly responsible for the increase in amino acids, due to its broad substrate specificity and because it accounts for more than 80% of the total aminopeptidase activity found in porcine skeletal muscle (Flores *et al.*, 1996a; Toldrá *et al.*, 1995).

In some cases, a decrease in concentration has been observed during the very latest stage of processing (McCain *et al.*, 1968; Toldrá and Aristoy, 1993). Buscailhon *et al.* (1994b) have reported changes in free amino acid concentrations during processing of French dry-cured ham. They observed a marked decrease in the concentration from 6 months of curing to 9 months. In this study (7 months to 12 months), a reduction in concentration was not detected except for asparagine and glutamine. These differences are thought to be due to the different processing technique employed when compared with the French process. Additionally, different origins of raw material may affect the generation of free amino acids since different patterns of muscle proteolytic and lipolytic enzymes have been shown among pig breed types (Flores *et al.*, 1994; Toldrá *et al.*, 1996).

Table 14.4 Amino acid and peptide concentrations in the short (7 months) and long (12 months) processes of Spanish 'Serrano' dry-cured ham

Amino acid	Short[a]	Long	Amino acid	Short	Long
Aspartic acid***[b]	98.8	267.4	Arginine***	186.6	295.7
Glutamic acid***	306.3	538.6	Proline***	113.6	207.7
Serine***	122.0	212.6	Anserine***	28.3	39.8
Asparagine[ns]	38.1	36.6	Tyrosine***	80.7	142.3
Glycine***	109.5	196.4	Valine***	145.8	272.9
Glutamine***	30.2	10.3	Methionine***	55.1	119.1
β-Alanine**	4.6	5.8	Isoleucine***	113.7	209.3
Taurine**	81.3	71.7	Leucine***	184.4	335.9
Histidine***	86.8	165.6	Phenylalanine***	108.3	191.6
γ-Aminobutyric*	5.5	14.6	Tryptophan***	24.2	46.1
Threonine***	119.9	226.1	Ornithine***	12.4	41.5
Alanine***	203.4	344.4	Lysine***	299.8	608.5
Carnosine***	499.4	664.8			

[a]Results are expressed as means of ten samples (mg free amino acid/100 g ham). [b]*P*: ns, nonsignificant, *significant at $P < 0.05$, **significant at $P < 0.01$, ***significant at $P < 0.001$.

The contents of dipeptides such as carnosine (β-alanin-histidine) and anserine (β-alanine-1-methylhistitine) seem to increase in the long process (Table 14.4), but these were reported on a wet weight basis and remained constant when expressed on a dry weight basis. Nishimura and Kato (1988) reported that both dipeptides have significant buffer action at a pH above 6.0, thereby playing an important role in the taste of foods, since the taste of ionic components of foods is strongly dependent on pH (Fuke and Konosu, 1991).

Capillary zone electrophoresis of the dry-cured ham extracts indicated significant changes in peptide mapping between long- and short-processed hams (Figure 14.2). Ten peaks exhibiting absorbance at 200 nm were detected (Flores *et al.*, 1997c). Peaks 2, 7, 8, 9 and 10 exhibited the same electrophoretic mobility as standards of carnosine, hypoxanthine, tryptophan, phenylalanine and tyrosine, respectively. All peaks were increased in the long process, except peak 6 which almost disappeared and peak 5 which showed a slight decrease in peak area. In order to study the possible contribution of peptides to the flavour of 'Serrano' dry-cured ham, the size of the peptides was determined. Peaks 1–5 and 7–10 had a molecular weight lower than 3000 Da. On the other hand, peak 6 showed a molecular weight between 20 and 30 kDa indicating that it could be responsible for the generation of lower molecular weight precursors. The increase in peak area of almost all the peptides was expected due to the longer time that hams were exposed to the action of proteolytic enzymes. On the other hand, the slight reduction of the area of peak 5 could be due to inactivation of the protease that produces it from a longer parent protein and/or further proteolysis by other enzymes.

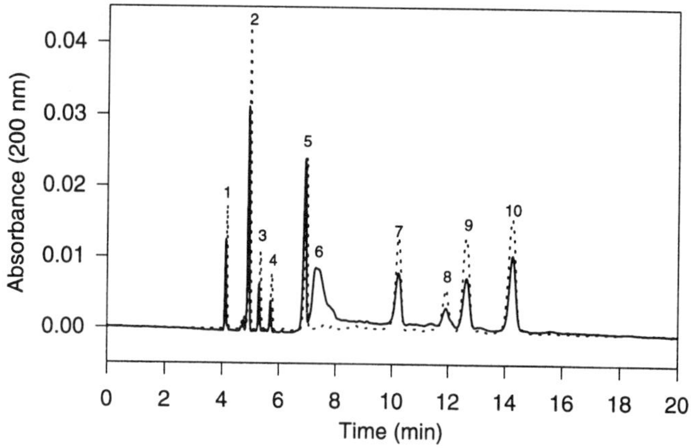

Figure 14.2 Capillary zone electropherogram of short (—) and long (- - -) processed Spanish 'Serrano' dry-cured ham.

14.5 Relations between sensory analysis and flavour components

The study of the contribution of volatile and nonvolatile components to the generation of dry-cured flavour was performed by analysing the data using the multivariate statistical method of factor analysis. One of the primary goals of multivariate factor analysis is to discover the minimum number of common factor axes that is needed to reproduce adequately the variation in the observed variables. Factors are usually considered interpretable when some observed variables load highly on them and the rest do not. This allows interpretation of the factor in terms of the high-loading variables. Loadings in excess of 0.30 are eligible for interpretation. Also, the factors are characterized by assigning them a name or a label having meaning in the science of application (Tabachnick and Fidell, 1983).

A principal factor analysis using an oblique (Promax) rotation was performed on the sensory and chemical data (SAS Institute, Inc., Cary, NC). Because the examination of all the components required an impractical amount of replication, we selected those volatile compounds that were responsible for specific aromas in the olfactory test to see their relation with the sensory descriptors. In addition, seven amino acids were selected based on their specific tastes, and peptides 3, 5 and 6, and peak 7, identified as hypoxanthine, according to their elution from the capillary electrophoresis system (the three latest peaks were proven to be amino acids).

The multivariate factor solution consisted of five factors (Table 14.5). We described Factor 1 as 'length of curing' because the response variables of those amino acids experienced a high increment in the process and exhibited the strongest loadings on the factor. Moreover, peptide 6 (PEP-6), which almost disappears in the long process, showed a strong negative correlation with the factor. We defined Factor 2 as 'pleasant aroma' because the response variables of peaks P-8 (2,3-butanedione, buttery), P-23 (methyl butanoate, sweet–caramel), P-35 (methylpyrazine, nutty),

Table 14.5 Variation in sensory, amino acid, peptide and GC-peak loadings onto common factors

Factor	1. Length	2. Pleasant	3. Pork	4. Off-flavour	5. Cured
+	Leu (0.94) Ile (0.93) Met (0.92) Glu (0.92) Lys (0.88) Asp (0.86) PEP-3 (0.75) PEP-7 (0.68)	P-68 (0.95) P-35 (0.83) P-50 (0.79) P-44 (0.72) P-8 (0.69) P-23 (0.69)	P-14 (0.80) P-34 (0.61) P-24 (0.55)	Pep-5 (0.78) Asn (0.63)	SOR (0.66) BTR (0.65) MTH (0.62)
–	PEP-6 (–0.79)		PRK (–0.57) FCX (–0.43)	BYD (–0.65) BOR (–0.52) SLT (–0.44)	

P-44 (*o*-xylene, sweet, fruit-candy), P-50 (2-butoxyethanol, dark toast–meaty) and P-68 (2-pentylfuran, ham-like) exhibited the strongest loadings on Factor 2. We named the third factor 'pork flavour' because it was defined by the sensory descriptors fat complex (FCX) and pork (PRK); furthermore it had a negative correlation with peaks P-14 (3-methylbutanal, cheesy–green), P-24 (dimethyl disulphide, dirty socks) and P-34 (hexanal, green–grassy). The fourth factor was named 'off-flavour' because it was defined by salty (SLT), boar taint (BOR) and barnyard (BYD) aromas and had a negative correlation with the amino acid asparagine (asn) and peptide-5 (PEP-5). It should be noted that 'pork flavour' and 'off-flavour' factors increased toward the negative direction of the factor axis. The last factor, Factor 5, was defined as 'cured flavour' because mouthfilling (MTH), sour (SOR) and bitter (BTR) exhibited the highest loadings on the factor.

As seen in Table 14.6, Factor 1 was negatively correlated with Factors 2, 3 and 4 and positively with Factor 5, meaning that the 'length of curing' is negatively correlated with the 'pleasant aroma', but due to a larger negative 'pork' and 'off-flavour' factor scores indicating stronger 'pork' and 'off-flavours', the 'length of curing' is positively related with the 'pork', 'off-flavour' and 'cured' flavours.

The average factor score of the long treatment had a positive relation with Factor 1 and 5, and negative with Factors 2, 3 and 4 (Table 14.7). The short average factor score showed a negative relation with Factor 1 and 5 and positive with Factors 2, 3 and 4 (Table 14.7). From the relationship of the factors with the treatments we attempt to explain the generation of the dry-cured ham flavour where the length of the drying stage produces an increase in 'pork', 'off-flavour', and 'cured' flavours, but masked the 'pleasant' aroma.

On the other hand, the relationship of the components with the treatment can be described in the short process (7 months) where compounds P-8 (2,3-butanodione, buttery), P-23 (methyl butanoate, sweet-caramel), P-35 (methylpyrazine, nutty), P-44 (*o*-xylene, sweet, fruit-candy), P-50 (2-butoxyethanol, dark toast-meaty), and P-68 (2-pentylfuran, ham-like) show the strongest correlation and gave a pleasant aroma; while compounds P-14 (3-methylbutanal, cheesy–green), P-24 (dimethyl disulphide, dirty socks) and P-34 (hexanal, green–grassy) gave a character of

Table 14.6 Inter-factor correlations

Factor	1. Length	2. Pleasant	3. Pork	4. Off-flavour	5. Cured
1. Length	1.000				
2. Pleasant	–0.165	1.000			
3. Pork	–0.269	0.163	1.000		
4. Off-flavour	–0.172	0.164	0.117	1.000	
5. Cured	0.177	0.164	–0.061	0.007	1.000

Table 14.7 Factor score average values for long and short dry-curing processes

Factor	1. Length	2. Pleasant	3. Pork	4. Off-flavour	5. Cured
+	Long (0.91 ± 0.07)[a]	Short (0.14 ± 0.13)	Short (0.41 ± 0.11)	Short (0.31 ± 0.13)	Long (0.11 ± 0.14)
–	Short (–0.91 ± 0.03)	Long (–0.14 ± 0.15)	Long (–0.41 ± 0.14)	Long (–0.31 ± 0.13)	Short (–0.11 ± 0.13)

[a]The number in brackets represents the mean ± sem.

fresh-cured pork flavour. Moreover, peptide 6 (PEP-6) had a strong correlation with the short process as also occurred with peptide 5 (PEP-5) and asparagine (asn). The long process had a strong correlation with amino acids glutamic acid (glu), aspartic acid (asp), methionine (met), isoleucine (ile), leucine (leu) and lysine (lys) and with peptide 3 (PEP-3) and peak 7 (PEP-7), identified as hypoxanthine.

These results agree, in part, with those of Careri *et al.* (1993) who found that esters, aromatic hydrocarbons and cyclic nitrogen compounds affected the aged odour of Italian-type dry-cured ham positively, although we did not find any positive contribution of three- and four-carbon alcohols to the aged aroma as they described. From the olfactory test, we did not detect any contribution of these alcohols to the aroma of the 'Serrano' dry-cured ham. When we compare our results with those found in French-type dry-cured ham (Buscailhon *et al.*, 1994a) we detected some similarities about the relation of several ketones with the pleasant aroma of the dry-cured ham and the relation of some aldehydes with the aroma of fresh-cured pork, but we did not find any contribution of 1-butanol to the aroma of 'Serrano' dry-cured ham. On the other hand, the volatile compounds detected in Spanish Iberian dry-cured ham were principally aldehydes, ketones, hydrocarbons and alcohols (Garcia *et al.*, 1991; Lopez *et al.*, 1992) but these authors did not study their contribution to the dry-cured ham flavour. All the differences in volatile composition and sensorial relations between the Italian, French and Spanish dry-cured hams could be due to the different manufacturing techniques employed and also to the pre-mortem factors that can influence the final sensory quality of the product.

Few studies have been done to establish the relation between nonvolatile components of dry-cured ham with its flavour. Careri *et al.* (1993) determined several relationships of nonvolatile components with sensory attributes of Italian-type dry-cured ham. They found that amino acids, such as lysine and tyrosine, were related with an improved quality on the aged taste of hams while asparagine affected it negatively. They also found a contribution to saltiness from glutamic acid, and to acid taste by phenylalanine and isoleucine; tyrosine was negatively related to acid taste. In French-type dry-cured ham the results were completely opposite because Buscailhon *et al.* (1994a) reported that the changes produced on the amino acid concentrations had little effect on the development of the aroma of cured meat and dry ham. From our study, we detected a strong relation of amino acids glutamic acid, aspartic acid, methionine, isoleucine, leucine and lysine, and peptide 3 and peak 7 (hypoxanthine) with the length of the process and with the 'cured' and 'pork' flavours. Moreover, the amino acid asparagine and peptide 5 were negatively related to 'off-flavours'.

14.6 Conclusions

The flavour of dry-cured ham is due to the interactions of the combination of flavour compounds originating from ham proteins, lipids and carbohydrates. The nonvolatile components, peptides and amino acids, constitute taste-active compounds that have a large impact on the final flavour of the dry-cured ham product, as seen by the increase glutamic acid, aspartic acid, methionine, isoleucine, leucine and lysine; these amino acids have been shown to contribute to the dry-cured ham flavour by their combinational interaction and not by individual interaction. The amino acids of dry-cured ham also contribute to flavour volatile formation as a result of Strecker degradations and include sulphide compounds, methyl-branched aldehydes and alcohols, and pyrazines. The remainder of the volatile compounds are formed by lipid oxidation that takes place during ham ripening. The volatile components of the headspace of 'Serrano' dry-cured ham contribute, both individually and in combination, to the distinctive aroma properties of the product. Ketones, esters, aromatic hydrocarbons and pyrazines are essentially the volatile compounds that correlate with the pleasant aroma of dry-cured ham, while hexanal, 3-methylbutanal and dimethyl disulphide are related with a short ripening–drying process. The study of dry-cured ham flavour should continue to establish the contribution of each individual component to the final flavour of the product in order to obtain a high quality dry-cured ham product.

Acknowledgements

Comprehensive studies such as this involve many individuals who deserve acknowledgement. Therefore, we thank the following: (1) for assistance in statistical analysis, Dr B. Vinyard, (2) for sensory analysis, Ms D.A. Ingram, Ms M.G. Franklyn and Dr K.L. Bett, (3) for chromatographic assistance Mr J.A. Miller, Mr S. Lloyd, Mr C. James, Dr C.C. Grimm and Dr M-C Aristoy, and (4) for assistance in the hams' processing, Mr Bolumar and Mr Lara from Jamones Segorbe (Castellón, Spain) and Dr J. Flores. Financinal support from FPI/MEC in the USA to M. Flores and grants CRG941159 from NATO Collaborative Research Grants Program and project SP05 from ICD/RESD (USDA) are gratefully acknowledged.

References

Baldini, P., Bellati, M., Camorali, G., Palmia, F., Parolari, G., Reverberi, M., Pezzani, G., Guerrieri, C., Racznsky, R. and Rivaldi, P. (1992). Caratterizzazione del prosciutto tipico Italiano in base a parametri chimici, fisici, microbiologici e organolettici. *Industria Conserve*, **67**, 149–159.

Baines, D.A. and Mlotkiewicz, J.A. (1984). The chemistry of meat flavour. In *Recent Advances in the Chemistry of Meat*, ed. A.J. Bailey. The Royal Society of Chemistry, London, pp. 119–164.

Barbieri, G., Bolzoni, L., Parolari, G., Virgili, R., Buttini, R., Careri, M. and Mangia, A. (1992). Flavour compounds of dry-cured ham. *J. Agric. Food Chem.* **40**, 2389–2394.

Berdague, J.-L., Denoyer, C., Le Quéré, J.-L. and Semon, E. (1991). Volatile components of dry-cured ham. *J. Agric. Food Chem.*, **39**, 1257–1261.

Berdague, J.-L., Bonnaud, N., Rousset, S. and Touraille, C. (1993). Influence of pig cross-breed on the composition, volatile compound content and flavour of dry cured ham. *Meat Sci.*, **34**, 119–129.

Bolzoni, L., Barbieri, G. and Virgili, R. (1996). Changes in volatile compounds of Parma ham during maturation. *Meat. Sci.*, **43**, 301–310.

Buscailhon, S., Berdague, J.-L. and Monin, G. (1993). Time-related changes in volatile compounds of lean tissue during processing of French dry-cured ham. *J. Sci. Food Agric.*, **63**, 69–75.

Buscailhon, S., Berdague, J.-L., Bousset, J., Cornet, M., Gandemer, G., Touraille, C. and Monin, G. (1994a). Relations between compositional traits and sensory qualities of French dry-cured ham. *Meat Sci.*, **37**, 229–243.

Buscailhon, S., Monin, G., Cornet, M. and Bousset, J. (1994b). Time-related changes in nitrogen fractions and free amino acids of lean tissue of French dry-cured ham. *Meat Sci.*, **37**, 449–456.

Careri, M., Mangia, A., Barbieri, G., Bolzoni, L., Virgili, R. and Parolari, G. (1993). Sensory property relationships to chemical data of Italian-type dry-cured ham. *J. Food Sci.*, **58**, 968–972.

Caul, J. F. (1957). The profile method of flavour analysis. *Adv. Food Res.*, **7**, 1–40.

Chang, S.S., Petersen, R.J. (1977). Recent developments in the flavour of meat. *J. Food Sci.*, **42**, 298–305.

Drumm, T.D. and Spanier, A.M. (1991). Changes in the content of lipid autoxidation and sulphur-containing compounds in cooked beef during storage. *J. Agric. Food Chem.*, **39**, 336–343.

Flores, J. and Toldrá, F. (1993). Curing: processes and applications. In *Encyclopaedia of Food Science, Food Technology and Nutrition.*, eds. R. Macrae, R. Robinson, M. Sadler and G. Fullerlove. Academic Press, London, pp. 1277–1282.

Flores, J., Nieto, P., Bermell, S. and Miralles, M.C. (1985). Cambios en los lípidos del jamón durante el proceso de curado, lento y rápido, y su relación con la calidad. *Rev. Agroquim. Tecnol. Aliment.*, **25**, 117–124.

Flores, M., Aristoy, M.-C. and Toldrá, F. (1993). HPLC purification and characterization of porcine muscle aminopeptidase B. *Biochimie*. **75**, 861–867.

Flores, M., Romero, J., Aristoy, M.-C., Flores, J. and Toldrá, F. (1994). Differences in muscle proteolytic activities among pork breed types. *Sci. Des Aliment.* **14**, 469–474.

Flores, M., Aristoy, M.-C. and Toldrá, F. (1996a). HPLC purification and characterization of soluble alanyl aminopeptidase from porcine skeletal muscle. *J. Agric. Food Chem.* **44**, 2578–2583.

Flores, M., Toldrá, F., and Spanier, A.M. (1996b). Impact of dry-curing process on meat flavour. *Meat Focus Int.*, **5**, 220–222.

Flores, M., Ingram, D.A., Bett, K.L., Toldrá, F. and Spanier, A.M. (1997a). Sensory characteristics of Spanish 'Serrano' dry-cured ham. *J. Sensory Studies.*, **12**, 169–179.

Flores, M., Grimm, C.C., Toldrá, F. and Spanier, A.M. (1997b). Correlations of sensory and volatile compounds of Spanish 'Serrano' dry-cured ham as a function of two processing times. *J. Agric. Food Chem.* **45**, 2178–2186.

Flores, M., Aristoy, M.C., Spanier, A.M. and Toldrá, F. (1997c). Correlations of nonvolatile components and sensory properties of Spanish 'Serrano' dry-cured ham. Effect of processing time. *J. Food Sci.*, In press.

Forss, D.A. (1972). Odor and flavour compounds from lipids. *Prog. Chem. Fats and Other Lipids*, **13**, 181–258.

Frankel, E.N. (1984). Recent advances in the chemistry of rancidity of fats. In *Recent Advances in the Chemistry of Meat*, ed. A.J. Bailey. The Royal Society of Chemistry, London, pp. 87–118.

Fuke, S. and Konosu, S. (1991). Taste-active components in some foods: A review of Japanese research. *Physiol. Behav.*, **49**, 863–868.

Garcia, C., Berdague, J.-L., Antequera, T., Lopez-bote, C., Cordoba, J.J. and Ventanas, J. (1991). Volatile components of dry-cured Iberian ham. *Food Chem.*, **41**, 23–32.

Hinrichsen, L.L. and Pedersen, S.B. (1995). Relationship among flavour, volatile compounds, chemical changes, and microflora in Italian-type dry-cured ham during processing. *J. Agric. Food Chem.*, **43**, 2932–2940.

Kato, H., Rhue, M.R. and Nishimura, T. (1989). Role of free amino acids and peptides in food taste. In *Flavour Chemistry. Trends and Development*, eds. R. Teranishi, R.G. Buttery and F. Shahidi. ACS Symposium Series 388, American Chemical Society, Washington, DC, pp. 158–174.

Lillard, D.A. and Ayres, J.C. (1969). Flavour compounds in country-cured hams. *Food Technol.*, **23**, 117–120.

Lopez, M.O., De la Hoz, L., Cambero, M.I., Gallardo, E., Reglero, G. and Ordoñez, J.A. (1992). Volatile compounds of dry hams from Iberian pigs. *Meat Sci.*, **31**, 267–277.

Lundström, K., Malmfors, B., Malmfors, G., Stern, S., Peterson H., Mortensen, A.B. and Sorensen, S.E. (1988). Skatole, androstenone and taint in boars fed two different diets. *Livestock Prod. Sci.*, **18**, 55–67.

Maga, J.A. (1982). Pyrazines in foods: An update. *CRC Crit. Rev. Food Sci. Nutr.*, **16**, 1–48.

Marriot, N.G., Graham, P.P., Shaffer, C.K. and Phelps, S.K. (1987). Accelerated production of dry cured hams. *Meat Sci.* **19**, 53–64.

Marriot, N.G., Graham, P.P. and Claus, J.R. (1992). Accelerated dry curing of pork legs (hams): A review. *J. Muscle Foods*, **3**, 159–168.

McCain, G.R., Blumer, T.N., Craig, H.B. and Steel, R.G. (1968). Free amino acids in ham muscle during succesive aging periods and their relation to flavour. *J. Food Sci.*, **33**, 142–146.

Meilgaard, M., Civille, G.V. and Carr, B.T. (1991). *Sensory Evaluation Techniques*, 2nd edn. CRC Press, Boca Raton, FL, pp. 174–175.

Min, D.B.S., Ina, K., Peterson, R.J., and Chang, S.S. (1979). Preliminary identification of volatile flavour compounds in the neutral fraction of roast beef. *J. Food Sci.* **44**, 639–642.

Molina, I. and Toldrá, F. (1992). Detection of proteolytic activity in microorganisms isolated from dry-cured ham. *J. Food Sci.*, **57**, 1308–1310.

Motilva, M.-J., Toldrá, F. and Flores, J. (1992). Assay of lipase and esterase activities in fresh pork meat and dry-cured ham. *Z. Lebensm. Unters. Forsch.*, **195**, 446–450.

Motilva, M.-J., Toldrá, F., Nieto, P. and Flores, J. (1993). Muscle lipolysis phenomena in the processing of dry-cured ham. *Food Chem.* **48**, 121–125.

Motilva, M.-J., Toldrá, F., Nadal, M.-I. and Flores, J. (1994). Pre-freezing hams affects lipolysis during dry-curing. *J. Food Sci.*, **59**, 303–305.

Mottram, D.S. (1985). The effect of cooking conditions on the formation of volatile heterocyclic compounds in pork. *J. Sci. Food Agric.*, **36**, 377–382.

Mottram, D.S. and Edwards, R.A. (1983). The role of triglycerides and phospholipids in the aroma of cooked beef. *J. Sci. Food Agric.* **34**, 517–522.

Muñoz, A.M. and Civille, G.V. (1992). The spectrum descriptive analysis method. In *Manual on Descriptive Analysis Testing for Sensory Evaluation*, ed. R.C. De Hootman. ASTM Manual Series, American Society for Testing and Materials, Philadelphia, PA, pp. 22–34.

Nishimura, T. and Kato, H. (1988). Taste of free amino acids and peptides. *Food Rev. Int.* **4**, 175–194.

Nishimura, T., Rhue, M.R., Okitani, A. and Kato, H. (1988). Components contributing to the improvement of meat taste during storage. *Agric. Biol. Chem.*, **52**, 2323–2330.

Nishimura, T., Okitani, A., Rhue, M.R. and Kato, H. (1990). Survey of neutral aminopeptidase in bovine, porcine and chicken skeletal muscles. *Agric. Biol. Chem.*, **54**, 2769–2775.

Ockerman, H.W., Blumer, T.N. and Craig, H.B. (1964). Volatile chemical compounds in dry-cured hams. *J. Food Sci.*, **29**, 123–129.

Parolari, G. (1996). Review: Achievements, needs and perspectives in dry-cured ham technology: The example of Parma ham. *Food Sci. Technol. Int.*, **2**, 69–78.

Pearson, A.M., Dickson, R.L., Hill, T.W. and Hilton, D.W. (1995). Boar taint: Androstenones or skatole? A review. In *41st International Congress of Meat Science and Technology*, San Antonio, TX, pp. 78–79.

Rosell, C.M. and Toldrá, F. (1996). Effect of curing agents on *m*-calpain activity through the curing process. *Z. Lebensm. Unters. Forsch.*, **203**, 320–325.

SAS Institute Inc. (1989). *SAS/STAT® User's Guide, Version 6*, 4th edn, Vols 1 and 2. SAS Institute Inc., Cary, NC.

Shahidi, F., Rubin, L.J. and D'Souza, L.A. (1986). Meat flavour volatiles: A review of the composition, techniques of analysis, and sensory evaluations. *CRC Crit. Rev. Food Sci. Nutr.*, **24**, 141–243.

Spanier, A.M. and Miller, J.A. (1993). Role of proteins and peptides in meat flavour. In *Food Flavour and Safety. Molecular Analysis and Design.*, eds. A.M. Spanier, H. Okai and M. Tamura. ACS Symposium Series 528, American Chemical Society, Washington, DC, pp. 78–97.

Spanier, A.M., Edwards, J.V. and Dupuy, H.P. (1988). The warmed-over flavour process in beef: A study of meat and peptides. *Food Technol.*, **42**, 110, 112–118.

Spanier, A.M., McMillin, K.W. and Miller, J.A. (1990). Enzyme activity levels in beef. Effect of postmortem aging and end-point cooking temperature. *J. Food Sci.*, **55**(2), 318–322, 326.

St. Angelo, A.J., Legendre, M.G. and Dupuy, H.P. (1980). Identification of lipoxygenase-linoleate decomposition products by direct gas chromatography–mass spectometry. *Lipids*, **15**, 45–49.

Tabachnick, B.G. and Fidell, L.S. (1983). Principal components analysis and factor analysis. In *Using Multivariate Statistics.* Harper & Row, New York, pp. 372–445.

Toldrá, F. and Etherington, B.J. (1988). Examination of cathepsin B, D, H and L activities in dry-cured hams. *Meat Sci.*, **23**, 1–7.

Toldrá, F. and Aristoy, M.-C. (1993). Availability of essential amino acids in dry-cured ham. *Int. J. Food Sci. Nutr.*, **44**, 215–219.

Toldrá, F., Aristoy, M.-C., Cervero, M.-C., Rico, E., Motilva, M.-J. and Flores, J. (1992). Muscle and adipose tissue aminopeptidase activities in raw and dry-cured ham. *J. Food Sci.*, **57**, 816–818, 833.

Toldrá, F., Rico, E. and Flores, J. (1993). Cathepsin B, D, H and L activities in the processing of dry-cured ham. *J. Sci. Food Agric.*, **62**, 157–161.

Toldrá, F., Flores, M. and Aristoy, M.-C. (1995). Enzyme generation of free amino acids and its nutritional significance in processed pork meats. In *Food Flavours: Generation, Analysis, and Process influence*, ed. G. Charalambous, Elsevier Science, Amsterdam, pp. 1303–1322.

Toldrá, F., Flores, M., Aristoy, M.-C., Virgili, R. and Parolari, G. (1996). Pattern of muscle proteolytic and lipolytic enzymes from light and heavy pigs. *J. Sci. Food Agric.*, **71**, 124–128.

Toldrá, F., Flores, M. and Sanz, Y. (1997a). Dry-cured ham flavour: enzymatic generation and process influence. *Food Chem.*, **59**, 523–530.

Toldrá, F., Flores, M., Navarro, J.-L., Aristoy, M-C. and Flores, J. (1997b). New developments in dry-cured ham. In *Chemistry of Novel Foods*, eds. H. Okai, O. Mills, A.M. Spanier and M. Tamura. Allured., Carol Stream, IL, pp. 259–272.

Virgili, R., Parolari G., Schivazappa C., Soresi Bordini, C., and Borri, M. (1995). Sensory and texture quality of dry-cured ham as affected by endogenous cathepsin B activity and muscle composition. *J. Food Sci.*, **60**, 1183–1186.

15 Smoke flavourings in processed meats*

J. ROZUM

15.1 Introduction

For centuries, people have been searching for ways to preserve their hunt to ensure a food supply that would last them through the leaner times of the year. Possibly the first method used to preserve foodstuffs was smoking. Not only did it provide a cooked product with a different aroma, flavour and colour, it also prevented the product from spoiling as fast. Whether this was an accidental discovery or a thought-out process will never be known, but it changed the way people lived and ate.

Wood smoke is composed of numerous chemical compounds and the possibilities of their reactions with food components are almost infinite. Over the course of the past 30 years or so, a great deal of research has been done to divide the components of smoke into major classes. Each class is considered the primary source for given functions in the smoking of foods. The major classes include acidic compounds which contribute to flavour and skin formation, phenolic compounds which provide flavour and preservation capabilities, and the carbonyls which react with proteins and other nitrogenous sources to give food a smoke colour. A fourth group, the polycyclic aromatic hydrocarbons (PAH), are undesirable fraction of smoke as they are known to be carcinogenic. The differences in natural vaporous and liquid smoke PAH content will be discussed later.

In early modern times researchers (Ostertag and Young, 1934) believed that the preservation of food by smoking was caused by the shrinkage of the muscle fibres and widening of the interstitial spaces. Other early investigators believed that the glossy appearance of meats after smoking meant that the proper resins, phenols and aldehydes were deposited onto the meat surface. The deposition of these compounds was termed the residual antiseptic effect in smoked meats.

During the pyrolysis of wood to produce smoke, there are many variables which complicate any generalization as to the source of the important components of wood smoke. However, it is generally assumed that cellulose and hemicellulose form the carbonyl and acid fractions and lignin forms the phenolic fractions. It is then the make-up and amount of these components in wood which lead to the various flavours from different

*This chapter is dedicated to the memory of Dr C.M. Hollenbeck.

wood species. Knowledge of the wood and smoke chemistry allows researchers to take the smoking process a step further than the ancient caveman. They are now able to control the pyrolysis and products of the pyrolysis of wood to provide a broad spectrum of flavours and colours to the food industry.

A look at the structure of these wood components, and some of the research that has been done on their breakdown products, seems to verify these generalizations (Maga, 1988). The chemical structures of the three major components of wood are shown in Figure 15.1. Some of the products and their possible pathways for formation during pyrolysis are discussed in the following sections.

15.2 Pyrolysis of cellulose

Cellulose is the major component of most wood species. The number of products produced by the pyrolysis of the cellulose component is large and

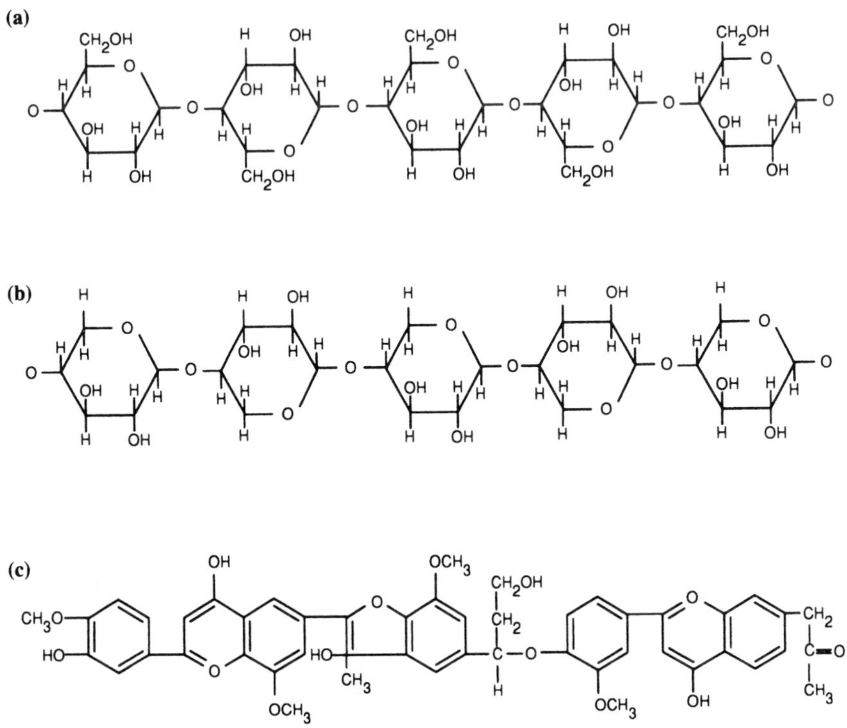

Figure 15.1 Structural inter-relationships of the three major components of wood: (a) cellulose; (b) hemicellulose; (c) lignin.

Table 15.1 Pyrolysis products of cellulose at 600°C

Compound	Relative %
Acetaldehyde	2.3
Furan	1.6
Acetone/propionaldehyde	1.5
Propanal	3.2
Methanol	2.1
2,3-Butanedione	2.0
1-Hydroxy-2-propanone	2.1
Glyoxal	2.2
Acetic acid	6.7
2-Furaldehyde	1.1
Formic acid	0.9
5-Methyl-2-furaldehyde	0.7
2-Furfuryl alcohol	0.5
Carbon dioxide	12.0
Water	18.0
Char	15.0
Tar	28.0

variable. The type and quantity of products are determined by the type of cellulose source and conditions of the pyrolysis performed. Shafizadeh (1984) has provided a list of some of the major products of cellulose pyrolysis at 600°C (Table 15.1). The aliphatic acids and the aldehydes are the important compounds contained in smoke flavourings produced by the breakdown of cellulose.

A series of pathways were proposed by Byrne *et al.* (1966) for the breakdown of cellulose to lower molecular weight aldehydes (Figure 15.2). It is these aldehydes which are responsible for the smoke colour formation in processed meats, including fish meat, and other foods.

15.3 Pyrolysis of hemicellulose

Hemicellulose does not represent only one compound but encompasses a mixture of various polysaccharides. It represents about 20–35% of the mass of wood. It is the first component to undergo thermal decomposition in pyrolysis.

Fengel and Wegener (1984) proposed some pathways for the decomposition of hemicellulose (Figure 15.3). The major degradation products are furan and its derivatives and a series of aliphatic carboxylic acids. These compounds contribute to the overall flavour and chemical properties of smoke but the exact reactions involved cannot be easily identified.

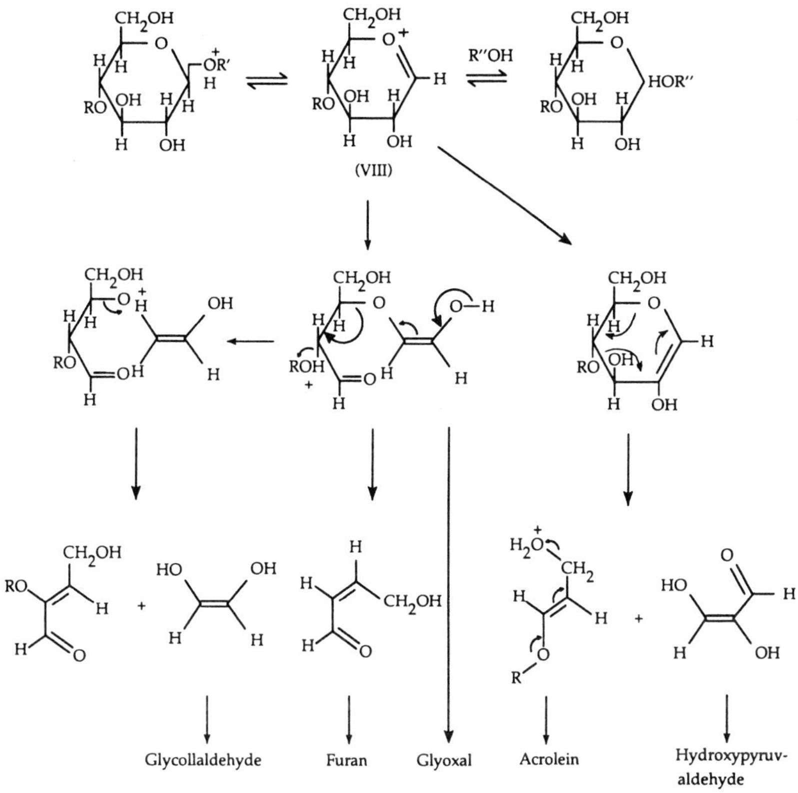

Figure 15.2 Mechanisms of the formation of carbonyl compounds.

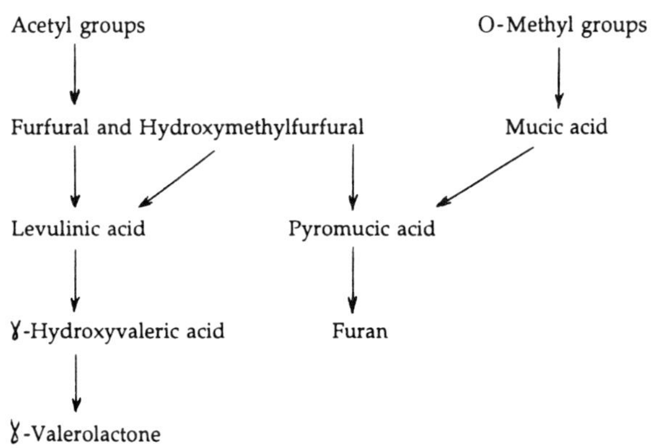

Figure 15.3 Thermal degradation products from hemicellulose.

15.4 Pyrolysis of lignin

Mature hardwoods contain 20–25% lignin. The pyrolysis of lignin produces mainly phenolic compounds. A pathway for the production of phenolics from lignin has been proposed by Gilbert and Knowles (1975) as depicted in Figure 15.4. Syringol, 2,6-dimethoxyphenol, is the phenol that is produced in the greatest quantity. Syringol, and other phenolics, play the greatest role in providing the 'smoked' flavour to smoked foods. Other phenols, contributing to the flavour of foods, found in large quantities in liquid smoke are eugenol, isoeugenol, guaiacol, phenol and cresols.

15.5 Formation of colour in smoked foods

The main colour-forming reaction in foods caused by smoke is the reaction of aldehydes with amino groups. This reaction, known as the Maillard reaction, and pathways involved have been described by Namiki and Hayashi (1983) as shown in Figure 15.5. Their pathway depicts the formation of a pyrazine polymer in the sugar–amino acid compound by way of a hydroxyacetaldehyde intermediate. Hydroxyacetaldehyde is the most active browning agent found in smoke. In the production of liquid smoke flavourings, every effort is made to maintain conditions for optimal aldehyde formation. This usually requires rapid quenching of the pyrolysis products to prevent their reaction with other smoke components. A recent trend in

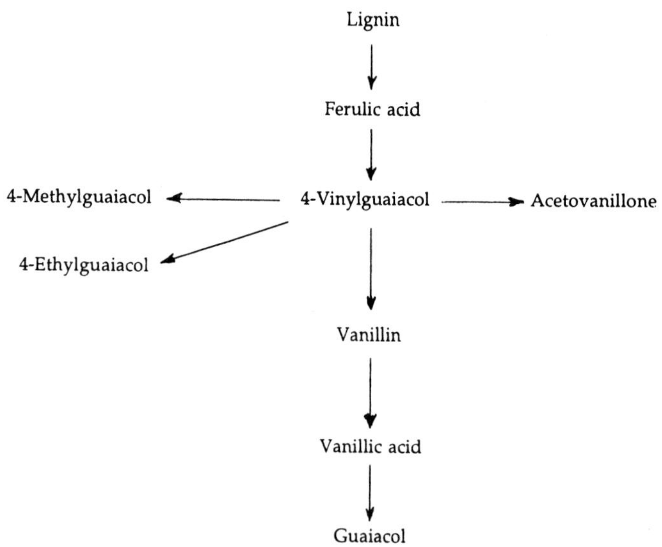

Figure 15.4 Thermal degradation products of ferulic acid.

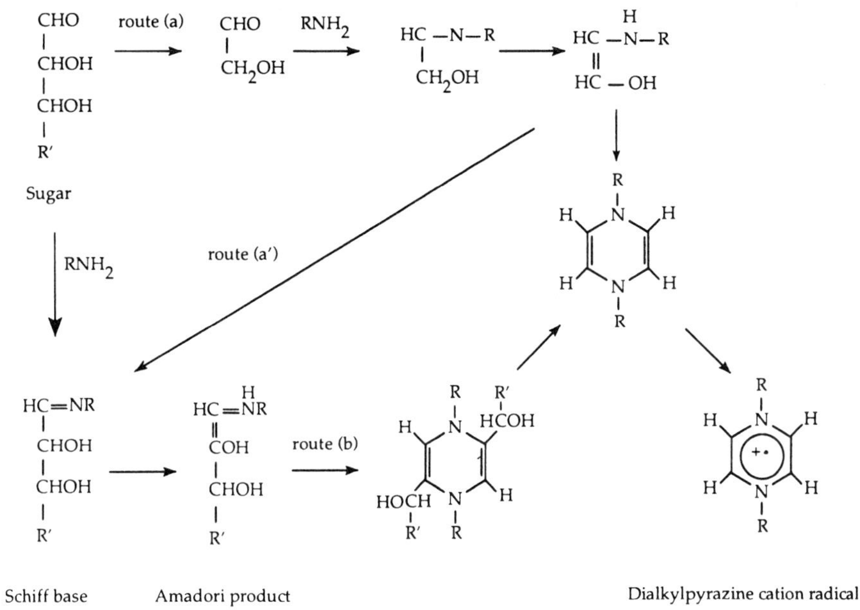

Figure 15.5 Two possible pathways for radical formation in the reaction of sugars with amino compounds.

liquid smoke production is in the development of a very rapid pyrolysis process to reduce the amount of time the smoke stays in the pyrolysis zone (Underwood and Graham, 1989). Current technology allows pyrolysis to occur in milliseconds instead of several seconds in older processes. Figure 15.6 shows how a reduced pyrolysis time affects the yield of hydroxy-acetaldehyde.

Table 15.2 shows the relative colour-forming potential of selected pure carbonyls when measured in three different ways. The browning index is a procedure used by Red Arrow Products Co. (Manitowoc, WI) in determining the amount of brown colour formation potential that is contained in their products. A glycine paper spot test involves coating a filter paper with a 1% glycine solution and allowing it to dry. A drop of the test solution is then dropped onto the glycine-coated filter paper. The filter paper is placed in an oven set at 200–300°F (93–149°C) for approximately 1.5 min or until dry. This will give a colour which is subjectively compared to other tests. Weiner dip tests are done to determine the carbonyls' reaction potential on a meat system. Normal weiners are stuffed and dipped for a certain time in a carbonyl solution, and then processed. Colour formation on the weiners is then compared. Rozum (1994) has indicated that various individual carbonyls provide colour to slow-cooked pork fat even though the nitrogen level is fairly low in fat cells. The study

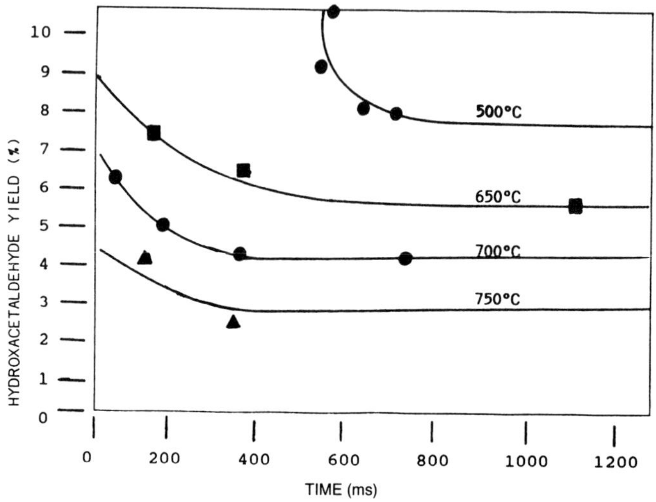

Figure 15.6 Fast pyrolysis of wood (hydroxyacetaldehyde versus residence time).

Table 15.2 Colour forming potential of selected carbonyls

Colour intensity	Browning index (pure carbonyls)[a]	Glycine paper (pure carbonyls)[b]	Wiener dips (pure carbonyls)[b]
Darkest	Glycoaldehyde (18.9)	Glyoxal	Glycoaldehyde
↓	Pyruvaldehyde (14.8)	Pyruvaldehyde	Glyoxal
	Glyoxal (6.4)	Glycoaldehyde	Formaldehyde
	Diacetyl (5.3)	Diacetyl	Pyruvaldehyde
	Furfural (4.9)	Acetol	Acetol
	Acetol (0)	Furfural	3-Hydroxy-2-butanone
Lightest	Formaldehyde (0)	3-Hydroxy-2-butanone	Furfural

[a]10% solutions.
[b]Glycoaldehyde 0–3% solution; all others 5%.

also found that none of the major phenolic compounds found in smoke coloured the pork fat.

In general, the reactions that occur to give the brown colour in smoked foods are time and temperature dependent. At higher temperatures the reactions tend to be fast, whereas at low temperatures the reactions tend to take longer. As an example, the glycine paper test takes about 1 min at 300°F (149°C) in the oven to form complete colour, but takes over a year at 0°F (–18°C) to form any colour.

15.6 Smoke flavour in processed meats

The flavour produced by smoking foods is a combination of unreacted smoke components and reacted smoke-protein components. The phenolic compounds are the major unreacted smoke components that contribute smokiness to the flavour. The unreacted and reacted acids, esters, lactones and carbonyls also contribute to the flavour. The flavour of the carbonyls comes from the Maillard reaction products. The flavour of individual smoke phenolics have been described by Kim (1974), as shown in Table 15.3. The profile of the phenolic components in a smoke can vary from one type of wood to another. The major independent wood sources are hickory and mesquite. To obtain a non-specific flavour a combination of hardwoods can be utilized.

15.7 Natural vaporous versus liquid smoke

The origin of smoke flavouring can be from two sources, the burning of wood in a smokehouse or from liquid smoke flavourings. Both provide all the necessary compounds needed to flavour and colour meats and other foods. In natural vaporous smokehouses, all compounds formed from the

Table 15.3 Flavour descriptions of various phenols isolated from wood smoke

Compound	Description
Phenol	Pungent
o-Cresol	Pungent
m- and *p*-Cresol	Pungent
2,3-Xylenol	Pungent
2,4-Xylenol	Pungent
2,6-Xylenol	Cresolic
3,4-Xylenol	Cresolic
3,5-Xylenol	Cresolic
2-Ethyl-5-methylphenol	Cresolic
3-Ethyl-5-methylphenol	Cresolic
2,3,5-Trimethylphenol	Cresolic
Guaiacol	Sweet, smoky, somewhat pungent
3-Methylguaiacol	Weak, phenolic
4-Methylguaiacol	Sweet, smoky
4-Ethylguaiacol	Sweet, smoky
4-Allyguaiacol	Woody
2,6-Dimethoxyphenol	Smoky
2,6-Dimethoxy-4-methylphenol	Mild, heavy, burnt
2,6-Dimethoxy-4-ethylphenol	Mild, heavy, burnt
2,6-Dimethoxy-4-propylphenol	Mild, heavy, burnt
2,6-Dimethoxy-4-propenylphenol	Mild, heavy, burnt
Pyrocatechol	Heavy, sweet, burnt
3-Methylpyrocatechol	Heavy, sweet, burnt
4-Methylpyrocatecol	Heavy, sweet, burnt
4-Ethylpyrocatechol	Heavy, sweet, burnt

burning of wood are deposited on the surface of the meat and the interior of the smokehouse. This includes all the wanted flavour and colouring components as well as the tars, volatile light organics and the carcinogenic PAH. It is these last three components that have caused many processors to go from the traditional natural vaporous smokehouse to liquid smoke. In the production of liquid smoke, the water insolubles (tars), light organics and PAHs are significantly reduced. This reduces the amount of time needed for clean-up (limited tar deposits), lowers light organic emissions, and lowers PAH levels in the final meat and fish products. The PAH levels may even be further reduced in some smokes by using resins to remove the PAH. Table 15.4 shows the reduction of PAH level in smoke over a period of time (Underwood and Rozum, 1995).

15.8 Evolution of smoke flavourings

The evolution of liquid smoke products has also allowed processors to obtain various flavours and colours that were not available to them before through natural vaporous smoking. They also allow the processor to add smoke flavourings to various recipes at a variety of steps in the processing.

Most liquid smoke flavourings originate from one main starting ingredient, a 10% acid liquid smoke produced by a smoke generator. From this product a wide variety of acid levels, additions and deletions can be made to the smoke to produce several products. One of the first derived flavours was an oil-based smoke flavouring developed by Hollenbeck (1969). This product is produced by extracting aqueous smoke flavouring with a vegetable oil. The oil extracts mainly phenols from the liquid to provide a flavouring agent with no colour forming properties. This product

Table 15.4 The removal of phenols from a 10% acid liquid smoke (Charsol C-10[a]) by adsorption on a resin column

Treated volume (gal)	Phenols (mg/ml)	Carbonyls (%)	Browning index	Brix
Starting feed	17.0	12.4	9.9	25.9
10	1.3	11.3	9.6	19.2
20	2.4	11.1	10.1	21.0
30	4.8	11.5	10.6	22.6
40	6.3	11.5	10.4	23.2
50	7.9	12.2	10.3	23.4
65	12.5	11.8	9.5	24.4
75	14.3	na	na	24.6
85	15.3	11.3	9.5	24.6
95	17.0	na	na	25.4
105	16.8	12.6	9.3	26.0

na = not analysed.
[a]Red Arrow Products Co., Maintowoc, WI.

is widely used in processed meats and is added into the emulsion prior to stuffing.

The next major development came 12 years later in the development of a brine-soluble flavouring. Underwood and Wendorff (1981) found that when a smoked oil was extracted with polysorbate 80, the phenolic compounds transferred into the polysorbate and became water soluble. This product is used as a smoke flavouring agent in almost all of the bacon currently produced in the USA, is used in other smoked meats, and is being added as a component of the pumping pickle.

When the polysorbate 80 product is used internally, processors would sometimes still smoke the outside of the bacon to provide the brown colour. Seeing the need for a smoke that colours but provides low flavour, Underwood (1990) developed an aqueous product that produced colour but possessed very little flavour. This was accomplished by treating the aqueous smoke flavour with a non-ionic resin adsorption column which removes a large portion of the phenolics from the smoke. Table 15.5 shows the removal of phenols by the resin without affecting the level of colour-forming carbonyls in a 10% acid liquid smoke solution (Underwood, 1990).

The low-flavour browning agent was then taken a step further to produce a product with no flavour with a high level of browning potential. This product developed by Underwood (1991) is used on a variety of products where only a minimum amount of smoke flavour is desired but a dark brown colour is wanted.

The production of liquid smoke produces two main by-products, both of which can be used in other applications. One, the ash, can be used as a fuel source or can be further processed to produce briquets. The other is tar, which is the insoluble fraction that remains after the liquid smoke production. Tar is a good fuel source and can be utilized as such. It can also be processed into food additives itself. The first use of tar as a food ingredient was proposed by Miler (1969) who used various extractions to produce a product for flavouring sausages. Another important discovery

Table 15.5 Antibacterial properties of smoke flavourings in culture media

Smoke flavouring	% inhibition			
	E. coli	*S. aureus*	*P. aeruginosa*	*L. viridescens*
Charsol C-6[a] (0.25% v/v, pH 2.4)	33	72	52	99
Charsol C-6[a] (0.25% v/v, pH 5.0)	11	31	51	97
Charsol C-6[a] (0.25% v/v, pH 7.0)	3	25	54	15
Acetic acid, 6.5% (0.25% v/v)	25	52	29	99
Chardex[a] (0.1% w/v)	0	77	46	21
Aro-Smoke P-50* (0.25% w/v)	20	60	62	10
Charoil[a] (0.25% v/v)	0	55	52	85

[a]Red Arrow Products Co., Manitowoc, WI.

was by Dainius *et al.* (1979) who used a distillation process to produce a low carcinogenic product for use in meats and pet food applications.

15.9 Food preservation with smoke

Not only does smoking provide flavour, colour and other sensory effects to foods, it also affects preservation of foods. Smoke flavourings are potential effective antibacterial and antifungal agents. Varying degrees of inhibition by different smoke types against several organism types have been found (Sofos *et al.*, 1988). The study showed that certain smoke varieties worked better than others against the same organisms. Wendorff (1981) measured the bacteriostatic and fungistatic activities of some smoke flavourings against some of the common food bacteria and fungi. Data from these tests are shown in Tables 15.6 and 15.7. It is believed that the acids and phenolics of smoke contribute to the inhibition of growth of both bacteria and fungi. However, Rozum (1995) found that when a polysorbate 80 smoke flavouring was used in cooked chicken, the phenolics were not able to control bacterial growth. It was hypothesized that

Table 15.6 Antifungal properties of smoke flavourings

Smoke flavouring	Zones of inhibition (mm^2)		
	Penicillium sp.	*A. niger*	*A. flavus*
Charsol C-6[a] (0.25% v/v, pH 2.5)	21	18	14
Charsol C-6[a] (0.25% v/v, pH 5.0)	19	16	13
Charsol C-6[a] (0.25% v/v, pH 7.0)	17	12	11
Acetic acid, 6.5% (0.25% v/v)	16	12	13
Chardex[a] (0.1% w/v)	12	9	8
Aro-Smoke P-50[a] (0.25% w/v)	9	9	8

[a]Red Arrow Products Co., Manitowoc, WI.

Table 15.7 Antioxidant properties of natural smoke flavourings

Antioxidant	Peroxide value (meq/kg fat)				
	Initial	Week 1	Week 2	Week 4	Week 26
Untreated control	0.8	3.5	8.5	18.4	43.1
Charoil[a] (0.4%)	0.8	1.2	2.0	2.4	3.7
Charoil[a] (0.2%)	0.8	1.3	2.1	3.0	4.6
Aro-Smoke P-50[a] (0.04%)	0.8	1.2	2.0	2.2	3.4
Aro-Smoke P-50[a] (0.02%)	0.8	1.4	2.1	2.9	4.9
BHA (0.02%)	0.8	1.4	2.1	3.3	12.5
BHT (0.02%)	0.8	1.3	1.9	2.8	4.8
Propyl gallate (0.02%)	0.8	1.4	2.1	3.0	5.8

BHA = butylated hydroxyanisole; BHT = butylated hydroxytoluene.
[a]Red Arrow Products Co., Manitowoc, WI.

when confronted with multiple organisms, the effectiveness of the antibacterial benefits is reduced.

The phenolics in oil and polysorbate 80 smoke flavourings tend to behave as strong antioxidants as shown by Wendorff (1981) with pork fat. Previous research and casual observations show that smoke-processed meats do not develop rancidity as rapidly as cured, unsmoked meats.

15.10 Summary

The pyrolysis of cellulose, hemicellulose and lignin is known to provide flavour and colour to smoked meats. The aldehydes produce the colour by reacting with nitrogenous compounds in meat, while flavour is mainly provided by the phenolics that are deposited on the meat surface.

Through years of research, several new and innovative ways have been developed to provide a wider range of smoke flavours to the meat and food industry. The flavours range from a basic liquid smoke to smoked oils to smoked emulsifiers plus a wide variety of liquids that have been treated to remove or add certain properties. What has already been developed will definitely not be the end of new smoke flavourings, but only a step towards future flavour development.

References

Byrne, G.A., Gardener, D. and Holmes, F.H. (1966). The pyrolysis of cellulose and the action of flame retardants. *J. Appl. Chem.*, **16**, 81–87.

Dainius, B., Dame, C. and O'Hara, J. (1979). Method of producing from wood tar a liquid smoke for use in food processing, and product of said method. US Patent 4 154 866.

Fengel, D. and Weener, F. (1984). *Wood: Chemistry Ultrastructure, Reactions*. Walter de Gruyter, Berlin, Chapter 12.

Gilbert, J. and Knowles, M.E. (1975). The chemistry of smoked foods. *J. Food Technol.*, **10**, 245–251.

Hollenbeck, C.M. (1969). Preparation and use of a smoke of a smoke flavoured edible oil. US Patent 3 480 446.

Kim, K., Kurata, T. and Fujimaki, M. (1974). Identification of flavour constituents in carbonyl, non-carbonyl, neutral and basic fractions of aqueous smoke condensates. *Agric. Biol. Chem.*, **38**(1), 53–63.

Maga, J.A. (1988). *Smoke in Food Processing*. CRC Press, Boca Raton, FL.

Miler, K. (1969). Method of producing a smoke preparation. US Patent 3 445 248.

Namiki, M. and Hayashi, T. (1983). A new mechanism of the Maillard reaction involving sugar fragmentation and free radical formation. In *The Maillard Reaction in Foods and Nutrition*, eds. G.R. Waller and M.S. Feather. ACS Symposium Series, American Chemical Society, Washington, DC, p. 45.

Ostertag, J.L. and Young, T.D. (1934). *Textbook of Meat Inspection*. Alexander Eger, Chicago, IL, pp. 529–530.

Rozum, J.J. (1994). Effects of various carbonyls and phenols on the colour of cooked pork fat. Unpublished work. Red Arrow Products Co., Manitowoc, WI.

Rozum, J.J. (1995). Microbiological quality of cooked chicken breasts containing commercially available shelf-life extenders. M.Sc. Thesis, University of Wisconsin.

Shafizadeh, F. (1984). The chemistry of pyrolysis and combustion. In *The Chemistry of Solid Wood*, ed. R. Rowell. American Chemical Society, Washington, DC, Chapter 13.

Sofos, J.N., Maga, J.A. and Boyle, D.L. (1988). Effect of ether extracts from condensed wood smokes on the growth of *Aeromonas hydrophila* and *Staphylococcus aureus*. *J. Food Sci.*, **53**, 1840–1843.

Underwood, G.L. (1990). Process making liquid smoke compositions and resin treated liquid smoke compositions. US Patent 4 959 232.

Underwood, G.L. (1991). High browning liquid smoke composition and method of making a high browning liquid smoke composition. US Patent 5 039 537.

Underwood, G.L. and Wendorff, W.L. (1981). Smoke flavoured hydrophilic liquid concentrate and process of producing same. US Patent 4 250 199.

Underwood, G.L. and Graham, R.G. (1989). Method of using fast pyrolysis liquids as liquid smoke. US Patent 4 876 108.

Underwood, G.L. and Rozum, J.J. (1995). Method of removing hydrocarbons from liquid smoke and flavouring compositions. US Patent pending 8 536 948.

Wendorff, W.L. (1981). Antioxidant and bacteriostatic properties of liquid smoke. In *Proceedings of Smoke Symposium*, Red Arrow Products Co., Manitowoc, WI, pp. 73–87.

16 Instrumental methods for analyzing the flavor of muscle foods

K.R. CADWALLADER and A.J. MACLEOD

16.1 Introduction

The first and often most important step in flavor analysis is the isolation of the target volatile solutes from the nonvolatile food components. After isolation, instrumental methods of analysis are used to separate, identify and quantify the various flavor components of the isolate. In some cases, sensory evaluation (e.g. gas chromatography–olfactometry) may be employed to indicate the important contributors to the characteristic aroma of the food. This chapter focuses on procedures for isolation and extraction of volatile flavor components as well as recent advances in analytical methodology for characterization of muscle food flavor.

16.2 Isolation of volatile flavor compounds

Analysis of volatile flavor components in foods is complicated due to the presence of only minute quantities of solutes in highly complex mixtures (especially in the case of muscle foods). These volatile solutes can be isolated from the nonvolatile material by taking advantage of their volatility or nonpolar nature. There are numerous methods for isolation of flavor volatiles from a food matrix. Those methods taking advantage of the volatility of the analytes include headspace techniques, distillation and direct injection. Solvent extraction and adsorption techniques rely on the relative nonpolar nature of the aroma compounds to isolate them from the food sample. There is no single 'perfect' method and each method will introduce a different type and degree of sampling bias into the resulting GC flavor profile (Reineccius, 1993; Etiévant, 1996). Often the best approach is to isolate the flavor volatiles by several complementary techniques that are based on different separation criteria. In this way, the biases of each method can be ascertained and accounted for in the final results. The methods most often employed in the analysis of volatiles in muscle foods will be discussed under the following headings:

1. headspace sampling and direct thermal desorption;
2. solvent extraction and distillation–extraction techniques.

16.2.1 Headspace sampling and direct thermal desorption

One property an aroma compound must inherently possess is volatility. Headspace sampling takes advantage of this property. Headspace sampling techniques, including static headspace, dynamic headspace, and purge-and-trap methodologies have been recently reviewed (Wampler, 1997). Static headspace sampling (SHS) is, in principle, the simplest among the headspace techniques. In SHS, the food sample is placed in a closed vessel and the volatile components are allowed to come to equilibrium between the sample matrix and the surrounding headspace. This equilibrium is affected by temperature of vessel, sample size, equilibration time, etc. Advantages of SHS include simple sample preparation, elimination of reagents and low risk of artifacts; however, the method is limited to analysis of highly volatile components. The method has been used to a limited extent in the analysis of the flavor of muscle foods, such as fish (Girard and Nakai, 1994; Milo and Grosch, 1995, 1996), beef (Guth and Grosch, 1994; Kerler and Grosch, 1996) and chicken (Elsner *et al.*, 1996).

The efficiency of headspace sampling can be greatly improved by use of an intermediate trapping step to enrich the volatiles prior to their analysis. This technique is generally referred to as dynamic headspace sampling (DHS) or purge-and-trap analysis. DHS currently is the most popular technique used in the study of the muscle food flavor. This method involves the continuous removal of the headspace volatile analytes from a thermostatted food sample using a stream of inert gas. These volatiles are then enriched by trapping onto adsorbent materials (generally porous polymers) or by cryogenic focusing. Alternatively, the volatiles may be sent directly to the analytical GC column for analysis. The use of adsorbent trapping–thermal desorption and direct thermal desorption techniques in flavor analysis has been reviewed (Hartman *et al.*, 1993; Grimm *et al.*, 1997). Direct thermal desorption is similar in principle to the external closed inlet device (ECID) that has been used extensively in flavor research (St. Angelo, 1996). Advantages of these headspace techniques include simple sample preparation, reduced sample size and low risk of artifact formation. Furthermore, volatiles isolated by DHS may more closely resemble the actual aroma composition that is perceived during smelling. A major disadvantage of DHS is that it is not efficient towards components of low volatility. DHS has been used by several researchers for the isolation of volatiles from muscle foods, such as chicken (Ang *et al.*, 1994; Patterson and Stevenson, 1995), bacon (Anderson and Hinrichsen, 1995), ham (Bolzoni *et al.*, 1996) and salmon (Refsgaard *et al.*, 1997).

An emerging technique, solid phase microextraction (SPME), is a rapid, solventless technique based on the partitioning of the volatile analytes between the sample or sample headspace and a polymer-coated fiber. For analysis, the adsorbed volatiles are thermally desorbed in the heated GC injector. SPME has been recently reviewed (Harmon, 1997). While this technique shows good potential for flavor analysis, its use in the study of muscle foods is limited.

16.2.2 Solvent extraction and distillation–extraction techniques

Most volatile flavor compounds are considerably less polar than the bulk aqueous food matrix material. Use of direct solvent extraction takes advantage of this difference in polarity. Additional clean-up of the extract is often required to separate the extracted volatile material from the nonvolatile residue. This can be accomplished by steam distillation, high vacuum distillation (Sen *et al.*, 1991) or by DHS (Buttery *et al.*, 1994). An alternative approach is to first distill the volatile material away from the sample followed by solvent extraction of the aqueous distillate. These processes may be combined to give simultaneous steam distillation–solvent extraction (SDE). In order to minimize thermal generation of artifacts SDE has been conducted under reduced pressure (Maignial *et al.*, 1992). Solvent extraction and distillation techniques have been discussed in great detail elsewhere (Parliment, 1997). These methods have been extensively used in the analysis of muscle foods. SDE has been recently applied for the study of chorizo (Mateo and Zumalacárregui, 1996) and lobster (Cadwallader *et al.*, 1995) flavor. Direct solvent extraction has been used for the determination of volatile flavor compounds by isotope dilution analysis in salmon and cod (Milo and Grosch, 1996) and stewed beef (Guth and Grosch, 1994).

16.3 Instrumental analysis of volatile flavor compounds

Since its inception, tandem gas chromatography–mass spectrometry (GC–MS) has been the technique of choice for the analysis of volatile food flavors. There are many reasons for the pre-eminence of GC–MS, including the fact that GC provides the best overall performance of all separation strategies. Furthermore, GC is ideally suited to deal with solutes in the vapor phase, such as volatile flavor components. Mass spectrometry is one of the most powerful techniques for identification of unknown compounds, and although nuclear magnetic resonance (NMR) spectroscopy is probably superior for most applications, NMR cannot easily operate in a tandem mode, on-line with a chromatographic device. Furthermore, its sensitivity, which is obviously very important in trace analysis, is generally inferior to that of mass spectrometry.

For nearly four decades, GC–MS has reigned supreme in flavor analysis, and continues to dominate. Most research conducted on muscle food flavor in the last few years has depended on GC–MS as the main analytical tool. The technique is so standard and so routine in flavor studies of muscle foods that there is no need to describe it any further here.

Despite the pre-eminence of GC–MS in flavor research, there are other ways of tackling the problem, which can, in certain circumstances, provide valuable additional and/or complementary information to GC–MS. These methods can be classified under the following headings:

1. refinements to routine GC in GC–MS;
2. refinements to routine MS in GC–MS;
3. alternatives to GC as a method of separation prior to identification;
4. alternatives to MS as a method of identification following separation.

Before dealing with these, it is appropriate to define what is considered to be the current, standard GC–MS used in flavor analysis, i.e. fused silica, capillary column GC with bonded phases, providing high resolution, combined with fast-scanning, high-sensitivity MS operating in the electron impact (EI) ionization mode (Bonelli, 1993; Hinshaw, 1994).

16.3.1 Refinements to routine GC in GC–MS

The main refinements to routine GC in GC–MS which have been used in flavor analysis are:

1. gas chromatography–olfactometry (GC–O);
2. multidimensional gas chromatography (MDGC);
3. chiral gas chromatography;
4. preparative gas chromatography.

Gas chromatography–olfactometry (GC–O). A high resolution GC column coupled with a standard GC detector is capable of separating and detecting hundreds of volatile compounds in a single run. However, it is likely that many of these components have little or no impact on the actual aroma of the food. The aroma-active components in the volatile isolate can be determined by combining GC with olfactometry. In GC–O, the analytes are first separated by GC and then delivered to an olfactometer (sniffer port) where they are mixed with humidified air. Human 'sniffers' through the nose continuously breathe the air emitted from the olfactometer, and record their perceptions, such as the odor intensity and description of the detected compounds. There are several extensive reviews dedicated to GC–O (Acree, 1993; Blank, 1997; Mistry *et al.*, 1997). Some common methods based on GC–O include aroma extract dilution analysis (AEDA) (Grosch, 1993), Charm (Acree, 1993) and Osme (McDaniel *et al.*, 1990). These methods mainly differ in how the GC–O data are recorded and analyzed. Given below are some examples of where GC–O was used in the study of the flavor of muscle foods.

AEDA and CharmAnalysis both rely on GC–O of a serially diluted series of a flavor extract. In AEDA, each odor-active compound is assigned a flavor dilution (FD) factor, which is based on the highest extract dilution at which the odorant was last detected by GC–O. FD factors are proportional to odor unit values (compound concentration/odor-detection threshold). CharmAnalysis differs from AEDA in that duration of perceived odor is taken into consideration in the calculation of odor unit values. AEDA has been used to determine potent odorants in beef (Guth and Grosch, 1994; Kerscher and Grosch, 1997) and other muscle foods (Cadwallader *et al.*, 1995; Milo and Grosch, 1996). A headspace technique based on the concept of AEDA called GC–O of headspace samples (GCO–H) has been used to indicate odorants responsible for warmed-over flavor in beef (Kerler and Grosch, 1996) and odor defects in cod and trout (Milo and Grosch, 1995). The use of CharmAnalysis for the evaluation muscle food flavor is limited (Langourieux and Escher, 1997). Other miscellaneous GC–O techniques also have been recently used in the study muscle food flavor (Anderson and Hinrichsen, 1995; Patterson and Stevenson, 1995; Farmer *et al.*, 1997; Guillard *et al.*, 1997).

Multidimensional gas chromatography (MDGC). With samples as complex as those encountered in a typical flavor analysis, even using the best high-resolution GC columns, components sometimes co-elute during GC–MS, producing mixed mass spectra which are difficult to interpret. MDGC, in which typically two different GC columns are used, a pre-column and an analytical column, can overcome this problem, and the technique can be applied in a number of ways, e.g. foreflushing, back-flushing and heartcutting. A thorough discussion of MDGC can be found elsewhere (Wright, 1997). Comprehensive two-dimensional GC, a type of MDGC, allows even greater separation efficiency than heartcut MDGC (Ledford *et al.*, 1996). Although MDGC has been used in flavor analysis (Wright *et al.*, 1986; Van Wassenhove *et al.*, 1988; Homatidou *et al.*, 1990; Arora *et al.*, 1995), the technique has not been extensively applied to the study of muscle food flavor.

Chiral gas chromatography. It is well known that different enantiomers of the same compound can impart difference aroma properties, therefore it is sometimes important to resolve these during flavor analysis. This can be accomplished by use of chiral stationary phases in GC (Mosandl, 1995). Currently, the most common chiral GC stationary phases are based on modified cyclodextrins. A common practice in flavor analysis has been to use chiral GC in combination with MDGC (Hener *et al.*, 1990; Bernreuther and Schreier, 1991). While the chiral GC technique is increasingly being used in general flavor analysis (Bernreuther *et al.*, 1997), it has not been used to any significant extent in the analysis of muscle food flavor.

Preparative gas chromatography. Although techniques of tandem analysis are almost universally employed in flavor analysis, preparative GC followed by off-line analysis is possible in certain simple situations. This provides the immense advantage of making it possible to analyze the collected solute at leisure by a variety of techniques, including the powerful NMR. Preparative GC is, however, extremely difficult, and requires great skill and technical expertise, which partly explains its limited use. Nevertheless it has been applied in studies of meat flavor compounds, but in the simpler model systems (Tressl *et al.*, 1985, 1986; Werkhoff *et al.*, 1989).

16.3.2 Refinements to routine MS in GC–MS

The main refinements to routine electron impact (EI) MS in GC–MS which have been used in flavor analysis are:

1. high resolution mass spectrometry;
2. selected ion monitoring mass spectrometry;
3. chemical ionization mass spectrometry;
4. negative ion chemical ionization mass spectrometry.

High resolution mass spectrometry. The ability of the modern high resolution mass spectrometer to yield precise elemental composition in the spectra of compounds separated by capillary GC has not yet been widely exploited in flavor analysis. However, with the continuing improvements in the performance of commercial magnetic sector mass spectrometers, especially with regard to sensitivity at high resolution, there is little doubt that this valuable facility will become more readily available and hence more widely used in the future.

Selected ion monitoring mass spectrometry. In selected monitoring (SIM), only selected ions representative of a specific compound, or group of compounds, are recorded during GC–MS. The technique is extremely useful in enabling a very high sensitivity assay for the known component or types of components in questions, but it does not contribute to the identification of unknown compounds, since full spectra are not recorded. This technique has been applied to the flavor analysis of muscle foods (Garcia-Regueiro and Diaz, 1989).

A related approach is 'mass chromatography', which is useful for deconvoluting co-eluted GC peaks (Thomas *et al.*, 1984). The difference here, however, is that complete mass spectra have been recorded throughout the GC–MS run, rather than selected ions as in SIM. The data analysis system can then be instructed to select appropriate specific ions from the full recorded spectra of the peak, with the objective of artificially resolving and recognizing the two (or more) components of the peak.

Mass chromatography also can be considered as retrospective selected ion monitoring. Thus the selected ion or ions from the full spectra can be output as single ion chromatograms throughout the GC run, rather than just one peak, hence pinpointing the compound or group of compounds of interest (e.g. selecting *m/z* 93 and 136 will isolate most monoterpene hydrocarbons from the full total ion current trace). This approach has the advantage over genuine SIM in that full spectra are available for detailed interpretation, but on the other hand it is, of course, far less sensitive.

Chemical ionization mass spectrometry. One of the main frustrations in conventional EI MS is when no molecular ion peak is obtained in the mass spectrum. This is often due to the instability of the molecular ion under the excessive energy imparted by electron impact (an energy of 70 eV is usually employed in EI MS). However, this is not so much a problem in muscle food flavor analysis as in other areas, since many of the aroma components are aromatic (in the chemical sense), including many of the heterocyclic compounds. Such compounds generally yield molecular ions of reasonable abundance due to aromatic stabilization. Nevertheless, not all meat flavor compounds fall into this category, and use of a 'softer' ionization, which imparts less energy to the molecular ion than EI and hence limits its fragmentation, can be very useful. Chemical ionization (CI) is the most common alternative, softer approach in GC–MS. In CI MS a reagent gas, such as methane, isobutane or ammonia, is introduced into the mass spectrometer source to be ionised by broadly conventional EI. A range of positive ions, such as $C_2H_5^+$ from methane, is produced. Sample molecules are then ionized by ion–molecule reactions with the reagent gas species. The result is that so-called pseudo-molecular ions are produced, such as $(M+H)^+$, by proton transfer. Typically an energy of only 5 eV is imparted to sample molecules, so usually very little fragmentation is observed under these conditions.

The value of CI MS in flavor analysis is to complement and supplement the data provided by EI MS. It is not often used on its own, although there are exceptions (Lange and Schultze, 1988a,b), for the very reason that few fragmentation data for interpretation are usually obtained.

Negative ion chemical ionization mass spectrometry. In addition to positive ion CI MS, it is possible to perform negative ion CI MS, in which negatively charged reagent gas ions, such as OH^-, undergo similar ion-molecule interactions with sample molecules, but with the result that negatively charged pseudo-molecular ions are obtained, such as $(M–H)^-$, for example, which is produced by proton abstraction. In many respects, negative ion CI MS can be superior to positive ion CI MS, both in terms of sensitivity and degree of 'softness'. A good example of the latter is that isobornyl acetate (MW 196) still fragments under positive ion CI MS to yield only one main peak in the spectrum at *m/z* 137 due to $(M\text{-}CH_3COO)^+$, whereas under negative ion CI MS an intense peak at *m/z* 195, due to $(M\text{-}H)^-$, is obtained (Bruins, 1986). Negative ion CI MS has not been widely used in flavor analysis (Bruins, 1979; Hendriks and Bruins, 1980, 1983; George, 1984), but it has great potential and is certainly an underutilized technique.

There are some other methods of 'soft' ionization in mass spectrometry, but they are either not applicable to GC–MS (e.g. fast atom bombardment or FAB) or they have been only slightly used in flavor analysis (e.g. photoionization MS; Adamczyk *et al.*, 1987).

16.3.3 Alternatives to GC as a method of separation prior to identification

There are four main alternatives to GC as the method of separation prior to identification:

1. high performance liquid chromatography (HPLC);
2. supercritical fluid chromatography (SFC);
3. capillary electrophoresis (CE);
4. mass spectrometry in MS–MS.

High performance liquid chromatography. An obvious question is why even consider HPLC, when GC offers superior performance, is a commercially more highly developed technique, and by definition is better suited for the analysis of volatile components? Answers to the specific points include the fact that modern HPLC is, in fact, a more efficient chromatographic process than even the best GC, routinely providing far greater numbers of theoretical plates per unit length and hence superior HETP values. GC, however, provides far better performance overall, by virtue of the much longer open tubular columns which can routinely be used, e.g. 50–60 m, whereas typical HPLC columns are only 25 cm in length. Although GC is undoubtedly the more developed technique, it did have a head start of about two decades, and HPLC is rapidly catching up, with recent developments in instrumentation and column packings. Finally, it must not be overlooked that HPLC also can cope with analysis of volatiles in the same way as GC, in addition to being able to deal with nonvolatile constituents. It is also better suited to the analysis of thermally labile flavor components, although clearly this is not such a significant advantage in meat flavor analysis as in some other areas.

The main problem that has held back HPLC as a viable alternative to GC has been the great difficulty in satisfactorily and efficiently interfacing HPLC instruments with mass spectrometers. There are, however, reasonably priced benchtop HPLC–MS instruments currently available (Imatani and Smith, 1996). A number of these instruments employ soft ionization techniques such as atmospheric pressure introduction, including electrospray and ion spray. Despite this, however, HPLC still has not been widely exploited in flavor analysis, although recently it has been used by some researchers (Kim *et al.*, 1994; Sägesser and Deinzer, 1996). A technique of recent interest is coupled HPLC–GC–MS (Mondello *et al.*, 1996).

Supercritical fluid chromatography. When a gas such as carbon dioxide is heated above its critical temperature at sufficiently high pressure, it becomes a fluid with liquid-like density and solvating power, which broadly has properties between those of a gas and a liquid. Use of such supercritical fluid as the mobile phase in chromatography provides a technique somewhat intermediate between GC and HPLC. GC provides better chromatographic performance, but supercritical fluid chromatography (SFC), like HPLC, is also capable of dealing with solutes not amenable to GC (e.g. those of low volatility, high polarity or thermal instability) as well as volatile components. SFC is superior to HPLC in a number of respects. For example, as already mentioned, the integration of HPLC with MS is somewhat problematical, whereas the combination of SFC with MS is easier. In particular, mobile phase elimination after chromatography before MS is clearly less of a problem with a supercritical fluid such as carbon dioxide than with a conventional liquid solvent. SFC has been successfully interfaced with standard benchtop mass spectrometers (Ramsey *et al.*, 1995). Both packed and capillary column SFC are possible; supercritical carbon dioxide is the most common mobile phase.

In many respects, SFC combines the best features of GC and HPLC, namely the excellent resolving power of the former and the mild operating conditions of the latter, but to date it has not been extensively used in flavor analysis. Some examples can be quoted (Flament *et al.*, 1987; Calvey *et al.*, 1994), but this undoubtedly is a technique that has been underutilized.

Interestingly, in addition to being combined with MS as an identification technique, SFC has also been coupled with Fourier transform infrared spectroscopy (Hellgeth *et al.*, 1986; Morin *et al.*, 1986, 1987b), and used in flavor analysis (Morin *et al.*, 1987a).

Capillary electrophoresis. In capillary electrophoresis (CE), a thin-walled fused silica capillary, about 1 m in length, is filled with a buffer solution and one end is held in a buffer reservoir at ground potential with the other end in the buffer at a high voltage potential, typically 30 kV. Under the influence of the applied electric field, ionic species have a tendency to migrate eletrophoretically to the appropriate electrode. The speed at which these species migrate is dependent on the strength of the applied electric field and the mass to charge ratio of analyte molecules. Thus, a small singly charged species migrates at a faster rate than a large multiply charged molecule, so that the latter will have a longer retention time.

In addition, however, there is also an overwhelming electro-osmotic flow that sweeps all solutes through the capillary from the positive to negative electrode, but without itself promoting any separation. This effect is caused by the interaction of the buffer and the negatively charged capillary wall, which arises from the ionization of surface silanol groups. As a result, all components actually flow in the same direction, and although not true chromatography, CE can then be used chromatographically (so-called 'capillary electrochromatography'). Thus, positively charged species migrate towards the cathode at a speed corresponding to the sum of the electro-osmotic flow and the electrokinetic migration experienced by the molecule. CE thus provides a complex and highly efficient separatory process, and close to 1 million theoretical plates per meter has been obtained. A thorough review of CE can be found elsewhere (Righetti, 1996).

Furthermore, CE also has been interfaced successfully with mass spectrometry (Smith *et al.*, 1994), usually either via electrospray ionization (Olivares *et al.*, 1987; Smith *et al.*, 1988a,b; Dunayevskiy *et al.*, 1996) or via CF–FAB (Caprioli *et al.*, 1989; Moseley *et al.*, 1989). CE–MS constitutes a powerful analytical technique which, although still very much in its infancy, is causing great enthusiasm and excitement amongst biological scientists. It is ideally suited to the analysis of a very wide range of labile biological molecules. It is included here mainly on the basis of its potential, since it has not been used to any great extent in flavor analysis. Impressive results have, however, recently been obtained using CE–MS to analyze products from Maillard model systems (Tomlinson, 1991).

Mass spectrometry in MS–MS. In mass spectrometry–mass spectrometry (MS–MS), there are effectively two mass spectrometers linked together. The first stage of the analysis achieves separation of a mixture on the basis of individual selection of constituents, and the second provides a conventional mass analysis. The mixture to be analyzed is introduced into MS–1, and the molecular ions (or pseudo-molecular ions from CI or FAB MS) which are produced are separated in the normal manner, according to their different masses. At the exit of MS–1 the separated molecular ion is subjected to collisionally activated dissociation (CAD) collision with a neutral gas in a high pressure region, as a result of which the molecular ion fragments into characteristic daughter ions. These are then mass analyzed conventionally in MS-2 to provide a pure spectrum. It should be noted that there are other forms of MS–MS, but the preceding concept is more relevant here.

Although this type of approach has been known for a long time, for example in the study of metastable ions and in 'mass analyzed ion kinetic energy spectrometry' (MIKES), it is only during the past decade that

MS–MS has been accepted as a simple and effective procedure for analyzing mixtures. In general, it is not necessary to buy two mass spectrometers, since a wide range of different types of multiple sector instruments is now commercially available for MS–MS. Triple-sector instruments provide an extra analyzer (electrostatic, magnetic or quadrupole) beyond the CAD cell after MS–1, and four-sector instruments with a variety of geometries also are possible. A common configuration is a hybrid triple sector, consisting of a quadrupole analyzer (as MS-2) beyond an otherwise conventional double-focusing instrument (as MS–1), although triple quadrupole instruments are also available for less demanding work. Ion trap mass spectrometers also are becoming popular for MS–MS. A discussion of MS–MS with ion trap instruments can be found elsewhere (Huston, 1997). Fay *et al.* (1997) have demonstrated use of a quadrupole tandem mass spectrometer for MS–MS in flavor studies.

Obviously MS–MS fails if isomers are present or if at low resolution other different components in the mixture give the same unit mass parent in MS–1. The latter can, of course, be overcome at high resolution. A problem which is more difficult to overcome is when a compound does not yield a molecular or pseudo-molecular ion in sufficient abundance for significant CAD, which itself is not always an efficient process.

The main advantages of MS–MS over GC–MS are its simplicity and the fact that, by its very nature, the necessity for extensive sample preparation and handling is reduced. Nevertheless, it is highly unlikely that MS–MS will ever replace GC–MS in flavor analysis, although it is certain that as commercial instrumentation becomes more widely available, its use will increase. At present, MS–MS is employed much more widely in other analytical fields, and has had limited application in flavor analysis (Sheehan, 1996; Huston, 1997), the majority has been in determining specific compounds or groups of compounds, which is one of the strengths of the technique.

16.3.4 Alternatives to MS as a method of identification following separation

Other than MS there is really only one instrumental method that can be satisfactorily combined, on-line in tandem with a separatory procedure for identification of unknowns, and that is Fourier transform infrared spectroscopy.

Fourier transform infrared spectroscopy (FTIR). Although the only viable possibility in this category, FTIR has been very widely and very commonly used in flavor analysis, more so that any of the variations previously discussed. However, this is not to imply that it is the most useful. Nor it is a rival to MS in GC–MS; rather it provides valuable, complementary information.

A major problem with GC–FTIR is that is possesses significantly lower sensitivity (a smaller dynamic range) than GC–MS, but on the other hand, when IR spectra can be obtained they can provide important structural information which is either lacking or less obvious in the mass spectrum. The most valuable attribute of IR spectroscopy, of course, is in providing information regarding functional groups and their environment, but it can also sometimes be extremely useful in flavor studies by enabling discrimination between isomers, especially geometric isomers. In addition, since the IR spectrum of a compound is a virtually unique 'fingerprint', the technique provides a powerful single method of identification, in combination with an appropriate library of reference spectra. In this latter context, however, another major limitation of GC–FTIR at present is the lack of adequate, extensive databases of vapor phase IR spectra.

Many excellent descriptions of the apparatus for GC–FTIR exist in the literature (Schreier and Idstein, 1985a,b), which relate specifically to its application in flavor analysis. Basically, the GC–FTIR functions as follows. The effluent from the GC is passed through a light-pipe, which is a heated capillary gas cell internally coated with a layer of gold and capped at both ends with KBr windows. IR irradiation, typically in the range of 4000–750 cm^{-1}, passes through the cell, interacting with the components as they elute from the GC, to be detected at the other end by a suitable device (e.g. a cooled mercury cadmium telluride (MCT) semiconductor detector). Interferograms are generated which are processed by computer to yield typical IR spectra. Full spectra can readily be obtained from 1 s 'scans' showing a resolution of 4 cm^{-1}. Many different commercial GC–FTIR instruments are available.

An alternative to light-pipe GC–FTIR is cryogenic matrix isolation GC–FTIR, which offers improved sensitivity of about 100-fold. Again, commercial equipment is available. In this system, the effluent from the capillary GC is mixed with a matrix gas (usually about 1% argon) and sprayed onto a cryogenic surface (often a slowly rotating disc) at about 12 K, where it immediately freezes on the surface. Helium carrier gas is removed by vacuum, while argon, which is then in excess, forms a solid matrix completely surrounding and trapping solute molecules. FTIR spectra can then be taken at leisure. If the cold surface is transparent to IR (e.g. a CsI or CsBr window) then the IR beam passes through the sample and surface to yield an absorption spectrum; the frozen matrix gas is transparent to IR and therefore does not interfere. If the surface is a mirror (e.g. gold coated) then the IR beam passes through the sample and is then reflected back off the mirror surface to give a reflectance spectrum. The use of cryogenic matrix isolation GC–FTIR in flavor analysis has been described elsewhere (Williams *et al.*, 1987; Croasmun and McGorrin, 1989).

A considerable number of papers describe the use of GC–FTIR in flavor analysis. Several have employed GC–FTIR in the flavor analysis of muscle foods (e.g. Cadwallader *et al.*, 1995; Baek and Cadwallader, 1997). Several reviews have dealt with use of this technique in flavor analysis (Herres, 1984; Schreier and Idstein, 1985b; Werkhoff *et al.*, 1990) and some publications have specifically dealt with the important problem of compiling and searching GC–FTIR library databases (Field and White, 1987). As previously mentioned, FTIR has also been successfully combined with SFC and used in flavor analysis (Morrin *et al.*, 1986, 1987a,b; Taylor and Jordan, 1995).

An interesting advance is to link together GC with both MS and FTIR, and hence to obtain both sets of spectral data at the same time rather than in two separate runs. In many respects this would seem relatively easy, since light-pipe GC–FTIR is a nondestructive system, and separated solutes can then be passed from the FTIR to the MS. However, most successful integration has utilized the more sensitive matrix isolation GC–FTIR system, which is, of course, destructive in this context, so GC effluents have to be split (e.g. 1:1) before being fed to both the FTIR and the MS (Croasmun and McGorrin, 1989). It is uncertain whether this 'in series' approach is more effective, but useful results from an analysis of chicken volatiles have been reported, illustrating the value of having data from both analytical techniques (Croasmun and McGorrin, 1989).

There are no other analytical instruments, which can be used on-line, in tandem, with a GC for identification of unknowns in flavor analysis, but it is worth commenting on a few sophisticated GC detectors which can provide more information than a routine FID or any other conventional detector. For example, modern atomic emission detectors provide detection of virtually all elements, as well as some stable isotopes, in a complex sample matrix (Johnson *et al.*, 1995). It has been used in a study of ham flavor to enable heterocyclic compounds to be pinpointed to facilitate GC–MS (Baloga *et al.*, 1990).

16.4 Conclusions

There are numerous methods for the isolation and analysis of the volatile flavor components of muscle foods. None of the alternative procedures discussed here is likely to prove superior to the classical GC–MS based methods for the analysis of muscle food flavor. However, many of the analytical procedures described can provide extremely valuable additional and/or complementary data to those obtained by GC–MS. Indeed, for certain specific problems, alternative approaches may sometimes be superior. In addition, with a problem as difficult and complex as studying and analyzing the flavor of muscle foods, it is absolutely essential to use all possible techniques and procedures which are available and which might yield constructive information.

References

Adamczyk, B., Genuit, W. and Boon, J.J. (1987). A simple approach to the dynamic headspace analysis of volatile flavors using a gas chromatography–photoionization mass spectrometer. *Biomed. Environ. Mass Spectrom.*, **161**, 373–375.

Acree, T.E. (1993). Gas chromatography–olfactometry. In *Flavor Measurement*, eds. C.T. Ho and C.H. Manley. Marcel Dekker, New York, pp. 77–94.

Anderson, H.J. and Hinrichsen, L.L. (1995). Changes in curing agents, microbial counts and volatile compounds during processing of green bacon using two different production technologies. *J. Sci. Food Agric.*, **68**, 477–487.

Ang, C.Y.W., Liu, F. and Sun, T. (1994). Development of a dynamic headspace GC method for assessing the influence of heating and end-point temperature on volatiles of chicken breast meat. *J. Agric. Food Chem.*, **42**, 2493–2498.

Arora, G., Cormier, F. and Lee, B. (1995). Analysis of odor-active volatiles in Cheddar cheese headspace by multidimensional GC/MS/sniffing. *J. Agric. Food Chem.*, **43**, 748–752.

Baek, H.H. and Cadwallader, K.R. (1997). Aroma volatiles in cooked alligator meat. *J. Food Sci.*, **62**, 321–325.

Baloga, D.W., Reineccius, G.A. and Miller, J.W. (1990). Characterization of ham flavor by an atomic emission detector. *J. Agric. Food Chem.*, **38**, 2021–2026.

Bernreuther, A. and Schreier, P. (1991). Multidimensional gas chromatography/mass spectrometry: a powerful tool for the direct chiral evaluation of aroma compounds in plant tissues. II. Linalool in essential oils and fruits. *Phytochem. Anal.*, **2**, 167.

Bernreuther, A., Epperlein, U. and Koppenhoefer, B. (1997). Enatiomers: Why they are important and how to resolve them. In *Techniques for Analyzing Food Aroma*, ed. R. Marsili. Marcel Dekker, New York, pp. 143–207.

Blank, I. (1997). Gas chromatography-olfactometry in food aroma analysis. In *Techniques for Analyzing Food Aroma*, ed. R. Marsili. Marcel Dekker, New York, pp. 265–292.

Bolzoni, L., Barbieri, G. and Virgilli, R. (1996). Changes in volatile compounds of Parma ham during maturation. *Meat Sci.*, **43**, 301–310.

Bonelli, E.J. (1993). A high-performance benchtop GC–MS. *Am Lab.*, **25**(1), 28C-28G.

Bruins, A.P. (1979). Negative ion chemical ionization mass spectrometry in the determination of components in essential oils. *Anal. Chem.*, **51**, 967–972.

Bruins, A.P. (1986). GC/MS employed with chemical ionisation sources for studies of positive and negative ions. In *Advances in Mass Spectrometry*, Part A, ed. J.F.J. Todd, Wiley, Chichester, UK, pp. 119–131.

Buttery, R.G., Takeoka, G.R., Krammer, G.E. and Ling L. (1994). Identification of 2,5-dimethyl-4-hydroxy-3(2*H*)-furanone (Furaneol) and 5-methyl-4-hydroxy-3(2*H*)-furanone in fresh and processed tomato. *Lebensm. Wiss. u. Technol.*, **27**, 592–594.

Cadwallader, K.R., Tan, Q., Chen, F. and Meyers, S.P. (1995). Evaluation of the aroma of cooked spiny lobster tail meat by aroma extract dilution analysis. *J. Agric. Food Chem.*, **43**, 2432–2437.

Calvey, E., Roach, J.A. and Block, E. (1994). Supercritical fluid chromatography of garlic (*Allium sativum*) extracts with mass spectrophotometric identification of allicin. *J. Chromatogr. Sci.*, **32**, 93–96.

Caprioli, R.M., Moore, W.T., Martin, M., DaGue, B.B., Wilson, K. and Moring, S. (1989). Coupling capillary zone electrophoresis and continuous-flow fast atom bombardment mass spectrometry for the analysis of peptide mixtures. *J. Chromatogr.*, **480**, 247–257.

Croasmun, W.R. and McGorrin, R.J. (1989). Gas chromatography–matrix isolation infrared spectroscopy for analysis of thermally generated aroma compounds. In *Thermal Generation of Aromas*, ed. T.H. Parliment, R.J. McGorrin and C.-T. Ho. American Chemical Society, Washington, DC, pp. 61–72.

Dunayevskiy, Y.M., Vouros, P., Wintner, E.A., Shipps, G.W., Carell, T. and Rebek, J. Jr (1996). Application of capillary electrophoresis–electrospray ionization mass spectrometry in the determination of molecular diversity. *Proc. Natl Acad. Sci.*, **93**, 6152–6157.

Elsner, T., Rothweiler, B., Marx, F. and Pfeilsticker, K. (1996). Detection of irradiated chicken breast by headspace gas chromatography and discriminant analytical evaluation. *Z. Lebensm. Unters. Forsch.*, **202**, 63–65.

Etiévant, P.X. (1996). Artifacts and contaminants in the analysis of food flavor. *CRC Crit. Rev. Food Sci. Nutr.*, **36**, 733–745.

Farmer, L.J., Hagan, T.D.J. and Paraskevas, O. (1997). A comparison of three sugars and inosine monophosphate as precursors of meat aroma. In *Flavor Science: Recent Developments*, eds. A.J. Taylor and D.S. Mottram. The Royal Society of Chemistry, Cambridge, pp. 225–230.

Fay, L.B., Blank, I. and Cerny, C. (1997). Tandem mass spectrometry in flavor research. In *Flavor Science: Recent Developments*, eds. A.J. Taylor and Mottram, D.S. The Royal Society of Chemistry, Cambridge, pp. 271–282.

Fields, R.E. and White, R.L. (1987). Real-time library search of vapor-phase spectra for gas chromatography/Fourier transform infrared spectrometry eluents. *Anal. Chem.*, **59**, 2709–2716.

Flament, I., Chevallier, C. and Keller, U. (1987). Extraction and chromatography of food constituents with supercritical CO_2. In *Flavor Science and Technology*, eds. M. Martens, G.A. Dalen and H. Russwurm Jr. Wiley, Chichester, pp. 151–163.

Garcia-Regueiro, J.A. and Diaz, I. (1989). Evaluation of the contribution of skatole, indole, androstenone and androstenols to boar-taint in back fat of pigs by HPLC and capillary gas chromatography. *Meat Sci.*, **25**, 307–316.

George, G. (1984) The negative ion technique in mass spectrometry. Application to problems of aroma. *Labo-Pharma – Probl. Tech.*, **343**, 479–481.

Girard, B. and Nakai, S. (1994). Sexual maturity stages of canned chum salmon segregated using static headspace volatile patterns. *Food Res. Int.* **27**, 469–476.

Grimm, C.C., Lloyd, S.W., Miller, J.A. and Spanier, A.M. (1997). The analysis of food volatiles using direct thermal desorption. In *Techniques for Analyzing Food Aroma*, ed. R. Marsili. Marcel Dekker, New York, pp. 59–80.

Grosch, W. (1993). Detection of potent odorants in foods by aroma extract dilution analysis. *Trends Food Sci. Technol.*, **4**, 68–73.

Guillard, A.S., Le Quéré, J.L. and Vendeuvre, J.L. (1997). A gas chromatographic–olfactometric study of cooked cured ham – impact of sodium nitrite. In *Flavor Science: Recent Developments*, eds. A.J. Taylor and D.S. Mottram. The Royal Society of Chemistry, Cambridge, pp. 231–238.

Guth, H. and Grosch, W. (1994). Identification of the character impact odorants of stewed beef juice by instrumental analyses and sensory studies. *J. Agric. Food Chem.*, **42**, 2862–2866.

Harmon, A.D. (1997). Solid phase microextraction for the analysis of flavors. In *Techniques for Analyzing Food Aroma*, ed. R. Marsili. Marcel Dekker, New York, pp. 81–112.

Hartman, T.G., Lech, J., Karmas, K., Salinas, J., Rosen, R.T. and Ho, C.-T. (1993). Flavor characterization using adsorbent trapping–thermal desorption or direct thermal desorption–gas chromatography and gas chromatography–mass spectrometry. In *Flavor Measurement*, eds. C.T. Ho and C.H. Manley, Marcel Dekker, New York, pp. 37–60.

Hellgeth, H., Jordan, J.W., Taylor, L.T. and Khorassani, M.A. (1986). Supercritical fluid chromatography of free fatty acids with on-line FTIR detection. *J. Chromatogr. Sci.*, **24**, 183–188.

Hendriks, H. and Bruins, A.P. (1980). Study of three types of essential oil of *Valeriana officialis* L. s.l. by combined gas chromatography–negative ion chemical ionization mass spectrometry. *J. Chromatogr.*, **190**, 321–330.

Hendriks, H. and Bruins, A.P. (1983). A tentative identification of components in the essential oil of *Cannabis sativa* L. by a combination of gas chromatography negative ion chemical ionization mass spectrometry and retention indices. *Biomed. Mass Spectrom.*, **10**, 377–381.

Hener, U., Hollnagel, A., Kreis, P., Maas, B., Schmarr, H.-G., Schubert, V., Rettinger, K., Weber, B. and Mosandl, A. (1990). Direct enatiomer separation of chiral volatiles from complex matrices by multidimensional gas chromatography. In *Flavor Science and Technology*, eds. Y. Bessiere and A.F. Thomas. Wiley, Chichester, pp. 25–28.

Herres, W. (1984). Capillary GC–FTIR analysis of volatiles: HRGC–FTIR. In *Analysis of Volatiles*, ed. P. Schreier. Walter de Gruyter, Berlin, pp. 183–217.

Hinshaw, J.V. (1994). Bench-top mass spectrometric detectors. *LCGC*, **12**, 906–912.

Homatidou, V., Karvouni, S. and Dourtoglou, V. (1990). Determination of characteristic aroma components of 'Cantaloupe' *Cucumis melo* using multidimensional gas chromatography (MDGC). In *Flavors and Off-flavors*, ed. G. Charalambous. Elsevier, Amsterdam, pp. 1011–1023.

Huston, C.K. (1997). Ion trap mass spectrophotometry for food aroma analysis. In *Techniques for Analyzing Food Aroma*, ed. R. Marsili. Marcel Dekker, New York, pp. 209–235.

Imatani, K. and Smith, C. (1996). An easy-to-use benchtop liquid chromatograph/mass selective detector. *Am. Lab.*, **28**(11), 29–33.

Johnson, D., Quimby, B. and Sullivan, J. (1995). An atomic emission detector for gas chromatography. *Am. Lab.*, **27**(10), 13–20.

Kerler, J. and Grosch, W. (1996). Odorants contributing to warmed-over flavor (WOF) of refrigerated cooked beef. *J. Food Sci.*, **61**, 1271–1284.

Kerscher, R. and Grosch, W. (1997). Comparative evaluation of potent odorants of boiled beef by aroma extract dilution and concentration analysis. *Z. Lebensm. Unters. Forsch. A*, **204**, 3–6.

Kim, H.-J., Ball, D., Giles, J. and White, F. (1994). Analysis of thermally produced compounds in foods by thermospray liquid chromatography–mass spectrometry and gas chromatography–mass spectrometry. *J. Agric. Food Chem.*, **42**, 2812–2816.

Lange, G. and Schultze, W. (1988a). Differentiation of isopulegol isomers by chemical ionization mass spectrometry. In *Bioflavor '87*, ed. P. Schreier. Walter de Gruyter, Berlin, pp. 115–122.

Lange, G. and Schultze, W. (1988b). Studies on terpenoids and non-terpenoid esters using chemical ionization mass spectrometry in GC/MS coupling. In *Bioflavor '87*, ed. P. Schreier. Walter de Gruyter, Berlin, pp. 105–114.

Langourieux, S. and Escher, F.E. (1997). Off-flavor formation in heat sterilised meat in trays. In *Flavor Science: Recent Developments*, eds. A.J. Taylor and D.S. Mottram. The Royal Society of Chemistry, Cambridge, pp. 194–199.

Ledford, E.B. Jr, Phillips, J.B., Xu, J., Gaines, R.B. and Blomberg, J. (1996). Ordered chromatograms: A powerful methodology in gas chromatography. *Am. Lab.*, **28**(6), 22–25.

Maignial, L., Pibarot, P., Bonetti, G., Chaintreau, A. and Marion, J.P. (1992). Simultaneous distillation–extraction under static vacuum: Isolation of volatile compounds at room temperature. *J. Chromatogr.*, **606**, 87–94.

Mateo, J. and Zumalacárregui, J.M. (1996). Volatile compounds in chorizo and their changes during ripening. *Meat Sci.*, **44**, 255–273.

McDaniel, M.R., Miranda-López, R., Watson, B.T., Micheals, N.J. and Libbey, L.M. (1990). Pinot noir aroma: a sensory/gas chromatographic approach. In *Flavors and Off-Flavors*, ed. G. Charalambous. Elsevier, Amsterdam, p. 23.

Milo, C. and Grosch, W. (1995). Detection of odor defects in boiled cod and trout by gas chromatography–olfactometry of headspace samples. *J. Agric. Food Chem.*, **43**, 459–462.

Milo, C. and Grosch, W. (1996). Changes in the odorants of boiled salmon and cod as affected by the storage of the raw material. *J. Agric. Food Chem.*, **44**, 2366–2371.

Mistry, B.S., Reineccius, T. and Olson, L. (1997). Gas chromatography–olfactometry for the determination of key odorants in foods. In *Techniques for Analyzing Food Aroma*. ed. R. Marsili. Marcel Dekker, New York, pp. 265–292.

Mondello, L., Dugo, G. and Bartle, K.D. (1996). Coupled HPLC–HRGC–MS: A new method for the on-line analysis of real samples. *Am. Lab.*, **28**(12), 41–49.

Morin, P., Claude, M., Richard, H. and Rossett, R. (1986). Carbon dioxide supercritical fluid chromatography–Fourier transform infrared spectrometry. *Chromatographia*, **21**, 523–530.

Morin, P., Claude, M., Richard, H. and Rossett, R. (1987a). On-line carbon dioxide SFC–FTIR in aroma research. Perspectives and limits. In *Flavor Science and Technology*, eds. M. Martens, G.A. Dalen and H. Russwurm Jr. Wiley, Chichester, pp. 165–174.

Morin, P., Claude, M., Richard, H. and Rossett, R. (1987b). Supercritical fluid chromatography–Fourier transform infrared spectrometry coupling applied to polycyclic aromatic and sesquiterpene hydrocarbon analysis. *Analysis*, **15**, 117–127.

Mosandl, A. (1995). Enantioselective capillary gas chromatography and stable isotope ratio mass spectrometry in the authenticity control of flavors and essential oils. *Food Rev. Int.*, **11**, 597–664.

Moseley, M.A., Deterding, L.J., Tomer, K.B. and Jorgenson, J.W. (1989). Coupling of capillary zone electrophoresis and capillary liquid chromatography with coaxial continuous-flow fast atom bombardment tandem sector mass spectrometry. *J. Chromatogr.*, **480**, 197–209.

Olivares, J.A., Nguyen, N.T., Yonker, C.R. and Smith, R.D. (1987). On-line mass spectrometric detection for capillary zone electrophoresis. *Anal. Chem.*, **59**, 1230–1232.

Parliment, T.H. (1997). Solvent extraction and distillation techniques. In *Techniques for Analyzing Food Aroma*, ed. R. Marsili. Marcel Dekker, New York, pp. 1–26.

Patterson, R.L.S. and Stevenson, M.H. (1995). Irradiation-induced off-odour in chicken and its possible control. *Br. Poult. Sci.*, **36**, 425–441.

Ramsey, E.D., Lawrence, S.D. and Games, D.E. (1995). Application of capillary supercritical fluid chromatography combined with benchtop mass spectrometry for the analysis of macrolide antibiotic mydecamycin Al. *Anal. Proc.*, **32**, 455–457.

Refsgaard, H.H.F., Brockhoff, P.M. and Jensen, B. (1997). Sensory and chemical changes of salmon during frozen storage. In *Flavor Science: Recent Developments*, eds. A.J. Taylor and D.S. Mottram. The Royal Society of Chemistry, Cambridge, pp. 344–347.

Reineccius, G. (1993). Biases in analytical flavor profiles introduced by isolation method. In *Flavor Measurement*, eds. C.T. Ho and C.H. Manley. Marcel Dekker, New York, pp. 61–76.

Righetti, P.G., ed. (1996). *Capillary Electrophoresis in Analytical Biotechnology*. CRC Press, Boca Raton, FL.

Sägesser, M. and Deinzer, M. (1996). HPLC–ion spray-tandem mass spectrometry of flavonol glycosides in hops. *J. Am. Soc. Brew. Chem.*, **54**, 129–134.

Schreier, P. and Idstein, H. (1985a). Advances in the instrumental analysis of food flavors. *Z. Lebensm. Unters. Forsch.*, **180**, 1–14.

Schreier, P. and Idstein, H. (1985b). High-resolution gas chromatography–Fourier transform infrared spectroscopy in flavor analysis. *Z. Lebensm. Unters. Forsch.*, **181**, 183–188.

Sen, A., Laskawy, G., Schieberle, P. and Grosch, W. (1991). Quantitative determination of (*E*)-β-damascenone in foods using a stable isotope dilution assay. *J. Agric. Food Chem.*, **39**, 757–759.

Sheehan, T.L. (1996). GC–MS–MS: The next logical step in benchtop GC–MS. *Am. Lab.*, **28**(11), 28V–28Y.

Smith, R.D., Baringa, C.J. and Udseth, H.R. (1988a). Improved electrospray ionization interface for capillary zone electrophoresis–mass spectrometry. *Anal. Chem.*, **60**, 1948–1952.

Smith, R.D., Olivares, J.A., Nguyen, N.T. and Udseth, H.R. (1988b). Capillary zone electrophoresis–mass spectrometry using an electrospray ionization interface. *Anal. Chem.*, **60**, 436–441.

Smith, R.D., Goodlet, D.R. and Wahl, J.H. (1994). *Capillary Electrophoresis–Mass Spectrometry*, CRC Press, Boca Raton, FL.

St. Angelo, A.J. (1996). Lipid oxidation in foods. *CRC Crit. Rev. Food Sci. Nutr.*, **36**, 175–224.

Taylor, L.T. and Jordan, S.L. (1995). Supercritical fluid extraction coupled directly with Fourier transform infrared spectrometry. *J. Chromatogr. A*, **703**(1–2), 537–548.

Thomas, A.F., Wilhalm, B. and Flament, I. (1984). Some aspects of GC–MS in the analysis of volatile flavors. In *Chromatography and Mass Spectrometry in Nutrition Science and Food Safety*, eds. A. Frigerio and H. Milon. Elsevier, Amsterdam, pp. 47–65.

Thomlinson, A.J. (1991). Analysis of the coloured products of the Maillard reaction. PhD Thesis, London University, London.

Tressl, R., Grunewald, K.G., Kersten, E. and Rewicki, D. (1985). Formation of pyrroles and tetrahydroindolizin-6-ones as hydroxyproline-specific Maillard products from erythrose and arabinose. *J. Agric. Food Chem.*, **33**, 1137–1142.

Van Wassenhove, F., Dirinck, P. and Schamp, N. (1988). Analysis of the key components of celery by two dimensional capillary gas chromatography. In *Bioflavor '87*, ed. P. Schreier. Walter de Gruyter, Berlin, pp. 137–144.

Wampler, T.P. (1997). Analysis of food volatiles using headspace-gas chromatographic techniques. In *Techniques for Analyzing Food Aroma*, ed. R. Marsili. Marcel Dekker, New York, pp. 27–58.

Werkhoff, P., Emberger, R., Guntert, M. and Kopsel, M. (1989). Isolation and characterization of volatile sulfur-containing meat flavor components in model systems. In *Thermal Generation of Aromas*, eds. T.H. Parliament, R.J. McGorrin and C.-T. Ho. American Chemical Society, Washington, DC, pp. 460–478.

Werkoff, P., Bretscheider, W., Herrman, H.J. and Schreiber, K. (1990). Progress in analysis of odorous substances. Part II. Techniques for structure determination of odorous and flavor compounds. *LaborPraxis*, **14**, 256–264.

Williams, A.A., Tucknott, O.G., Lewis, M.J., May, H.V. and Wachter, L. (1987). Examples of cryogenic matrix isolation GC/IR in the analysis of flavour extracts. In *Flavour Science and Technology*, eds. M. Martens, G.A. Dalen and H. Russwurm Jr. Wiley, Chichester, pp. 259–270.

Wright, D.W. (1997). Application of multidimensional gas chromatography techniques to aroma analysis. In *Techniques for Analyzing Food Aroma*, ed. R. Marsili. Marcel Dekker, New York, pp. 113–141.

Wright, D.W., Mahler, K.O. and Ballard, L.B. (1986). The application of an expanded multidimensional GC system to complex fragrance evaluations. *J. Chromatogr. Sci.*, **24**, 60–65.

17 Assessment of lipid oxidation and off-flavour development in meat, meat products and seafoods

F. SHAHIDI

17.1 Introduction

Lipid oxidation is a major cause of flavour deterioration of muscle foods. Many methods for its evaluation have been employed and these measure: (1) primary changes such as those in the structure of original fatty acids and formation of free radical intermediates, oxygen uptake and weight gain; (2) formation of lipid hydroperoxides, the primary products of oxidation; (3) formation of secondary products of lipid oxidation (secondary changes) such as aldehydes production; and (4) overall changes that occur in food and indirect methods. However, the ultimate criterion for the suitability of any of these markers as an index of lipid oxidation is its adequate correlation with sensory data. This chapter discusses some of the commonly used methods of assessment of lipid oxidation and off-flavour development in meat and meat products.

A major cause of quality deterioration of foods, in general, and of muscle foods in particular, is due to changes in their lipid components by both enzymatic and nonenzymatic lipid oxidation processes. Oxidation of lipids results not only in the loss of essential fatty acids but it generally brings about undesirable changes in flavour, colour, texture and functional properties, as well as causing the destruction of fat-soluble vitamins and formation of cholesterol oxides. The susceptibility to and rate of oxidation of fatty acids in the lipids depend on the degree of their unsaturation (Belitz and Grosch, 1987). Thus, autoxidation of major fatty acids of meat follows the order given below.

$$C18{:}0 < C18{:}1 < C18{:}2 < C18{:}3$$

Seafoods, containing a large proportion of C20:5 and C20:6 fatty acids (eicosapentaenoic acid, EPA, and docosahexaenoic acid, DHA, respectively) in their lipids, are even more prone to autoxidation than the C18 acids listed above. Tichivangana and Morrissey (1985) have shown that the autoxidation of muscle foods occurs in the following order:

$$\text{fish} > \text{poultry (chicken and turkey)} > \text{pork} > \text{beef} \gg \text{lamb}$$

Autoxidation of lipids proceeds via a free-radical chain mechanism and is catalysed by many factors such as the presence of heat, light, ionizing radiation, metal ions and metalloporphyrins (Wong, 1989). It involves initiation, propagation and termination steps, as given below.

$RH \longrightarrow R^{\bullet} + H^{\bullet}$		Initiation
$R^{\bullet} + O_2 \longrightarrow ROO^{\bullet}$	}	Propagation
$ROO^{\bullet} + RH \longrightarrow ROOH + R^{\bullet}$		
$R^{\bullet} + R^{\bullet} \longrightarrow R\text{–}R$	}	Termination
$R^{\bullet} + ROO^{\bullet} \longrightarrow ROOR$		
$ROO^{\bullet} + ROO^{\bullet} \longrightarrow ROOR + O_2$		

Hydroperoxides (ROOH) are the primary products of lipid autoxidation. Each fatty acid may produce several hydroperoxides upon oxidation; however, hydroperoxides do not possess any flavour. Their breakdown into secondary oxidation products produces a wide array of low molecular weight organic compounds. Some of these degradation products are potently flavour-active and impart off-flavour to cooked, stored muscle foods (Figure 17.1) (Chang and Peterson, 1977). One way to retard lipid oxidation is by using natural or synthetic antioxidants.

Primary antioxidants act as scavengers or terminators of free radicals by an electron- or a hydrogen-donating mechanism (Labuza, 1971). Phenolic antioxidants such as butylated hydroxyanisole (BHA), butylated hydroxytoluene (BHT), propyl gallate (PG) and *t*-butyl hydroquinone (TBHQ) are common food additives used in processed products. Secondary antioxidants are compounds that act as deactivators of catalysts of autoxidation such as chelating agents or those which show a synergistic effect when used in combination with common antioxidants (Labuza, 1971; Roozen, 1987). Thus, for extending the shelf-life of muscle foods, elimination or inactivation of pro-oxidants and/or oxygen from the system and termination of free radicals are common methods employed. Polyphosphates, citric acid and ethylenediaminetetraacetic acid (EDTA) are examples of typical secondary antioxidants used in the food industry.

In meats, nitrite curing is an effective means of prevention of flavour deterioration (Gray *et al.*, 1981; Gray and Pearson, 1984). The exact role of nitrite in inhibiting lipid autoxidation and chelating properties, as well as its possible role in stabilizing meat membrane lipids, have been suggested (MacDonald *et al.*, 1980; Zubillaga *et al.*, 1984; Igene *et al.*, 1985).

The techniques employed for assessing lipid oxidation in muscle foods include analysis of the following: changes in the fatty acids, presence and type of free radical intermediates, conjugated dienes and trienes, polar components, polymers and *trans*-isomers. Determination of oxygen uptake, total or individual carbonyl compounds, peroxide value (PV), 2-thiobarbituric acid (TBA) value and anisidine value (AnV) are also commonplace. Furthermore, the Kries test, the oxirane test, fluorescence tests, and polarographic and chromatographic determinations are employed.

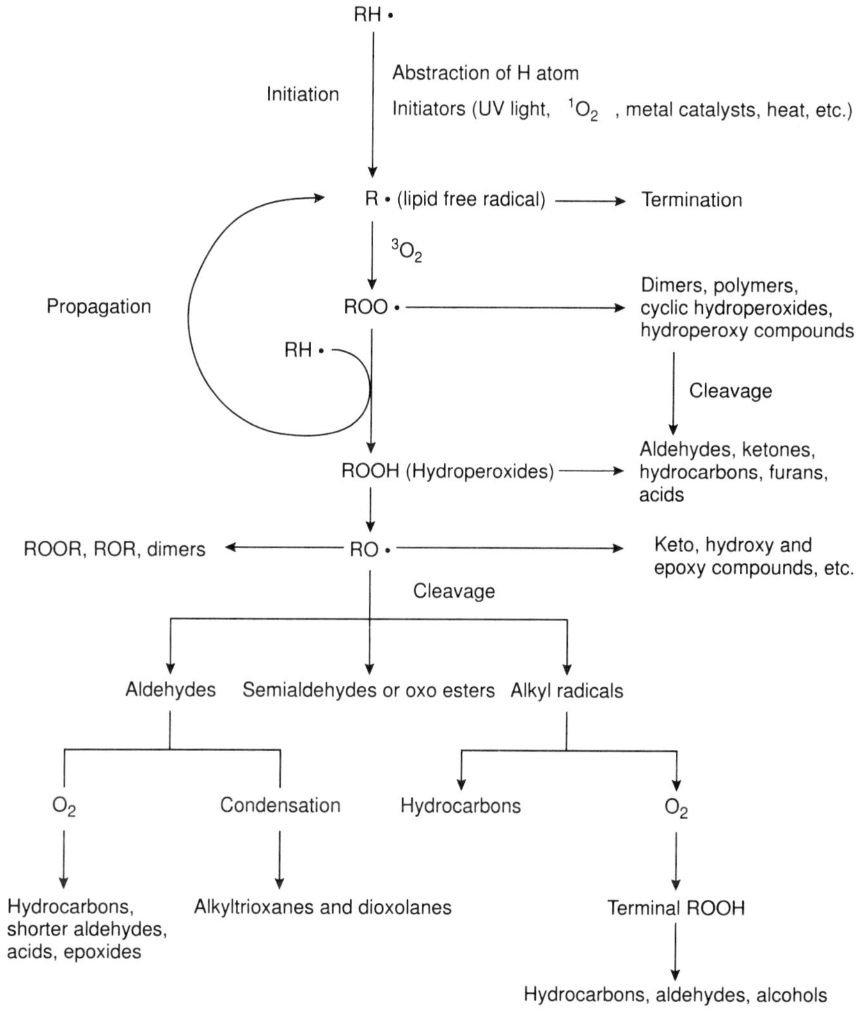

Figure 17.1 Oxidation of muscle food lipids.

In order to assess the quality of muscle foods, one must be able to quantify the extent of oxidation. Different methods for measurement of the oxidative state of foods have been employed as listed above. These methods may measure one of the following parameters.

1. Measurement of changes in the concentration of one or more substrates. Such as methods include analysis of lipid fatty acids, formation of conjugated dienes/trienes and/or *trans*-isomers, and oxygen uptake. Formation of lipid free radicals, or their derivatives, may also be measured.

2. Measurement of the primary products of lipid oxidation. In this respect, estimation of PV is practised.
3. Measurement of the secondary products of lipid oxidation. Such methods commonly employed include analysis of the content of malonaldehyde (MA), 2-thiobarbituric acid reactive substances (TBARS), and total carbonyls or selected volatile constituents.
4. Measurement of the overall changes in lipids such as those predicted by total oxidation (TOTOX) value, among others.
5. Indirect methods. These methods may include measurements of texture, functional properties and colour.
6. Sensory analysis. This is a method of great importance and generally all objective methods of quantification of lipid oxidation must be correlated with it.

Apart from sensory evaluation for estimation of the oxidative state of muscle foods, other methods are categorized as either chemical or physical. Sensory evaluation techniques are simple, but are time-consuming, costly and often poorly reproducible. To avoid human bias and error, several measurements are performed and statistical analysis is used to process the results. The present chapter reviews methods of assessment of the oxidative changes of meat and meat products.

17.2 Fatty acid analysis

An indirect method to measure the susceptibility of meat lipids to oxidation is to monitor changes that occur in their fatty acid composition (Keller and Kinsella, 1973). A 25% decrease in the concentration of C20:4 of phosphatidylethanolamine in hamburgers upon cooking was noticed and this corresponded well with an increase in the TBA values and total content of carbonyl compounds. However, measurement of fatty acids is not a very sensitive method as they must possess three or more double bonds to be very susceptible to oxidation (Moerck and Ball, 1974). In meats, these fatty acids reside primarily in the phospholipid fraction. Thus, separation of phospholipids from other lipid components prior to their determination is required.

Dimick and MacNeil (1970) and Moerck and Ball (1974) reported that the level of polyunsaturated fatty acids (PUFA) in chicken phospholipids decreases rapidly during refrigerated or frozen storage. Similar results have been reported by Igene *et al.* (1980) in beef and poultry meat model systems and by Gokalp *et al.* (1981) for beef patties. However, in intact muscles, lipids were less susceptible to changes in their content of PUFA. Furthermore, variations in the content of α-tocopherol as well as the level of haemoproteins may influence the degree of changes that might occur in the use of chromatographic methods such as HPLC procedures for the separation and quantification of individual phospholipids may present a fast and accurate method for assessing lipid autoxidation of muscle foods (Bolton *et al.*, 1985; Yeo and Horrocks, 1985).

17.3 Oxygen uptake

The uptake of oxygen has been used as a method of assessing the susceptability of muscle tissues to lipid oxidation (Fisher and Deng, 1977; Lee *et al.*, 1975; Silberstein and Lillard, 1978). Although oxidation of proteins may contribute to oxygen uptake results by muscle tissue lipids, existing differences in the rate of their oxidation makes effective utilization of this method viable (Rhee, 1978b). Our own work on seal muscle tissues lends support to this latter finding (Shahidi, unpublished results):

17.4 Conjugated dienes

Sklan *et al.* (1983) determined the content of conjugated dienes, trienes and tetraenes, referred to as total conjugated products of oxidation, in total lipid extracts of turkey meat during a 60-day storage at 4°C and more than 18 months' storage at –18°C. The content of conjugated products resulting from oxidation was found to increase at both temperatures. Similar studies were made by Ahmad and Augustin (1985) for fried fish during a 40-day storage at –60°C. These authors indicated that the level of both dienes and trienes increased with increasing storage time. Therefore, this method may offer a practical procedure for assessing lipid oxidation of meat and meat products.

17.5 Peroxide value

Hydroperoxides are primary products of lipid autoxidation. A common method for monitoring oxidative changes in muscle foods is via determination of peroxide value (PV) of their lipids. Peroxides in meat and meat products may be measured by a variety of techniques as described by Gray (1978). Peroxide value is commonly reported in milliequivalents of iodine per kilogram of fat and is generally determined by the use of an iodometric method (Fiedler, 1974; AOCS, 1989). Although interferences from the presence of oxygen in the assay sample and uptake of iodine by lipid double bonds are possible, this method nonetheless has been used extensively for assessing the oxidative state of fatty foods.

The time to reach a certain PV may be used as an index of oxidative stability of meat lipids. The effect of antioxidants and food processing on lipids is often monitored in this way. Thus, a longer period to reach a certain PV generally indicates a better antioxidant activity for the additive under examination.

Bailey *et al.* (1973) have used PV for evaluation of pork fat quality during frozen storage, and the oxidative state of beef muscle tissues during a 10-week storage at –10°C was reported by Owen *et al.* (1975). Jeremiah (1980) reported the keeping quality of frozen pork samples in different packaging materials. In general, a significant relationship between the PV and sensory flavour scores was noted. However, breakdown of hydroperoxides to secondary oxidation products may result in a decrease in PV during the storage period (Awad *et al.*, 1968; Noble, 1976). Due to the aforementioned considerations, PV has not been used extensively in evaluation of the oxidative state of meat and meat products (Pearson *et al.*, 1977; Melton, 1983).

17.6 2-Thiobarbituric acid test

The 2-thiobarbituric acid (TBA) test is the most widely used method for assessing oxidative state of muscle foods (Tarladgis *et al.*, 1960; Gray, 1978; Melton, 1983). Malonaldehyde (MA) is a decomposition product of lipid peroxides formed in meats. It has been extensively studied due to its reactivity with biological molecules such as amino groups of amino acids, proteins, nucleic acids and with sulphydryl groups (Chio and Tappel, 1969a,b; Draper *et al.*, 1986). Malonaldehyde itself is generally bound to biological materials and therefore, prior to determination, it must be released from muscle tissues by acid treatment.

Kohn and Liversedge (1944) were the first to use the 2-thiobarbituric acid (TBA) test for evaluation of the oxidative state of meats. The specificity of this method for estimating malonaldehyde, as its TBA derivative, was then reported by Sinnhuber and Yu (1958), as formulated in Figure 17.2. The pink-coloured chromophore formed between the TBA reagent and malonaldehyde in a 2:1 ratio (Figure 17.2) (Nair and Turner, 1984) has an absorption maximum at 532 nm. However, researchers have shown that alkenals and 2,4-alkadienals may also react with the TBA reagent to form a coloured complex with absorption at 532 nm (Marcuse and Johansson, 1973; Kosugi *et al.*, 1987, 1988). Therefore, the TBA test is generally used to quantify the content of TBA reactive substances (TBARS).

The TBA test may be performed directly on a food product (Wills, 1965), and this may be followed by extraction of the coloured pigment into butanol or a butanol–pyridine mixture (Placer *et al.*, 1966; Uchiyama and Mihara, 1978; Ohkawa *et al.*, 1979). The test may also be carried out on an aliquot of an acid extract of food (usually in 7.5–28% trichloroacetic acid (TCA) solution) (Siu and Draper, 1978) or on a portion of a

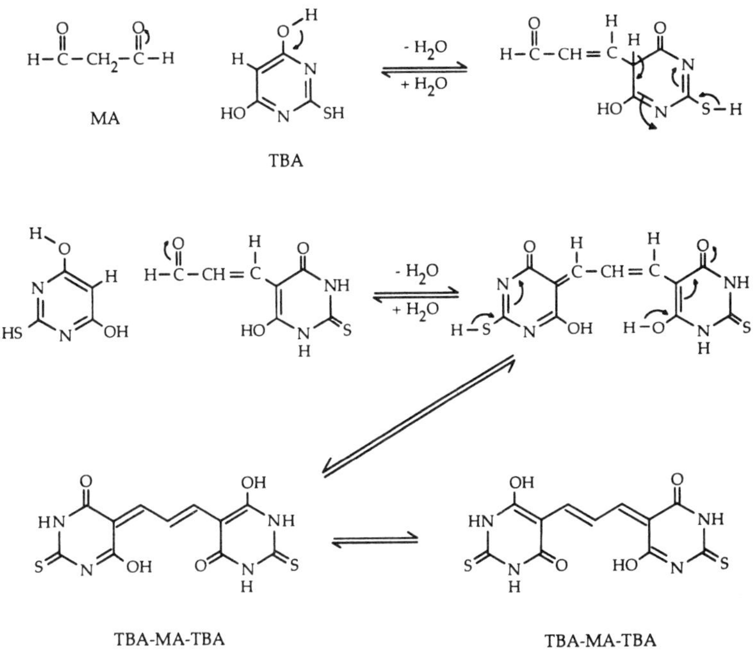

Figure 17.2 Mechanism of formation of 2-thiobarbituric acid (TBA)–malonaldehyde (MA) adduct.

steam distillate of the sample under investigation (Tarladgis *et al.*, 1960, 1964). The distillation procedure developed by Tarladgis *et al.* (1960) is the method frequently used and the TCA extraction procedure has also been used by many researchers (Shahidi and Hong, 1991).

Quantification of MA and its precursor(s), using TBA colorimetry, has been followed by heating the assay sample with the TBA reagent in an acidic solution. However, the various procedures used differ from one another. Therefore, sample preparation, types of acidifying agents and their concentration in the reaction mixture, pH of the reaction mixture, composition of the TBA reagent, length of the TBA reaction time and possible use of antioxidants and chelators in the systems, are factors which may influence results reported by researchers from various laboratories. For example, Moerck and Ball (1974) suggested that Tenox II should be added to the distillation mixture prior to heating in order to retard further oxidation and consequently artifact formation during this step, while Ke *et al.* (1977) reported the use of propyl gallate (PG) and ethylenediaminetetraacetic acid (EDTA) during distillation for this purpose. However, researchers have pointed out that some phenolic antioxidants such as butylated hydroxyanisole (BHA) used in order to retard further oxidation of samples may in fact enhance the decomposition of lipid peroxides during distillation (Rhee, 1978a).

Recently Shahidi and Hong (1991) showed that addition of commonly used antioxidants and chelators has a marginal effect in the prevention of further lipid oxidation of meat lipids during the distillation step. It was further demonstrated that while the distillation procedure generally affords results that are numerically higher than those obtained by the acid extraction procedure, each method has its own advantages and drawbacks. Nonetheless, the relative, rather than the absolute TBA values, should be compared against one another in such determinations. The extraction procedure has a further disadvantage in that the presence of coloured ingredients, such as the cooked cured-meat pigment (CCMP), may cause overestimation of the numerical TBA values. Furthermore, a similar effect may be observed due to the occurrence of turbidity in the extraction solution owing to dissolution of certain proteins or the presence of fat emulsions in the systems.

Many modifications to improve the TBA test, proposed by Tarladgis *et al.* (1960), have been reported by researchers. The use of whole-tissue homogenates without deproteinization (Wills, 1965) for the TBA test is commonplace; however, some deproteinization prior to analysis may afford more realistic results. Placer *et al.* (1966) modified the TBA test for whole-tissue homogenates and used an alkaline pyridine–butanol mixture to extract the TBA–MA complex. The TBA–MA adduct was extracted into butanol and the absorbance of the solution was measured at both 520 and 535 nm. The difference in absorbance values at these two wavelengths was then taken as the TBA value. Quantification of malonaldehyde alone by a high-performance liquid chromatographic (HPLC) method has been reported (Kakuda *et al.*, 1981; Csallany *et al.*, 1984; Bull and Marnett, 1985).

17.6.1 Advantages and limitations of the TBA test

Lipid oxidation in muscle foods is conveniently assessed by the classical TBA methodology as mentioned earlier. This method is simple and generally correlates well with sensory data. However, there are some limitations associated with it. TBA values of cooked muscle foods tend to increase over the storage period, reach a maximum and then start to decline. Interaction of MA with available amino groups of meat constituents forming 1-amino-3-aminopropene structures has been suggested to occur (Chio and Tappel, 1969b). Consequently, a given TBA value may correspond with two points during the storage period (i.e. one value during the early stages of storage corresponding to an increase in MA concentration and another equal value during its decrease over a longer storage period). Nonetheless, TBA values do correspond well with a single point during the early stages of storage as studied by many researchers (Spanier and Taylor, 1991).

In nitrite-cured products, use of sulphanilamide (SA) is recommended (Zipser and Watts, 1962). SA reacts with residual nitrite present in the sample, leading to the formation of a diazonium salt, thus preventing the nitrosation of malonaldehyde which would be otherwise unreactive towards the TBA reagent (Figure 17.3). However, in the absence of residual nitrite, SA itself may react with MA to form an amino-imino-propene derivative. Formation of this latter complex has been positively detected by fluorescence spectroscopy (Figure 17.4) (Shahidi *et al.*, 1991). Furthermore, in the presence of TBA, a new complex of SA–MA–TBA (SMT, 1:1:1, mole basis) was detected and elucidated (Figure 17.5) (Pegg *et al.*, 1992).

Use of an extraction, rather than a distillation procedure, as well as its advantages and disadvantages in the TBA test such as extraction of coloured components and dissolution of some proteins or dispersion of fat droplets which may interfere with the determination, has been discussed earlier. Bound malonaldehyde also may not be released in the absence of heat treatment or acidic conditions. Therefore, the TBA methodology, although simple and adequate for general estimation of oxidative state of meats, is not without its drawbacks when accurate quantitation is required. Nonetheless, the TBA test, due to its simplicity and acceptability, remains the most widespread procedure for estimation of the oxidative state of meat and meat products.

17.7 The Kries test

This is a rapid test which may be performed on lipid extracts from meat (see Robards *et al.*, 1988). It is one of the first methods of evaluation of the oxidative state of lipids and involves the formation of a red-coloured adduct when phloroglucinol reacts with oxidized lipids under acidic conditions. Although the Kries test is simple, it does not always parallel the development of rancidity in foods. Nonetheless, it has been reported that careful experimentation may provide both qualitative and quantitative information (Rossell, 1991). Therefore, lipids may be reacted with phloroglucinol in diethyl ether followed by extraction with a hydrochloric

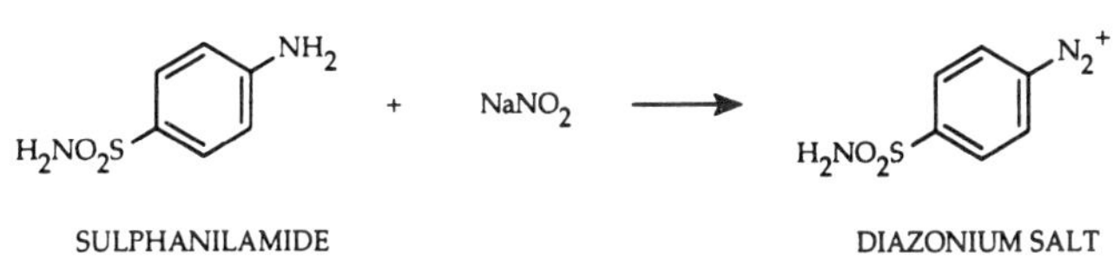

Figure 17.3 Interaction of sulphanilamide with nitrite.

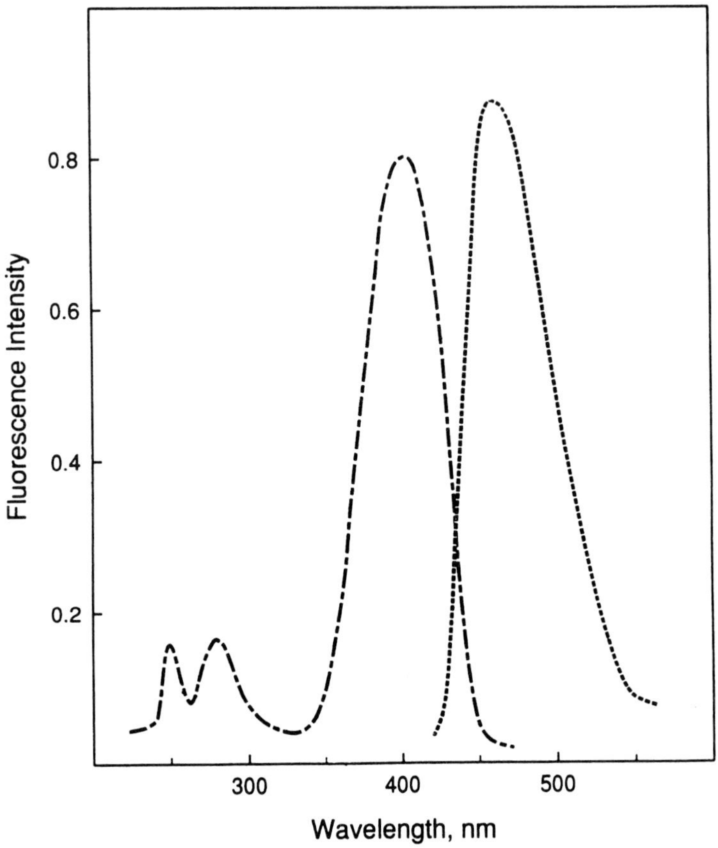

Figure 17.4 Fluorescence of malonaldehyde (MA)–sulphanilamide (SA) adduct. Excitation, – – – – – –, and fluorescence, - - - - - - .

acid solution. The red colour so produced may be measured with a Lovibond colorimeter. However, interference from certain colorants and food additives may be noted (Rossell, 1991).

17.8 Anisidine value

The peroxides in food lipids are transitory species that decompose into various carbonyl and other products. Reaction of aldehydes of oxidized lipids with the *p*-anisidine reagent in isooctane is followed by measuring the concentration of reaction products spectrophotometrically at 350 nm (Figure 17.6). Various factors influencing the accuracy and reproducibility of the method have been reviewed by Rossell (1986) and Robards *et al.*

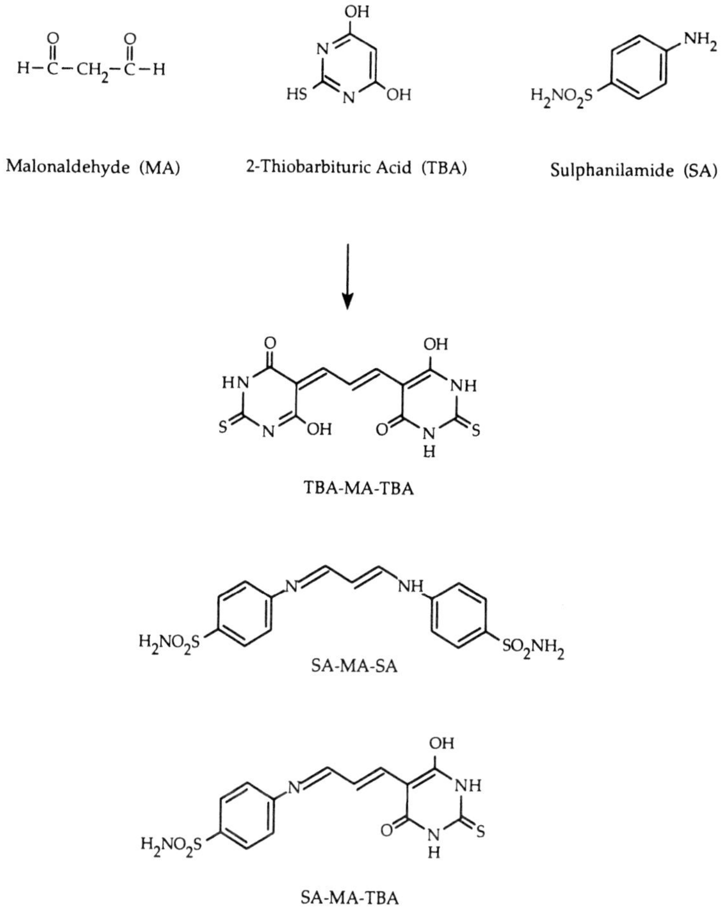

Figure 17.5 Interaction of malonaldehyde (MA) with sulphanilamide (SA) and 2-thiobarbituric acid (TBA).

(1988). This method has not usually been employed for meat products but has been used for evaluation of the quality of fish meal as feed for aquacultured species (Cho, 1990).

17.9 Totox value

The anisidine value (AnV) is often used together with the PV to calculate the so-called total oxidation value or totox value. Totox value is defined as 2PV + AnV. It is generally considered that extracted lipids with a totox value of less than 10 are of good quality. This parameter has been used

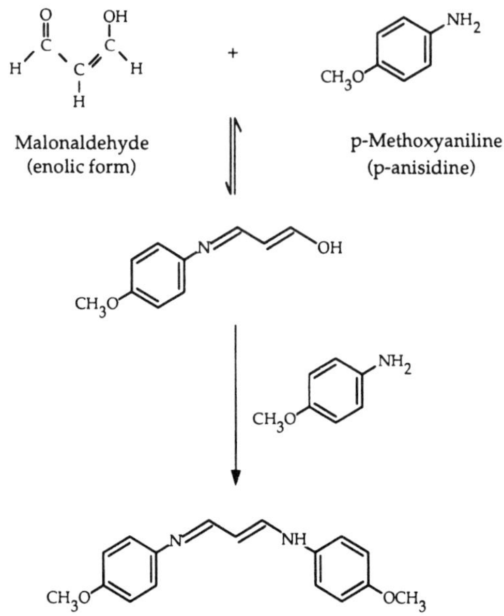

Figure 17.6 Interaction of *p*-anisidine reagent with lipid oxidation products.

extensively in the evaluation of fish meal quality. Nonetheless, its theoretical significance is questionable as parameters with different dimensions are added together.

17.10 Carbonyl compounds

Carbonyl compounds in meat and meat products may be determined by converting them to their 2,4-dinitrophenylhydrazones. Carbonyl compounds together with meat triacylglycerols are generally extracted first from muscle tissue into hexane and then derivitized on a column of activated celite impregnated with 2,4-dinitrophenylhydrazones, prior to measurement at 340 nm (Schwartz *et al.*, 1963). Alternatively, total carbonyls may be converted to their 2,4-dinotrophenylhydrazones in an aqueous medium and then extracted with hexane prior to their measurement at 340 nm (Lawrence, 1965; Keller and Kinsella, 1973). The general reaction involved is given in Figure 17.7.

Evaluation of the oxidative state of muscle foods from beef, lamb, poultry and fish by this procedure has been reported extensively (Schwartz *et al.*, 1963; Dimick and MacNeil, 1970; Dimick *et al.*, 1972; Sink and Smith, 1972; Keller and Kinsella, 1973; Caporaso and Sink, 1978; Kunsman *et al.*, 1978; Mai and Kinsella, 1979; Misock *et al.*, 1979). In the column separation

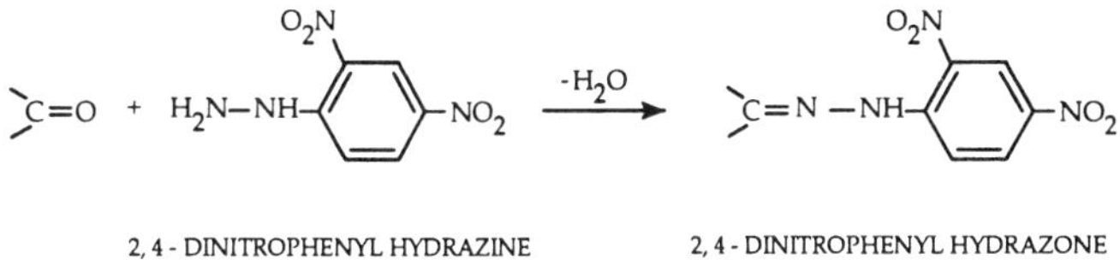

Figure 17.7 Interaction of carbonyl compounds with 2,4-dinitrophenylhydrazine.

of hydrazones from lipid components, hexane was used to elute the lipid fraction whereas chloroform was employed to elute the hydrazone derivatives. Ketoglycerides unintentionally derivatized may interfere with the analysis and their removal on an alumina column with benzene–hexane as eluent is necessary. Monocarbonyl compounds from fresh muscle foods may be isolated by similar techniques (Thomas *et al.*, 1971). Monocarbonyls may also be separated from unreacted lipids and dicarbonyls and their content determined spectrophotometrically at 365 nm. The monocarbonyl fraction may be separated further into alkanals, 2-alkanones, 2-alkenes and 2,4-alkadienals by the chromatographic procedure of Schwartz *et al.* (1963) or its modification (Caporaso and Sink, 1978). It has been reported that the total monocarbonyls served as a better indicator of lipid oxidation in mechanically deboned meats than the total carbonyl content (Kunsman *et al.*, 1978). However, individual constituents of the monocarbonyls did not serve as good indicators of oxidative deterioration of muscle foods (Dimick and MacNeil, 1970; Dimick *et al.*, 1972).

Regeneration of the original carbonyl compounds from their hydrazone derivatives is another interesting aspect (Dartey and Kinsella, 1971). However, regeneration efficiency of carbonyl compounds decreased as their chain length increased. Hydrolysis of hydrazones with 4N HCl followed by the extraction of the liberated carbonyl compounds in pentane is recommended (Buttery, 1973).

17.11 Hexanal, propanal and other carbonyl compounds

Hexanal is a seemingly ubiquitous component of food, both fresh and stored. This stems from the fact that practically all foods have some percentage of linoleate, the fatty acid which undergoes oxidation to produce hexanal (Frankel *et al.*, 1981). Linoleic acid may form a large number of volatile compounds during its autoxidation (Ullrich and Grosch, 1987). Hexanal, pentanal and 2,4-decadienal have been proposed as indicators for the development of off-flavours in vegetable oils (Dupuy *et al.*, 1976; Warner *et al.*, 1978).

Initially the autoxidation of linoleic acid produces 9-hydroperoxy-10, 12-octadecadienoic (9-HPOD) and 13-hydroperoxy-9,11-octodecadienoic

(13-HPOD) acids. Hexanal may be formed from both of these hydroperoxides (Schieberle and Grosch, 1981). It has also been shown that hexanal may be formed via degradation of secondary oxidation products of lipids such as 2-octenal and 2,4-decadienal. Hexanal itself may be oxidized to hexanoic acid. Nonetheless, the content of hexanal in muscle foods has been shown to be a good indicator for evaluation of the oxidative state of meat lipids (Shahidi *et al.*, 1987).

The protocols used for hexanal quantification may involve the isolation of carbonyls, together with all other volatile compounds by distillation and/or solvent extraction, followed by their analysis by gas chromatography or gas chromatography–mass spectrometry without any derivitization (Shahidi *et al.*, 1987). A direct headspace analysis may also be used. Alternatively, the carbonyl compounds may be reacted with 2,4-dinitro-phenylhydrazine. The hydrazones so obtained may then be separated by a reversed-phase high-performance liquid chromatographic method and the hexanal content quantified as its hydrozone derivative (Morrissey and Apte, 1988).

Hexanal has been successfully used for evaluation of the oxidative state of red meat from different species (Cross and Ziegler, 1965; Ruenger *et al.*, 1978; Shahidi *et al.*, 1987; Morrissey and Apte, 1988; Torres *et al.*, 1989; Lamikanra and Dupuy, 1990; Drum and Spanier, 1991). Shahidi *et al.* (1987) reported that a linear relationship existed between the content of hexanal and both the TBA values and sensory acceptability scores of cooked ground pork (Figure 17.8). Similar results were reported for fresh beef, pork and fish (Morrissey and Apte, 1988), for chevon (Lamikanra

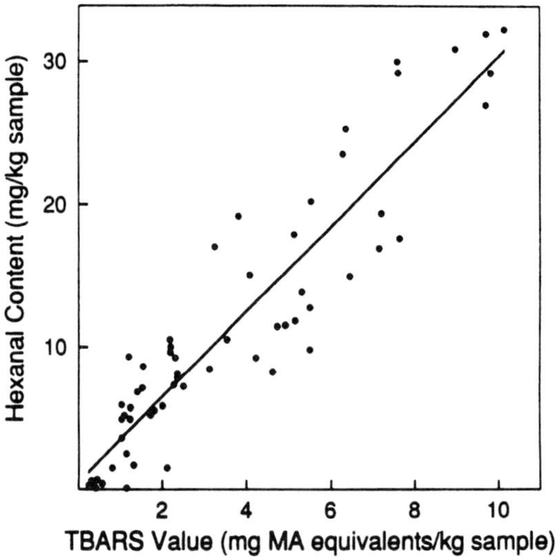

Figure 17.8 Relationship between hexanal content and 2-thiobarbituric acid reactive substances (TBARS) in cooked pork (Wettasinghe and Shahidi, 1997).

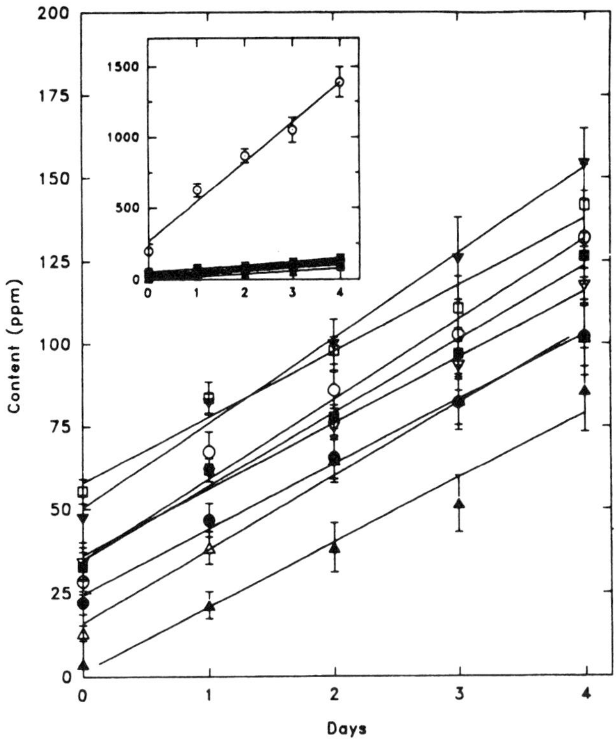

Figure 17.9 Relationship between hexanal (main plot), and hexanal and other volatiles (inset plot), and storage period of beef patties (Drumm and Spanier, 1991).

and Dupuy, 1990), for salted and dried beef (Torres *et al.*, 1989) and for cooked beef (Drumm and Spanier, 1991). The latter authors have also shown that hexanal concentration increases much faster than any other aldehyde during the storage of cooked beef (Figure 17.9). A similar conclusion was reached by Shahidi (1991) in a recent study. Therefore, hexanal appears to be a sensitive and reliable indicator for evaluation of the oxidative state and flavour quality of meat and meat products. Furthermore, it should be noted that hexanal has a distinctive odour described by many researchers as 'green' or grassy' (MacLeod and Ames, 1986; Ullrich and Grosch, 1987; Gasser and Grosch, 1988). It has a very low odour threshold concentration and is detectable at levels as low as 100 ppm (Bengtsson *et al.*, 1967; Schieberle and Grosch, 1987).

In muscle foods rich in omega-3 fatty acids such as fish and poultry meat from birds fed extensively on fish meal, ethanal, propanal and propenal as indicators of oxidation may be useful. These compounds are typical breakdown products of hydroperoxides of omega-3 fatty acids.

Nonetheless, the fact that linoleic acid occurs to some degree in all muscle foods means that measurement of hexanal could provide a rapid, sensitive and reliable method for evaluation of the oxidative state of muscle foods. Use of propanal as an indicator for quality deterioration of fish meat has been reported (He and Shahidi, 1997).

17.12 Pentane and other alkanes

Formation of short-chain hydrocarbons, particularly ethane and pentane, from oxidation of lipids is documented (Horvat *et al.*, 1964; Evans *et al.*, 1967). Measurement of pentane, another dominant breakdown product of linoleic acid, as an indicator of lipid oxidation in freeze-dried muscle foods was reported by Seo (1976) and Seo and Joel (1980). Bigalli (1977) showed that the pentane content relates well with the development of off-flavours in several food systems.

17.13 Recent developments for quantitation of lipid oxidation

Lipid oxidation has conventionally been studied by monitoring either primary of secondary oxidation products. However, more recent advances in pulse radiolysis (Simic, 1980) and electron spin resonance (ESR; Schaich and Borgi, 1980) techniques have facilitated the detection and study of short-lived radical intermediates. ESR spectroscopy allows the detection of species with an unpaired electron such as free radicals, but a major requirement is that the radical concentrations remain higher than 10^{-8} M. Since radical lifetimes in solutions are very short and generally less than 1 ms and steady-state concentrations generally remain well below 10^{-7} M, either the rate of production of radicals should be enhanced or their rate of disappearance decreased. Thus, rapid freezing, lyophilization or spin trapping are employed to meet this requirement (Davies, 1987). Although application of ESR to study lipid oxidation in biological systems is commonplace, its use in food-related studies is relatively new. Use of ESR technique to determine the number of nitric oxide moieties on ferrous iron of cooked cured meat pigment was recently reported (Pegg *et al.*, 1996). Significance of meat pigments in inducing lipid oxidation in muscle foods has been discussed elsewhere in this monograph.

Infrared (IR) spectroscopy has also been used for measurement of rancidity. For example, hydroperoxides absorb at about 2.93 μm and appearance of a peak at 5.72 μm is due to C=O stretching as a result of formation of aldehydes, ketones and acids. Application of Fourier transform infrared (FTIR) for monitoring oxidation of food lipids has been reviewed by Sedman *et al.* (1997).

Recently, Burkow *et al.* (1992) reported that lypochlorite-activated chemiluminescence could provide a useful method for evaluation of antioxidants in edible oils. Chemiluminescence method has now been tested for monitoring the deterioration of edible oils (Burkow *et al.*, 1995) and shelf-life dating of fish samples (Miyazawa *et al.*, 1991).

Autoxidation of unsaturated lipids may also be evaluated using high resolution nuclear magnetic resonance spectroscopy (NMR). Changes that occur in the relative proportions of various types of hydrogen atoms in triacylglycerol (TAG) molecules during oxidation are monitored. This method has been used for evaluation of oxidation of marine oils, fish and fish meal samples (Saito and Udagawa, 1992; Wanasundara and Shahidi, 1993; Shahidi *et al.*, 1994; Saito 1997). For this purpose the oil or extracted oil is dissolved in $CDCl_3$ and the NMR spectrum of the lipid is recorded. The ratios of aliphatic to olfinic (R_{ao}) and aliphatic to diallyl-methylene (R_{ad}) protons are easily calculated from the integration data. Numerical values of R_{ao} and R_{ad} may be plotted against storage time or against PV or TOTOX ($2PV + p$-An) value to establish the relevant relationships. However, this method is more suitable for muscle foods containing a high proportion of polyunsaturated lipids.

17.14 Conclusions

Although a large number of procedures for assessing lipid oxidation in meat and seafood products are available, both the TBA test and hexanal and/or propanal assays seem to provide the best means for monitoring oxidative deterioration of muscle foods during storage. Absolute TBA values obtained by the test do not always provide accurate results; nonetheless in uncured meats the TBA test always offers a reliable trend when comparing several systems under investigation. In cured meat products, however, the presence of residual nitrite in samples may require addition of sulphanilamide to the assay in order to prevent the nitrosation of malonaldehyde, which results in its underestimation. Use of hexanal and propanal as indicators of lipid oxidation during the early stages of oxidative changes in meats and seafoods, respectively, has proved beneficial.

Acknowledgements

Financial support from the Natural Sciences and Engineering Research Council (NSERC) of Canada is gratefully acknowledged.

References

Ahmad, S. and Augustin, M.A. (1985). Effect of triarylbutylhydroquinone on lipid oxidation in fish crackers. *J. Sci. Food Agric.*, **36**, 393–401.

AOCS (1989). Official Methods and Recommended Practices of the American Oil Chemists' Society, 4th edn, Champaign, IL.

Awad, A., Powrie, W.D. and Fennema, O. (1968). Chemical deterioration of frozen bovine muscle at –4°C. *J. Food Sci.*, **33**, 227–235.

Bailey, C., Cutting, C.K., Enser, M.B. and Rhodes, D.N. (1973). The influence of slaughter weight on the stability of pork sides in frozen storage. *J. Sci. Food Agric.*, **24**, 1299–1304.

Belitz, H.-D. and Grosch, W. (1987). Lipids. In *Food Chemistry*. Springer-Verlag, Berlin, pp. 128–200.

Bengtsson, B.L., Bosund, I. and Rasmussen, I. (1967). Hexanal and ethanol formation in peas in relation to off-flavour development. *Food Technol.*, **21**(3), 160A–164A.

Bigalli, G. (1977). Determination of pentane formed during autoxidation of oils contained in solid samples. *J. Am. Oil Chem.* Soc., **54**, 229–232.

Bolton, J.C., Melton, S.L., Riemann, M.J. and Backus, W.R. (1985). An isocratic HPLC method for analysis of bovine phospholipids. Presented at the 77th Annual Meeting of the American Society of Animal Sciences, University of Georgia, August 13–16.

Bull, A.W. and Marnett, L.J. (1985). Determination of malondialdehyde by ion-pairing high-performance liquid chromatography. *Anal. Biochem.*, **149**, 284–290.

Burkow, I.C., Moen, P. and Overbo, K. (1992). Chemiluminescence as a method for oxidative rancidity assessment of marine oils. *J. Am. Oil Chem. Soc.*, **69**, 1108–1111.

Burkow, I.C., Vikersveen, L. and Saarem, K. (1995). Evaluation of antioxidants for cod liver oil by chemiluminescence and ranciant methods. *J. Am. Oil Chem. Soc.*, **72**, 553–557.

Buttery, R.G. (1973). Some unusual volatile carbonyl components of potato chips. *J. Agric. Food Chem.*, **21**, 31–33.

Caporaso, F. and Sink, J.D. (1978). Lipid-soluble carbonyl components of ovine adipose tissue. *J. Food Sci.*, **43**, 1379–1381.

Chang, S.S. and Peterson, R.J. (1977). Symposium: The basis of quality in muscle foods. Recent developments in the flavour of meat. *J. Food Sci.*, **42**, 298–305.

Chio, K.S. and Tappell, A.L. (1969a). Inactivation of ribonuclease and other enzymes by peroxidizing lipids and by malonaldehyde. *Biochemistry*, **8**, 2827–2832.

Chio, K.S. and Tappel, A.L. (1969b). Synthesis and characterization of the fluorescent products derived from malonaldehyde and amino acids. *Biochemistry*, **8**, 2821–2827.

Cho, C.Y. (1990). Fish nutrition, feeds, and feeding: with special emphasis on salmonid aquaculture. *Food Rev. Int.*, **6**, 333–357.

Cross, C.K. and Ziegler, P. (1965). A comparison of the volatile fractions from cured and uncured meat. *J. Food Sci.*, **30**, 610–614.

Csallany, A.S., Guan, M.D., Manwaring, J.D. and Addis, P.B. (1984). Free malonaldehyde determination in tissues by high-performance liquid chromatography. *Anal. Biochem.*, **142**, 277–283.

Darty, C.K. and Kinsella, J.E. (1971). Rate of formation of methyl ketones during blue cheese ripening. *J. Agric. Food Chem.*, **19**, 771–774.

Davies, M.J. (1987). Application of electron spin resonance spectroscopy to the identification of radicals produced during lipid peroxidation. *Chem. Phys. Lipids*, **44**, 149–173.

Dimick, P.S. and MacNeil, J.H. (1970). Poultry product quality. 2. Storage time–temperature effects on carbonyl composition of cooked turkey and chicken skin fractions. *J. Food Sci.*, **35**, 186–190.

Dimick, P.S., MacNeil, J.H. and Grunden, L.P. (1972). Poultry product quality. Carbonyl composition and organoleptic evaluation of mechanically deboned poultry meat. *J. Food Sci.*, **37**, 544–546.

Draper, H.H., McGirr, L.G. and Hadley, M. (1986). The metabolism of malondialdehyde. *Lipids*, **21**, 305–307.

Drumm, T.D. and Spanier, A.M. (1991). Changes in the content of lipid autoxidation and sulphur-containing compounds in cooked beef during storage. *J. Agric. Food Chem.*, **39**, 336–343.

Dupuy, H.P., Rayner, E.T. and Wadsworth, J.I. (1976). Correlations of flavour scores with volatiles of vegetable oils. *J. Am. Oil Chem. Soc.*, **53**, 628–631.
Evans, C.D., List, G.R., Dolev, A., McConnell, D.G. and Hoffman, R.L. (1967). Pentane from thermal decomposition of lipoxidase-derived products. *Lipids*, **2**, 432–434.
Fielder, U. (1974). A coulometric method for the determination of low peroxide values of fats and oils. *J. Am. Oil Chem., Soc.*, **51**, 101–103.
Fischer, J. and Deng, J.C. (1977). Catalysis of lipid oxidation: a study of mullet (*Mugil cephalus*) dark flesh and emulsion model system. *J. Food Sci.*, **42**, 610–614.
Frankel, E.N., Neff, W.E. and Selke, E. (1981). Analysis of autoxidized fats by gas chromatography–mass spectrometry: VII. Volatile thermal decomposition products of pure hydroperoxides from autoxidized and photosensitized oxidized methyl oleate, linoleate and linolenate. *Lipids*, **16**, 279–285.
Gasser, U. and Grosch, W. (1988). Flavour deterioration of soya bean oil: identification of intense odour compounds during flavour reversion. *Fett. Wiss. Technol.*, **90**, 332–336.
Gokalp, H.Y., Ockerman, H.W., Plimpton, R.F. and Peng, A.C. (1981). Qualitative alteration of phospholipid in beef patties cooked after vacuum and nonpackaged, frozen storage. *J. Food Sci.*, **46**, 19–22.
Gray, J.I. (1978). Measurement of lipid oxidation: a review. *J. Am. Oil Chem. Soc.*, **55**, 539–546.
Gray, J.I. and Pearson, A.M. (1984). Cured meat flavor. *Adv. Food Res.*, **29**, 1–86.
Gray, J.I., MacDonald, B., Pearson, A.M. and Morton, I.D. (1981). Role of nitrite in cured meat flavor: a review. *J. Food Prot.*, **44**, 302–312.
He, Y. and Shahidi, F. (1997). Antioxidant activity of green tea and its catechins in a fish meat model system. *J. Agric. Food Chem.*, **45**, 4262–4266.
Horvat, R.J., Lane, W.G., Ng., H. and Shepherd, A.D. (1964). Saturated hydrocarbons from autoxidizing methyl linoleate. *Nature*, **203**, 523–524.
Igene, J.O., Pearson, A.M., Dugan, L.R., Jr. and Price, J.F. (1980). Role of triglycerides and phospholipids on development of rancidity in model meat systems during frozen storage. *Food Chem.*, **5**, 263–276.
Igene, J.O., Yamauchi, K., Pearson, A.M., Gray, J.I. and Aust, S.D. (1985). Mechanisms by which nitrite inhibits the development of warmed-over flavor (WOF) in cured meat. *Food Chem.*, **18**, 1–18.
Jeremiah, L.E. (1980). Effect of frozen storage and protective wrap upon the cooking losses, palatability, and rancidity of fresh and cured pork cuts. *J. Food Sci.*, **45**, 187–196.
Kakuda, Y., Stanley, D.W. and Van de Voort, F.R. (1981). Determination of TBA number of high performance liquid chromatography. *J. Am. Oil Chem. Soc.*, **58**, 773–775.
Ke, P.J., Ackman, R.G., Linke, B.A. and Nash, D.M. (1977). Differential lipid oxidation in various parts of frozen mackerel. *J. Food Technol.*, **12**, 37–47.
Keller, J.D. and Kinsella, J.E. (1973). Phospholipid changes and lipid oxidation during cooking and frozen storage of raw ground beef. *J. Food Sci.*, **38**, 1200–1204.
Kohn, H.I. and Liversedge, M. (1944). On a new aerobic metabolite whose production by brain is inhibited by apomorphine, emetine, epinephrine and menadione. *J. Pharmacol.*, **82**, 292–295.
Kosugi, H., Kato, T. and Kikugawa, K. (1987). Formation of yellow, orange, and red pigments in the reaction of alk-2-enals with 2-thiobarbituric acid. *Anal. Biochem.*, **165**, 456–464.
Kosugi, H., Kato, T. and Kikugawa, K. (1988). Formation of red pigment by a two-step 2-thiobarbituric acid reaction of alka-2,4-dienals. Potential products of lipid oxidation. *Lipids*, **23**, 1024–1031.
Kunsman, J.E., Field, R.A. and Kazantzis, D. (1978). Lipid oxidation in mechanically deboned red meat. *J. Food Sci.*, **43**, 1375–1378.
Labuza, T.P. (1971). Kinetics of lipid oxidation in foods. *CRC Crit. Rev. Food Technol.*, **2**, 355–405.
Lamikanra, V.T. and Dupuy, H.P. (1990). Analysis of volatiles related to warmed over flavor of cooked chevon. *J. Food Sci.*, **55**, 861–862.
Lawrence, R.C. (1965). Use of 2,4-dinitrophenylhydrazine for the estimation of micro amounts of carbonyls. *Nature*, **205**, 1313–1314.
Lee, Y.B., Hargus, G.L., Kirkpatrick, J.A., Berner, D.L. and Forsythe, R.H. (1975). Mechanism of lipid oxidation in mechanically deboned chicken meat. *J. Food Sci.*, **40**, 964–967.

MacDonald, B., Gray, J.I. and Gibbins, L.N. (1980). Role of nitrite in cured meat flavour: Antioxidant role of nitrite. *J. Food Sci.*, **45**, 893–897.

MacLeod, G. and Ames, J. (1986). The effect of heat on beef aroma: comparisons of chemical composition and sensory properties. *Flavour Fragrance J.* **1**, 91–104.

Mai, J. and Kinsella, J.E. (1979). Changes in lipid composition of cooked mince carp (*Cyprinus carpio*) during frozen storage. *J. Food Sci.*, **44**, 1619–1624.

Marcuse, R. and Johansson, L. (1973). Studies on the TBA test for rancidity grading: II. TBA reactivity of different aldehyde classes. *J. Am. Oil Chem. Soc.*, **50**, 387–391.

Melton, S.L. (1983). Methodology for following lipid oxidation in muscle foods. *Food Technol.*, **37**(7), 105–111, 116.

Misock, J.P., Kunsman, J.E. and Field, R.A. (1979). Lipid oxidation in bologna containing mechanically deboned beef. *J. Food Sci.*, **44**, 151–153.

Miyazawa, T., Kikuch, M., Fujimoto, K., Endo, Y., Cho, S.-Y., Usuki, R. and Keneda, T. (1991). Shelf-life dating of fish meats in terms of oxidative rancidity as measured by chemiluminescence. *J. Am. Oil Chem. Soc.*, **68**, 39–43.

Moerck, K.E. and Ball, H.R., Jr. (1974). Lipid autoxidation in mechanically deboned chicken meat. *J. Food Sci.*, **39**, 876–879.

Morrissey, P.A. and Apte, S. (1988). Influence of species, haem and non-haem iron fraction and nitrite on hexanal production in cooked muscle systems. *Sci. Alimentes*, **8**, 3–14.

Nair, V. and Turner, G.A. (1984). The thiobarbituric acid test for lipid peroxidation: structure of the adduct with malondialdehyde. *Lipids*, **19**, 804–805.

Noble, A.C. (1976). Effect of carbon dioxide and sodium chloride on oxidative stability of frozen mechanically deboned poultry meat. *Can. Inst. Food Sci. Technol. J.*, **9**, 105–107.

Ohkawa, H., Ohishi, N. and Yagi, K. (1979). Assay for lipid peroxides in animal tissues by thiobarbituric acid reaction. *Anal. Biochem.*, **95**, 351–358.

Owen, J.E., Lawrie, R.A. and Hardy, B. (1975). Effect of dietary variation, with respect to energy and crude protein levels, on the oxidative rancidity exhibited by frozen porcine muscles. *J. Sci. Food Agric.*, **26**, 31–41.

Pearson, A.M., Love, J.D. and Shorland, F.B. (1977). 'Warmed-over' flavor in meat, poultry, and fish. *Adv. Food Res.*, **23**, 1–74.

Pegg, R.B., Shahidi, F. and Jablonski, C.R. (1992). Interactions of sulfanilamide and 2-thiobarbituric acid with malonaldehyde: structure of adducts and implications in determination of oxidative state of nitrite-cured meats. *J. Agric. Food Chem.*, **40**, 1826–1832.

Pegg, R.B., Shahidi, F., Gogan, N.J. and DeSilva, S.J. (1996). Elucidation of the chemical structure of preformed cooked, cured-meat pigment by electron paramagnetic resonance spectroscopy. *J. Agric. Food Chem.*, **44**, 416–421.

Placer, Z.A., Cushman, L.L. and Johnson, B.C. (1966). Estimation of product of lipid peroxidation (malonyl dialdehyde) in biochemical systems. *Anal. Biochem.*, **16**, 359–364.

Rhee, K.S. (1978a). Minimization of further lipid peroxidation in the distillation 2-thiobarbituric acid test of fish and meat. *J. Food Sci.*, **43**, 1776–1778, 1781.

Rhee, K.S. (1978b). Factors affecting oxygen uptake in model systems used for investigating lipid peroxidation in meat. *J. Food Sci.*, **43**, 6–9.

Robards, K., Kerr, A.F. and Patsalides, E. (1988). Rancidity and its measurement in edible oils and snack foods. *Analyst*, **113**, 213–224.

Roozen, J.P. (1987). Effects of types I, II and III antioxidants on phospholipid oxidation in a meat model for warmed over flavour. *Food Chem.*, **24**, 167–185.

Rossell, J.B. (1986). Measurement of rancidity. In *Rancidity in Foods*, 2nd edn, eds. J.C. Allen and R.J. Hamilton. Elsevier Science, London, pp. 23–52.

Rossell, J.B. (1991). How to measure oxidative rancidity in fats and fatty foods. *Lipid Technol.*, **3**, 122–126.

Ruenger, E.L., Reineccius, G.A. and Thompson, D.R. (1978). Flavor compounds related to the warmed-over flavor of turkey. *J. Food Sci.*, **43**, 1198–1200.

Saito, H. (1997). Evaluation method for lipid oxidation by nuclear magnetic resonance. In *Flavor and Lipid Chemistry of Seafoods*, eds. F. Shahidi and K. Cadwallader, ACS Symposium Series 674, American Chemical Society, Washington, DC, pp. 218–239.

Saito, H. and Udagawa, M. (1992). Use of NMR to evaluate the oxidative deterioration of Niboshi, boiled and dried fish. *Biosci. Biotechnol. Biochem.*, **56**, 831–832.

Schaich, K.M. and Borgi, D.C. (1980). EPR studies in autoxidation. In *Autoxidation in Food and Biological Systems*, eds. M.G. Simic and M. Karel. Plenum Press, New York, pp. 45–70.

Schieberle, P. and Grosch, W. (1981). Model experiments about the formation of volatile carbonyl compounds. *J. Am. Oil Chem. Soc.*, **58**, 602–607.

Schieberle, P. and Grosch, W. (1987). Evaluation of the flavour of wheat and rye bread crusts by aroma extra dilution analysis. *Z. Lebensm. Unters. Forsch.*, **185**, 111–113.

Schwartz, D.P., Haller, H.S. and Keeney, M. (1963). Direct quantitative isolation of monocarbonyl compounds from fats and oils. *Anal. Chem.*, **35**, 2191–2194.

Sedman, J., Ismail, A.A., Nicodema, A., Kubow, S. and Van de Voort, F.R. (1997). Application of FTIR. ATR differential spectroscopy for monitoring oil oxidation and antioxidant efficacy. In *Natural Antioxidants: Chemistry, Health Effects and Applications*, ed. F. Shahidi. AOCS Press, Champaign, IL, pp. 358–378.

Seo, C.W. (1976). Hydrocarbon production from freeze-dried meats. *J. Food Sci.*, **41**, 594–597.

Seo, C.W. and Joel, D.L. (1980). Pentane production as an index of rancidity in freeze-dried pork. *J. Food Sci.*, **45**, 26–28, 92.

Shahidi, F. (1991). Prevention of lipid oxidation in muscle foods by nitrite and nitrite-free compositions. In *Lipid Oxidation in Food*. ed. A.J. St. Angelo. ACS Symposium Series 500, American Chemical Society, Washington, DC, pp. 161–182.

Shahidi, F. and Hong, C. (1991). Evaluation of malonaldehyde as a marker of oxidative rancidity in meat products. *J. Food Biochem.*, **15**, 97–105.

Shahidi, F., Yun, J., Rubin, L.J. and Wood, D.F. (1987). The hexanal content as an indicator of oxidative stability and flavour acceptability in cooked ground pork. *Can. Inst. Food Sci. Technol. J.*, **20**, 104–106.

Shahidi, F., Pegg, R.B. and Harris, R. (1991). Effects of nitrite and sulfanilamide on the 2-thiobarbituric acid (TBA) values in aqueous model and cured meat systems. *J. Muscle Foods*, **2**, 1–9.

Shahidi, F., Wanasundara, U.N. and Brunet, N. (1994). Oxidative stability of oil from blubber of harp seal (*Phoca groenlandica*) as assessed by NMR and standard procedures. *Food Res. Int.*, **27**, 555–562.

Silberstein, D.A. and Lillard, D.A. (1978). Factors affecting the autoxidation of lipids in mechanically deboned fish. *J. Food Sci.*, **43**, 764–766.

Simic, M.G. (1980). Kinetic and mechanistic studies of peroxy, vitamin E and antioxidant free radicals by pulse radiolysis. In *Autoxidation in Food and Biological Systems*, eds. M.G. Simic and M. Karel. Plenum Press, New York, pp. 17–26.

Sink, J.D. and Smith, P.W. (1972). Changes in lipid soluble carbonyls of beef muscle during aging. *J. Food Sci.*, **37**, 181–182.

Sinnhuber, R.O. and Yu, T.C. (1958). 2-Thiobarbituric acid method for the measurement of rancidity in fishery products. II. The quantitative determination of malonaldehyde. *Food Technol.*, **12**(1), 9–12.

Siu, G.M. and Draper, H.H. (1978). A survey of the malonaldehyde content of retail meats and fish. *J. Food Sci.*, **453**, 1147–1149.

Sklan, D., Tenne, Z. and Budowski, P. (1983). The effect of dietary fat and tocopherol on lipolysis and oxidation of turkey meat stored at different temperatures. *Poult. Sci.*, **62**, 2017–2021.

Spanier, A.M. and Taylor, R.D. (1991). A rapid, direct chemical assay for the quantitative determination of thiobarbituric acid reactive substances in raw, cooked, and cooked/stored muscle foods. *J. Muscle Foods*, **2**, 165–176.

Tarladgis, B.G., Watts, B.M., Younathan, M.T. and Dugan, L.R., Jr. (1960). A distillation method for the quantitative determination of malonaldehyde in rancid foods. *J. Am. Oil Chem. Soc.*, **37**, 44–48.

Tarladgis, B.G., Pearson, A.M. and Dugan, L.R., Jr. (1964). Chemistry of the 2-thiobarbituric acid test for determination of oxidative rancidity in foods. II. Formation of the TBA–malonaldehyde complex without acid-heat treatment. *J. Sci. Food Agric.*, **15**, 602–607.

Thomas, C.P., Dimick, P.S. and MacNeil, J.H. (1971). Poultry product quality. 4. Levels of carbonyl compounds in fresh, uncooked chicken and turkey skin. *J. Food Sci.*, **36**, 527–531.

Tichivangana, J.Z. and Morrissey, P.A. (1985). Metmyoglobin and inorganic metals as prooxidants in raw and cooked muscle systems. *Meat Sci.*, **15**, 107–116.

Torres, E., Pearson, A.M., Gray, J.I., Ku, P.K. and Shimokomaki, M. (1989). Lipid oxidation in charqui (salted and dried beef). *Food Chem.*, **32**, 257–268.

Uchiyama, M. and Mihara, M. (1978). Determination of malonaldehyde precursor in tissues by thiobarbituric acid test. *Anal. Biochem.*, **86**, 271–278.

Ullrich, F. and Grosch, W. (1987). Identification of the most intense volatile flavour compounds formed during autoxidation of linoleic acid. *Z. Lebensm. Unters. Forsch.*, **184**, 277–282.

Wanasundara, U.N. and Shahidi, F. (1993). Application of NMR spectroscopy to assess oxidative stability of canola and soybean oils. *J. Food Lipids*, **1**, 15–24.

Warner, K., Evans, C.D., List, G.R., Dupuy, H.P., Wadsworth, J.I. and Goheen, G.E. (1978). Flavor score correlation with pentanal and hexanal contents of vegetable oil. *J. Am. Oil Chem. Soc.*, **55**, 252–256.

Wettasinghe, M. and Shahidi, F. (1997). Oxidative stability of cooked comminuted lean pork as affected by alkali and alkali-earth halides. *J. Food Sci.*, **61**, 1160–1164.

Wills, E.D. (1965). Mechanisms of lipid peroxide formation in tissues. Role of metals and haematin proteins in the catalysis of the oxidation of unsaturated fatty acids. *Biochim. Biophys. Acta*, **98**, 238–251.

Wong, D.W.S. (1989). Lipids. In *Mechanism and Theory in Food Chemistry*. Van Nostrand Reinhold, New York, pp. 1–47.

Yamauchi, K., Nagai, Y. and Ohashi, T. (1980). Quantitative relationship between alpha-tocopherol and polyunsaturated fatty acids and its connection to development of oxidative rancidity in porcine skeletal muscle. *Agric. Biol. Chem.*, **44**, 1061–1067.

Yeo, Y.K. and Horrocks, L.A. (1985). Decrease in plasmalogin in beef muscle by electrical stimulation. Presented at the 45th Annual Meeting of the Institute of Food Technologists, Atlanta, GA, June 9–12.

Zipser, M.W. and Watts, B.M. (1962). A modified 2-thiobarbituric acid (TBA) method for the determination of malonaldehyde in cured meats. *Food Technol.*, **16**, 102–104.

Zubillaga, M.P., Maerker, G. and Foglia, I.A. (1984). Antioxidant activity of sodium nitrite in meat. *J. Am. Oil Chem. Soc.*, **61**, 772–776.

18 Sensory and statistical analyses in meat flavour research

A.M. SPANIER, B.T. VINYARD, K.L. BETT and
A.J. ST. ANGELO

18.1 Introduction

In 1985, the Southern Regional Research Center (SRRC) in New Orleans, LA, established a new research unit, Food Flavour Quality (FFQ), whose primary mission was to establish sensory and objective definitions of food flavours in meat, catfish, eggs and peanut products. Included in this mission was research on the chemical and biological origins of flavour compounds as well as causes of flavour deterioration and undesirable flavour development. In 1995 the programme was reorganized to include fresh-cut fruits, rice and rice products. Of course the ultimate goal, as established by the United States Department of Agriculture in founding the FFQ Research Unit, was to serve both the consumer and the food industry through quality assessment.

One part of the original research mission dealt with the etiology and control of undesirable flavours of meat, predominantly beef, more specifically the 'warmed-over flavour' or WOF phenomon. Tims and Watts (1958) described WOF as the rapid development of oxidized flavour in cooked, uncured meat that has been stored refrigerated over a period of a few days. Actually, the flavour can deteriorate after only a few hours of refrigeration. Since first described by Tims and Watts in 1958, the original definition has been expanded to include raw meat that is ground and exposed to air (Green, 1969; Sato and Hegarty, 1971). The process in which meat develops WOF has recently been called 'meat flavour deterioration', or MFD (Spanier *et al.*, 1990; St. Angelo *et al.*, 1992) to describe more appropriately the process that was occurring. To investigate WOF, or MFD as hereafter referred to in this chapter, we set out to establish a sensory panel of highly 'trained' experts that could judge flavour intensities by descriptive analysis, and to correlate those findings with chemical and instrumental analyses. These correlations were made by appropriate use of statistical methods, such as multivariant analysis by principal components analysis (PCA). The purpose of this chapter is to present an overview of the multidisciplinary approach used to examine and solve this problem.

18.2. Sensory evaluation

The sensory laboratory for evaluating meat samples requires different considerations than those for evaluating nonheated or noncooked samples as will be explained in this section. There are several different requirements that a laboratory must meet to minimize bias among panellists and among samples.

18.2.1 Odour control

A meat sensory laboratory should consist of a minimum of two rooms, one for preparation of samples and the other for evaluation of samples. The preparation of meat samples generates odours that will interfere with flavour evaluation; therefore, the two processes need to be separated. More rooms for other functions may be necessary, such as an office, storage room, training room, entrance and exit areas, or computer space. The air conditioning system in the evaluation room should have positive pressure compared to surrounding rooms to keep odours flowing out of the room. Activated carbon filters installed in the air conditioning system prevent outside odours from entering into the evaluation room. Care should be taken when setting up the evaluation room that no highly odorous materials are used. For example, vinyl floors should be used instead of carpet.

18.2.2 Lighting

The room should have an adequate level of illumination, such as that provided by fluorescent lights. To hide colour differences in the samples, a special lighting system is required. This system can include incandescent lighting fixtures with red, green and/or blue bulbs or theatre-type light filters in frames over recessed lighting (Meilgaard *et al.*, 1987). Care should be taken to provide adequate lighting when coloured lights are used, because persons with poor eyesight will have difficulty. Pangborn (1967) suggested presenting samples singly to prevent comparison of colour. This works well in descriptive flavour analysis, but not for forced choice methods.

18.2.3 General comfort

The temperature of the evaluation room should be comfortable for the panellists. If it is too warm or too cool the panelists will be distracted by their discomfort. The ideal temperature should be 72°F (22°C) and 45–50% humidity (Meilgaard *et al.*, 1987). In New Orleans, LA, the relative humidity (RH) is normally between 80 and 100%. We find that 21.5°C (71°F) with 67% RH is comfortable for our panellists and readily controlled with our installed air conditioning system in the humid climate of southeastern Louisiana.

The evaluation room should be large enough to accommodate comfortably the number of panellists present. Booths can be constructed to reduce auditory and visual distractions. They should be 68.5–81 cm wide with dividers that extend approximately 46 cm beyond the counter top. Sinks should not be installed in booths because of inherent odour problems. Counter tops should be 46–56 cm deep and 74–79 cm from the floor. Comfortably padded operator chairs (that swivel and roll) are easiest for panellists to manoeuvre and turn around to listen to instructions.

If a computerized ballot system is used, the design (placement of monitor, keyboard, etc.) should be comfortable for the panellists to use. If light pens or touch screens are used, the video screen should be low enough that panellists do not have to hold their arm in the air for long periods of time, yet at a level that makes it easy to see. A software package that is user friendly should be selected since it allow the panellist to concentrate on the sample rather than the computer system.

18.2.4 Preparation area

The preparation area should include appropriate cooking equipment. A survey by Cross (1977) indicated that most researchers broil or roast steaks. This may be true today also. Convection ovens are most appropriate for roasting. Broiling can be done in a conventional oven or on an electric grill. Braising (cooking slowly in a moist atmosphere) is another common method of cooking meat.

Intact meat samples are cooked to an internal temperature to determine 'doneness'. Thermocouples with wire diameters less than 0.02 cm connected to a monitoring device are recommended. The thermocouple should not be encased in metal sheaths, because the metal conducts heat into the sample. This causes the sample in contact with the metal sheath to reach the end-point temperature before the rest of the sample. A constant time of cooking method can be used for ground or flaked samples when a temperature monitoring system is not available (AMSA, 1987).

The preparation area should be constructed of easy-to-clean materials. Equipment and counter tops should be made of stainless steel or a comparable material that will not transfer volatiles to the sample. Plastic materials are unsuitable for this reason. Porous materials are not suitable because they absorb odours, which subsequently can be transferred to the sample, and they harbour bacteria that can contaminate the sample (Meilgaard *et al.*, 1987).

18.2.5 Sample preparation and serving

The objectives of a given experiment determine the best preparation method. If one wishes to compare the effect of two or more treatments on flavour descriptors, then a ground meat model (3 oz (80 g) ground patties) results in a more uniform sample. If one wants to compare the descriptive analysis data with consumer data, then a model that is patterned after the consumer samples is necessary. The method of cooking, whether roasted, broiled, braised or grilled, depends on the objectives of the experiment and the serving model. Ground patties can be grilled or broiled. Whole roasts are roasted or braised. Meat should be cooked to a specified end-point temperature since temperature has a profound effect on meat flavour (Spanier, 1992; Spanier *et al.*, 1994a,b; Spanier and Drumm-Boylston, 1994; Spanier and Miller, 1996).

Reheating of meat samples can be accomplished by several methods. Johnsen and Civille (1986) used three methods: baking, broiling and boiling. To bake, patties were rewarmed in a 150°C preheated oven until they reached an internal temperature of 65°C. To broil, patties were rewarmed in a preheated broiler for 5 min on each side. To steam-cook, patties were sealed in polyethylene pouches and boiled for 10 min in 2 quarts (0.95 l) of water.

Microwaving can also be used to rewarm samples. Cremer microwaved beef samples on high power until the internal temperature reached 74°C (Cremer, 1982). Length of time of exposure was shown to be dependent upon the weight of sample microwaved. Times needed to be determined before the start of the experiment. Cremer found that beef patties reheated in a convection oven had higher sensory quality ratings for flavour, appearance and general acceptability than those reheated in a microwave. From these data, one can conclude that the objectives of any experiment should determine the reheating method, otherwise a convection oven is preferred.

At SRRC, we trim the visible fat, grind the meat and form 85 g patties. Making ground patties increases the uniformity of the samples. We use a Farbarware electric grill (Model 455ND, subsidiary of W. Kidde Co., Bronx, New York) to cook the patties 7 min per side to a well-done state. Frequent monitoring of internal temperature assures that the end-point remains constant. Patties are cut into wedges and placed in covered glass petri plates (60 × 15 mm) for serving to panellists. To keep them warm for the few minutes between placing in petri plates and serving to panellists, petri plates are placed in Hobart warming drawers (Troy, Ohio) set at 51.7°C. Cooking or reheating of samples is staggered so they can be served as soon as possible after cooking or reheating to minimize time in the warming drawers and flavour changes.

After receiving a sample, a panellist will cut the wedge of ground meat into two pieces, lengthwise, to ensure having two equal portions. This allows for a second chance of evaluation. Meat samples are identified with a three-digit random number. The order of serving is randomly selected before cooking, so no clues can be reaped from the order of presentation.

Panellists receive between four and six experimental samples at a session. Each session begins with a warm-up (nonexperimental) freshly grilled sample. Consensus (mean scores) are taken on that sample and the panellists use those sample scores to compare the other samples to, instead of comparing between samples. The interval of time between presentation of samples is 5 min. This allows time to evaluate the sample and to let the senses recover. Unsalted crackers and reverse osmosis-treated, deionized water are available at all times to cleanse the panellists' palate.

18.3 Sensory analysis of meat

18.3.1 Descriptive flavour panel

Descriptive flavour panellists should be selected on the basis of having 'normal' abilities to taste and smell and on their availability. Testing for normal abilities to taste and smell should include tests for rating or ranking intensity, tests that determine a person's ability to discriminate between aromas and between tastes, and tests to determine the ability to describe what is being perceived (Meilgaard *et al.*, 1987).

18.3.2 Descriptor development

Once the panellists have been selected and oriented in the basic principles of descriptive flavour analysis, they can start descriptor development. A range of samples that include the flavours and off-flavours similar to those of interest are presented, and the panellists describe the flavours that they perceive. A trained panel leader then guides the discussion to eliminate redundant descriptors. The panellists need to come to a consensus on the descriptors that eventually will be placed on the ballot.

Johnsen and Civille (1986) worked with seven experts in the meat industry to develop descriptors for warmed-over flavour in meats. Love (1988) reported on the development of beef descriptors at SRRC. The panel at SRRC also developed descriptors for lamb (St. Angelo *et al.*, 1991), for aging beef (Spanier *et al.*, 1997) and for dry-cured ham (Flores *et al.*, 1998).

The descriptor beefy/brothy, defined by Love (1988), was divided into beefy and brothy by the SRRC descriptive-analysis panel to aid in studies on desirable beef flavours. It was hypothesized that these were two unique descriptors, but it was to be determined if the descriptive panel could evaluate them. At the first session, the panel was presented with drippings from roasted chicken, roasted pork, roasted veal and roasted top round beef, and broth from boiled chicken, boiled pork, boiled veal and boiled beef (top round). They also received a sample of grilled ground beef in which to discern these two flavours. At this session, the panel decided that two terms were feasible and developed definitions. 'Beefy' was defined as the aromatics commonly associated with matured cooked beef muscle products. The reference sample is prepared by boiling cubes of top round roast (semimembranosus muscle) in water until well done. The liquid is drained off and served to the panellists. It contains a distinct beefy flavour. The broth from chicken tasted distinctly like chicken and the broth from pork tasted distinctly like pork, etc. 'Brothy' was defined as the aromatics associated with the drippings from roasted meat which is characteristic of all meats (i.e. poultry, beef and pork). The reference is the drippings from top round roasted in a 177°C (350°F) oven until done. It has a brothy note that is characteristic in the drippings from beef, poultry, veal and pork. In the second session, panellists evaluated beefy and brothy along with the other descriptors in ground patties made of top round, chuck, veal and T-bone. This further tested the feasibility of these revised descriptors.

An experiment was designed to test panellists' ability to evaluate flavour with the revised descriptors. Three cuts of beef and veal were individually ground and formed into 3 oz (80 g) patties. One-half of the patties was frozen at –11°C for 3 days and the other half was refrigerated at 4°C for 3 days. At the end of 3 days, the patties were grilled and served to the panel as described above. The analysis of variance was accomplished on means across panellists of samples within a session (Table 18.1). Analysing the means instead of individual panellist's scores removes most of the panellist effect. The results of the analysis of variance showed that there was a significant session effect for beefy and brothy, as well as browned/caramel and cardboardy. Different meat was purchased for each session, so a significant session effect covers differences in meat from week to week and differences in the panellists' mean response from session to session. This indicated that the brothy flavour intensity did not significantly differ between cuts. Conversely, beefy flavour was highest in intensity in T-bone and top round, lowest in veal, and chuck was in between (Table 18.2). The other descriptors that vary among beef cuts are painty, cooked liver, serum/raw, brown/caramel, salty, sour, and bitter. Differences in flavour notes between cuts are also seen upon refrigerated storage (Spanier and Miller, 1996; Table 18.2).

18.4 Chemical and instrumental parameters

18.4.1 Thiobarbituric acid reactive substances (TBARS)

The test for the presence of thiobarbituric acid reactive substances or TBARS is the most popular and widely accepted methods for measuring

Table 18.1 Sums of squares and coefficients of variability from the analysis of variance

Source	df	STY	BEF	BTH	PTY	SER	BRC	CKL	CBD	SOU	SWT	BTR
Cut	2	0.06	5.65**	0.03	0.40*	1.06*	4.66**	0.78*	0.10	0.89**	0.41	0.48**
Storage	1.00	0.11*	0.23	0.26	0.20	0.15*	0.23	0.01	0.12	0.03	0.03	0.07
Cut, storage*	0.01	0.26	0.23	0.05	0.04	0.10	0.03	0.03	0.10	0.03	0.06	
Error	12	0.20	1.18	0.67	0.54	0.36	0.90	0.70	0.69	0.29	0.23	0.02
Total	23	0.52	11.71	2.27	1.71	1.73	8.05	2.79	2.17	1.68	1.13	0.40
C.V.		9.60	9.80	11.60	34.50	18.90	11.80	12.60	31.40	16.40	12.30	26.70
Grand mean		1.30	3.20	2.00	0.60	0.90	2.30	1.90	0.80	0.90	1.10	1.30

Abbreviations: STY, salty; BEF, beefy; BTH, brothy; PTY, painty; SER, serumy; BRC, browned/caramel; CKL, cooked liver; CBD, cardboardy; SOU, sour; SWT, sweet; BTR, bitter. **means are significantly different at $p < 0.01$; *means are significantly different at $p < 0.05$.

Table 18.2 Table of means of descriptive flavour analysis by beef cuts

Descriptor	Veal	T-bone	Top round	Chuck
Salty	1.5[a]	1.3[ab]	1.3[ab]	1.2[b]
Beefy	2.4[c]	3.6[a]	3.7[a]	3.1[b]
Brothy	2.1[a]	2.1[a]	2.0[a]	1.9[a]
Painty	0.6[a]	0.4[a]	0.5[a]	0.8[a]
Serum/raw	1.1[a]	0.6[c]	0.9[b]	1.1[a]
Browned/caramel	1.5[c]	2.7[a]	2.7[a]	2.3[b]
Cooked liver	1.5[b]	2.0[a]	2.1[a]	2.0[a]
Cardboardy	0.7[a]	0.7[a]	0.7[a]	0.9[a]
Sour	1.3[a]	0.7[b]	0.9[b]	0.9[b]
Sweet	1.0[b]	1.4[a]	1.1[b]	1.0[b]
Bitter	0.9[a]	0.5[b]	0.6[b]	0.7[ab]

[a,b,c]Means with the same letter are not significantly ($p < 0.05$) different based on LSD mean comparison test.

lipid oxidation in food products. The extent of rancidity is usually expressed in terms of TBA number (μg malonaldehyde/g sample). One mole of malondialdehyde (MDA), a secondary decomposition product from peroxidized unsaturated fatty acids, can complex with 2 moles of TBA reagent to form a chromaphore measurable by spectrophotometry. However, there have been problems related to the TBA tests. Its sensitivity has been questioned (Tsoukala and Grosch, 1977); its reliability has been questioned (Witty *et al.*, 1970); interfering compounds have been identified, which led to a modification of the method to overcome their effect (Sinnhuber and Yu, 1977). More recently, limitation of the TBA test for oxidized lipids containing alka-2,4-dienals was claimed since the test was found to be nonspecific to malonaldehyde (Kosugi *et al.*, 1988). Nevertheless, in spite the problems related to the TBA–MDA reaction, the method is still used throughout the food industry. The distillation method of Tarladgis (1960) seems to be the most popular. In addition to measuring rancidity by the Tarladgis procedure, samples were also analysed by a second objective method, i.e. gas chromatography (see below).

18.4.2 Direct gas chromatography

As lipids oxidize, they form hydroperoxides, which decompose to produce volatile flavour components, such as aldehydes, ketones, alcohols, etc. Over the years, enrichment techniques have been developed to obtain a sufficient concentration of these volatile compounds for analysis by gas chromatography (GC).

An early novel approach to enrichment of flavour compounds for analysis by GC was developed by Dupuy *et al.* (1971). They described their method as a simple and very sensitive technique to analyse volatiles in vegetable oils by direct gas chromatography (DGC). Briefly, the method involved adding a sample (500 mg) of oil directly onto glass wool, which is in a glass liner. The glass liner was then inserted into a heated injection port. A carrier gas then sweeps the volatiles onto a column for analysis. The method requires no extractions, distillations or formation of derivatives. Later, Legendre and co-workers modified the procedure by inventing an external closed inlet device (ECID that can be attached to almost any gas chromatograph) (Legendre *et al.*, 1979). Whereas the original method was first used with packed columns, Dupuy *et al.* later modified the method to utilize capillary columns (Dupuy *et al.*, 1985). This method is neither an example of the classical static headspace technique nor the dynamic headspace method as defined by Wampler *et al.* (1985). Hence, the name 'direct' is assigned to the 'Dupuy' GC method.

Although first employed to assess vegetable oil quality, this methodology has been utilized during the past two decades to assess the quality of many foods, including meats, e.g. bacon (Dupuy *et al.*, 1978), beef, ham and pepperoni (Bailey *et al.*, 1980), beef, chicken and turkey (Dupuy *et al.*, 1987), and more recently lamb (St. Angelo *et al.*, 1991). A typical profile of freshly cooked ground beef patties that had undergone MFD using packed columns is shown in Figure 18.1. A capillary column profile is shown in Figure 18.2 for fresh cooked ground beef patties and Figure 18.3 for WOF patties stored for 24 h at 4°C. The primary lipid oxidation markers are the aldehydes, such as hexanal, pentanal, nonanal and the decadienals. As oxidation progresses, the intensity of these compounds increases proportionally.

18.5 Correlations among sensory, chemical and instrumental analyses

When comparisons were made between volatile compounds obtained from DGC and TBARS numbers, from both fresh and cooked/stored ground beef patties, the results correlated very well (St. Angelo *et al.*, 1987; Bailey *et al.*, 1987). The major volatile compounds (for example, butanal, pentanal, hexanal, heptanal, 2,3-octanedione, 2,4-heptadienal, nonanal, and *trans*, *cis*- and *trans*, *trans*-2,4-decadienal) that were found in fresh and MFD/WOF meat samples by DGC were also found in the distillate prepared for the TBARS reaction. Similar correlations were also made between DGC data or TBARS numbers and that obtained from sensory panels for fresh and MFD ground beef patties (St. Angelo *et al.*, 1987; Spanier *et al.*, 1988). Subsequent research used multivariate principal components analysis to show statistical correlations among the sensory (cooked beef/brothy, browned caramel, serumy, sweet, painty, cardboardy, sour, bitter, and salty) chemical (TBA) and instrumental attributes (namely, hexanal, 2,3-octanedione, nonanal and pentanal). Similar correlation among sensory, instrumental, and chemical data using multivariate

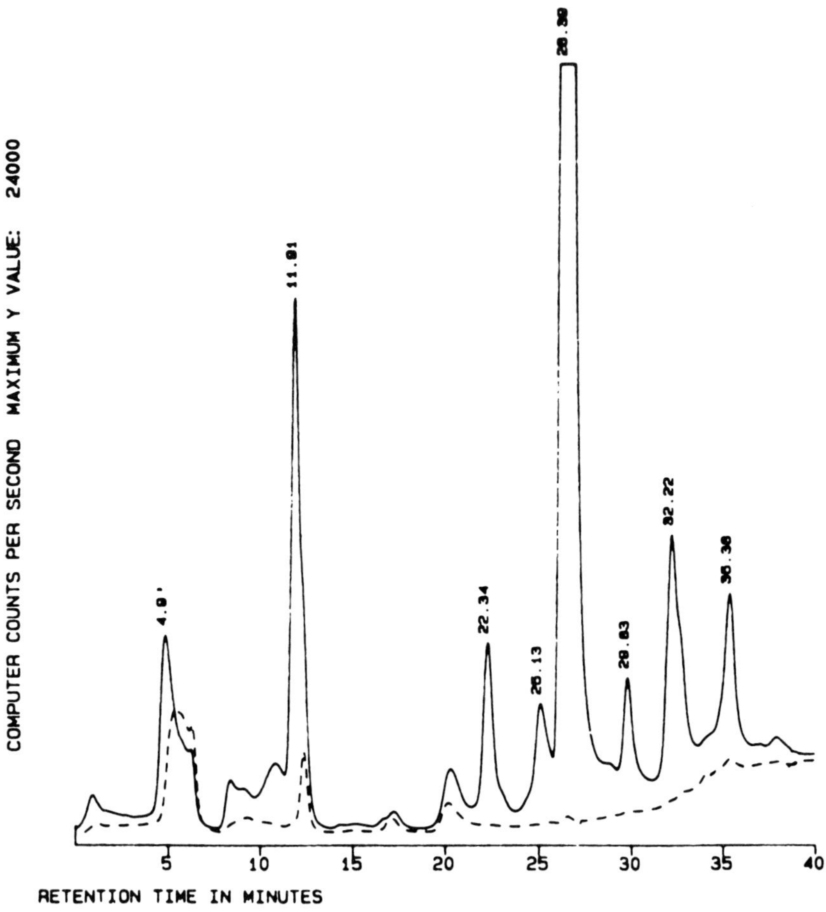

Figure 18.1 Volatile profiles of cooked beef obtained with packed GC. The broken line graph represents freshly cooked beef; the solid line graph represents cooked beef after storage at 4°C for 8 h. Compounds identified were propanol (11.9 min retention time), butanal (17), pentanal (22.3), 2,3-hydroxybutanone (25.1), hexanal (26.4), heptanal (29.8), 2,3-octanedione (32.2) and nonanal (35.4).

principal components analysis showed that the MFD process was sensitive and showed a dose response to added antioxidant–chelator mixtures and to the additives in conjunction with vacuum (Spanier *et al.*, 1992a,b).

The collection of sensory, chemical and instrumental data on meat samples, as discussed above, provides a profile of each meat sample that can facilitate identification of relationships among these characteristics. Such relationships can subsequently provide insight regarding the mechanisms at work in meat samples that do or do not contribute to meat flavour deterioration (MFD). Two experiments, a replicated lamb study

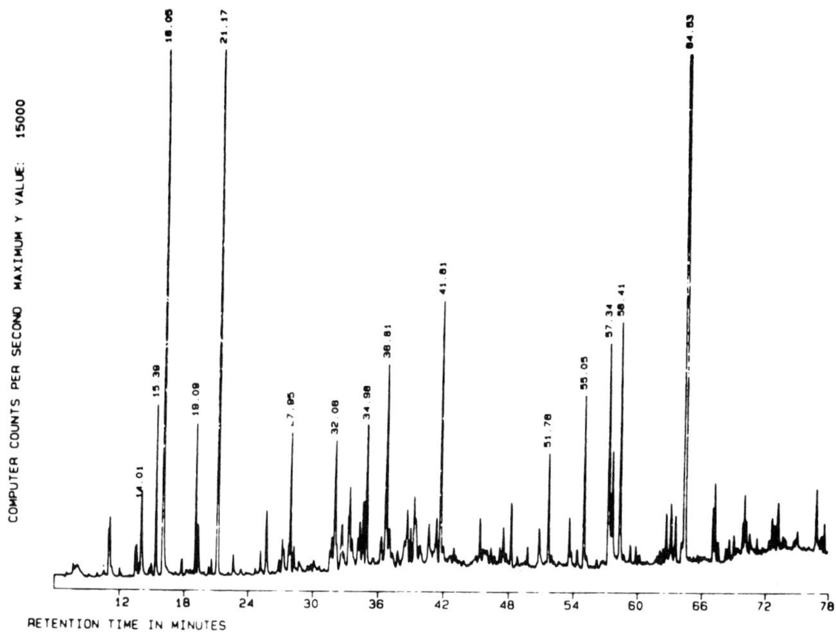

Figure 18.2 Volatile profile of freshly cooked beef obtained with capillary column GC. Compounds identified were pentanal (15.4 min retention time), 2,3-hydroxybutanone (19.1), hexanal (21.2), heptanal (27.9), 2,3-octanedione (32.1), 2-pentylfuran (34.9), nonanal (41.8), *trans*-2, *cis*-4-decadienal (57.3) and *trans*-2, *trans*-4-decadienal (58.4).

and an beef additive study, will be discussed to illustrate the appropriate statistical techniques for examining these relationships.

18.5.1 Experimental designs

An experimental design should be developed for each experiment. If collection of data is accomplished without regard to a design, then relationships of interest may become confounded (i.e. confused) with day-to-day variation. Designing an experiment requires (1) the construction of a schedule by which the experimental treatments are applied to a design structure (i.e. panel sessions), as discussed below, and (2) a decision regarding how much replication can be conducted, as discussed later.

Application of treatments to a design structure. The manner by which treatments are appropriately applied to panel sessions depends mainly upon the experimental objective(s) and, to a lesser degree, upon the number of treatments being considered in the experiment. It is also essential that at least one identically treated sample should be administered

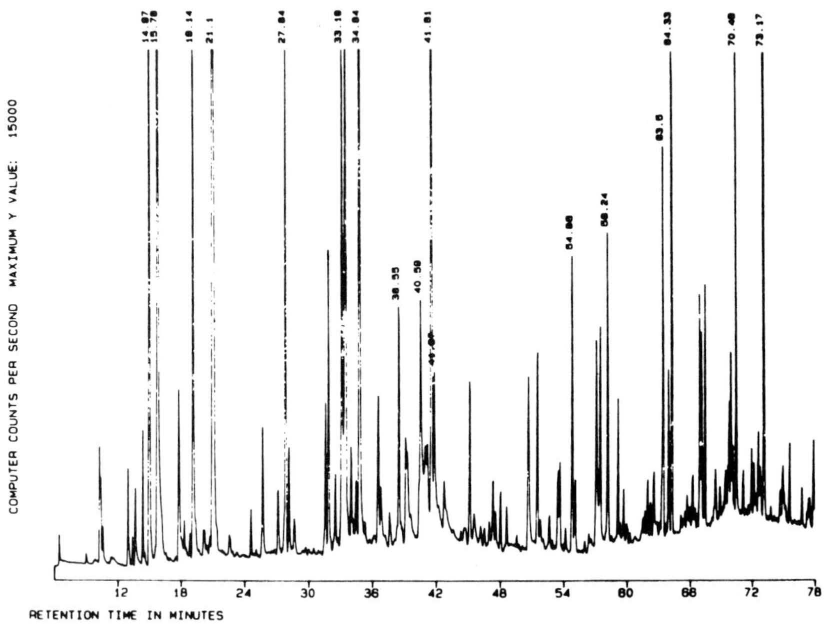

Figure 18.3 Volatile profile of cooked/stored WOF patties obtained with capillary GC; compounds identified had comparable retention times to those listed in Figure 18.2.

during every panel session to serve as a blind control for use in removing session-to-session variability.

Consideration of experimental objective and treatments. Ideally, it is recommended that the number of experimental treatments chosen for consideration in an experiment should be limited to the number of samples that can be presented in one sensory panel session. This restriction prevents the effect of treatments from becoming confounded (i.e. confused) with the day-to-day variation in meat samples and panel performance. Each sensory panel session is limited to a maximum of six experimental treatments and a control treatment to calibrate the panellists.

Practically, an experiment limited to the examination of six experimental treatments lacks scope for providing sufficiently useful information. Hence, the experimental factors are assigned a degree of importance so that levels of the experimental factor of primary interest are administered to panellists in the same panel session.

For example, St. Angelo *et al.* (1991) compared the effects of five distinct tenderization treatments applied to lamb carcasses immediately after slaughter on fresh frozen and 2-day stored samples. These treatments were as follows: Sample 1, Control (standard slaughter); Sample 2, ES (electrical stimulation); Sample 3, ES + Ca (ES plus calcium chloride); Sample 4, ES + Ca + M (as Sample 3 plus maltol); and Sample 5, ES + Ca + SA (as Sample 3 plus sodium ascorbate). Freshly ground leg of lamb was used as a standard.

Assessing the tenderization of meat was the primary objective of this experiment and the consistency of response to the tenderization treatments with storage was secondary. Hence, a design was constructed to randomly administer samples from lamb treated with all five tenderization treatments in the same panel session. Each panel session was then completely composed of either fresh frozen samples or 2-day stored samples totally confounding the variability of secondary interest (variability due to storage), with session-to-session variability. Reduction of session effects to allow examination of the variability due to storage only is discussed in the next section. One replicate was comprised of the two sessions described above.

In the current beef additive study, we initially investigated the effect of six chemical additives (each administered to beef samples at three concentrations, e.g. 0 ppm, 50 ppm and 125 ppm) on the sensory, chemical and instrumental responses of fresh frozen, and 2-day stored beef samples. The objective of primary importance was for each of the six additives to characterize the MFD trends occurring in stored samples relative to fresh frozen samples and to determine if these trends exhibited a significant reduction in rate of MFD development with the application of varying concentrations of chemical additive to the beef samples. Secondary to the study was a comparison among the six additives.

A design was constructed so that each session was composed of both fresh frozen and 2-day stored samples, each treated with all three concentrations of a single additive. The schedule of sample administration for one such panel (i.e. Session 1) constitutes a single replicate of all six storage times (zero and 2-day) by concentration (zero, higher percentage and highest percentage) combinations.

Analysis of data collected using this design indicated no significant reduction in MFD with increased additive concentrations for any of the six additives. Hence, the experiment was repeated using fresh frozen and 4-day stored beef samples. Session 2 included the use of a sample presentation schedule identical to that of the first session, except for the replacement of 2-day samples with 4-day samples. One replicate was comprised of Sessions 1 and 2.

A design constructed to initially consider all three levels of (fresh frozen, 2-day and 4-day) stored samples would have been very similar to the design comprised of Sessions 1 and 2. Based on the experimental objective, a preferred design would administer samples stored for all three storage times and only two of the three concentrations of an additive during the same session. The differences between this preferred design and the design resulting from the combination of the two experiments actually conducted can, under certain circumstances, be assumed negligible, as discussed in the next section.

Including a blind control. The session effect, caused by session-to-session variability, can be effectively eliminated from the sensory data prior to its analysis by insuring that the experiment be designed to include a blind control in every panel session.

Before discussing the blind controls for our two experiments, we will describe how the session-to-session variability (i.e. session effect) can be removed from the sensory data. The average rating assigned to a sample for a particular flavour characteristic by the panellists is considered to be one replicate measurement of the treatment assigned to that sample. Hence, each panel session provides one replicate of each treatment represented in that session. If an identical blind control treatment is applied to one sample in each session then data collected from all panel sessions can be adjusted to one another respective to the ratings received by this common sample. The session effect can be practically eliminated under the assumptions that none of the blind control samples are outliers in any manner and that any variability observed from session-to-session can be indentified by differences among the blind control samples. The elimination of session effect is accomplished by comparing the average panellist rating of this blind control to the overall average rating of the blind control from all panel sessions. If a session's blind control was rated above (or below) the overall average rating then the average rating observed for each treatment in the session is adjusted downward (or upwards) by this difference.

St. Angelo *et al.* (1991) required an adjustment for session effect to allow comparison of fresh frozen samples to samples stored for 2 days. Owing to the limited amount of experimental sample material no blind control sample was administered. Hence, the 'standard' samples used to calibrate the panellists at the beginning of each session were used, less effectively, to eliminate session effects.

In the current beef additive study, interest lay in direct comparisons only lay in direct comparison between the storage time trends occurring at three concentrations of each particular additive. Hence, for analytical purposes, a separate experiment was conducted for each additive, and elimination of session effect was conducted separately for each additive. The blind control for this experiment was actually a collection of the experimental treatments. One replicate, as described above, indicates that each session contains six experimental samples and that three of the six experimental samples are fresh frozen (zero-day stored) samples, each representing one of the three additive concentrations. Each of these three fresh frozen samples receives an average rating with respect to a particular flavour characteristic. An average of these three sample averages then provides a 'comprehensive' blind control for eliminating session effect.

Session effect would have been eliminated in data collected according to the 'preferred' design, discussed above, by using an average of fresh frozen, 2-day and 4-day stored samples with 0% additive. Hence, under the assumption that this adjustment technique effectively eliminates session effects, the design that was actually used in conducting the experiment provides information comparable to that provided by the preferred design.

Summary – applications of treatments to a design structure. Techniques for applying treatments to panel sessions have been discussed. These techniques emphasize the application of treatments directly related to the primary objective of the experiment to the same panel session, and mandate the inclusion of an identical blind control sample in each panel session. These concepts are equally applicable to replicated and unreplicated experiments.

Replication – how much is possible? The amount of replication that can be conducted in any experiment is primarily dependent upon the material, physical, monetary, time and human resources available. Experimental material is usually a valuable and limited commodity. However, if sufficient experimental material is available to conduct five replicate panel sessions for each treatment (considered optimal based on the sensory history at SRRC), then it must be determined if the sensory panel can be convened for the required number of sessions within time frame limitations. If these requirements are feasible, then there must 'ideally' remain a sufficient amount of experimental material, time, physical and human resources to conduct all chemical and instrumental analyses on a sample of the same experimental material, representing each treatment, administered to the panel during each session.

For instance, St. Angelo *et al.* (1991) conducted four complete replicates of the five tenderization treatments evaluated in fresh frozen and 2-day stored lamb samples. The experiment required eight panel sessions, each session consisting of samples representing one of the two storage times and all five experimental treatments.

Practically, owing to the time-intensive nature of many chemical analyses, the researcher must sometimes settle for a single or duplicate chemical analysis for each treatment. If this is the case, the resulting data cannot be combined with sensory data for use in a multivariate analysis such as that discussed below. Rather, a separate univariate analysis must be conducted for each chemical response variable to identify any differences among experimental treatments.

Ideally, each experimental treatment should be replicated as many times as there are sensory, chemical and physical characteristics of interest. Practically, if relationships among these characteristics are to be identified using multivariate statistical analyses, then (1) each treatment must have been replicated; and (2) the number of treatments times the amount of treatment replication must be at least three times the total number of sensory, chemical and instrumental responses of interest (Tabachnick and Fidell, 1983). Hence, the number of sensory, chemical and instrumental measures that can be considered in a multivariate statistical analysis is limited by the amount of replication possible.

For example, in the lamb study (St. Angelo *et al.*, 1991) four replicates were conducted on each of the ten tenderization and storage time treatments to obtain a data set containing a total of 40 observations. Hence, a multivariate analysis can be conducted using an absolute maximum of 13 response variables collected in the experiment.

The scope of or the exploratory nature of an experiment may sometimes limit each treatment to a single replicate. Although informative univariate statistical analyses can always be performed on the data collected from unreplicated experiments, no multivariate statistical analyses can be conducted. The current beef additive study provided duplicates only for fresh frozen (zero-day stored) samples at each concentration of additive. This partial replication provided sufficient information to conduct a univariate analysis of variance (ANOVA; Steel and Torrie, 1980) for each sensory, chemical and instrumental characteristic measured. Milliken and Johnson (1989) present techniques for analysing completely unreplicated experiments.

Summary – experimental design. Techniques have been discussed for applying experimental and blind control treatments to panel sessions based on the primary experimental objective(s). The dependency of various types of statistical analyses on the amount of replication conducted and the dependency of replication on resource and practical constraints were examined. Additional reading may be found in Bett (1993) and Bett and Grimm (1994).

18.5.2 Statistical analysis

The data collected according to an experiment's design can easily be placed in a format to facilitate the univariate and multivariate statistical analyses of the sensory, chemical and instrumental responses of interest.

Data preparation. Prior to its analysis, sensory data can be subjected to noise reduction techniques of Bett *et al.* (1993) to identify panellists who are either super-sensitive or cannot discern the presence of a particular flavour characterisitic and to normalize such data to clarify the information contained in the data. The averages of panellist responses for each treatment in a session are then used as data for univariate and multivariate statistical analyses, as shown in Table 18.3.

Table 18.3 Data format for univariate and multivariate analyses of sensory and chemical data from lamb study

Session	STOR	TRT	HEX	TVOL	TBA	CBD	MTH	MTY	PTY	SWT
1	2	Control	347.0	841.0	12.96	2.6	1.5	3.7	2.5	1.3
1	2	ES	66.0	161.5	8.63	1.9	1.4	3.9	2.0	1.2
1	2	ES+Ca	218.0	456.0	12.30	2.5	1.5	4.2	2.6	1.0
1	2	ES+Ca+M	25.0	216.0	7.84	0.9	1.7	4.1	0.8	1.1
1	2	ES+Ca+SA	44.0	87.0	1.91	1.8	1.1	4.6	1.6	0.8
2	0	Control	2.0	48.0	2.37	1.4	0.9	4.3	1.1	1.0
2	0	ES	46.0	201.0	2.39	0.3	1.1	4.7	0.1	1.5
2	0	ES+Ca	30.0	134.0	3.61	0.5	1.5	4.9	0.3	1.4
2	0	ES+Ca+M	32.0	94.0	1.34	0.3	1.4	4.7	0.1	1.3
2	0	ES+Ca+SA	1.0	30.0	0.57	1.0	1.3	4.8	0.7	1.1
7	0	Control	1.0	57.0	–	2.2	0.4	4.5	2.8	0.6
7	0	ES	10.0	59.0	2.43	0.3	1.3	4.9	0.1	1.3
7	0	ES+Ca	48.0	117.0	5.35	0.3	1.6	4.6	0.4	1.3
7	0	ES+Ca+M	13.0	100.0	1.18	0.7	1.7	4.3	0.4	1.5
7	0	ES+Ca+SA	4.0	65.0	0.66	1.2	1.2	4.3	0.9	1.3
8	2	Control	260.0	451.0	8.06	1.6	2.0	3.2	2.3	2.0
8	2	ES	187.0	300.0	7.05	2.3	1.4	3.5	2.1	1.1
8	2	ES+Ca	415.0	700.0	22.51	2.1	1.3	3.7	2.3	1.1
8	2	ES+Ca+M	107.0	235.0	7.78	0.9	1.6	4.3	0.6	1.2
8	2	ES+Ca+SA	–	–	–	0.9	1.1	3.6	0.7	2.1

Abbreviations: STOR, days storage; TRT, treatments; HEX, hexanal; TVOL, total volatiles; TBA, thiobarbituric acid; CBD, cardboardy; MTH, muttony/herby; MTY, meaty; PTY, painty, SWT, sweet; ES, electrical stimulation; Ca, calcium chloride; M, maltol; SA, sodium ascorbate.

If the chemical and/or instrumental measurements were not conducted for every replicate sample administered to the sensory panel, such as the TBA value for the control treatment in Session 7 of Table 18.3 and the values for all three chemical variables for the ES + Ca + SA treatment in Session 8, then one of two options can be utilized. The first option simply omits these two lines of data (i.e. treatment replicates) from use in the multivariate analysis and conducts the analysis based on 38 rather than 40 observations. The second option requires a separate analysis for the sensory and the chemical response variables. Because of interest in relationships among chemical and sensory variables, the first option is the most desirable. The collected data are now ready for individual and simultaneous analysis of sensory, chemical and instrumental responses.

Univariate analyses of variance. A univariate analysis of variance (ANOVA; Steel and Torrie, 1980) is conducted using data collected for a single response variable. An ANOVA can identify experimental factors (such as storage time, additive and additive concentration) and combinations of these factors that affect statistically distinct behaviour with respect to a particular sensory, chemical or instrumental response. Once all statistically significant treatment effects have been identified for a particular response variable, appropriate mean comparisons (i.e. contrasts) or orthogonal polynomial contrasts (Steel and Torrie, 1980) can be conducted. These contrasts identify, specifically, which experimental treatments or combinations of treatments are statistically distinct with respect to a response variable.

St. Angelo *et al.* (1991) conducted contrasts, some of which are shown in Table 18.4, to compare results from specific pairs and specific linear combinations of the five tenderizing treatments, respective to treatment effects found significant in the ANOVA for each individual sensory, chemical and instrumental response. This approach, however, for comparing responses measured on distinct treatments could not be applied to the analysis of data from the current beef additive study.

In the current beef additive study, interest focused on a response's trend across zero, 2- and 4-day storage times at various concentrations of a particular additive. Hence, the objective was to compare the effect on a response resulting from a numeric change in the amount of a single treatment applied to the experimental material. Alternatively, the objective of the lamb study was to compare the effect on a response of the application of several distinct treatments.

The appropriate technique for identifying differences among the numeric levels of treatments such as storage times and additive concentrations is to fit a response surface, i.e. a regression model (Milliken and Johnson, 1989), to the data observed at all combinations of storage times and concentrations considered. The response surface model and plot for the data collected on the sensory descriptor 'painty' using the maltol additive is shown in Figure 18.4. This figure illustrates that an increased

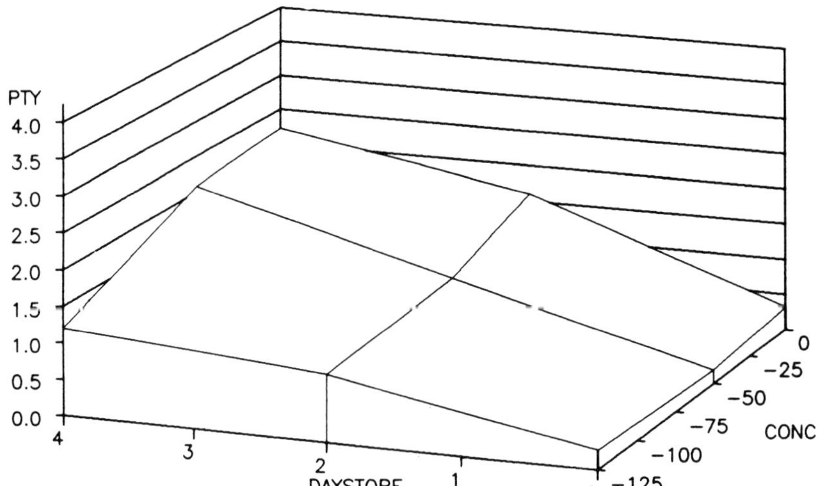

Figure 18.4 Univariate factor analysis; effect of maltol on the intensity of 'painty'. Model: Painty = 0.29 + (0.53 − 0.0022 × concentration) daystore.

Table 18.4 Means, standard errors and *p*-values of treatment contrasts for responses with significant treatment by storage time interaction

Response	Day	Means/standard error treatment			*p* value contrast		
		ES (1)	ES + $CaCl_2$ (2)	(2) + SA[a] (3)	(1) versus (2)	(2) versus (3)	(1) versus (3)
Hexanal	0	18.25/6.97	47.25/6.97	5.25/6.97	0.0101*	0.007**	0.2068
(area counts × 10^3)	2	126.50/38.13	261.75/38.13	24.67/44.03	0.0251*	0.0011**	0.1023
Total volatiles	0	84.50/21.67	133.25/21.67	70.25/21.67	0.0576	0.0576	0.6486
(area counts × 10^3)	2	230.63/66.27	481.25/66.27	123.67/76.52	0.0033**	0.0033**	0.3986
TBARS	0	2.02/0.44	4.50/0.44	1.18/0.44	0.0013**	0.0001**	0.1940
(mg MDA/kg sample)	2	10.10/1.68	19.15/1.68	1.56/1.95	0.0020**	0.0001**	0.0020**
Meaty[b]	0	4.88/0.11	4.73/0.12	4.56/0.11	0.3502	0.2852	0.0682
	2	3.74/0.15	3.72/0.15	3.97/0.14	0.9417	0.2277	0.2935
Cardboardy[b]	0	0.51/0.15	0.57/0.15	1.42/0.14	0.7631	0.0001**	0.0018**
	2	1.96/0.19	2.33/0.19	1.54/0.18	0.1590	0.0028	0.0484
Painty[b]	0	0.31/0.15	0.35/0.16	1.13/0.14	0.8495	0.0002**	0.0135
	2	2.05/0.20	2.59/0.02	1.52/0.02	0.0557	0.0002**	0.0485

[a]Sodium ascorbate.
[b]Intensity values: 0, lowest; 15, highest.
*Mean differences significant at the 0.05 level
**Mean differences significant at the 0.01 level.

concentration of maltol significantly reduces the rate at which the off-flavour descriptor, painty, increases with sample storage time.

At minimum, partial replication (such as the zero-day duplicates observed for all three concentrations of an additive) is strongly recommended for any experiment. Without replication, there is no source by which to estimate error variability in the data for the subsequent testing for significant differences among experimental treatments. How much replication is actually conducted can be determined as discussed above.

Multivariate principal factor analysis. Multivariate statistical analyses utilize the relationships among the response variables included in the analysis to better describe differences occurring among the experimental treatments. Specifically, a principal factor analysis utilizes the linear correlation structure existing among the observed responses to create a new set of 'factors'. Each factor identifies a unique source of common variability observed among the response variables respective to the experimental treatments.

Before conducting a principal factor analysis, it is essential that any of the sensory, chemical and/or instrumental analysis variables are omitted from this analysis which, by means of a mathematical relationship among one or more of the other response variables, theoretically reproduce identical information. This problem can be avoided by choosing a set of sensory, chemical and instrumental analysis variables that are of key importance to the experiment and are known to provide unique information.

The data from St. Angelo *et al.* (1991) can be used to illustrate a principal factor analysis. Five key sensory variables are chosen from a set of 15 measured variables and all three chemical responses are included for a total of eight variables to be utilized in the analysis. The 38 data observations, for which all eight response variables had values, provided sufficient replication, as discussed above, to conduct the factor analysis.

The first step in choosing an appropriate factor solution (i.e. result from a principal factor analysis) is determining the number of factors that can be extracted from the variability among treatments exhibited in the data set. This choice depends upon the variance of each extracted factor. The variance (i.e. eigenvalue) of each principal factor indicates its ability to describe variation occurring in the data. If the eigenvalue of a principal factor is less than 1 then an individual response variable is capable of explaining more of the data variability than is this particular principal factor (Tabachnick and Fidell, 1983). Hence, only those principal factors with eigenvalues of size at least 1 need be considered in examining the relationship among the experimental treatments.

A principal factor analysis is only one of several techniques for extracting factors from a set of data. The factor solution, i.e. factors extracted, resulting from a principal factor analysis is often readily interpretable (Tabachnick and Fidell, 1983). In general, a desirable factor solution will exhibit three characteristics. First, each factor must 'load' (i.e. be highly correlated with) a minimum of two response variables. Second, each response variable used in the analysis should have a much higher loading (i.e. stronger) correlation with one factor than with any of the other factors. Finally, the factor solution should be rotated to produce mutually uncorrelated factors.

The factor solution resulting from conducting a principal factor analysis on the data collected in the lamb study consisted of two factors. Of the total treatment variance of 6.436, the variances of Factor 1 and Factor 2 were 4.587 and 1.504, respectively. All other extracted factors had variances less than unity. Hence, the two factors in this factor solution were able to explain 94.6% of the total variability among the tenderization and storage treatment combinations.

The rotated factor pattern shown in Table 18.5 illustrates which of the eight response variables exhibits the strongest correlation with each factor. Rotation of a factor pattern is simply a technique for removing any correlation among the factors in a factor solution. The response variables hexanal, total volatiles, TBA, cardboardy and painty all exhibit a high positive 'loading' (i.e. correlation with) Factor 1 while the sensory descriptor meaty exhibits a high negative loading onto Factor 1. This indicates that hexanal, total volatiles, TBA, cardboardy and painty are all directly related but are inversely related to meaty. The common variability represented by Factor 1 can be labelled as 'desirability' of the meat samples. The two response variables, muttony–herby and sweet, exhibit positive loadings onto Factor 2 of similar magnitude. However, there is no apparent interpretation of the common variability represented by Factor 2.

The factor pattern (Table 18.5) can be used to graphically examine the relationship among the experimental treatments. Each principal factor produces a score for each observation in the data set used for the analysis. The principal factor scores resulting from the observations associated with a particular treatment and principal factor can be averaged. These factor score averages for each treatment can be plotted (Figure 18.5), together

Table 18.5 Rotated factor pattern from a principal factor analysis for the ham tenderization study

Response variable	Factor 1	Factor 2
Hexanal	0.868[a]	0.302
Total volatiles	0.847[a]	0.312
TBA	0.823[a]	0.184
Cardboardy	0.906[a]	–0.317
Muttony/herby	0.310	0.702[a]
Meaty	–0.797	–0.221
Painty	0.937[a]	–0.193
Sweet	–0.100	0.776[a]

[a]Response variable loads onto this factor.

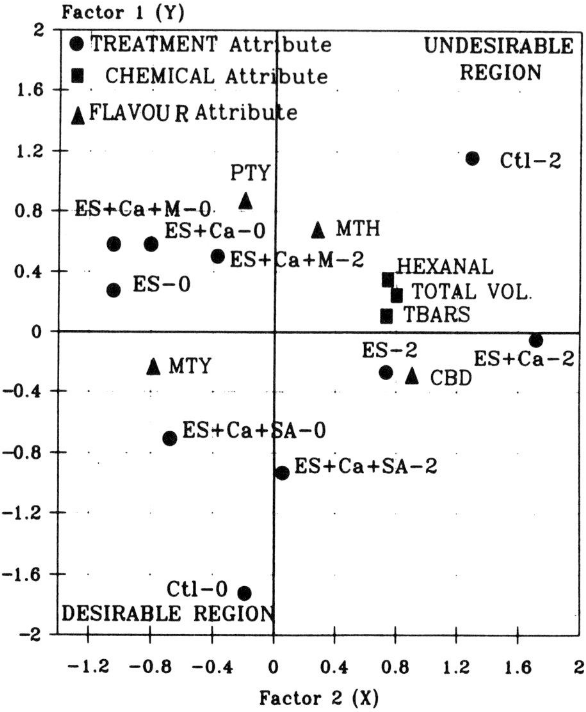

Figure 18.5 Multivariate factor analysis; infusion of lamb muscle experiment.

with the factor loadings for each response variable, to facilitate graphical interpretation of relationships existing among the response variables and the experimental treatments. Notice in Figure 18.5 that the undesirable sensory descriptor and the chemical volatile loadings appear in the upper portion of the plot while the loadings for the more desirable sensory descriptors appear in the lower portion of the plot. Hence, scores for tenderization and storage treatments appearing in the upper or lower portion of the plot associate each treatment, respectively, with either undesirable or desirable sensory and chemical characteristics.

The researcher should note that a factor solution is specific to the data from which it was created. Hence, the validity of any generalization of relationships between response variables and common behaviour, i.e. principal factors, is a direct function of the amount of replication conducted. Most factor analyses are considered 'exploratory' rather than 'confirmatory' analyses (Tabachnick and Fidell, 1983).

18.6 Summary

Aspects of appropriate experimental designs and statistical analyses for both replicated and unreplicated experiments have been discussed. The design of an experiment applies the treatments under consideration to experimental material in a manner to allow statistically correct analyses of the collected data. The types of treatments, whether distinct treatments or a single treatment observed at various levels, together with the amount of replication allowed by the resources and practical limitations of the experiment determine an appropriate statistical analysis for the collected data.

Acknowledgements

We thank the following Ms D.A. Ingram and Ms M.G. Franklyn for their assistance in sensory analysis and Mr J.A. Miller and Mr C. James for chromatographic assistance.

References

AMSA (1987). *Guidelines for Cookery and Sensory Evaluation of Meat.* American Meat Science Association in cooperation with the National Livestock and Meat Board, Chicago, IL.

Bailey, M.E., Dupuy, H.P. and Legendre, M.G. (1980). In *The Analysis and Control of Less Desirable Flavours in Foods and Beverages*, ed. G. Charalambous. Academic Press, Orlando, FL, pp. 31–52.

Bailey, M.E., Shin-Lee, S.Y., Dupuy, H.P., St. Angelo, A.J. and Vercellotti, J.R. (1987). In *Warmed-Over Flavour of Meat*, eds. A.J. St. Angelo and M.E. Bailey. Academic Press, Orlando, FL, pp. 237–266.

Bett, K.L. (1993). Measuring sensory properties of meat in the laboratory. *Food Technol.*, **47**, 121–126, 134.

Bett, K.L. and Grimm, C.C. (1994). Flavour and aroma – its measurement. In *Quality Attributes and Their Measurement in Meat, Poultry and Fish Products*, eds. A.M. Pearson and T.R. Dutson. Blackie Academic & Professional, New York, pp. 202–221.

Bett, K.L., Shaffer, G.P., Vercellotti, J.R., Sanders, T.H., and Blankenship, P.D. (1993). Reducing the noise contained in descriptive sensory data. *J. Sens. Studies*, **8**, 13–29.

Cremer, M.L. (1982). Sensory quality and energy use for scrambled eggs and beef patties heated in institutional microwave and convection ovens. *J. Food Sci.*, **47**, 871–874.

Cross, H.R. (1977). A survey of meat cookery and sensory evaluation methods among AMSA meat scientists. Recip. Meat Conf. Auburn, Alabama.

Dupuy, H.P., Fore, S.P. and Goldblatt, L.A. (1971). Elution and analysis of volatiles in vegetable oils by gas chromatography. *J. Am. Oil Chem. Soc.*, **48**, 876–884.

Dupuy, H.P., Bailey, M.E., St. Angelo, A.J., Legendre, M.G. and Vercellotti, J.R. (1987). In *Warmed-Over Flavour of Meat*, eds. A.J. St. Angelo and M.E. Bailey. Academic Press, Orlando, FL, pp. 165–191.

Dupuy, H.P., Brown, M.L., Legendre, M.G., Wadsworth, J.I. and Rayner, W.T. (1978). In *Lipids as a Source of Flavour*, ed. M.K. Supran, ACS Symposium Series 75, American Chemical Society, Washington, DC, pp. 60–67.

Dupuy, H.P., Flick, G.J. Jr, Bailey, M.E., St. Angelo, A.J. and Legendre, M.G. (1985). Direct sampling capillary gas chromatography of volatiles in vegetable oils. *J. Am. Oil Chem. Soc.*, **62**, 1690–1693.

Flores, M., Ingram, D.A., Bett, K.L., Toldrá, F. and Spanier, A.M. (1997). Sensory characteristics of Spanish 'Serrano' dry-cured ham. *J. Sensory Studies,* **12**, 169–179.

Greene, B.E. (1969). Lipid oxidation and pigment changes in raw beef. *J. Food Sci.*, **34**, 110–113.

Johnsen, P.B. and Civille, G.V. (1986). A standardized lexicon of meat WOF descriptors. *J. Sens. Studies*, **1**, 99–104.

Kosugi, H., Kato, T. and Kikugawa, K. (1988). Formation of red pigment by a two-step 2-thiobarbituric acid reaction of alka-2,4-dienals. Potential products of lipid oxidation. *Lipids*, **23**, 1024–1031.

Legendre, M.G., Fisher, G.S., Schuller, W.H., Dupuy, H.P. and Rayner, E.T. (1979). Novel technique for the analysis of volatiles in aqueous and nonaqueous systems. *J. Am. Oil Chem. Soc.*, **56**, 552–555.

Love, J. (1988). Sensory analysis of warmed-over flavour in meat. *Food Technol.* **42**, 140–143.

Meilgaard, M.C., Civille, G.V. and Carr, B.T. (1987). In *Sensory Evaluation Techniques*, Vol. II. CRC Press Inc., Boca Raton, FL, pp. 47–49.

Milliken, G.A. and and Johnson, D.E. (1989). *Analysis of Messy Data – Nonreplicated Experiments*, Vol. 2. Van Nostrand Reinhold, New York.

Pangborn, R.M. (1967). Use and misuse of sensory methodology. *J. Food Qual.*, **15**, 7–40.

Sato, K. and Hegarty, G.R. (1971). Warmed-over flavour in cooked meat. *J. Food Sci.*, **36**, 1098–1102.

Sinnhuber, R.O. and Yu, T.C. (1977). The 2-thiobarbituric acid reaction, an objective measure of the oxidative deterioration occurring in fats and oils. *J. Jpn. Oil Chem. Soc.*, **26**, 259–267.

Spanier, A.M. (1992). Current approaches to the study of meat flavour. In *Food Science and Human Nutrition*, ed. G. Charalambous. Elsevier Science, Amsterdam pp. 696–709.

Spanier, A.M. and Drumm-Boylston, T. (1994). Effect of temperature on the analyses of beef flavour volatiles: A look at carbonyl and sulphur-containing compounds. *Food Chem.*, **50**, 251–259.

Spanier, A.M. and Miller, J.A. (1996). Effect of temperature on the quality of muscle food. *J. Muscle Foods*, **7**, 355–375.

Spanier, A.M., Edwards, J.V., and Dupuy, H.P. (1988). The W.O.F. process in beef: A study of meat proteins and polypeptides. *Food Technol.*, **42**, 110–118.

Spanier, A.M., McMillin, K.W. and Miller, J.A. (1990). Enzyme activity levels in beef: Effect of postmortem aging and end-point cooking temperature. *J. Food Sci.*, **55**, 318–322.

Spanier, A.M., St. Angelo, A.J. and Shaffer, G.P. (1992a). Response of beef flavour to oxygen depletion and an antioxidant/chelator mixture. *J. Agric. Food Chem.* **40**, 1656–1662.

Spanier, A.M., Vercellotti, J.R. and James, C. Jr. (1992b). Correlation of sensory, instrumental and chemical attributes of beef as influenced by meat structure and oxygen exclusion. *J. Food Sci.*, **57**, 10–15.

Spanier, A.M., St. Angelo, A.J., Grimm, C.C. and Miller, J.A. (1994a). The relationship of temperature to the production of lipid volatiles from beef. In *Lipids in Food Flavours*, eds. C.T. Ho and J. Hartman. ACS Symposium Series 558, American Chemical Society, Washington, DC, pp. 78–97.

Spanier, A.M., Grimm, C.C. and Miller, J.A. (1994b). The sulphur-containing compounds in beef: Are they really present or are they artifacts? In *Sulphur Compounds in Foods*, eds. C.J. Mussinan, and M.E. Keelan. ACS Symposium Series 564, American Chemical Society, Washington, DC, pp. 49–62.

Spanier, A.M., Flores, M., McMillan, K.W. and Bidner, T.D. (1997). The effect of postmortem aging on meat flavour quality. Correlation of treatments, sensory, instrumental, and chemical descriptors. *Food Chem.*, **59**, 531–538.

St. Angelo, A.J., Vercellotti, J.R., Legendre, M.G., Vinnett, C.H., Kuan, J.W., James, C. Jr and Dupuy, H.P. (1987). Chemical and instrumental analysis of warmed-over flavour in beef. *J. Food Sci.*, **52**, 1163–1168.

St. Angelo, A.J., Koohmaraie M., Crippen, K.L. and Crouse, J. (1991). Acceleration of postmortem tenderization/inhibition of warmed-over flavour by calcium chloride–antioxidant infusion into lamb carcasses. *J. Food Sci.*, **56**, 359–362.

St. Angelo, A.J., Spanier, A.M. and Crippen, K.L. (1992). Chemical and sensory evaluation of flavour in untreated and antioxidant-treated meat. In *Lipid Oxidation in Foods*, ed.

A.J. St. Angelo. ACS Symposium Series 500, American Chemical Society, Washington, DC, pp. 140–160.

Steel, R.G.D. and Torrie, J.H. (1980). *Principles and Procedures of Statistics – A Biometrical Approach*, 2nd edn. McGraw-Hill, New York.

Tabachnick, B.G. and Fidell, L.S. (1983). *Using Multivariate Statistics*. Harper & Row, New York.

Tarladgis, B.G., Watts, B.M., Younathan, M.T. and Dugan, L. (1960). A distillation method for the quantitative determination of malonaldehyde in rancid foods. *J. Am. Oil Chem. Soc.* **37**, 44–48.

Tims, M.J. and Watts, B.M. (1958). Protection of cooked meats with phosphates. *Food Technol.*, **12**, 240–243.

Tsoukala, B. and Grosch, W. (1977). Analysis of fat deterioration – comparison of some photometric tests. *J. Am. Oil Chem. Soc.*, **54**, 490–493.

Wampler. T.P., Bowe, W.A. and Levy, E.J. (1985). Splitless capillary GC analysis of herbs and spices using cryofocusing. *Am. Lab.*, **October**, 76–81.

Witty, V.C., Krause, G.F. and Bailey, M.E. (1970). A new extraction method for determining 2-thiobarbituric acid values of pork and beef during storage. *J. Food Sci.*, **35**, 582–585.

Index